STATISTICS FOR RESEARCH

THIRD EDITION

STATISTICS FOR RESEARCH

THIRD EDITION

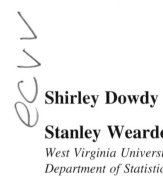

Shirley Dowdy

Stanley Weardon
West Virginia University
Department of Statistics and Computer Science
Morgantown, WV

Daniel Chilko
West Virginia University
Department of Statistics and Computer Science
Morgantown, WV

WILEY-INTERSCIENCE

A JOHN WILEY & SONS, INC. PUBLICATION

For ordering and customer service, call 1-800-CALL-WILEY.

Library of Congress Cataloging-in-Publication Data:

Dowdy, S. M.
   Statistics for research / Shirley Dowdy, Stanley Weardon, Daniel Chilko.
      p. cm. – (Wiley series in probability and statistics; 1345)
   Includes bibliographical references and index.
   ISBN 0-471-26735-X (cloth : acid-free paper)
      1. Mathematical statistics. I. Wearden, Stanley, 1926– II. Chilko, Daniel M. III. Title. IV. Series.

QA276.D66 2003
519.5–dc21

                                                                        2003053485

Printed in the United States of America.

10 9 8 7 6 5 4 3 2 1

CONTENTS

PREFACE TO THE THIRD EDITION

In preparation for the third edition, we sent an electronic mail questionnaire to every statistics department in the United States with a graduate program. We wanted modal opinion on what statistical procedures should be addressed in a statistical methods course in the twenty-first century. Our findings can readily be summarized as a seeming contradiction. The course has changed little since R. A. Fisher published the inaugural text in 1925, but it also has changed greatly since then. The goals, procedures, and statistical inference needed for good research remain unchanged, but the nearly universal availability of personal computers and statistical computing application packages make it possible, almost daily, to do more than ever before. The role of the computer in teaching statistical methods is a problem Fisher never had to face, but today's instructor must face it, fortunately without having to make an all-or-none choice.

We have always promised to avoid the black-box concept of computer analysis by showing the actual arithmetic performed in each analysis, and we remain true to that promise. However, except for some simple computations, with every example of a statistical procedure in which we demonstrate the arithmetic, we also give the results of a computer analysis of the same data. For easy comparison we often locate them near each other, but in some instances we find it better to have a separate section for computer analysis. Because of greater familiarity with them, we have chosen the SAS® and JMP®, computer applications developed by the SAS Institute.[†] SAS was initially written for use on large main frame computers, but has been adapted for personal computers. JMP was designed for personal computers, and we find it more interactive than SAS. It is also more visually oriented, with graphics presented in the output before any numerical values are given. But because SAS seems to remain the computer application of choice, we present it more frequently than JMP.

Two additions to the text are due to responses to our survey. In the preface to the first edition, we stated our preference for discussing probability only when it is needed to explain some aspect of statistical analysis, but many respondents felt a course in statistical methods needs a formal discussion of probability. We have attempted to "have it both ways" by including a very short presentation of probability in the first chapter, but continuing to discuss it as needed. Another frequent response was the idea that a statistical analysis course now should include some minimal discussion of logistic regression. This caused us almost to surrender to black-box instruction. It is fairly easy to understand the results of a computer analysis of logistic regression, but many of our students have a mathematical background a bit shy of that needed for performing logistic regression analysis. Thus we discuss it, with a worked example, in the last section to make it available for those with the necessary

---

[†]SAS and JMP are registered trademarks of SAS Institute Inc., Cary, NC, USA.

mathematical background, but to avoid alarming other students who might see the mathematics and feel they recognize themselves in Stevie Smith's poem[†]:

Nobody heard him, the dead man,
But still he lay moaning:
I was much further out than you thought
And not waving but drowning.

Consulting with research workers at West Virginia University has caused us to add some topics not found in earlier editions. Many of our examples and exercises reflect actual research problems for which we provided the statistical analysis. That has not changed, but the research areas that seek our help have become more global. In earlier years we assisted agricultural, biological, and behavioral scientists who can design *prospective* studies, and in our text we tried to meet the needs of their students. After helping researchers in areas such as health science who must depend on *retrospective* studies, we made additions for the benefit of their students as well. We added examples to show how statistics is applied to health research and now discuss risks, odds and their ratios, as well as repeated-measures analysis. While helping researchers prepare manuscripts for publication, we learned that some journals prefer the more conservative Bonferroni procedures, so we have added them to the discussion of mean separation techniques in Chapter 10. We also have a discussion of ratio and difference estimation. However, that inclusion may be self-serving to avoid yet another explanation of "Why go to the all the trouble of least squares when it is so much easier to use a ratio?" Now we can refer the questioner to the appropriate section in Chapter 9.

There are additions to the exercises as well as the body of the text. We believe our students enjoy hearing about the research efforts of Sir Francis Galton, that delightfully eccentric but remarkably ingenious gentleman scientist of Victorian England. To make them suitable exercises, we have taken a few liberties with some of his research efforts, but only to demonstrate the breadth of ideas of a pioneer who thought everything is measurable and hence tractable to quantitative analysis. In respect for a man who—dare we say?—"thought outside the black box," many of the exercises that relate to Galton will require students to think on their own as he did. We hope that, like Galton himself, those who attempt these exercises will accept the challenge and not be too concerned when they do not succeed.

We are pleased that Daniel M. Chilko, a long-time colleague, has joined us in this endeavor. His talents have made it easier to update sections on computer analysis, and he will serve as webmaster for the web site that will now accompany the text.

We wish to acknowledge the help we received from many people in preparation of this edition. Once again, we thank SAS Institute for permission to discuss their SAS and JMP software.

We want to express our appreciation to the many readers who called to our attention a flaw in the algorithm used to prepare the Poisson confidence intervals in Table A8. Because they alerted us, we made corrections and verified all tables generated by us for this edition.

To all who responded to our survey, we are indeed indebted. We especially thank Dr. Marta D. Remmenga, Professor at New Mexico State University. She provided us with a detailed account of how she uses the text to teach statistics and gave us a number of helpful suggestions for this edition. All responses were helpful, and we do appreciate the time taken by so many to answer our questionnaire.

[†]Not Waving But Drowning, *The Top 500 Poems*, Columbia University Press, New York.

Even without this edition, we would be indebted to long-time colleagues in the Department of Statistics at West Virginia University. Over the years, Erdogan Gunel, E. James Harner, and Gerald R. Hobbs have provided the congenial atmosphere and enough help and counsel to make our task easy and joyful.

Shirley M. Dowdy
Stanley Wearden
Daniel M. Chilko

PREFACE TO THE SECOND EDITION

From its inception, the intent of this text has been to demystify statistical procedures for those who employ them in their research. However, between the first and second editions, the use of statistics in research has been radically affected by the increased availability of computers, especially personal computers which can also serve as terminals for access to even more powerful computers. Consequently, we now feel a new responsibility also to try to demystify the computer output of statistical analyses.

Wherever appropriate, we have tried to include computer output for the statistical procedures which have just been demonstrated. We have chosen the output of the SAS® System* for this purpose. SAS was chosen not only for its relative ubiquity on campus and research centers, but also because the SAS printout shares common features with many other statistical analysis packages. Thus if one becomes familiar with the SAS output explained in this text, it should not be too difficult to interpret that of almost any other analysis system. In the main, we have attempted to make the computer output relatively unobtrusive. Where it was reasonable to do so, we placed it toward the end of each chapter and provided output of the computer analysis of the same data for which hand-calculations had already been discussed. For those who have ready access to computers, we have also provided exercises containing raw data to aid in learning how to do statistics on computers.

In order to meet the new objective of demystifying computer output, we have included the programs necessary to obtain the appropriate output from the SAS System. However, the reader should not be mislead in believing this text can serve as a substitute for the SAS manuals. Before one can use the information provided here, it is necessary to know how to access the particular computer system on which SAS is available, and that is likely to be different from one research location to another. Also, to keep the discussion of computer output from becoming too lengthy, we have not discussed a number of other topics such as data editing, storage, and retrieval. We feel the reader who wants to begin using computer analysis will be better served by learning how to do so with the equipment and software available at his or her own research center.

At the request of many who used the first edition, we now include nonparametric statistics in the text. However, once again with the intent of keeping these procedures from seeming to be too arcane, we have approached each nonparametric test as an analog to a previously discussed parametric test, the difference being in the fact that data were collected on the nominal or ordinal scale of measurement, or else transformed to either of these scales of measurement. The test statistics are presented in such a form that they will appear as similar as possible to their parametric counterparts, and for that reason, we consider only large samples

---

*SAS is a registered trademark of SAS Institute Inc., Cary, NC, USA.

for which the central limit theorem will apply. As with the coverage of computer output, the sections on nonparametric statistics are placed near the end of each chapter as material supplementary to statistical procedures already demonstrated.

Finally, those who have reflected on human nature realize that when they are told "no one does that any more," it is really the speaker who doesn't want to do it any more. It is in accord with that interpretation that we say "no one does multiple regression by hand calculations any more," and correspondingly present considerable revision in Chapter 14. Consistent with our intention of avoiding any appearance of mystery, we use a very small sample to present the computations necessary for multiple regression analysis. However, more space is devoted to examination and explanation of the computer analyses available for multiple regression problems.

We are indebted to the SAS Institute for permission to discuss their software. Output from SAS procedures is printed with the permission of SAS Institute Inc., Cary NC, USA, Copyright © 1985.

We want to thank readers of the first edition who have so kindly written to us to advise us of misprints and confusing statements and to make suggestions for improvement. We also want to thank our colleagues in the department, especially Donald F. Butcher, Daniel M. Chilko, E. James Harner, Gerald R. Hobbs, William V. Thayne and Edwin C. Townsend. They have read what we have written, made useful suggestions, and have provided data sets and problems. We feel fortunate to have the benefit of their assistance.

Shirley Dowdy
Stanley Wearden

Morgantown, West Virginia
November 1990

PREFACE TO THE FIRST EDITION

This textbook is designed for the population of students we have encountered while teaching a two-semester introductory statistical methods course for graduate students. These students come from a variety of research disciplines in the natural and social sciences. Most of the students have no prior background in statistical methods but will need to use some, or all, of the procedures discussed in this book before they complete their studies. Therefore, we attempt to provide not only an understanding of the concepts of statistical inference but also the methodology for the most commonly used analytical procedures.

Experience has taught us that students ought to receive their instruction in statistics early in their graduate program, or perhaps, even in their senior year as undergraduates. This ensures that they will be familiar with statistical terminology when they begin critical reading of research papers in their respective disciplines and with statistical procedures before they begin their research. We frequently find, however, that graduate students are poor with respect to mathematical skills; it has been several years since they completed their undergraduate mathematics and they have not used these skills in the subsequent years. Consequently, we have found it helpful to give details of mathematical techniques as they are employed, and we do so in this text.

We should like our students to be aware that statistical procedures are based on sound mathematical theory. But we have learned from our students, and from those with whom we consult, that research workers do not share the mathematically oriented scientists' enthusiasm for elegant proofs of theorems. So we deliberately avoid not only theoretical proofs but even too much of a mathematical tone. When statistics was in its infancy, W. S. Gosset replied to an explanation of the sampling distribution of the partial correlation coefficient by R. A. Fisher:[†]

> ...I fear that I can't conscientiously claim to understand it, but I take it for granted that you know what you are talking about and thankfully use the results!

> It's not so much the mathematics, I can often say "Well, of course, that's beyond me, but we'll take it as correct, but when I come to 'Evidently' I know that means two hours hard work at least before I can see why.

Considering that the original "Student" of statistics was concerned about whether he could understand the mathematical underpinnings of the discipline, it is reasonable that today's students have similar misgivings. Lest this concern keep our students from appreciating the importance of statistics in research, we consciously avoid theoretical mathematical discussions.

[†]From letter No. 6, May 5, 1922, in *Letters From W. S. Gosset to R. A. Fisher 1915–1936*, Arthur Guinness Sons and Company, Ltd., Dublin. Issued for private circulation.

We want to show the importance of statistics in research, and we have taken two specific measures to accomplish this goal. First, to explain that statistics is an integral part of research, we show from the very first chapter of the text how it is used. We have found that our students are impatient with textbooks that require eight weeks of preparatory work before any actual application of statistics to relevant problems. Thus, we have eschewed the traditional introductory discussion of probability and descriptive statistics; these topics are covered only as they are needed. Second, we try to present a practical example of each topic as soon as possible, often with considerable detail about the research problem. This is particularly helpful to those who enroll in the statistical methods course before the research methods course in their particular discipline. Many of the examples and exercises are based on actual research situations that we have encountered in consulting with research workers. We attempt to provide data that are reasonable but that are simplified for each of computation. We realize that in an actual research project a statistical package on a computer will probably be used for the computations, and we considered including printouts of computer analyses. But the multiplicity of the currently available packages, and the rapidity with which they are improved and revised, makes this infeasible.

It is probable that every course has an optimum pace at which it should be taught; we are convinced that such is the case with statistical methods. Because our students come to us unfamiliar with inductive reasoning, we start slowly and try to explain inference in considerable detail. The pace quickens, however, as soon as the students seem familiar with the concepts. Then when new concepts, such as bivariate distributions, are introduced, it is necessary to pause and reestablish the gradual acceleration. Testing helps to maintain the pace, and we find that our students benefit from frequent testing. The exercises at the end of each section are often taken directly from these tests.

A textbook can never replace a reference book. But, many people, because they are familiar with the text they used when they studied statistical methods, often refer to that book for information during later professional activities. We have kept this in mind while designing the text and have included some features that should be helpful: Summaries of procedures are clearly set off, references to articles and books that further develop the topics discussed are given at the end of each chapter, and explanations on reading the statistical tables are given in the table section.

We thank Professor Donald Butcher, Chairman of the Department of Statistics and Computer Science at West Virginia University, for his encouragement of this project. We are also grateful for the assistance of Professor George Trapp and computer science graduate students Barry Miller and Benito Herrera in the production of the statistical methods with us during the preliminary version of the text.

Shirley Dowdy
Stanley Wearden

Morgantown, West Virginia
December 1982

1 The Role of Statistics

In this chapter we informally discuss how statistics is used to attempt to answer questions raised in research. Because probability is basic to statistical decision making, we will also present a few probability rules to show how probabilities are computed. Since this is an overview, we make no attempt to give precise definitions. The more formal development will follow in later chapters.

1.1. THE BASIC STATISTICAL PROCEDURE

Scientists sometimes use statistics to describe the results of an experiment or an investigation. This process is referred to as *data analysis* or *descriptive statistics*. Scientists also use statistics another way; if the entire population of interest is not accessible to them for some reason, they often observe only a portion of the population (a sample) and use statistics to answer questions about the whole population. This process is called *inferential statistics*. Statistical inference is the main focus of this book.

Inferential statistics can be defined as the science of using *probability* to make decisions. Before explaining how this is done, a quick review of the "laws of chance" is in order. Only four probability rules will be discussed here, those for (1) simple probability, (2) mutually exclusive events, (3) independent events, and (4) conditional probability. For anyone wanting more than covered here, Johnson and Kuby (2000) as well as Bennett, Briggs, and Triola (2003) provide more detailed discussion.

Early study of probability was greatly influenced by games of chance. Wealthy games players consulted mathematicians to learn if their losses during a night of gaming were due to bad luck or because they did not know how to compute their chances of winning. (Of course, there was always the possibility of chicanery, but that seemed a matter better settled with dueling weapons than mathematical computations.) Stephen Stigler (1986) states that formal study of probability began in 1654 with the exchange of letters between two famous French mathematicians, Blaise Pascal and Pierre de Fermat, regarding a question posed by a French nobleman about a dice game. The problem can be found in Exercise 1.1.5.

In games of chance, as in experiments, we are interested in the outcomes of a random phenomenon that cannot be predicted with certainty because usually there is more than one outcome and each is subject to chance. The probability of an outcome is a measure of how likely that outcome is to occur. The random outcomes associated with games of chance should be equally likely to occur if the gambling device is *fair*, controlled by chance alone. Thus the probability of getting a head on a single toss of a fair coin and the probability of getting an even number when we roll a fair die are both $1/2$.

Statistics for Research, *Third Edition*, Edited by Shirley Dowdy, Stanley Weardon, and Daniel Chilko. ISBN 0-471-26735-X © 2004 John Wiley & Sons, Inc.

Because of the early association between probability and games of chance, we label some collection of equally likely outcomes as a success. A collection of outcomes is called an event. If success is the event of an even number of pips on a fair die, then the event consists of outcomes 2, 4, and 6. An event may consist of only one outcome, as the event head on a single toss of a coin. The probability of a success is found by the following probability rule:

$$\text{probability of success} = \frac{\text{number of successful outcomes}}{\text{total number of outcomes}}$$

In symbols

$$P(\text{success}) = P(S) = \frac{n_s}{N}$$

where n_S is the number of outcomes in the event designated as success and N is the total number of possible outcomes. Thus the simple probability rule for equally likely outcomes is to count the number of ways a success can be obtained and divide it by the total number of outcomes.

Example 1.1. Simple Probability Rule for Equally Likely Outcomes

There is a game, often played at charity events, that involves tossing a coin such as a 25-cent piece. The quarter is tossed so that it bounces off a board and into a chute to land in one of nine glass tumblers, only one of which is red. If the coin lands in the red tumbler, the player wins \$1; otherwise the coin is lost. In the language of probability, there are $N = 9$ possible outcomes for the toss and only one of these can lead to a success. Assuming skill is not a factor in this game, all nine outcomes are equally likely and $P(\text{success}) = 1/9$.

In the game described above, $P(\text{win}) = 1/9$ and $P(\text{loss}) = 8/9$. We observe there is only one way to win \$1 and eight ways to lose 25¢. A related idea from the early history of probability is the concept of *odds*. The odds for winning are $P(\text{win})/P(\text{loss})$. Here we say, "The odds for winning are one to eight" or, more pessimistically, "The odds against winning are eight to one." In general,

$$\text{odds for success} = \frac{P(\text{success})}{1 - P(\text{success})}$$

We need to stress that the simple probability rule above applies only to an experiment with a discrete number of equally likely outcomes. There is a similarity in computing probabilities for *continuous variables* for which there is a distribution curve for measures of the variable. In this case

$$P(\text{success}) = \frac{\text{area under the curve where the measure is called a success}}{\text{total area under the curve}}$$

A simple example is provided by the "spinner" that comes with many board games. The spinner is an arrow that spins freely around an axle attached to the center of a circle. Suppose that the circle is divided into quadrants marked 1, 2, 3, and 4 and play on the board is determined by the quadrant in which the spinner comes to rest. If no skill is involved in spinning the arrow, the outcomes can be considered uniformly distributed over the 360° of the circle. If it is a success to land in the third quadrant of the circle, a spin is a success when the arrow stops anywhere in the $90°$ of the third quadrant and

$$P(\text{success}) = \frac{\text{area in third quadrant}}{\text{total area}} = \frac{90}{360} = \frac{1}{4}$$

While only a little geometry is needed to calculate probabilities for a *uniform distribution*, knowledge of calculus is required for more complex distributions. However, finding probabilities for many continuous variables is possible by using simple tables. This will be explained in later chapters.

The next rule involves events that are *mutually exclusive*, meaning one event excludes the possibility of another. For instance, if two dice are rolled and the event is that the sum of spots is $y = 7$, then y cannot possibly be another value as well. However, there are six ways that the spots, or pips, on two dice can produce a sum of 7, and each of these is mutually exclusive of the others. To see how this is so, imagine that the pair consists of one red die and one green; then we can detail all the possible outcomes for the event $y = 7$:

Red die:	1	2	3	4	5	6
Green die:	6	5	4	3	2	1
Sum:	7	7	7	7	7	7

If a success depends only on a value of $y = 7$, then by the simple probability rule the number of possible successes is $n_S = 6$; the number of possible outcomes is $N = 36$ because each of the six outcomes of the red die can be paired with each of the six outcomes of the green die and the total number of outcomes is $6 \times 6 = 36$. Thus $P(\text{success}) = n_S/N = 6/36 = 1/6$. However, we need a more general statement to cover mutually exclusive events, whether or not they are equally likely, and that is the addition rule.

If a success is any of k mutually exclusive events $E_1, E_2, \ldots, E_k$, then the *addition rule for mutually exclusive events* is $P(\text{success}) = P(E_1) + P(E_2) + \cdots + P(E_k)$. This holds true with the dice; if E_1 is the event that the red die shows 1 and the green die shows 6, then $P(E_1) = 1/36$. Then, because each of the $k = 6$ events has the same probability,

$$P(\text{success}) = \left(\frac{1}{36}\right) + \left(\frac{1}{36}\right) + \left(\frac{1}{36}\right) + \left(\frac{1}{36}\right) + \left(\frac{1}{36}\right) + \left(\frac{1}{36}\right) = \frac{6}{36} = \frac{1}{6}$$

Here $1/36$ is the common probability for all events, but the addition rule for mutually exclusive events still holds true even when the probability values are not the same for all events.

Example 1.2. Addition Rule for Mutually Exclusive Events

To see how this rule applies to events that are not equally likely, suppose a coin-operated gambling device is programmed to provide, on random plays, winnings with the following probabilities:

Event	$P(\text{Event})$
Win 10 coins	0.001
Win 5 coins	0.010

Event	P(Event)
Win 3 coins	0.040
Win 1 coin	0.359
Lose 1 coin	0.590

Because most players consider it a success if any coins are won, $P(\text{success}) = 0.0001 + 0.010 + 0.040 + 0.359 = 0.410$, and the odds for winning are $0.41/0.59 = 0.695$, while the odds against a win are $0.59/0.41 = 1.44$.

We might ask why we bother to add $0.0001 + 0.010 + 0.040 + 0.359$ to obtain $P(\text{success}) = 0.41$ when we can obtain it just from knowledge of $P(\text{no success})$. On a play at the coin machine, one either wins of loses, so there is the probability of a success, $P(S) = 0.41$, and the probability of no success, $P(\text{no success}) = 0.59$. The opposite of a success, is called its *complement*, and its probability is symbolized as $P(\bar{S})$. In a play at the machine there is no possibility of neither a win nor a loss, $P(S) + P(\bar{S}) = 1.0$, so rather than counting the four ways to win it is easier to find $P(S) = 1.0 - P(\bar{S}) = 1.0 - 0.59 = 0.41$. Note that in the computation of the odds for winning we used the ratio of the probability of a win to its complement, $P(S)/P(\bar{S})$.

At games of chance, people who have had a string of losses are encouraged to continue to play with such remarks as "Your luck is sure to change" or "Odds favor your winning now," but is that so? Not if the plays, or events, are independent. A play in a game of chance has no memory of what happened on previous plays. So using the results of Example 1.2, suppose we try the machine three times. The probability of a win on the first play is $P(S_1) = 0.41$, but the second coin played has no memory of the fate of its predecessor, so $P(S_2) = 0.41$, and likewise $P(S_3) = 0.41$. Thus we could insert 100 coins in the machine and lose on the first 99 plays, but the probability that our last coin will win remains $P(S_{100}) = 0.41$. However, we would have good reason to suspect the honesty of the machine rather than bad luck, for with an honest machine for which the probability of a win is 0.41, we would expect about 41 wins in 100 plays.

When dealing with independent events, we often need to find the *joint probability* that two or more of them will all occur simultaneously. If the total number of possible outcomes (N) is small, we can always compile tables, so with the $N = 52$ cards in a standard deck, we can classify each card by color (red or black) and as to whether or not it is an honor card (ace, king, queen, or jack). Then we can sort and count the cards in each of four groups to get the following table:

Color

Honor	Black	Red	Total
No	18	18	36
Yes	8	8	16
Total	26	26	52

If a card is dealt at random from such a deck, we can find the *joint probability* that it will be red and an honor by noting that there are 8 such cards in the deck of 52; hence $P(\text{red and honor}) = P(RH) = 8/52 = 2/13$. This is easy enough when the total number of outcomes is small or when they have already been tabulated, but in many cases there are too many or there is a process such as the slot machine capable of producing an infinite number of outcomes. Fortunately there is a probability rule for such situations.

The multiplication rule for finding the joint probability of k independent events E_1, $E_2, \ldots, E_k$ is

$$P(E_1 \text{ and } E_2 \text{ and } \ldots E_k) = P(E_1) \times P(E_2) \times \cdots \times P(E_k)$$

With the cards, k is 2, E_1 is a red card, and E_2 is an honor card, so $P(E_1E_2) = P(E_1) \times P(E_2) = (26/52) \times (16/52) = (1/2) \times (4/13) = 4/26 = 2/13$.

Example 1.3. The Multiplication Rule for Independent Events

Gender and handedness are independent, and if $P(\text{female}) = 0.50$ and $P(\text{left handed}) = 0.15$, then the probability that the first child of a couple will be a left-handed girl is

$$P(\text{female and left handed}) = P(\text{female}) \times P(\text{left handed}) = 0.50 \times 0.15 = 0.075$$

If the probability values $P(\text{female})$ and $P(\text{left handed})$ are realistic, the computation is easier than the alternative of trying to tabulate the outcomes of all first births. We know the biological mechanism for determining gender but not handedness, so it was only estimated here. However, the value we would obtain from a tabulation of a large number of births would also be only an estimate. We will see in Chapter 3 how to make estimates and how to say scientifically, "The probability that the first child will be a left-handed girl is likely somewhere around 0.075."

The multiplication rule is very convenient when events are independent, but frequently we encounter events that are not independent but rather are at least partially related. Thus we need to understand these and how to deal with them in probability. When told that a person is from Sweden or some other Nordic country, we might immediately assume that he or she has blue eyes, or conversely dark eyes if from a Mediterranean country. In our encounters with people from these areas, we think we have found that the probability of eye color $P(\text{blue})$ is not the same for both those geographic regions but rather depends, or *is conditioned*, on the region from which a person comes. *Conditional probability* is symbolized as $P(E_2|E_1)$, and we say "The probability of event 2 given event 1." In the case of eye color, it would be the probability of blue eyes given that one is from a Nordic country.

The *conditional probability rule* for finding the conditional probability of event 2 given event 1 is

$$P(E_2|E_1) = \frac{P(E_1E_2)}{P(E_1)}$$

In the deck of cards, the probability a randomly dealt card will be red and an honor card is $P(\text{red and honor}) = 8/52$, while the probability it is red is $P(R) = 26/52$, so the probability that it will be an honor card, given that it is a red card is $P(RH)/P(R) = 8/26 = 4/13$, which is the same as $P(H)$ because the two are independent rather than related. Hence independent events can be defined as satisfying $P(E_2|E_1) = P(E_2)$.

Example 1.4.   The Conditional Probability Rule

Suppose an oncologist is suspicious that cancer of the gum may be associated with use of smokeless tobacco. It would be ideal if he also had data on the use of smokeless tobacco by those free of cancer, but the only data immediately available are from 100 of his own cancer patients, so he tabulates them to obtain the following:

Smokeless Tobacco

Cancer Site	No	Yes	Total
Gum	5	20	25
Elsewhere	60	15	75
Total	65	35	100

There are 25 cases of gum cancer in his database and 20 of those patients had used smokeless tobacco, so we see that his best estimate of the probability that a randomly drawn gum cancer patient was a user of smokeless tobacco is $20/25 = 0.80$. This probability could also be found by the conditional probability rule. If $P(\text{gum}) = P(G)$ and $P(\text{user}) = P(U)$, then

$$P(U|G) = \frac{P(GU)}{P(G)} = \frac{(20/100)}{(25/100)} = \frac{20}{25} = 0.80$$

Are gum cancer and use of smokeless tobacco independent? They are if $P(U|G) = P(U)$, and from the data set, the best estimate of users among all cancer patients is $P(U) = 35/100 = 0.35$. The discrepancy in estimates is 0.80 for gum cancer patients compared to 0.35 for all patients. This leads us to believe that gum cancer and smokeless tobacco usage are related rather than independent. In Chapter 5, we will see how to test to see whether or not two variables are independent.

Odds obtained from medical data sets similar to but much larger than that in Example 1.4 are frequently cited in the news. Had the odds been the same in a data set of hundreds or thousands of gum cancer patients, we would report that the odds were $0.80/0.20 = 4.0$ for smokeless tobacco, and $0.35/0.65 = 0.538$ for smokeless tobacco among all cancer patients. Then, for sake of comparison, we would report the *odds ratio*, which is the ratio of the two odds, $4.0/0.538 = 7.435$. This ratio gives the relative frequency of smokeless tobacco users among gum cancer patients to smokeless tobacco users among all cancer patients, and the medical implications are ominous. For comparison, it would be helpful to have data on the usage of smokeless tobacco in a cancer-free population, but first information about an association such as that in Example 1.4 usually comes from medical records for those with a disease.

Caution is necessary when trying to interpret odds ratios, especially those based on very low incidences of occurrence. To show a totally meaningless odds ratio, suppose we have two data sets, one containing 20 million broccoli eaters and the other of 10 million who do not eat the vegetable. Then, if we examine the health records of those in each group, we find there are two in each group suffering from chronic bladder infections. The odds ratio is 2.0, but we would garner strange looks rather than prestige if we attempted to claim that the odds for

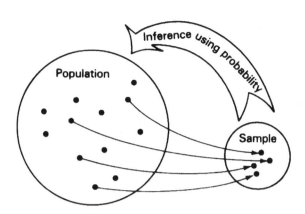

FIGURE 1.1. Statistical inference.

chronic bladder infection is twice as great for broccoli eaters when compared to those who do not eat the vegetable. To use statistics in research is happily more than just to compute and report numbers.

The basic process in inferential statistics is to assign probabilities so that we can reach conclusions. The inferences we make are either decisions or estimates about the population. The tool for making inferences is probability (Figure 1.1).

We can illustrate this process by the following example.

Example 1.5.   Using Probabilities to Make a Decision

A sociologist has two large sets of cards, set A and set B, containing data for her research. The sets each consist of 10,000 cards. Set A concerns a group of people, half of whom are women. In set B, 80% of the cards are for women. The two files look alike. Unfortunately, the sociologist loses track of which is A and which is B. She does not want to sort and count the cards, so she decides to use probability to identify the sets. The sociologist selects a set. She draws a card at random from the selected set, notes whether or not it concerns a woman, replaces the card, and repeats this procedure 10 times. She finds that all 10 cards contain data about women. She must now decide between two possible conclusions:

1. This is set B.
2. This is set A, but an unlikely sample of cards has been chosen.

In order to decide in favor of one of these conclusions, she computes the probabilities of obtaining 10 cards all for females:

$$P(10 \text{ females}) = P(\text{first is female})$$
$$\times P(\text{second is female}) \times \cdots \times P(\text{tenth is female})$$

The multiplication rule is used because each choice is independent of the others. For the set A, the probability of selecting 10 cards for females is $(0.50)^{10} = 0.00098$ (rounded to two significant digits). For set B, the probability of 10 cards for females is $(0.80)^{10} = 0.11$ (again rounded to two significant digits). Since the probability of all 10 of the cards being for women if the set is B is about 100 times the probability if the set is A, she decides that the set is B, that is, she decides in favor of the conclusion with the higher probability.

When we use a strategy based on probability, we are not guaranteed success every time. However, if we repeat the strategy, we will be correct more often than mistaken. In the above example, the sociologist could make the wrong decision because 10 cards chosen at random from set A could all be cards for women. In fact, in repeated experiments using set A, 10 cards for females will appear approximately 0.098% of the time, that is, almost once in every thousand 10-card samples.

The example of the files is artificial and oversimplified. In real life, we use statistical methods to reach conclusions about some significant aspect of research in the natural, physical, or social sciences. Statistical procedures do not furnish us with proofs, as do many mathematical techniques. Rather, statistical procedures establish probability bases on which we can accept or reject certain hypotheses.

Example 1.6.   Using Probability to Reach a Conclusion in Science

A real example of the use of statistics in science is the analysis of the effectiveness of Salk's polio vaccine.

A great deal of work had to be done prior to the actual experiment and the statistical analysis. Dr. Jonas Salk first had to gather enough preliminary information and experience in his field to know which of the three polio viruses to use. He had to solve the problem of how to culture that virus. He also had to determine how long to treat the virus with formaldehyde so that it would die but retain its protein shell in the same form as the live virus; the shell could then act as an antigen to stimulate the human body to develop antibodies. At this point, Dr. Salk could conjecture that the dead virus might be used as a vaccine to give patients immunity to paralytic polio.

Finally, Dr. Salk had to decide on the type of experiment that would adequately test his conjecture. He decided on a *double-blind* experiment in which neither patient nor doctor knew whether the patient received the vaccine or a saline solution. The patients receiving the saline solution would form the *control group*, the standard for comparison. Only after all these preliminary steps could the experiment be carried out.

When Dr. Salk speculated that patients inoculated with the dead virus would be immune to paralytic polio, he was formulating the *experimental hypothesis*: the expected outcome if the experimenter's speculation is true. Dr. Salk wanted to use statistics to make a decision about this experimental hypothesis. The decision was to be made solely on the basis of probability. He made the decision in an indirect way; instead of considering the experimental hypothesis itself, he considered a statistical hypothesis called the *null hypothesis*—the expected outcome if the vaccine is ineffective and only chance differences are observed between the two sample groups, the inoculated group and the control group. The null hypothesis is often called the hypothesis of no difference, and it is symbolized H_0. In Dr. Salk's experiment, the null hypothesis is that the incidence of paralytic polio in the general population will be the same whether it receives the proposed vaccine or the saline solution. In symbols[†]

$$H_0: \pi_I = \pi_C.$$

[†] The use of the symbol π has nothing to do with the geometry of circles or the irrational number $3.1416\dots$.

in which π_I is the proportion of cases of paralytic polio in the general population if it were inoculated with the vaccine and π_C is the proportion of cases if it received the saline solution. If the null hypothesis is true, then the two sample groups in the experiment should be alike except for chance differences of exposure and contraction of the disease.

The experimental results were as follows:

	Proportion with Paralytic Polio	Number in Study
Inoculated Group	0.0001603	200,745
Control Group	0.0005703	201,229

The incidence of paralytic polio in the control group was almost four times higher than in the inoculated group, or in other words the odds ratio was $0.0005703/0.0001603 = 3.56$.

Dr. Salk then found the probability that these experimental results or more extreme ones could have happened with a true null hypothesis. The probability that $\pi_I = \pi_C$ and the difference between the two experimental groups was caused by chance was less than 1 in 10,000,000, so Salk rejected the null hypothesis and decided that he had found an effective vaccine for the general public.[†]

Usually when we experiment, the results are not as conclusive as the result obtained by Dr. Salk. The probabilities will always fall between 0 and 1, and we have to establish a level below which we reject the null hypothesis and above which we accept the null hypothesis. If the probability associated with the null hypothesis is small, we reject the null hypothesis and accept an alternative hypothesis (usually the experimental hypothesis). When the probability associated with the null hypothesis is large, we accept the null hypothesis. This is one of the basic procedures of statistical methods—to ask: What is the probability that we would get these experimental results (or more extreme ones) with a true null hypothesis?

Since the experiment has already taken place, it may seem after the fact to ask for the probability that only chance caused the difference between the observed results and the null hypothesis. Actually, when we calculate the probability associated with the null hypothesis, we are asking: If this experiment were performed over and over, what is the probability that chance will produce experimental results as different as are these results from what is expected on the basis of the null hypothesis?

We should also note that Salk was interested not only in the samples of 401,974 people who took part in the study; he was also interested in *all* people, then and in the future, who could receive the vaccine. He wanted to make an inference to the entire population from the portion of the population that he was able to observe. This is called the *target population*, the population about which the inference is intended.

Sometimes in science the inference we should like to make is not in the form of a decision about a hypothesis; but rather it consists of an estimate. For example, perhaps we want to estimate the proportion of adult Americans who approve of the way in which the president is handling the economy, and we want to include some statement about the amount of error possibly related to this estimate. Estimation of this type is another kind of inference, and it also depends on probability. For simplicity, we focus on tests of hypotheses in this

---

[†]This probability is found using a chi-square test (see Section 5.3).

introductory chapter. The first example of inference in the form of estimation is discussed in Chapter 3.

EXERCISES

1.1.1. A trial mailing is made to advertise a new science dictionary. The trial mailing list is made up of random samples of current mailing lists of several popular magazines. The number of advertisements mailed and the number of people who ordered the dictionary are as follows:

	Magazine				
	A	B	C	D	E
Mailed:	900	810	1100	890	950
Ordered:	18	15	10	30	45

    a. Estimate the probability and the odds that a subscriber to each of the magazines will buy the dictionary.

    b. Make a decision about the mailing list that will probably produce the highest percentage of sales if the entire list is used.

1.1.2. In Examples 1.5 and 1.6, probability was used to make decisions and odds ratios could have been used to further support the decisions. To do so:

    a. For the data in Example 1.5, compute the odds ratio for the two sets of cards.

    b. For the data in Example 1.6, compute the odds ratio of getting polio for those vaccinated as opposed to those not vaccinated.

1.1.3. If 60% of the population of the United States need to have their vision corrected, we say that the probability that an individual chosen at random from the population needs vision correction is $P(C) = 0.60$.

    a. Estimate the probability that an individual chosen at random does not need vision correction. *Hint:* Use the complement of a probability.

    b. If 3 people are chosen at random from the population, what is the probability that all 3 need correction, $P(CCC)$? *Hint:* Use the multiplication law of probability for independent events.

    c. If 3 people are chosen at random from the population, what is the probability that the second person does not need correction but the first and the third do, $P(CNC)$?

    d. If 3 people are chosen at random from the population, what is the probability that 1 out of the 3 needs correction, $P(CNN$ or NCN or $NNC)$? *Hint:* Use the addition law of probability for mutually exclusive events.

    e. Assuming no association between vision and gender, what is the probability that a randomly chosen female needs vision correction, $P(C|F)$?

1.1.4. On a single roll of 2 dice (think of one green and the other red to keep track of all outcomes) in the game of craps, find the probabilities for:

    a. A sum of 6, $P(y = 6)$

b. A sum of 8, $P(y = 8)$

c. A win on the first roll; that is, a sum of 7 or 11, $P(y = 7$ or $11)$

d. A loss on the first roll; that is, a sum of 2, 3, or 12, $P(y = 2, 3,$ or $12)$

1.1.5. The dice game about which Pascal and de Fermat were asked consisted in throwing a pair of dice 24 times. The problem was to decide whether or not to bet even money on the occurrence of at least one "double 6" during the 24 throws of a pair of dice. Because it is easier to solve this problem by finding the complement, take the following steps:

a. What is the probability of not a double 6 on a roll, $P(E) = P(y \neq 12)$?

b. What is the probability that $y = 12$ on all 24 rolls, $P(E_1 E_2, \ldots, E_{24})$?

c. What is the probability of at least one double 6?

d. What are the odds of a win in this game?

1.1.6. Sir Francis Galton (1822–1911) was educated as a physician but had the time, money, and inclination for research on whatever interested him, and almost everything did. Though not the first to notice that he could find no two people with the same fingerprints, he was the first to develop a system for categorizing fingerprints and to persuade Scotland Yard to use fingerprints in criminal investigation. He supported his argument with fingerprints of friends and volunteers solicited through the newspapers, and for all comparisons $P(\text{fingerprints match}) = 0$. To compute the number of events associated with Galton's data:

a. Suppose fingerprints on only 10 individuals are involved.

i. How many comparisons between individuals can be made? *Hint:* Fingerprints of the first individual can be compared to those of the other 9. However, for the second individual there are only 8 additional comparisons because his fingerprints have already been compared to the first.

ii. How many comparisons between fingers can be made? Assume these are between corresponding fingers of both individuals in a comparison, right thumb of one versus right thumb of the other, and so on.

b. Suppose fingerprints are available on 11 individuals rather than 10. Use the results already obtained to simplify computations in finding the number of comparisons among people and among fingers.

1.2. THE SCIENTIFIC METHOD

The natural, physical, and social scientists who use statistical methods to reach conclusions all approach their problems by the same general procedure, the *scientific method*. The steps involved in the scientific method are:

1. State the problem.
2. Formulate the hypothesis.
3. Design the experiment or survey.
4. Make observations.
5. Interpret the data.
6. Draw conclusions.

We use statistics mainly in step 5, "interpret the data." In an indirect way we also use statistics in steps 2 and 3, since the formulation of the hypothesis and the design of the experiment or survey must take into consideration the type of statistical procedure to be used in analyzing the data.

The main purpose of this book is to examine step 5. We frequently discuss the other steps, however, because an understanding of the total procedure is important. A statistical analysis may be flawless, but it is not valid if data are gathered incorrectly. A statistical analysis may not even be possible if a question is formulated in such a way that a statistical hypothesis cannot be tested. Considering all of the steps also helps those who study statistical methods before they have had much practical experience in using the scientific method. A full discussion of the scientific method is outside the scope of this book, but in this section we make some comments on the five steps.

STEP 1. STATE THE PROBLEM. Sometimes, when we read reports of research, we get the impression that research is a very orderly analytic process. Nothing could be further from the truth. A great deal of hidden work and also a tremendous amount of intuition are involved before a solvable problem can even be stated. Technical information and experience are indispensable before anyone can hope to formulate a reasonable problem, but they are not sufficient. The mediocre scientist and the outstanding scientist may be equally familiar with their field; the difference between them is the intuitive insight and skill that the outstanding scientist has in identifying relevant problems that he or she can reasonably hope to solve.

One simple technique for getting a problem in focus is to formulate a clear and explicit statement of the problem and put the statement in writing. This may seem like an unnecessary instruction for a research scientist; however, it is frequently not followed. The consequence is a vagueness and lack of focus that make it almost impossible to proceed. It leads to the collection of unnecessary information or the failure to collect essential information. Sometimes the original question is even lost as the researcher gets involved in the details of the experiment.

STEP 2. FORMULATE THE HYPOTHESIS. The "hypothesis" in this step is the experimental hypothesis, the expected outcome if the experimenter's speculations are true. The experimental hypothesis must be stated in a precise way so that an experiment can be carried out that will lead to a decision about the hypothesis. A good experimental hypothesis is comprehensive enough to explain a phenomenon and predict unknown facts and yet is stated in a simple way. Classic examples of good experimental hypotheses are Mendel's laws, which can be used to explain hereditary characteristics (such as the color of flowers) and to predict what form the characteristics will take in the future.

Although the null hypothesis is not used in a formal way until the data are being interpreted, it is appropriate to formulate the null hypothesis at this time in order to verify that the experimental hypothesis is stated in such a way that it can be tested by statistical techniques.

Several experimental hypotheses may be connected with a single problem. Once these hypotheses are formulated in a satisfactory way, the investigator should do a literature search to see whether the problem has already been solved, whether or not there is hope of solving it, and whether or not the answer will make a worthwhile contribution to the field.

STEP 3. DESIGN THE EXPERIMENT OR SURVEY. Included in this step are several decisions. What *treatments* or conditions should be placed on the objects or subjects of the investigation in order to test the hypothesis? What are the *variables* of interest, that is, what variables should be measured? How will this be done? With how much precision? Each of these decisions is complex and requires experience and insight into the particular area of investigation.

Another group of decisions involves the choice of the *sample*, that portion of the population of interest that will be used in the study. The investigator usually tries to utilize samples that are:

(a) Random

(b) Representative

(c) Sufficiently large

In order to make a decision based on probability, it is necessary that the sample be *random*. Random samples make it possible to determine the probabilities associated with the study. A sample is random if it is just as likely that it will be picked from the population of interest as any other sample of that size. Strictly speaking, statistical inference is not possible unless random samples are used. (Specific methods for achieving random samples are discussed in Section 2.2.)

Random, however, does not mean haphazard. Haphazard processes often have hidden factors that influence the outcome. For example, one scientist using guinea pigs thought that time could be saved in choosing a treatment group and a control group by drawing the treatment group of animals from a box without looking. The scientist drew out half of the guinea pigs for testing and reserved the rest for the control group. It was noticed, however, that most of the animals in the treatment group were larger than those in the control group. For some reason, perhaps because they were larger, or slower, the heavier guinea pigs were drawn first. Instead of this haphazard selection, the experimenter could have recorded the animals' ear-tattoo numbers on plastic disks and drawn the disks at random from a box.

Unfortunately, in many fields of investigation random sampling is not possible, for example, meteorology, some medical research, and certain areas of economics. Random samples are the ideal, but sometimes only nonrandom data are available. In these cases the investigator may decide to proceed with statistical inference, realizing, of course, that it is somewhat risky. Any final report of such a study should include a statement of the author's awareness that the requirement of randomness for inference has not been met.

The second condition that an investigator often seeks in a sample is that it be *representative*. Usually we do not know how to find truly representative samples. Even when we think we can find them, we are often governed by a subconscious bias.

A classic example of a subconscious bias occurred at a Midwestern agricultural station in the early days of statistics. Agronomists were trying to predict the yield of a certain crop in a field. To make their prediction, they chose several 6-ft × 6-ft sections of the field which they felt were representative of the crop. They harvested those sections, calculated the arithmetic average of the yields, then multiplied this average by the number of 36-ft^2 sections in the field to estimate the total yield. A statistician assigned to the station suggested that instead they should have picked random sections. After harvesting several random sections, a second average was calculated and used to predict the total yield. At harvest time, the actual yield of the field was closer to the yield predicted by the statistician. The agronomists had predicted a much larger yield, probably because they chose sections that looked like an ideal crop. An entire field, of course, is not ideal. The unconscious bias of the agronomists prevented them from picking a representative sample. Such unconscious bias cannot occur when experimental units are chosen at random.

Although representativeness is an intuitively desirable property, in practice it is usually an impossible one to meet. How can a sample of 30 possibly contain all the properties of a population of 2000 individuals? The 2000 certainly have more characteristics than can possibly be proportionately reflected in 30 individuals. So although representativeness seems necessary for proper reasoning from the sample to the population, statisticians do not rely on representative samples—rather, they rely on random samples. (Large random samples will very likely be representative). If we do manage to deliberately construct a sample that is representative but is not random, we will be unable to compute probabilities related to the sample and, strictly speaking, we will be unable to do statistical inference.

It is also necessary that samples be *sufficiently large*. No one would question the necessity of repetition in an experiment or survey. We all know the danger of generalizing from a single observation. Sufficiently large, however, does not mean massive repetition. When we use statistics, we are trying to get information from relatively small samples. Determining a reasonable sample size for an investigation is often difficult. The size depends upon the magnitude of the difference we are trying to detect, the variability of the variable of interest, the type of statistical procedure we are using, the seriousness of the errors we might make, and the cost involved in sampling. (We make further remarks on sample size as we discuss various procedures throughout this text.)

STEP 4. MAKE OBSERVATIONS. Once the procedure for the investigation has been decided upon, the researcher must see that it is carried out in a rigorous manner. The study should be free from all errors except random measurement errors, that is, slight variations that are due to the limitations of the measuring instrument.

Care should be taken to avoid *bias*. Bias is a tendency for a measurement on a variable to be affected by an external factor. For example, bias could occur from an instrument out of calibration, an interviewer who influences the answers of a respondent, or a judge who sees the scores given by other judges. Equipment should not be changed in the middle of an experiment, and judges should not be changed halfway through an evaluation.

The data should be examined for unusual values, *outliers*, which do not seem to be consistent with the rest of the observations. Each outlier should be checked to see whether or not it is due to a recording error. If it is an error, it should be corrected. If it cannot be corrected, it should be discarded. If an outlier is not an error, it should be given special attention when the data are analyzed. For further discussion, see Barnett and Lewis (2002).

Finally, the investigator should keep a complete, legible record of the results of the investigation. All original data should be kept until the analysis is completed and the final report written. Summaries of the data are often not sufficient for a proper statistical analysis.

STEP 5. INTERPRET THE DATA. The general statistical procedure was illustrated in Example 1.6, in which the Salk vaccine experiment was discussed. To interpret the data, we set up the null hypothesis and then decide whether the experimental results are a rare outcome if the null hypothesis is true. That is, we decide whether the difference between the experimental outcome and the null hypothesis is due to more than chance; if so, this indicates that the null hypothesis should be rejected.

If the results of the experiment are unlikely when the null hypothesis is true, we reject the null hypothesis; if they are expected, we accept the null hypothesis. We must remember, however, that statistics does not prove anything. Even Dr. Salk's result, with a probability of less than 1 in 10,000,000 that chance was causing the difference between the experimental outcome and the null hypothesis, does not prove that the null hypothesis is false. An extremely small probability, however, does make the scientist believe that the difference is not due to chance alone and that some additional mechanism is operating.

Two slightly different approaches are used to evaluate the null hypothesis. In practice, they are often intermingled. Some researchers compute the probability that the experimental results, or more extreme values, could occur if the null hypothesis is true; then they use that probability to make a judgment about the null hypothesis. In research articles this is often reported as the *observed significance level*, or the *significance level*, or the *P value*. If the *P* value is large, they conclude that the data are consistent with the null hypothesis. If the *P* value is small, then either the null hypothesis is false or the null hypothesis is true and a rare event has occurred. (This was the approach used in the Salk vaccine example.)

Other researchers prefer a second, more decisive approach. Before the experiment they decide on a *rejection level*, the probability of an unlikely event (sometimes this is also called the *significance level*). An experimental outcome, or a more extreme one, that has a probability below this level is considered to be evidence that the null hypothesis is false. Some research articles are written with this approach. It has the advantage that only a limited number of probability tables are necessary. Without a computer, it is often difficult to determine the exact *P* value needed for the first approach. For this reason the second approach became popular in the early days of statistics. It is still frequently used.

The sequence in this second procedure is:

(a) Assume H_0 is true and determine the probability P that the experimental outcome or a more extreme one would occur.

(b) Compare the probability to a preset rejection level symbolized by α (the Greek letter alpha).

(c) If $P \leq \alpha$, reject H_0. If $P > \alpha$, accept H_0.

If $P > \alpha$, we say, "Accept the null hypothesis." Some statisticians prefer not to use that expression, since in the absence of evidence to reject the null hypothesis, they choose simply to withhold judgment about it. This group would say, "The null hypothesis may be true" or "There is no evidence that the null hypothesis is false."

If the probability associated with the null hypothesis is very close to α, more extensive testing may be desired. Notice that this is a blend of the two approaches.

An example of the total procedure follows.

Example 1.7.   Using a Statistical Procedure to Interpret Data

A manufacturer of baby food gives samples of two types of baby cereal, A and B, to a random sample of four mothers. Type *A* is the manufacturer's brand, type *B* a competitor's. The mothers are asked to report which type they prefer. The manufacturer wants to detect any preference for their cereal if it exists.

The null hypothesis, or the hypothesis of no difference, is H_0: $\pi = 1/2$, in which π is the proportion of mothers in the general population who prefer type *A*. The experimental hypothesis, which often corresponds to a second statistical hypothesis called the *alternative hypothesis*, is that there is a preference for cereal A, H_a: $\pi > 1/2$.

Suppose that four mothers are asked to choose between the two cereals. If there is no preference, the following 16 outcomes are possible with equal probability:

AAAA	*AAAB*	*ABBA*	*BBAB*
BAAA	*BBAA*	*ABAB*	*BABB*
ABAA	*BABA*	*AABB*	*ABBB*
AABA	*BAAB*	*BBBA*	*BBBB*

The manufacturer feels that only 1 of these 16 cases, *AAAA*, is very different from what would be expected to occur under random sampling, when the null hypothesis of no preference is true. Since the unusual case would appear only 1 time out of 16 times when the null hypothesis is true, α (the rejection level) is set equal to $1/16 = 0.0625$.

If the outcome of the experiment is in fact four choices of type *A*, then $P = P(AAAA) = 1/16$, and the manufacturer can say that the results are in the region of rejection, or the results are significant, and the null hypothesis is rejected. If the outcome is three choices of type *A*, however, then $P = P(3 \text{ or more } A\text{'s}) = P(AAAB \text{ or } AABA \text{ or } ABAA \text{ or } BAAA \text{ or } AAAA) = 5/16 > 1/16$, and he does not reject the null hypothesis. (Notice that P is the probability of this type of outcome or a more extreme one in the direction of the alternative hypothesis, so *AAAA* must be included.)

The way in which we set the rejection level α depends on the field of research, on the seriousness of an error, on cost, and to a great degree on tradition. In the example above, the sample size is 4, so an α smaller than $1/16$ is impossible. Later (in Section 3.2), we discuss using the seriousness of errors to determine a reasonable α. If the possible errors are not serious and cost is not a consideration, traditional values are often used.

Experimental statistics began about 1920 and was not used much until 1940, but it is already tradition bound. In the early part of the twentieth century Karl Pearson had his students at University College, London, compute tables of probabilities for reasonably rare events. Now computers are programmed to produce these tables, but the traditional levels used by Pearson persist for the most part. Tables are usually calculated for α equal to 0.10, 0.05, and 0.01. Many times there is no justification for the use of one of these values except tradition and the availability of tables. If an α close to but less than or equal to 0.05 were desired in the example above, a sample size of at least 5 would be necessary, then $\alpha = 1/32 = 0.03125$ if the only extreme case is *AAAAA*.

STEP 6. DRAW CONCLUSIONS. If the procedure just outlined is followed, then our decisions will be based solely on probability and will be consistent with the data from the experiment. If our experimental results are not unusual for the null hypothesis, $P > \alpha$, then the null hypothesis seems to be right and we should not reject it. If they are unusual, $P \leq \alpha$, then the null hypothesis seems to be wrong and we should reject it. We repeat that our decision could be incorrect, since there is a small probability α that we will reject a null hypothesis when in fact that null hypothesis is true; there is also a possibility that a false null hypothesis will be accepted. (These possible errors are discussed in Section 3.2.)

In some instances, the conclusion of the study and the statistical decision about the null hypothesis are the same. The conclusion merely states the statistical decision in specific terms. In many situations, the conclusion goes further than the statistical decision. For example, suppose that an orthodontist makes a study of malocclusion due to crowding of the adult lower front teeth. The orthodontist hypothesizes that the incidence is as common in males as in females, H_0: $\pi_M = \pi_F$. (Note that in this example the experimental hypothesis coincides with the null hypothesis.) In the data gathered, however, there is a preponderance of males and $P \leq \alpha$. The statistical decision is to reject the null hypothesis, but this is not the final statement. Having rejected the null hypothesis, the orthodontist concludes the report by stating that this condition occurs more frequently in males than in females and advises family dentists of the need to watch more closely for tendencies of this condition in boys than in girls.

EXERCISES

1.2.1. Put the example of the cereals in the framework of the scientific method, elaborating on each of the six steps.

1.2.2. State a null and alternative hypotheses for the example of the file cards in Section 1.1, Example 1.5.

1.2.3. In the Salk experiment described in Example 1.6 of Section 1.1:

    a. Why should Salk not be content just to reject the null hypothesis?

    b. What conclusion could be drawn from the experiment?

1.2.4. Two college roommates decide to perform an experiment in extrasensory perception (ESP). Each produces a snapshot of his home-town girl friend, and one snapshot is placed in each of two identical brown envelopes. One of the roommates leaves the room and the other places the two envelopes side by side on the desk. The first roommate returns to the room and tries to pick the envelope that contains his girl friend's picture. The experiment is repeated 10 times. If the one who places the envelopes on the desk tosses a coin to decide which picture will go to the left and which to the right, the probabilities for correct decisions are listed below.

Number of Correct Decisions	Probability	Number of Correct Decisions	Probability
0	1/1024	6	210/1024
1	10/1024	7	120/1024
2	45/1024	8	45/1024
3	120/1024	9	10/1024
4	210/1024	10	1/1024
5	252/1024		

    a. State the null hypothesis based on chance as the determining factor in a correct decision. (Make the statement in words and symbols.)

    b. State an alternative hypothesis based on the power of love.

    c. If α is set as near 0.05 as possible, what is the region of rejection, that is, what numbers of correct decisions would provide evidence for ESP?

    d. What is the *region of acceptance*, that is, those numbers of correct decisions that would not provide evidence of ESP?

    e. Suppose the first roommate is able to pick the envelope containing his girl friend's picture 10 times out of 10; which of the following statements are true?

        i. The null hypothesis should be rejected.

        ii. He has demonstrated ESP.

        iii. Chance is not likely to produce such a result.

        iv. Love is more powerful than chance.

        v. There is sufficient evidence to suspect that something other than chance was guiding his selections.

        vi. With his luck he should raise some money and go to Las Vegas.

1.2.5. The mortality rate of a certain disease is 50% during the first year after diagnosis. The chance probabilities for the number of deaths within a year from a group of six persons with the disease are:

Number of deaths:	0	1	2	3	4	5	6
Probability:	1/64	6/64	15/64	20/64	15/64	6/64	1/64

A new drug has been found that is helpful in cases of this disease, and it is hoped that it will lower the death rate. The drug is given to 6 persons who have been diagnosed as having the disease. After a year, a statistical test is performed on the outcome in order to make a decision about the effectiveness of the drug.

a. What is the null hypothesis, in words and symbols?

b. What is the alternative hypothesis, based on the prior evidence that the drug is of some help?

c. What is the region of rejection if α is set as close to 0.10 as possible?

d. What is the region of acceptance?

e. Suppose that 4 of the 6 persons die within one year. What decision should be made about the drug?

1.2.6. A company produces a new kind of decaffeinated coffee which is thought to have a taste superior to the three currently most popular brands. In a preliminary random sample, 20 consumers are presented with all 4 kinds of coffee (in unmarked containers and in random order), and they are asked to report which one tastes best. If all 4 taste equally good, there is a 1-in-4 chance that a consumer will report that the new product tastes best. If there is no difference, the probabilities for various numbers of consumers indicating by chance that the new product is best are:

Number picking new product:	0	1	2	3	4
Probability:	0.003	0.021	0.067	0.134	0.190
Number picking new product:	5	6	7	8	9
Probability:	0.202	0.169	0.112	0.061	0.027
Number picking new product:	10	11	12	13–20	
Probability:	0.010	0.003	0.001	<0.001	

a. State the null and alternative hypotheses, in words and symbols.

b. If α is set as near 0.05 as possible, what is the region of rejection? What is the region of acceptance?

c. Suppose that 6 of the 20 consumers indicate that they prefer the new product. Which of the following statements is correct?

    i. The null hypothesis should be rejected.

    ii. The new product has a superior taste.

    iii. The new product is probably inferior because fewer than half of the people selected it.

    iv. There is insufficient evidence to support the claim that the new product has a superior taste.

1.3. EXPERIMENTAL DATA AND SURVEY DATA

An *experiment* involves the collection of measurements or observations about populations that are treated or controlled by the experimenter. A *survey*, in contrast to an experiment, is an examination of a system in operation in which the investigator does not have an opportunity to assign different conditions to the objects of the study. Both of these methods of data collection may be the subject of statistical analysis; however, in the case of surveys some cautions are in order.

We might use a survey to compare two countries with different types of economic systems. If there is a significant difference in some economic measure, such as per-capita income, it does not mean that the economic system of one country is superior to the other. The survey takes conditions as they are and cannot control other variables that may affect the economic measure, such as comparative richness of natural resources, population health, or level of literacy. All that can be concluded is that at this particular time a significant difference exists in the economic measure. Unfortunately, surveys of this type are frequently misinterpreted.

A similar mistake could have been made in a survey of the life expectancy of men and women. The life expectancy was found to be 74.1 years for men and 79.5 years for women. Without control for risk factors—smoking, drinking, physical inactivity, stressful occupation, obesity, poor sleeping patterns, and poor life satisfaction—these results would be of little value. Fortunately, the investigators gathered information on these factors and found that women have more high-risk characteristics than men but still live longer. Because this was a carefully planned survey, the investigators were able to conclude that women biologically have greater longevity.

Surveys in general do not give answers that are as clear-cut as those of experiments. If an experiment is possible, it is preferred. For example, in order to determine which of two methods of teaching reading is more effective, we might conduct a survey of two schools that are each using a different one of the methods. But the results would be more reliable if we could conduct an experiment and set up two balanced groups within one school, teaching each group by a different method.

From this brief discussion it should not be inferred that surveys are not trustworthy. Most of the data presented as evidence for an association between heavy smoking and lung cancer come from surveys. Surveys of voter preference cause certain people to seek the presidency and others to decide not to enter the campaign. Quantitative research in many areas of social, biological, and behavioral science would be impossible without surveys. However, in surveys we must be alert to the possibility that our measurements may be affected by variables that are not of primary concern. Since we do not have as much control over these variables as we have in an experiment, we should record all concomitant information of pertinence for each observation. We can then study the effects of these other variables on the variable of interest and possibly adjust for their effects.

EXERCISES

1.3.1. In each of the research situations described below, determine whether the researcher is conducting an experiment or a survey.

    a. Traps are set out in a grain field to determine whether rabbits or raccoons are the more frequently found pests.

    b. A graduate student in English literature uses random 500-word passages from the writings of Shakespeare and Marlowe to determine which author uses the conditional tense more frequently.

    c. A random sample of hens is divided into 2 groups at random. The first group is given minute quantities of an insecticide containing an organic phosphorus compound; the second group acts as a control group. The average difference in eggshell thickness between the 2 groups is then determined.

    d. To determine whether honeybees have a color preference in flowers, an apiarist mixes a sugar-and-water solution and puts equal amounts in 2 equal-sized sets of vials of different colors. Bees are introduced into a cage containing the vials, and the frequency with which bees visit vials of each color is recorded.

1.3.2. In each of the following surveys, what besides the mechanism under study could have contributed to the result?

    a. An estimation of per-capita wealth for a city is made from a random sample of people listed in the city's telephone directory.

    b. Political preference is determined by an interviewer taking a random sample of Monday morning bank customers.

    c. The average length of fish in a lake is estimated by:

        i. The average length of fish caught, reported by anglers

        ii. The average length of dead fish found floating in the water

    d. The average number of words in the working vocabulary of first-grade children in a given county is estimated by a vocabulary test given to a random sample of first-grade children in the largest school in the country.

    e. The proportion of people who can distinguish between two similar tones is estimated on the basis of a test given to a random sample of university students in a music appreciation class.

1.3.3. *Time* magazine once reported that El Paso's water was heavily laced with lithium, a tranquilizing chemical, whereas Dallas had a low lithium level. *Time* also reported that FBI statistics showed that El Paso had 2889 known crimes per 100,000 population and Dallas had 5970 known crimes per 100,000 population. The article reported that a University of Texas biochemist felt that the reason for the lower crime rate in El Paso lay in El Paso's water. Comment on the biochemist's conjecture.

1.4. COMPUTER USAGE

The practice of statistics has been radically changed now that computers and high-quality statistical software are readily available and relatively inexpensive. It is no longer necessary to spend large amounts of time doing the numerous calculations that are part of a statistical analysis. We need only enter the data correctly, choose the appropriate procedure, and then have the computer take care of the computational details.

Because the computer can do so much for us, it might seem that it is now unnecessary to study statistics. Nothing could be further from the truth. Now more than ever the researcher needs a solid understanding of statistical analysis. The computer does not choose the statistical procedure or make the final interpretation of the results; these steps are still in the hands of the investigator.

Statistical software can quickly produce a large variety of analyses on data regardless of whether these analyses correspond to the way in which the data were collected. An inappropriate analysis yields results that are meaningless. Therefore, the researcher must learn the conditions under which it is valid to use the various analyses so that the selection can be made correctly.

The computer program will produce a numerical output. It will not indicate what the numbers mean. The researcher must draw the statistical conclusion and then translate it into the concrete terms of the investigation. Statistical analysis can best be described as a search for evidence. What the evidence means and how much weight to give to it must be decided by the researcher.

In this text we have included some computer output to illustrate how the output could be used to perform some of the analyses that are discussed. Several exercises have computer output to assist the user with analyzing the data. Additional output illustrating nearly all the procedures discussed is available on an Internet website.

Many different comprehensive statistical software packages are available and the outputs are very similar. A researcher familiar with the output of one package will probably find it easy to understand the output of a different package. We have used two particular packages, the SAS system and JMP, for the illustrations in the text. The SAS system was designed originally for batch use on the large mainframe computers of the 1970's. JMP was originally designed for interactive use on the personal computers of the 1980's. SAS made it possible to analyze very large sets of data simply and efficiently. JMP made it easy to visualize smaller sets of data. Because the distinction between large and small is frequently unclear, it is useful to know about both programs.

The computer could be used to do many of the exercises in the text; however, some calculations by the reader are still necessary in order to keep the computer from becoming a magic box. It is easier for the investigator to select the right procedure and to make a proper interpretation if the method of computation is understood.

REVIEW EXERCISES

Decide whether each of the following statements is true or false. If a statement is false, explain why.

1.1. To say that the null hypothesis is rejected does not necessarily mean it is false.

1.2. In a practical situation, the null hypothesis, alternative hypothesis, and level of rejection should be specified before the experimentation.

1.3. The probability of choosing a random sample of 3 persons in which the first 2 say "yes" and the last person says "no" from a population in which $P(\text{yes}) = 0.7$ is $(0.7)(0.7)(0.3)$.

1.4. If the experimental hypothesis is true, chance does not enter into the outcome of the experiment.

1.5. The alternative hypothesis is often the experimental hypothesis.

1.6. A decision made on the basis of a statistical procedure will always be correct.

1.7. The probability of choosing a random sample of 3 persons in which exactly 2 say "yes" from a population with $P(\text{yes}) = 0.6$ is $(0.6)(0.6)(0.4)$.

1.8. In the total process of investigating a question, the very first thing a scientist does is state the problem.

1.9. A scientist completes an experiment and then forms a hypothesis on the basis of the results of the experiment.

1.10. In an experiment, the scientist should always collect as large an amount of data as is humanly possible.

1.11. Even a specialist in a field may not be capable of picking a sample that is truly representative, so it is better to choose a random sample.

1.12. If in an experiment $P(\text{success}) = 1/3$, then the odds against success are 3 to 1.

1.13. One of the main reasons for using random sampling is to find the probability that an experiment could yield a particular outcome by chance if the null hypothesis is true.

1.14. The α level in a statistical procedure depends on the field of investigation, the cost, and the seriousness of error; however, traditional levels are often used.

1.15. A conclusion reached on the basis of a correctly applied statistical procedure is based solely on probability.

1.16. The null hypothesis may be the same as the experimental hypothesis.

1.17. The "α level" and the "region of rejection" are two expressions for the same thing.

1.18. If a correct statistical procedure is used, it is possible to reject a true null hypothesis.

1.19. The probability of rolling two 6's on two dice is $1/6 + 1/6 = 1/3$.

1.20. A weakness of many surveys is that there is little control of secondary variables.

SELECTED READINGS

Anscombe, F. J. (1960). Rejection of outliers. *Technometrics*, **2**, 123–147.

Barnard, G. A. (1947). The meaning of a significance level. *Biometrika*, **34**, 179–182.

Barnett, V., and T. Lewis (2002). *Outliers in Statistical Data*, 3rd ed. Wiley, New York.

Bennett, J. O., W. L. Briggs, and M. F. Triola (2003). *Statistical Reasoning for Everyday Life*, 2nd ed. Addison-Wesley, New York.

Berkson, J. (1942). Tests of significance considered as evidence. *Journal of the American Statistical Association*, **37**, 325–335.

Box, G. E. P. (1976). Science and statistics. *Journal of the American Statistical Association*, **71**, 791–799.

Cox, D. R. (1958). *Planning of Experiments*. Wiley, New York.

Duggan, T. J., and C. W. Dean (1968). Common misinterpretation of significance levels in sociology journals. *American Sociologist*, **3**, 45–46.

Edgington, E. S. (1966). Statistical inference and nonrandom samples. *Psychological Bulletin*, **66**, 485–487.

Edwards, W. (1965). Tactical note on the relation between scientific and statistical hypotheses. *Psychological Bulletin*, **63**, 400–402.

Ehrenberg, A. S. C. (1982). Writing technical papers or reports. *American Statistician*, **36**, 326–329.

Gibbons, J. D., and J. W. Pratt (1975). *P*-values: Interpretation and methodology. *American Statistician*, **29**, 20–25.

Gold, D. (1969). Statistical tests and substantive significance. *American Sociologist*, **4**, 42–46.

Greenberg, B. G. (1951). Why randomize? *Biometrics*, **7**, 309–322.

Johnson, R., and P. Kuby (2000). *Elementary Statistics*, 8th ed. Duxbury Press, Pacific Grove, California.

Labovitz, S. (1968). Criteria for selecting a significance level: A note on the sacredness of .05. *American Sociologist*, **3**, 220–222.

McGinnis, R. (1958). Randomization and inference in sociological research. *American Sociological Review*, **23**, 408–414.

Meier, P. (1990). Polio trial: an early efficient clinical trial. *Statistics in Medicine*, **9**, 13–16.

Plutchik, R. (1974). *Foundations of Experimental Research*, 2nd ed. Harper & Row, New York.

Rosenberg, M. (1968). *The Logic of Survey Analysis*. Basic Books, New York.

Royall, R. M. (1986). The effect of sample size on the meaning of significance tests. *American Statistician*, **40**, 313–315.

Selvin, H. C. (1957). A critique of tests of significance in survey research. *American Sociological Review*, **22**, 519–527.

Stigler, S. M. (1986). *The History of Statistics*. Harvard University Press, Cambridge.

2 Populations, Samples, and Probability Distributions

In Chapter 1 we showed that statistics often plays a role in the scientific method; it is used to make inference about some characteristic of a population that is of interest. In this chapter we define some terms that are needed to explain more formally how inference is carried out in various situations.

2.1. POPULATIONS AND SAMPLES

We use the term *population* rather broadly in research. A population is commonly understood to be a natural, geographical, or political collection of people, animals, plants, or objects. Some statisticians use the word in the more restricted sense of the set of measurements of some attribute of such a collection; thus they might speak of "the population of heights of male college students." Or they might use the word to designate a set of categories of some attribute of a collection, for example, "the population of religious affiliations of U.S. government employees."

In statistical discussions, we often refer to the physical collection of interest as well as to the collection of measurements or categories derived from the physical collection. In order to clarify which type of collection is being discussed, in this book we use the term *population* as it is used by the research scientist: The population is the physical collection. The derived set of measurements or categories is called the set of *values of the variable of interest*. Thus, in the first example above, we speak of "the set of all values of the variable *height* for the population of male college students."

This distinction may seem overly precise, but it is important because in a given research situation more than one variable may be of interest in relation to the population under consideration. For example, an economist might wish to learn about the economic condition of Appalachian farmers. He first defines the population. Involved in this is specifying the geographical area "Appalachia" and deciding whether a "farmer" is the person who owns land suitable for farming, the person who works on it, or the person who makes managerial decisions about how the land is to be used. The economist's decision depends on the group in which he is interested. After he has specified the population, he must decide on the variable or variables, that characteristic or set of characteristics of these people, that will give him information about their economic condition. These characteristics might be money in savings accounts, indebtedness in mortgages or farm loans, income derived from the sale of livestock, or any of a number of other economic variables. The choice of variables will depend on the objectives of his study, the specific questions he is trying to answer. The problem of choosing

Statistics for Research, *Third Edition*, Edited by Shirley Dowdy, Stanley Weardon, and Daniel Chilko. ISBN 0-471-26735-X © 2004 John Wiley & Sons, Inc.

characteristics that pertain to an issue is not trivial and requires a great deal of insight and experience in the relevant field.

Once the population and the related variable or variables are specified, we must be careful to restrict our conclusions to this population and these variables. For example, if the above study reveals that Appalachian farm managers are heavily in debt, it cannot be inferred that owners of Kansas wheat farms are carrying heavy mortgages. Nor if Appalachian farm workers are underpaid can it be inferred that they are suffering from malnutrition, poor health, or any other condition that was not directly measured in the study.

After we have defined the population and the appropriate variable, we usually find it impractical, if not impossible, to observe all the values of the variable. For example, all the values of the variable *miles per gallon in city driving for this year's model of a certain type of car* could not be obtained since some of the cars probably are yet to be produced. Even if they did exist, the task of obtaining a measurement from each car is not feasible. In another example, the values of the variable *condition of all packaged bandages (sterile or contaminated) produced on a particular day by a certain firm* could be obtained, but this is not desirable since the bandages would be made useless in the process of testing. Instead, we consider a *sample* (a portion of the population), obtain measurements or observations from this sample (the *sample data*), and then use statistics to make an inference about the entire set of values. To carry out this inference, the sample must be random. We discussed the need for randomness in Chapter 1; in the next section we outline the mechanics.

EXERCISES

2.1.1. In each of the following examples identify the population, the sample, and the research variable.

    a. To determine the total amount of error in all students' bills, a large university selects 150 accounts for a special check of accuracy.

    b. A wildlife biologist collects information on the sex of the 28 surviving California condors.

    c. An organic chemist repeats the synthesis of a certain compound 5 times using the same procedure and each time determines the percentage of yield.

    d. The Census Bureau distributes a special questionnaire to 1 out of every 20 households in the census and among other questions inquires about the number of rooms in the dwelling.

    e. A manufacturer examines the records of each of its employees to determine how long each one has worked for the company.

2.1.2. Identify 3 different research variables that might be investigated for each of the following populations.

    a. All adults living in Colorado

    b. All patients of a certain opthalmologist

    c. All farms in Oklahoma

    d. All veterans' hospitals

2.1.3. For two years Francis Galton explored unmapped areas of South Africa. Thereafter, he tried to explore unmapped areas of science. In both Africa and science, however, he made some wrong turns. One of them was in the sampling procedure he used in his study of the inheritance of genius. To simplify his study, he evaluated the number and quality of academic, artistic, musical, and other worthy "abilities" a notable person displayed in his life, and the variable of interest was the man's score on the scale Galton used (see Exercise 2.3.5). He would then examine the life of that man's father and score his abilities in the same fashion. After gathering data on a number of son-and-father pairs, he wanted to see if sons with high scores had fathers with high scores.

a. To obtain data, Galton used information from obituaries.

    i. What is the target population, the population about which Galton wanted to make inference?

    ii. Tell why his data selection process meets the definition of a sample. Since it is a sample, why is it of questionable use for making reliable inference?

    iii. Give some ways in which his process could lead to biased results.

b. How would you have sampled the target population and what variable of interest would you have used?

2.2. RANDOM SAMPLING

Most statistics departments have entire courses in which different sampling techniques and their efficiencies are studied; only a brief description of sampling can be given here. If we have a population of N items from which a sample of n is to be drawn and we choose the n items in such a way that every combination of n items has an equally likely chance of being chosen, then this is called a *simple random sample*.

In an attempt to ensure that all combinations are equally likely, we often use a lottery or other gambling technique in drawing a sample. Thus, if we have 5 pairs of human twins in whom we wish to compare 2 methods of teaching speed reading, we may toss a coin to decide which twin is assigned to a particular method. Or a physiologist may have 35 frogs and want a sample of 10 for use in testing an antispasmodic drug. In one technique, he paints with vegetable dye the numerals 1 through 35 on the backs of the frogs and numbers 35 index cards with the same numerals. He then shuffles the cards and draws 10 cards. The 10 numbers determine which frogs will be in the treatment group.

Such methods are only as reliable as the gambling or lottery device used. A notably poor method was used in the 1970 military draft, when young men were being called to fight in the Vietnam War. Each date of the year was placed in a capsule, but the capsules were separated by month to ensure that every day of every month was included. The first month's capsules were checked and placed in a container. The second month's capsules were checked and added to the container, and both groups were mixed together. Then the third month was checked, added, and mixed. This process was repeated for each of the succeeding months. Thus January was mixed 11 times, February 10 times, March 9 times, and so on. Finally, the capsules were poured into a different container and the lottery began. Young men of draft age were to be called into service in the order in which their birth dates were drawn. However, later analysis of the order indicated that those born in certain months were much more likely to be drafted than those born in other months. The Selective Service System was criticized and was unable to defend the randomness of its procedure. In 1971 the procedure was modified; it made use of two containers, one holding a capsule for every date of the year and the other the numbers from 1 to 365. Two capsules were picked at each draw, one from each container, and the number drawn indicated the order of call-up for the date drawn. This order was acceptably random.

Instead of a gambling device, the use of *random numbers* is usually advisable. If we have access to a computer, it probably has a random-number generator. From this, we can obtain a random listing of n of the available N numbered items. Some hand-held calculators produce random numbers. If a computer or a random-number generator is not available, many tables of random numbers are in existence. Table A.1 in the Appendix of Useful Tables at the back of this book is an example of a small table of random numbers. There are various ways to use a table of random numbers; the example that follows illustrates one method.

Example 2.1.   Using a Table of Random Numbers to Choose a Simple Random Sample

The physiologist who wants a random sample of 10 of his 35 frogs might use Table A.1 in the following fashion:

1. He begins anywhere in the table, for example, at row 39 and column 14 (columns are composed of single digits, the 5-digit groups are to aid in reading the table). He can read the table in any direction, and he chooses to read it horizontally.
2. He reads the table as pairs of digits because the largest number for a frog (35) requires a 2-digit number. To save time, he may want to use not only 01 through 35 but also 36 through 70. To use this latter group, he subtracts 35 from each of its members, and the difference indicates the number of the frog to be included in the sample. He does not use values between 71 and 00 (100) because this group does not have 35 members. If he used them similarly to 36 through 70, there would then be three ways in which frogs 1 through 30 could be in the sample but only two ways that frogs 31 through 35 could be included, and the probability of selecting 1 through 30 would be higher than the probability of selecting 31 through 35.
3. The pairs of digits as he finds them in Table A.1 are as follows, with parentheses around the pairs that cannot be used:

$$04, (85), 50, 62, 67, (62), 24, (84), 14, (72), 26, 34, (74), 69, 03, 02$$

The frogs to be included in the sample are

$$04, 50 - 35 = 15, 62 - 35 = 27, 67 - 35 = 32, 24$$
$$14, 26, 34, (69 - 35 = 34), 03, 02$$

If only one random sample is going to be used in a study, the investigator can begin reading the random-number table at any place. However, if several random samples are to be used in the same study, it is important that different parts of the table are used so that the same set of random numbers is not used more than once. One way to accomplish this is to mark the table at the end of the first random sample, then begin at that point when the second sample is selected, and so on, for all the necessary samples.

Table A.1 in the Appendix is suitable for most small or moderate-sized samples. Should a very large sample be required, however, one would need a list of random digits generated by a computer program or would need to refer to a published listing such as *A Million Random Digits with 100,000 Normal Deviates* by the Rand Corporation.

Sometimes it is not possible to sample from the entire population of interest because part of the population is not available for sampling. A geologist may be interested in the heavy minerals in a certain layer of sandstone in a sequence of shale but the layer of sandstone is only available at a few exposed ledges. The rest is buried and hidden from view. Similarly, a sociologist may be interested in a characteristic of all of the families in a certain city but the only feasible list of families for sampling purposes is a current commercially published city directory. Some families have moved into the city since the directory was compiled, and some have left. Using the directory makes it impossible to include any of the new families in the sampling process. In situations such as these, the researcher often modifies the description of the population so that it coincides with the population available for sampling. Statistical inference from the sample is made only to the available population, then a judgment is made from within the specialized area whether or not the conclusion can be applied to the entire population of interest.

There are other methods of sampling besides simple random sampling. One is *stratified random sampling*. This consists in dividing the population into groups, or *strata*, and then taking a simple random sample from each stratum. This is done to improve the accuracy of estimates, to reduce cost, or to make it possible to compare strata. The sampling is often proportional so that the sizes of the samples from the strata are proportional to the sizes of the strata.

In this book, unless specified otherwise, all random samples are simple random samples. If a sampling design other than simple random sampling is employed, then adjustments of the techniques we describe are usually necessary. For more information about such adjustments, one should consult a text on sampling such as those listed in Selected Readings at the end of this chapter.

EXERCISES

2.2.1. Use Table A.1 to find the following.

    a. Select 3 of 8 items if the starting point is row 35 and column 20 and you read vertically.

    b. Give the first 2 random digits if the starting point is row 38 and column 30 and you read vertically.

    c. Five of 45 items are to be selected at random. What are they if the starting point is row 13, column 42, and you read vertically?

    d. Select 4 of 25 items when the starting point is row 2, column 15, and you read horizontally.

2.2.2. Use Table A.1 to pick a random sample of 15 people out of a group of 100 beginning at row 41, column 31, and reading horizontally.

2.2.3. Use Table A.1 to pick a random sample of 5 mice out of a collection of 25 mice beginning at row 1, column 1, and reading vertically.

2.2.4. Heights (in Inches) of 50 Male Students

Student Number	(Units)									
(Tens)	00	01	02	03	04	05	06	07	08	09
00		64	65	65	66	66	67	67	67	68
10	68	68	69	69	69	69	69	69	69	69
20	70	70	70	70	70	70	70	70	70	70
30	71	71	71	71	71	71	71	72	72	72
40	72	72	72	72	73	73	73	74	74	74
50	75									

    a. The accompanying table represents the values of the variable *height* for a population of 50 male students. Use the table of random digits to draw a random sample of 10 men from this population and record the corresponding sample data.

b. Compute the arithmetic average of your sample data and compare it to 70, which is the mean of the variable height for the entire population.

2.2.5. Body mass index (BMI) takes into account both the height and weight of individuals, so large numbers represent those who are heavy for their height. It is a useful measure for orthopedists when treating patients with pain in a weight-bearing joint such as the knee. Suppose an orthopedist has been treating 40 patients with such severe knee pain that all have agreed to submit to a form of experimental surgery, but prudence dictates that the surgery be performed only on $n = 10$, and in case of duplicates a computer-generated random sample of 15 numbers between 1 and 40 is obtained. The random digits are

$$8 \ 39 \ 16 \ 11 \ 37 \ 39 \ 22 \ 22 \ 2 \ 3 \ 33 \ 21 \ 35 \ 3 \ 39$$

The number of the 40 patients, their genders, and BMI values in a comma-delimited format are

1,F,46	2,M,18	3,F,22	4,M,28	5,M,39
6,M,41	7,F,25	8,F,29	9,F,43	10,F,18
11,F,29	12,M,48	13,F,23	14,F,14	15,F,25
16,F,19	17,M,18	18,M,20	19,F,28	20,F,46
21,M,33	22,F,38	23,F,29	24,M,32	25,M,12
26,F,26	27,M,34	28,M,18	29,F,19	30,F,31
31,F,42	32,M,40	33,F,40	34,F,27	35,F,45
36,M,49	37,F,19	38,F,26	39,M,10	40,F,20

a. Use the computer-generated set of random digits to select the numbers of the 10 patients to receive the experimental surgery.

b. To evaluate the representativeness of the sample:

    i. Compute the percentage of females and compare that to fact that 25 of the original 40 are females.

    ii. Compute the sample BMI average and compare it to the mean of 28.875 for all 40 patients.

c. Tell why you think the 10 chosen for surgery are (or are not) representative of the original 40?

2.3. LEVELS OF MEASUREMENT

When we make observations about a sample from some population of interest, we are collecting the sample data. These data may consist of lists of measurements, tallies of particular categories, answers to questions, and so on. The attribute we are observing will take on different values, or will vary, from observation to observation, so we have been calling these attributes *variables*. Thus, collecting sample data consists in recording the various values the variables assume for each member of the sample. We call this process *measurement*.

We often have a choice of levels when we are measuring. For example, a proctologist collecting data on cancer of the colon could record information about polyps in patients using different levels of measurement. She might simply record that polyps are present or not present in the colon of a patient—a rough categorization involving a low level of measurement. She might choose a higher level of measurement and rank her patients from the one with the most polyps to the patient with the fewest. Another approach would be to record the actual number of polyps, a higher level of measurement than ranks. There is an even higher level of measurement; she could determine the percentage of the area of the colon which is affected by polyps; this would locate the degree of invasion on a continuous scale. A different level of measurement is used in each of these cases. These levels are called the nominal scale, the ordinal scale, the discrete numerical scale, and the continuous numerical scale, respectively.

Levels of Measurement	Example
Numerical scales	
Continuous	Percentage of invasion
Discrete	Number of polyps
Ordinal scale	Rank among patients
Nominal scale	Present/not present

We are using the *nominal scale* when we put observations into categories that have no natural numerical relationship to each other. Examples are sex, occupation, color of eyes, and state of residence. When choosing categories for a nominal scale, it is necessary that there be a class for each observation and that no observation belong to more than one class.

The *ordinal scale* is a higher level of measurement than the nominal scale. We are using the ordinal scale if we rank the observations. For example, we could rank the pelts of 10 foxes from the lightest color to the darkest. When the ordinal scale is used, the ranks give some numerical information about the categories, but the underlying classification need not be numerical, as in this case of the color of the pelts. If the underlying categories are numerical, the difference between any two consecutive ranks need not be constant. For example, if we rank the weights of 5 research animals, the difference between the first and second weight might be 3 ounces, while the difference between the second and third weight might be only 1 ounce. In this example there is more precise underlying information, but we choose not to record it. If the only information available is on the ordinal scale, then it is not possible to specify the underlying difference between any two ranks.

We are using the *discrete numerical scale* when the observations are naturally numerical, the scale is uniform, and there is a built-in limit to how precisely the measurements can be taken. If data are on a discrete numerical scale, there are only a finite number of values possible, or possibly a countable infinity—as many as the counting numbers.[†] Examples are the number of offspring in a litter, the number of rooms in a house, the number of quarts of milk ordered by a supermarket (the count here could be in 1/4 quarts, but no more precise measurement is usually possible), the values of various coins, shoe sizes (for a fixed width), and the number of wells drilled until oil is found.

The *continuous numerical scale* is the highest level of measurement. A variable is continuous when its values are "measurements" in the common meaning of that term; that is,

---

[†]The nominal and ordinal scales are also discrete.

the scale is uniform and observations are as precise as we choose. Continuous variables theoretically can assume as many values as there are real numbers. In practice, we measure in whole numbers or to a few decimal places so the data are collected on the discrete numerical scale, but theoretically there is a more precise underlying scale of measurement. Examples are weight, blood pressure, age, length, and temperature.

If we have collected data using either numerical scale, it is possible to decrease the level of measurement to the ordinal scale. For example, if the measurements are the heights in inches of 5 men, these measurements can be reduced to ranks. The scale could even be reduced to a nominal scale by classifying the men as tall or short.

Although we can reduce the scale from a higher to a lower level of measurement, it is impossible for us to move the other way. If it is known that a certain number of men are tall and another number short, there is no way of determining how many men are 69 in. tall. It is important to be aware of this during the planning of an experiment. We must be sure to make our observations at a level high enough to give us pertinent information. If data are collected at too low a level of measurement, it is impossible to recover more precise information. On the other hand, no one should go to extreme efforts to obtain a very fine measurement if this information is not necessary or if it is distracting. For example, it is sufficient to know that an insecticide kills termites within a 24-hour period. There is no advantage to knowing whether it attains 100% mortality in 17 hours, 13 minutes, 49 seconds compared with another insecticide that attains 100% mortality in 18 hours, 31 minutes, 11 seconds.

Knowledge of the different levels of measurement not only enables us to make decisions about the desired level of precision but also helps us to choose the statistical procedures appropriate for analyzing the data. One set of procedures applies only to the nominal scale, another set to the ordinal scale, and still others are applicable to the discrete or continuous numerical scale. Unless we can recognize the level of measurement being used, we will be unable to choose an appropriate analysis. Chapters 3 through 5 deal mainly with procedures for data collected on the nominal scale or reduced to the nominal scale after collection. The remaining chapters deal with numerical data; however, at various points where appropriate, procedures are also provided for data which were collected on the ordinal scale or reduced to it. These alternative procedures will be identified as *nonparametric statistics*, with the term defined in Section 3.4. For more extensive coverage of such procedures, the reader is referred to one of the texts on nonparametric statistics in the Selected Readings, namely Conover (1998), Daniel (1990), or Hollander and Wolfe (1999).

EXERCISES

2.3.1. Which is the highest level of measurement possible for each of the following variables?

    a. Daily high temperature for a given year in Chicago

    b. Marital status of the applicants for a particular job

    c. Class standings at a university (freshman, sophomore, etc.)

    d. Colors of roses

    e. Weights of all American-made cars

    f. Number in attendance per day at a particular high school

    g. Birthdays of people in a certain group

2.3.2. Which of the following sets of categories are suitable for a nominal scale when classifying persons? (There must be a unique category for each observation.)

  a. Female, only child, under 66 in. tall
  b. Only child, has only brothers, has only sisters, has both brothers and sisters
  c. Less than three children in a family, more than three children in a family
  d. Left handed, right handed
  e. Blue eyed, female, blond

2.3.3. Correct each of the unsuitable sets in Exercise 2.3.2.

2.3.4. In Exercise 2.2.4:
  a. The level of measurement used to record height for this population is the numerical scale. Is it discrete or continuous?
  b. Could a higher level of measurement have been employed to record the data?
  c. Could height have been measured more accurately?

2.3.5. Sir Francis Galton believed that manual skills are among the many abilities that are inherited. Hence, even the young children of skilled laborers should show greater manual dexterity than those of unskilled laborers. For evidence, suppose he watched 20 children of the age of 3 at play with toys requiring some manual ability. Ten of the children are children of skilled laborers and the other 10 of unskilled laborers, but at the time of measurement, he would not know to which group a child belongs. When making subjective measures, Galton used the scale

$$x \; g \; f \; e \; d \; c \; b \; a \; A \; B \; C \; D \; E \; F \; G \; X$$

in which a lower-case x is the lowest possible measurement and an upper-case X the highest. Assume this is used to measure the abilities of the 20 children and the following data were obtained:

Father	Children's Scores									
Skilled	e	b	a	B	C	D	F	G	G	X
Unskilled	x	g	f	d	d	c	A	B	E	F

  a. What is the scale of measurement? Explain.
  b. Galton would see evidence that the children of skilled laborers have greater dexterity. Explain why.
  c. How would you summarize the data, graphically or numerically, to support the idea of greater ability for the group with skilled-laborer fathers?

2.4. RANDOM VARIABLES AND PROBABILITY DISTRIBUTIONS

In Example 1.7, a test of hypothesis is carried out to determine if there is a preference for type *A* baby cereal over type *B*. The sample is a randomly chosen group of 4 mothers and the variable is recorded on the nominal scale (*A* or *B*). The test of hypothesis amounts to comparing the empirical results of sampling and recording outcomes in the real world with a theoretical *model* of what happens if the null hypothesis is true. The theoretical model is called a *probability distribution*. In this section we discuss the nature of probability distributions and how they act as models for studies that involve random sampling.

To develop the theoretical model for the test in Example 1.7 the possible outcomes of the study are associated with numbers, the number of mothers out of the 4 in the sample who prefer cereal A. The outcomes of this study are associated with 0, 1, 2, 3, or 4 (Figure 2.1). Numbers of this type, that is, those that are associated with the possible outcomes of an experiment or survey, are called the *values of the random variable y*. The random variable is the process of association. The random variable in this example is a *discrete random variable* because it has a countable number of values: 0, 1, 2, 3, 4.

To build the model, we assume that the null hypothesis is true and we determine the probability of each of the values of the random variable. Since the null hypothesis in this example is that the mothers have no preference between A and B (i.e., a randomly chosen mother will prefer A with probability $1/2$ and B with probability $1/2$), the 16 outcomes in Figure 2.1 are equally likely. The value of the random variable is 0 if no mothers prefer A; thus the probability of 0 is $1/16$ since there is only 1 outcome of this type ($BBBB$) among the 16 equally likely outcomes. We write $p(0) = 1/16$ to indicate that the probability that the value of the random variable will be 0 is $1/16$.

To find $P(y = 1) = p(1)$, we note that there are four cases in which exactly 1 mother out of 4 prefers A; thus $p(1) = 4/16$. As we saw in Chapter 1, the general rule for calculating the *probability of an event* when all outcomes are equally likely is

$$\text{probability of success} = \frac{\text{number of successful outcomes}}{\text{total number of outcomes}}$$

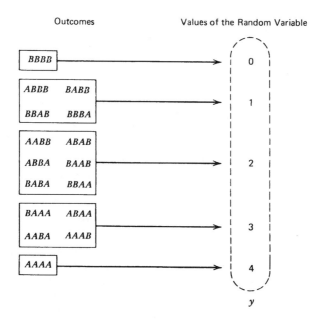

FIGURE 2.1. Associating numbers with nominal data.

In more general terms we can say:

$$\text{probability of an event} = \frac{\text{number of outcomes giving the event}}{\text{total number of outcomes}}$$

All of the probabilities are summarized in the table of Figure 2.2a and in the graph of Figure 2.2b.

The values of a discrete random variable y together with their associated probabilities are called a *probability distribution*, and $p(y)$ is called the *probability function*. In order for $p(y)$ to be a probability function, two conditions are necessary:

1. $0 \le p(y) \le 1$ for all values of y.
2. $\sum_{y} p(y) = 1$, that is, the sum of $p(y)$ over all values of y is 1.

Note that in the baby cereal example these two conditions are satisfied.

There are many functions that satisfy these two conditions. In Table 2.1, examples A through D represent discrete probability distributions. In example D the random variable has a countable infinity of values, and $p(y)$ can be given by the formula $p(y) = (1/2)^y$. In many cases it is possible to represent the probability function by a formula.

It is not difficult to find functions with the two properties required for a probability function. However, a probability distribution will only be of value statistically if it represents—models—a real-life situation. Some examples of probability distributions used as models occur in Exercises 1.2.4 through 1.2.6. The method for determining the probabilities in these examples is explained in Chapter 3. An example of a test of hypothesis that uses a different type of discrete probability distribution follows.

Example 2.2.   Testing a Hypothesis Using a Discrete Probability Distribution

A new salesperson for a company is told that the probability of making a sale on a single call is $1/4$. The salesperson calls on 7 people and makes no sales. Finally, on the eighth attempt, a

y	$p(y)$
0	$p(0) = 1/16$
1	$p(1) = 4/16$
2	$p(2) = 6/16$
3	$p(3) = 4/16$
4	$p(4) = 1/16$

(*a*)

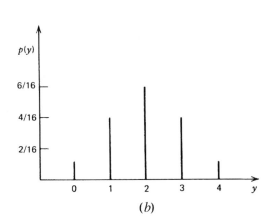

(*b*)

FIGURE 2.2. A discrete probability distribution. (a) Tabular form. (b) Graph.

TABLE 2.1. Four Discrete Probability Distributions

A		B		C		D	
y	$p(y)$	y	$p(y)$	y	$p(y)$	y	$p(y)$
0	1/4	5	1/5	0.5	0.125	1	1/2
1	1/2	6	1/5	1.0	0.125	2	1/4
2	1/4	7	1/5	1.5	0.125	3	1/8
		8	1/5	2.0	0.625	4	1/16
		9	1/5			5	1/32
						$\vdots$	$\vdots$
						N	$1/2^N$

sale is completed. The salesperson wonders if there is any evidence (at the 0.05 level of significance) that the probability of $1/4$ for a sale is too high.

The null hypothesis is H_0: $\theta = 1/4$; that is, the probability of a sale is $1/4$ on a single attempt.[†] The alternative is H_a: $\theta < 1/4$ because the salesperson is looking for evidence that the figure is too high.

If the probability of a sale is $1/4$, then the probability of no sale on a single trial is $3/4$. Using these values, the probability model can be found. The probability of a sale on the first call is

$$p(1) = \frac{1}{4}$$

and the probability that the first sale occurs on the second call is

$$p(2) = \left(\frac{3}{4}\right)\left(\frac{1}{4}\right) = \frac{3}{16}$$

since there is no sale on the first call and there is a sale on the second call. The probabilities are multiplied because the calls are assumed to be *independent* of each other; that is, we assume the customers are randomly chosen and do not influence each other and the salesperson behaves the same way on each call.

Similarly,

$$p(3) = \left(\frac{3}{4}\right)\left(\frac{3}{4}\right)\left(\frac{1}{4}\right) = \frac{9}{64}$$

and

$$p(y) = \left(\frac{3}{4}\right)^{y-1}\left(\frac{1}{4}\right)$$

is the general formula for the probability that the first sale occurs on the yth call. This probability distribution is known as a *geometric distribution*.

[†]The Greek letter θ is read "theta".

The beginning of the geometric distribution that is the model of this study can be summarized as follows:

y		$p(y)$	
1	1/4	= 0.2500	
2	3/16	= 0.1875	
3	9/65	= 0.1406	
4	27/256	= 0.1055	0.8665
5	81/1,024	= 0.0791	
6	243/4,096	= 0.0593	
7	729/16,384	= 0.0445	
8	2,187/65,536	= 0.0334	
⋮		⋮	

If $\theta < 1/4$, a larger number of calls will be necessary before the first sale than if $\theta = 1/4$. Thus the P value associated with this study is

$$P = P(8 \text{ or more calls needed for the first sale})$$
$$= 1 - P(1 \text{ through 7 calls needed for the first sale})$$
$$= 1 - 0.8665$$
$$= 0.1335$$

Since $P = 0.1335 > \alpha = 0.05$, the null hypothesis is accepted. There is no statistically significant evidence that the figure given to the salesperson is too high.

If the data are recorded on a continuous scale, the variable of interest corresponds to a *continuous random variable*. In this type of model it is not possible to represent the related probabilities by a table or a line graph; instead, a smooth curve is used to indicate the *continuous probability distribution* that is the model for the study.

Example 2.3.   A Continuous Probability Distribution

One of the major problems in coal mining is roof collapse. Any procedure which will increase the probability of a roof collapse must be used with great caution. A mining engineer questions whether the drilling of air shafts affects the stability of the roof. In one area of the mine, two air shafts are located 360 ft apart along a straight tunnel (Figure 2.3). The engineer reasons that if the roof's stability is unaffected by the air shafts, then the amount of debris from the roof that falls to the floor will be uniformly distributed between the shafts. If, however, the air shafts are causing instability, larger amounts of roof debris will appear close to the air shafts.

A uniform distribution of debris can be modeled by the graph in Figure 2.4. The random variable y is the location along the floor between the shafts, a number on a continuous scale between 0 and 360. The curve is a horizontal line which indicates that the debris is uniformly deposited on the floor. This line, $f(y) = 1/360$, is called the *probability density function* of the random variable y. The curve (the horizontal line) is placed at $1/360$ on the vertical axis so that the area of the rectangle under the line and between 0 and 360 is equal to 1. The proportion of debris between location 90 and 180 is represented by the area between 90 and 180 and under the

FIGURE 2.3. Cross section of mine tunnel.

curve; the proportion, or probability, is $1/4$. The probability of debris between 0 and 95 is given by the area under the curve and to the left of 95. The probability is $95/360 = 19/72$ (Figure 2.5).

FIGURE 2.4. Continuous uniform probability distribution.

Notice that the density function, unlike a probability function for a discrete random variable, does not indicate a probability directly; rather the density function is used to find an area that corresponds to the probability. Because areas correspond to probabilities, the probability of debris at a particular point, say $y = 95$, is 0. This becomes clear by noticing that, rather than a region, there is only a vertical line segment at 95 and that a line segment has no area. It follows that $P(y \le 95) = P(y < 95)$ in a continuous probability distribution, but this is not true in a discrete distribution.

In many models for continuous random variables, the continuous probability distribution is given by a curve that is neither a straight line nor a figure formed from straight lines. In these cases, areas are difficult to determine and calculus must be used. Fortunately, tables are available for most of the commonly encountered distributions, and thus even those who are not familiar with calculus are able to use continuous probability distributions that are represented by curves. The first distribution of this type is discussed in Chapter 5.

EXERCISES

2.4.1.

y:	2	4	6	8	10
$p(y)$:	$1/6$	$2/6$	$1/6$	—	$1/6$

FIGURE 2.5. Shaded area indicates $P(0 \le y \le 95)$.

    a. If the table above represents a probability distribution, what is the value of $p(8)$?

    b. Graph the probability distribution.

    c. Find $P(y \le 6)$, $P(y < 6)$, $p(y = 6)$, and $p(y \ge 6)$.

2.4.2. If $p(y) = 1/5$ for $y = 1, 2, 3, 4, 5$:

    a. Show that this is a probability distribution.

    b. Draw the graph.

    c. Find $P(y > 3)$, $P(y = 3)$, $P(y \le 3)$, and $P(y < 3)$.

2.4.3. Given the continuous probability distribution in Figure 2.6, imagine that the distribution represents the probability that a certain expert dart thrower will hit a 1-ft target within a certain distance y from the center 0.

    a. What is the total area within the triangle?

    b. What is the area of the shaded portion of the distribution?

    c. What is the probability that the dart will hit at a point that is from 6 in. to 1 ft from the center of the target?

    d. What is the area of the unshaded portion of the distribution?

    e. What is the probability that the dart will hit at a point that is less than 6 in. from the center of the target?

2.4.4. An oil company believes that the probability of striking oil on a single random drilling in a certain field is $1/3$. They drill and hit oil on the sixth attempt. Is there any evidence that the probability of a strike is less than $1/3$?

2.5. EXPECTED VALUE AND VARIANCE OF A PROBABILITY DISTRIBUTION

Since probability distributions are the key to statistical inference, it is helpful to study some of their characteristics. Two useful characteristics of a probability distribution are its *expected value* and its *variance*. Expected value is a measure of the location of the distribution, while variance is a measure of its spread.

    To introduce the idea of *expected value*, let us consider a certain electronic game that involves hitting a random target. To make the game sufficiently challenging to hand-eye coordination, it has been programmed so that the position of the target, the moment that the target appears, and the number of targets that appear during the period of play all vary. The number of targets to appear can be 11, 12, 13, 14, 15, or 16. They occur randomly and with equal frequency over a large

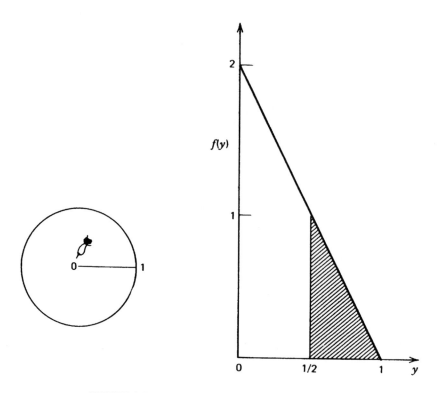

FIGURE 2.6. Continuous triangular probability distribution.

number of periods of play. A player of the game is unable to predict the number of targets that will appear during any one playing period, but the player can determine the expected number of targets, that is, the average number per playing session if the game is played many times.

The number of targets can be modeled by a discrete uniform probability distribution in which the values of the random variable y are 11, 12, 13, 14, 15, and 16 and the probability function $p(y)$ is $1/6$ for each of the values because they occur with equal frequency.

y	$p(y)$
11	1/6
12	1/6
13	1/6
14	1/6
15	1/6
16	1/6

The expected number of targets, $E(y)$, per playing period is

$$E(y) = \frac{11 + 12 + 13 + 14 + 15 + 16}{6} = \frac{81}{6} = 13.5$$

that is, the arithmetic average of the 6 equally frequent numbers. If many games are played, on the average 13.5 targets will appear per session. Note that the expected value need not be one of the possible values of the random variable; 13.5 targets never appear in a playing session.

Another way to compute the expected value is to use the formula

$$E(y) = \sum yp(y)$$

that is, the expected value of y is the sum of the products of the values of y times their corresponding probabilities. The following table illustrates how this formula is used:

y	$p(y)$	$yp(y)$
11	1/6	11/6
12	1/6	12/6
13	1/6	13/6
14	1/6	14/6
15	1/6	15/6
16	1/6	16/6
		$E(y) = \sum yp(y) = 81/6 = 13.5$

A third column is computed from the probability distribution. This third column is obtained by finding the product of the corresponding elements in the first two columns. The expected value of y is the sum of the products in the third column. The advantage of this second approach is that it can be used to find an expected value even if the probabilities are not all the same. The following example illustrates this general type of problem.

Example 2.4.   The Expected Value of a Discrete Probability Distribution

A teacher gives frequent short quizzes that consist of 2 multiple-choice questions. Each question is followed by 4 answers, and only 1 is correct. Because these quizzes are so short, the teacher wonders if they are useful for determining which students have learned the material. The teacher decides to find out how many questions a student can be expected to answer correctly if the student has no knowledge of the material and is choosing answers in a random fashion.

On a single question, the probability of a correct guess is $1/4$ because each answer is equally likely to be chosen and only 1 answer is correct. For 2 questions, the number of correct responses y can be 0, 1, or 2, and the probability distribution, which is a model of the number of correct responses under guessing, is

y	$p(y)$
0	9/16
1	6/16
2	1/16

The probabilities in this distribution are obtained by computing $p(0) = P(\text{two incorrect}) = (3/4)(3/4) = 9/16$ and $p(2) = P(\text{two correct}) = (1/4)(1/4) = 1/16$; then $p(1)$ must equal $6/16$ so that the sum of the probabilities is equal to 1.

If a large number of quizzes of this type are given, then the expected number of correct answers per quiz is

$$E(y) = \sum yp(y)$$

In tabular form:

y	$p(y)$	$yp(y)$
0	9/16	0
1	6/16	6/16
2	1/16	2/16
	$E(y) = \sum yp(y) = 8/16 = 0.5$	

On the average, the student will guess correctly only 0.5 of an answer per quiz. Although it is impossible to get 0.5 of an answer correct on a single quiz, the expected value is meaningful for a large number of quizzes.

The teacher decides that the quizzes are useful for distinguishing those who are guessing from those who have knowledge of the material. For example, if 40 such 2-question quizzes are given, then the student who is guessing is expected to answer correctly about 20 out of the 80 questions asked. A student who answers many more correctly, for example, 60 out of the 80 questions, demonstrates some knowledge of the material.

The expected value can be thought of as the location, or center, of the probability distribution. This seems reasonable if we visualize a uniform calibrated bar on which we place weights (all of equal heaviness): nine at 0, six at 1, and one at 2 (Figure 2.7). The bar will balance at 0.5, the expected value.

Another useful characteristic of a probability distribution is its *variance*. Variance is a measure of the spread of a distribution relative to its expected value. In the electronic game example, the random variable y had values 11, 12, 13, 14, 15, and 16 with equal frequency. The deviations of these values from the expected value of 13.5 are

y	$y - E(y)$
11	$11 - 13.5 = -2.5$
12	$12 - 13.5 = -1.5$
13	$13 - 13.5 = -0.5$
14	$14 - 13.5 = 0.5$
15	$15 - 13.5 = 1.5$
16	$16 - 13.5 = 2.5$

The deviations are shown graphically in Figure 2.8.

We might expect to measure spread by averaging these deviations. However, since the sum of the deviations from the expected value is always 0, this is not a useful measure. To obtain a meaningful average, we use the squares of the deviations. The variance of a probability distribution is the average squared deviation from its expected value. Using the probabilities,

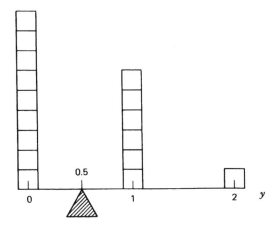

FIGURE 2.7. Expected value as the balancing point.

the formula for the variance of y is

$$V(y) = \sum [y - E(y)]^2 p(y)$$

In tabular form (using fractions to avoid rounding error), the computations are

y	$p(y)$	$y - E(y)$	$[y - E(y)]^2$	$[y - E(y)]^2 \, p(y)$
11	1/6	$-2.5 = -5/2$	25/4	25/24
12	1/6	$-1.5 = -3/2$	9/4	9/24
13	1/6	$-0.5 = -1/2$	1/4	1/24
14	1/6	$0.5 = 1/2$	1/4	1/24
15	1/6	$1.5 = 3/2$	9/4	9/24
16	1/6	$2.5 = 5/2$	25/4	25/24
				$V(y) = \overline{70/24}$

This formula is used even if the probabilities are not all equal.

FIGURE 2.8. Deviations from the expected value.

Variance measures the spread of a distribution. The larger the variance, the larger the spread. If we take the positive square root of the variance, we obtain the *standard deviation* of the random variable, sd(y). In this example

$$\text{sd}(y) = \sqrt{V(y)} = \sqrt{70/24} = 1.71$$

If we are told only the expected value and standard deviation of a probability distribution, we know a surprising amount about the nature of the distribution. Values of the random variables that are more than two or three standard deviations from the mean have very low probabilities associated with them. For example, in the case of the electronic game

$$E(y) = 13.50$$
$$\text{sd}(y) = 1.71$$

and

$$2[\text{sd}(y)] = 3.42$$

Two standard deviations below the expected value is

$$E(y) - 2[\text{sd}(y)] = 13.50 - 3.42 = 10.08$$

and the probability of 10 or fewer targets in a single playing period is very low; in fact, it is 0. Two standard deviations above the expected value is

$$E(y) + 2[\text{sd}(y)] = 13.50 + 3.42 = 16.92$$

and the probability of 17 or more targets is 0.

In practice, the computation of the variance from the formula

$$V(y) = \sum [y - E(y)]^2 p(y)$$

is sometimes tedious because of the subtractions and squaring. A mathematically equivalent formula may be used:

$$V(y) = \sum y^2 p(y) - [E(y)]^2$$

We illustrate this formula for the probability distribution of the 2-question multiple-choice quizzes.

Example 2.5.   The Variance of a Probability Distribution

For the short quizzes, a fourth column $y^2 p(y)$ is computed and summed after the computation of the expected value. The fourth column is obtained by multiplying the elements in the first column by the corresponding elements in the third column:

y	$p(y)$	$yp(y)$	$y^2 p(y)$
0	9/16	0	0
1	6/16	6/16	6/16
2	1/16	2/16	4/16
			$\sum y^2 p(y) = \overline{10/16}$

Then

$$V(y) = \sum y^2 p(y) - [E(y)]^2$$

$$= \frac{10}{16} - \left(\frac{1}{2}\right)^2$$

$$= \frac{6}{16}$$

Note that in this example

$$E(y) = 0.5$$

$$\text{sd}(y) = \sqrt{6/16} = 0.61$$

and 2 standard deviations below and above the expected value are

$$E(y) - 2[\text{sd}(y)] = 0.5 - 2(0.61) = -0.72$$
$$E(y) + 2[\text{sd}(y)] = 0.5 + 2(0.61) = 1.72$$

There is 0 probability that the value of the random variable is below -0.72 and $1/16$ probability that the random variable will have a value above 1.72. Using only these facts, if a student frequently answered both questions correctly, the teacher decides that the model based on guessing does not fit this student and the student probably has knowledge of the material.

The main use of the variance (or standard deviation) is for purposes of inference. This application is developed more fully in later chapters. The discussion in this section is restricted to discrete random variables. It is also possible to consider the expected value and variance of a continuous random variable; in such cases, calculus is usually needed to find the values.

Procedure. Expected Value and Variance of a Probability Distribution

Expected value: $E(y) = \sum yp(y)$

Variance: $V(y) = \sum [y - E(y)]^2 p(y)$

Standard deviation: $\text{sd}(y) = \sqrt{V(y)}$

EXERCISES

2.5.1. Find the mean and the variance of the probability distributions A to C in Table 2.1.

2.5.2. In Mendel's experiments on pea plants, he found that the trait of being tall is dominant over being short. His theory indicates that if pure-line tall and pure-line short plants are cross-pollinated and then the hybrids in the next generation are cross-pollinated, in the resulting population approximately $3/4$ of the plants will appear tall and $1/4$ will appear short. If 4 plants are chosen at random from such a population, the best model for the number of tall plants in 4 is

y:	0	1	2	3	4
p(y):	1/256	12/256	54/256	108/256	81/256

a. Find the expected value of this probability distribution.

b. Find the variance of the probability distribution.

c. What is the probability that the value of the random variable will be more than 2 standard deviations below the expected value?

d. What is the probability that the value of the random variable will be more than 2 standard deviations above the expected value?

2.5.3. A gambling game is played in which there is a group of 100 cards with one $25 winning card, two $10 winning cards, and three $5 winning cards. After paying a certain fee, a player selects one card at random. If it is one of the winning cards, the player receives the designated amount. If it is one of the other cards, the player wins nothing. The card is returned to the deck, the cards shuffled, and they are ready for the next play.

a. Find the probability distribution for y, the number of dollars won (use the rule for equally likely events).

b. If a large number of plays are purchased, what are the expected winnings per play, or in statistical terms, what is the expected value of y?

c. Would it be reasonable to pay $1 to play this game?

d. Find the variance of this probability distribution.

e. What proportion of the time will the winnings be within two standard deviations of the expected value?

2.5.4.

y:	1	2	3	4	5
p(y):	1/5	1/5	1/5	1/5	1/5

a. Find the expected value of y.

b. Find $V(y)$.

c. Compare your answers with those found in Exercise 2.5.1 for Table 2.1, distribution B. Explain why there is a difference in the expected values but the variances are the same.

2.5.5.

y:	1	2	3	4
p(y):	1/4	1/4	1/4	1/4

a. Find $E(y)$.

b. Compare this result with that of Exercise 2.5.4; find a simple general formula for the expected value of a discrete uniform distribution of successive integers from a to b.

REVIEW EXERCISES

Decide whether each of the following statements is true or false. If a statement is false, explain why.

2.1. The objective of statistics is to make inference about a population based on information contained in a sample from that population.

2.2. A single population may have several variables of interest to the investigator.

2.3. A lottery device may be an acceptable way to obtain a completely random sample.

2.4. When using a random-number table to select a sample, always begin at the beginning of the table.

2.5. The choice of sampling design has no effect on the choice of the procedure used for statistical analysis.

2.6. When choosing categories for the nominal scale, the only condition is that there is a category for each piece of data.

2.7. Data on the numerical scale can be easily changed to the nominal scale.

2.8. The ordinal scale is sometimes used even though more precise numerical information is available.

2.9. Data on an ordinal scale can be easily changed to the numerical scale.

2.10. Barometric pressure is usually recorded on the ordinal scale.

2.11. Yearly wages to the nearest dollar are recorded on the discrete numerical scale.

2.12. In a continuous probability distribution, the total area between the curve representing the distribution and the horizontal axis is 1.

2.13. In a continuous probability distribution, the probability of any particular value is the vertical distance at the value between the horizontal axis and the curve representing the distribution.

2.14. In a discrete probability distribution, the length of a vertical line at a certain value can be interpreted as the probability that such a value will result from random sampling.

2.15. If a population is infinite in size, the variable of interest is continuous.

2.16. Random variables always have numerical values.

2.17. The expected value of a probability distribution can be thought of as the center of balance.

2.18. The variance of a probability distribution is a measure of location, and the expected value indicates the spread.

2.19. If 2 probability distributions have equal variances, then their expected values are equal also.

2.20. The variance of a probability distribution can be defined symbolically as $E[y - E(y)]^2$.

SELECTED READINGS

Anderson, N. H. (1961). Scales and statistics: Parametric and nonparametric. *Psychological Bulletin*, **58**, 305–316.

Cochran, W. G. (1977). *Sampling Techniques*, 3rd ed. Wiley, New York.

Conover, W. J. (1998). *Practical Nonparametric Statistics*, 3rd ed. Wiley, New York.

Daniel, W. W. (1990). *Applied Nonparametric Statistics*, 2nd ed. Duxbury Press, Pacific Grove, California.

Hollander, M., and D. S. Wolfe (1999). *Nonparametric Statistical Methods*, 2nd ed. Wiley, New York.

Kish, L. (1965). *Survey Sampling*. Wiley, New York.

Lohr, S. L. (1999), *Sampling: Design and Analysis*. Duxbury Press, Pacific Grove, California.

Rand Corporation (1955). *A Million Random Digits with 100,000 Normal Deviates*. Free Press, Glencoe, Illinois.

Scheaffer, R.L., W. Mendenhall, and L. Ott (1998). *Elementary Survey Sampling*, 5th ed. Duxbury Press, Pacific Grove, California.

3 Binomial Distributions

In many experiments and surveys in which the variable of interest is being recorded at the nominal level, there are only 2 possible values or outcomes for the variable. For example, a salesman either makes a sale or does not make a sale, a newborn child is either a girl or a boy, and an insecticide may kill an insect or fail to kill it. Under certain conditions, samples involving dichotomous variables of this type can be represented by a theoretical probability distribution called a *binomial distribution*, binomial because of the two possible outcomes. In this chapter we look at the statistical interpretation of experimental results that can be modeled by binomial distributions.

3.1. THE NATURE OF BINOMIAL DISTRIBUTIONS

The population of human beings can be classified as "having type O blood" or "not having type O blood." There is no way that we can get exact information about the entire population, since this group is so large. It has been estimated that the proportion of people with type O blood is 0.40. Assume that the estimate is correct. If we observe a single person selected at random, the probability that the person will have type O blood is 0.40 and the probability that the person will not have type O blood is 0.60.

Now let us imagine that a large metropolitan hospital has a list of several thousand people willing to donate blood. If 4 people are chosen at random from the list, how likely is it that none have type O blood? One has type O? Two? Three? Four?

We first list the different possible outcomes for a sample of 4 people. Let O mean that a person has type O blood, and let N mean that the person does not have type O blood. The sequence of symbols indicates the results in the order in which they occur in the experiment, so *NNON* is a different outcome from *ONNN*.

Number with Type O Blood	Possible Outcomes
0	*NNNN*
1	*ONNN NONN NNON NNNO*
2	*OONN ONON ONNO NOON NONO NNOO*
3	*NOOO ONOO OONO OOON*
4	*OOOO*

When we ask a question like "How likely is it that 2 persons out of 4 have type O blood?" we have shifted our focus from the underlying variable of blood type (O or not-O) on the

Statistics for Research, *Third Edition*, Edited by Shirley Dowdy, Stanley Weardon, and Daniel Chilko.
ISBN 0-471-26735-X © 2004 John Wiley & Sons, Inc.

nominal scale to a count that is on the discrete numerical scale. Since it is numerical, the count can be thought of as a random variable, and we are looking for the probability distribution of this discrete random variable. We have already seen an example like this in the baby cereal preference study (Example 1.7 and Section 2.4), except in that case the probabilities were all equal.

Since not all of the 16 outcomes in this example are equally likely, to find the probabilities associated with 0, 1, 2, 3, and 4, we must use binomial probability rules based on the probability rules discussed in Chapter 1.

Binomial Probability Rules

1. If p is a probability, $0 \leq p \leq 1$.
2. If A and $\bar{A}$ are two mutually exclusive events that together include all possible outcomes, then $P(A) + P(\bar{A}) = 1$. [Two events A and B are *mutually exclusive* if they are nonoverlapping, that is, if $P(AB) = 0$.]
3. *Addition Rule.* The probability of a specified outcome is the sum of the probabilities of the mutually exclusive events making up that outcome.
4. *Multiplication Rule.* The probability of an event that is the simultaneous occurrence of two or more independent events is the product of the probabilities of the events. [Two events A and B are *independent* if the occurrence or nonoccurrence of A has no effect on the probability of B and vice versa.]

We already used the second rule when we stated that $P(N) = 0.60$. We reasoned that $P(N) = 1 - P(O) = 1 - 0.40 = 0.60$. Now we find that the probability of zero out of four having type O blood is

$$p(0) = P(NNNN) = [P(N)]^4 = (0.60)^4 = 0.1296$$

and the probability that 1 out of 4 will have type O blood is

$$p(1) = P(ONNN \text{ or } NONN \text{ or } NNON \text{ or } NNNO)$$
$$= P(ONNN) + P(NONN) + P(NNON) + P(NNNO)$$
$$= (0.40)(0.60)^3 + (0.60)(0.40)(0.60)^2 + (0.60)^2(0.40)(0.60) + (0.60)^3(0.40)$$
$$= 4(0.40)(0.60)^3 = 0.3456$$

In a similar way, we find that

$$p(2) = 6(0.40)^2(0.60)^2 = 0.3456$$

$$p(3) = 4(0.40)^3(0.60) = 0.1536$$

$$p(4) = (0.40)^4 = 0.0256$$

In summary, for this example the probability distribution is as appears in Figure 3.1. The discrete random variable with values 0, 1, 2, 3, 4 represents the number of people with type O blood in a random sample of 4 people, and $p(y)$ is the probability function of y. This probability distribution is called a *binomial probability distribution*. Note that a binomial probability distribution is a model of an experiment with only 2 possible outcomes. We

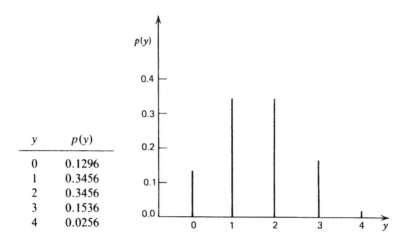

y	$p(y)$
0	0.1296
1	0.3456
2	0.3456
3	0.1536
4	0.0256

FIGURE 3.1. The binomial distribution with $n = 4$, $\pi = 0.40$.

concentrate on one of the outcomes, type O blood, and count the number of occurrences (successes) in the sample. The probability of type O blood does not change from observation to observation,[†] and the observations are independent of each other. We call such a survey or experiment a *binomial experiment*.

A *binomial experiment* is an experiment in which

1. there are only 2 possible outcomes, success S or failure F, with $P(S) = \pi$ and $P(F) = 1 - \pi$;
2. the experiment is repeated n times, that is, there are n trials;
3. $P(S) = \pi$ is constant from trial to trial;
4. the trials are independent of each other; and
5. we are interested in y, the number of successes, with $y = 0, 1, 2, \ldots, n$.

The probability of success π is called the *binomial parameter*. A parameter is a numerical characteristic of a population and the distribution which is used to model random sampling from the population. In the blood-type example, $\pi = 0.40$ is the proportion of the population with type O blood. The parameter π also specifies the theoretical model for the experiment, the binomial distribution with $n = 4$ trials and $P(S) = \pi = 0.40$.

In the seventeenth century, members of the Bernoulli family found a formula to calculate the binomial distribution for any number of trials and any probability of success. Before examining their formula, it may be best to explain the notation that occurs in it.

The symbol π^y means $(\pi)(\pi) \cdots (\pi)$, that is, the product when π is used as a factor y times. For example,

$$\left(\frac{3}{4}\right)^5 = \left(\frac{3}{4}\right)\left(\frac{3}{4}\right)\left(\frac{3}{4}\right)\left(\frac{3}{4}\right)\left(\frac{3}{4}\right) = \frac{243}{1024}$$

[†]Each time we remove a person from the population the probability of type O blood does in fact change slightly. However, since we are selecting only 4 people from several thousand, the changes are negligible.

Similarly,

$$(1 - \pi)^{n-y} = \underbrace{(1 - \pi)(1 - \pi) \cdots (1 - \pi)}_{n-y \text{ times}}$$

so that

$$\left(1 - \frac{3}{4}\right)^{7-5} = \left(\frac{1}{4}\right)\left(\frac{1}{4}\right) = \frac{1}{16}$$

The symbol $\binom{n}{y}$ is read "the number of combinations of n things taken y at a time." For example, if there are 4 slips of paper marked A, B, C, and D in a box and 2 slips are drawn at random, the possible combinations are

$$AB, AC, AD, BC, BD, CD$$

In this case $\binom{4}{2} = 6$. We are not interested in which letter is drawn first, so AB and BA are the same combination.

The symbol $\binom{n}{y}$ can also be applied to the blood-type example. Here $\binom{4}{2}$ means the number of different places that two O's can appear in a sequence of 4 symbols, that is, we are picking 2 positions out of the 4 possible positions. If first, second, third, and fourth are the positions, O can occur

1st and 2nd	1st and 3rd	1st and 4th
2nd and 3rd	2nd and 4th	3rd and 4th

or

OONN	ONON	ONNO
NOON	NONO	NNOO

In general,

$$\binom{n}{y} = \frac{n!}{y!(n-y)!}$$

where $n! = n(n-1)(n-2) \cdots (2)(1)$, and $n!$ is read "n factorial." Some examples are

$$\binom{4}{2} = \frac{4!}{2!(4-2)!} = \frac{4 \cdot 3 \cdot 2 \cdot 1}{(2 \cdot 1)(2 \cdot 1)} = 6$$

and

$$\binom{4}{0} = \frac{4!}{0!(4-0)!} = \frac{4 \cdot 3 \cdot 2 \cdot 1}{1(4 \cdot 3 \cdot 2 \cdot 1)} = 1$$

because $0! = 1$ by definition.

Table A.2 in the Appendix of Useful Tables is a table for $n!$, and Table A.3 is a table for $\binom{n}{y}$, the *binomial coefficients*. It should be noted that $\binom{n}{y} = \binom{n}{n-y}$ since this will often shorten calculations.

The Bernoulli formula for calculating binomial probabilities will now be understandable. To find $b(y; n, \pi)$, the probability in the binomial distribution of y successes when the number of trials is n and the probability of success on a single trial is π, we use the following formula:

$$b(y; n, \pi) = \binom{n}{y} \pi^y (1 - \pi)^{n-y}$$

Thus the mathematical model in the blood-type example is the random variable y having values 0, 1, 2, 3, 4 and probability function $b(y; 4, 0.40)$. The probabilities are computed in Table 3.1. This is the same result we previously computed by listing all possible experimental outcomes.

Since the Bernoulli formula can be used for any sample size and any probability of success, there is no need to go back to the list of all possible outcomes. If the number of trials is 20 and $\pi = 0.30$, then the probability of 7 successes out of 20 trials is

$$b(7; 20, 0.30) = \binom{20}{7}(0.30)^7(1 - 0.30)^{20-7}$$

$$= 77{,}520(0.30)^7(0.70)^{13}$$

$$= 0.16$$

Most of the time it is not necessary to use this formula since tables are available for many sample sizes and probabilities. Computers can easily be programmed to produce other tables of binomial distributions. The website for this text presents an example of this. It is useful, however, to know the formula so that the tables are meaningful.

Table 3.2 is an example of a table for 4 binomial distributions. The value of $b(7; 20, 0.30)$, which was calculated earlier in this section, can be found in the eighth row of the second column.

Note that there are entries of 0.000 in some positions, for example, $b(1; 20, 0.50)$. This does *not* mean that there is zero probability of getting 1 successful outcome in a sample of 20 when $\pi = 0.50$; rather it means that the probability of 1 successful outcome is smaller than $1/1000$.

TABLE 3.1. Computing Binomial Probabilities

y	$b(y; 4, 0.4)$
0	$\binom{4}{0}(0.4)^0(1 - 0.4)^{4-0} = (1)(0.4)^0(0.6)^4 = 0.1296$
1	$\binom{4}{1}(0.4)^1(1 - 0.4)^{4-1} = (4)(0.4)^1(0.6)^3 = 0.3456$
2	$\binom{4}{2}(0.4)^2(1 - 0.4)^{4-2} = (6)(0.4)^2(0.6)^2 = 0.3456$
3	$\binom{4}{3}(0.4)^3(1 - 0.4)^{4-3} = (4)(0.4)^3(0.6)^1 = 0.1536$
4	$\binom{4}{4}(0.4)^4(1 - 0.4)^{4-4} = (1)(0.4)^4(0.6)^0 = 0.0256$

The most likely outcome(s) for each value of π can be read from this table. If $\pi = 0.30$, the most likely outcome is 6 because it has the greatest probability. Similarly, for $\pi = 0.50$, the most likely outcome is 10; for $\pi = 0.70$ it is 14; and for $\pi = 0.75$ it is 15.

Since a binomial distribution is a probability distribution, we can find its expected value, $E(y)$, and variance, $V(y)$, by using the formulas introduced in Section 2.5. However, because of the special nature of the binomial distribution, shorter formulas exist. For a binomial distribution

$$E(y) = n\pi$$
$$V(y) = n\pi(1 - \pi)$$

Thus, for $b(y; 20, 0.50)$

$$E(y) = 20(0.5) = 10$$
$$V(y) = 20(0.5)(0.5) = 5$$

$$sd(y) = \sqrt{5} = 2.24$$

If we consider an interval from two standard deviations below the expected value to 2 standard deviations above the expected value, that is,

$$10 \pm 2(2.24)$$

TABLE 3.2. Four Binomial Distributions

y	$b(y; 20, 0.30)$	$b(y; 20, 0.50)$	$b(y; 20, 0.70)$	$b(y; 20, 0.75)$	y
0	0.001	0.000	0.000	0.000	0
1	0.007	0.000	0.000	0.000	1
2	0.028	0.000	0.000	0.000	2
3	0.072	0.001	0.000	0.000	3
4	0.130	0.005	0.000	0.000	4
5	0.179	0.015	0.000	0.000	5
6	0.192	0.037	0.000	0.000	6
7	0.164	0.074	0.001	0.000	7
8	0.114	0.120	0.004	0.001	8
9	0.065	0.160	0.012	0.003	9
10	0.031	0.176	0.031	0.010	10
11	0.012	0.160	0.065	0.027	11
12	0.004	0.120	0.114	0.061	12
13	0.001	0.074	0.164	0.112	13
14	0.000	0.037	0.192	0.169	14
15	0.000	0.015	0.179	0.202	15
16	0.000	0.005	0.130	0.190	16
17	0.000	0.001	0.072	0.134	17
18	0.000	0.000	0.028	0.067	18
19	0.000	0.000	0.007	0.021	19
20	0.000	0.000	0.001	0.003	20

or

$$5.52 \text{ to } 14.48$$

we find a probability of 0.958 that a value of the random variable will be within this interval and only a 0.042 probability that the value will be outside this interval.

In the next two sections we see how binomial distributions can help interpret the results of experiments.

EXERCISES

3.1.1. In a certain large college course, past records show that grades of A, B, C, D, and F are equally likely. If 1 student is chosen at random, find the following probabilities:

a. $P(C)$

b. $P(A \text{ or } B)$

c. $P(\text{a grade higher than D})$

d. $P(A, B, C, D, \text{ or } F)$

e. $P(B \text{ and } D)$

f. $P(E)$

g. $P(\text{not-A})$

h. $P(\text{not-A and not-F})$

3.1.2. If 2 people who do not study together take the course described in Exercise 3.1.1, find:

a. $P(2 \text{ A's})$

b. $P(\text{same grade})$

c. $P(\text{different grades})$

d. $P(\text{both higher than D})$

e. $P(\text{both fail})$

f. $P(\text{one passes and one fails})$

3.1.3. In a certain city, a fourth of the families take their children to the doctor for regular checkups. Five families are chosen at random.

a. What is the probability that exactly 3 families out of the 5 take their children to the doctor for regular checkups?

b. What is the probability that at most 2 families out of the 5 take their children for regular checkups?

c. What is the probability that more than 1 family out of the 5 take their children?

3.1.4. Assume a standard deck of 52 cards is used in the following problems.

a. Find the probability of drawing a heart or a picture card when selecting 1 card at random. Explain why $P(\text{heart or picture card}) \neq P(\text{heart}) + P(\text{picture card})$.

b. Find the probability of drawing 2 cards of the same color if the first card is randomly selected and kept out of the deck and the second card is then selected at random. Explain why $P(2 \text{ red cards}) \neq (1/2)(1/2)$.

3.1.5. In the game of Yahtzee, 5 ordinary dice are tossed.

  a. How likely is it that a player will get exactly four 2's on a random roll of the dice?

  b. In this game, 50 points are awarded if all 5 dice show the same number. How likely is this to happen on a random toss?

3.1.6. Find:

  a. 4!

  b. 0!

  c. 5!

  d. 1!3!

  e. 2!(6 − 2)!

  f. (10 − 2)!

3.1.7. Compute:

  a. $\binom{4}{4}$

  b. $\binom{3}{2}$

  c. $\binom{5}{0}$

  d. $\binom{5}{3}$

  e. $\binom{5}{1}$

  f. $\binom{4}{3}$

3.1.8. Use Exercise 3.1.7 to find the following without doing any further computations:

  a. $\binom{5}{5}$

  b. $\binom{3}{1}$

  c. $\binom{5}{2}$

  d. $\binom{5}{4}$

  e. $\binom{4}{1}$

  f. $\binom{4}{0}$

3.1.9. Compute:

  a. $\binom{7}{3}(0.20)^3(0.80)^4$

  b. $\binom{8}{0}(0.70)^0(0.30)^8$

  c. $\binom{10}{8}(0.10)^8(0.90)^2$

3.1.10. Compute the following binomial probabilities:
    a. $b(y; 3, 0.25)$ for $y = 0, 1, 2, 3$
    b. $b(y; 4, 0.30)$ for $y = 0, 1, 2, 3, 4$
    c. $b(y; 5, 0.10)$ for $y = 0, 1, 2, 3, 4, 5$
    d. Use part b to find the binomial distribution $b(y; 4, 0.70)$ without doing any further computations.

3.1.11. Find the expected value and variance for the blood-type example.
    a. Using the formulas given in Section 2.4
    b. Using the special formulas for the expected value and variance of a binomial distribution that are given in this section

3.1.12. An experimental psychologist has 20 volunteers for a sensory perception experiment and wishes to draw a random sample of 10 of these volunteers. Suppose that he decides to write all combinations of 10 names on index cards and then draw 1 of the cards at random. How many combinations will there be?

3.1.13. A geneticist studying dairy cattle has 4 bulls and 8 cows that can be used in an experiment. How many different matings are possible?

3.1.14. There are 6 teams in a baseball conference.
    a. How many games are necessary before each team plays every other team once?
    b. If there are no ties in standings, how many ways can the teams be ranked on the basis of number of games won?

3.1.15. Twelve school photographs (all the same size) are placed in random order face down on a table. Two of them are of identical twin boys. One of the twins is brought into the room and asked to select a photograph.
    a. What is the probability that he will select his own by chance?
    b. What is the probability that he will select his own or his brother's?
    c. If he is asked to select 2 photographs, what is the probability that he will select his own and his brother's?

3.1.16. There is evidence that among lower forms of animal life behavioral characteristics can be transferred from one individual to another along with the transfer of the chemical substance known as RNA. In an experimental study of this transfer behavior, 8 salamanders are divided at random into 2 equal-sized groups of 4. One group will be the experimental group and the other the control group.
    a. Show that there are 70 different ways the 2 groups can be formed.
    b. What is the probability that the 4 fastest swimmers are all in the same group?
    c. What is the probability that 3 of the 4 fastest swimmers are in the same group?
    d. All of the salamanders in one group (called the experimental group) received RNA from a salamander that has been trained to swim fast. The other group (called the control group) receives RNA from an untrained salamander. Before one could believe that behavior is transferred with RNA, what should the number of fastest swimmers in the experimental group be? Explain.

3.1.17. Many candy manufacturers who use artificial chocolate claim that their customers cannot tell it from real chocolate. Suppose 5 customers are selected at random and each is allowed to taste a candy bar made with real chocolate and the same kind of bar made with artificial chocolate. They are not told which contains real chocolate, and they are asked which one it is.

   a. If the manufacturer is correct about their inability to tell real from artificial chocolate, find the probability that a taster will correctly choose the one that is the real chocolate.

   b. What is the probability that all 5 tasters will choose correctly?

3.1.18. A certain basketball player has a success record of 1 in 3 for making attempted field goals. Suppose she attempts 7 field goals in a game.

   a. What conditions must be true in order to use the binomial distribution to produce reliable probability statements?

   b. Assuming the necessary conditions are met, compute the probability that the player will make exactly 4 field goals.

   c. What is the expected number of field goals she will make?

3.1.19. A night watchman must check in at 9 stations in a warehouse during each round of inspection. He decides to try all possible sequences of the 9 stations and use the shortest of these as his routine round of inspection. There are 9! possible different sequences of the stations.

   a. Why are there 9! different sequences?

   b. How many sequences must he try?

   c. If he walks 4 rounds of inspection each night, how many nights will he require to try all possible sequences?

3.1.20. A sociologist examines 6 northern cities that have the same percentage of racial minorities. He is able to rank the cities according to employment opportunities for high-school graduates from the minority groups. He then orders the cities on the basis of truancy among minority high-school students.

   a. How many ways is it possible to order 6 cities on the basis of truancy among minority students?

   b. If ordering by truancy and by job opportunities are unrelated, how likely is it that truancy will have a perfect reverse ordering to job opportunities?

   c. If the truancy ordering is the exact reverse ordering of that for job opportunities, should the sociologist decide that this happened by chance and that there is no relationship between the two?

3.1.21. A person claims the extrasensory ability of looking at a photograph and telling whether the subject of the photograph is still living or has died. In an experiment to test her claimed ability, she is shown 10 photographs of people unknown to her. (To improve the experiment, the subjects should be of the same age and the photographs taken at the same time; a high-school yearbook would meet both conditions.) She is asked to point out the 5 subjects who are now dead.

   a. How many ways can she select 5 of the 10 photographs?

   b. How many ways can she select the photographs of the 5 dead subjects?

   c. What is the probability of selecting the correct 5 photographs by guessing rather than by extrasensory ability?

   d. Why should this be a double-blind experiment?

3.1.22. The grading of laboratory reports is tedious, so a laboratory instructor decides that he will grade only a randomly chosen 2 of the 5 reports that each student has submitted.

If both are acceptable, the student will be given an A as his laboratory grade; if 1 is acceptable, he will receive a B; a C will be given if neither is acceptable.

a. How likely is a student to receive an A when he has submitted 5 acceptable reports? 4? 3? 2? 1? 0?

b. How likely is a student to receive a C when he has submitted 5 acceptable reports? 4? 3? 2? 1? 0?

3.1.23. In Exercise 1.1.6 the number of ways that all pairwise comparisons could be made among 10 people was determined by counting all of the events, and the answer was $9 + 8 + 7 + 6 + 5 + 4 + 3 + 2 + 1 = 45$.

a. Use combinations to verify that answer.

b. Why do both procedures produce the same answer? *Hint:* Add the integers from the ends toward the middle, $(9 + 1) + (8 + 2) + \cdots$.

3.2. TESTING HYPOTHESES

We return to the basic statistical problem of using probability to make decisions about populations that are not totally accessible. The following example shows how the probabilities in a theoretical binomial distribution can help to interpret the results of an experiment. (We have already seen an example in the baby cereal preference study, Example 1.7 and Section 2.4.)

Example 3.1.   Using a Binomial Distribution to Test a Hypothesis

Because dairy farmers need more cows than bulls, it would be advantageous for them if a method could be found to change the approximately 1-to-1 sex ratio found in nature. Many biological experiments have been performed in an attempt to alter sex ratio, either by trying to separate the sperm cells which produce male offspring or by finding some way to inactivate them so that they cannot fertilize an egg cell.

A reproductive physiologist believes that by treating the semen of the bull with a mild acid and using artificial insemination he can change the sex ratio of calves. (This is the scientific hypothesis.) He decides to perform an experiment and observe 20 calves that have been produced by this method. He is going to use statistics in order to generalize the result from these 20 calves to the entire population of calves that could be produced by this method. Thus, the statistical procedure begins at this point, prior to the actual experiment.

The steps in the statistical procedure are:

1. State the null hypothesis.
2. State the alternative hypothesis.
3. Establish α, the level of rejection, and the region of rejection.
4. Perform the experiment and observe the outcome.
5. Draw conclusions.

Step 1. State the Null Hypothesis. In this experiment, H_0: $\pi = 0.5$, that is, under chance alone, the probability of a newborn calf being female is 0.5. In other words, the treatment has no effect on the sex ratio. The theoretical probability distribution if the null hypothesis is true is $b(y; 20, 0.50)$. This experiment can be done in such a way that it satisfies the 5 conditions of a binomial experiment: There are only 2 possible outcomes, a male calf or a female calf. There will be a repeated number of trials, 20. If the null hypothesis is true, P(female calf) $= 0.5$ for each trial. The 20 cows can be selected at random, and the semen can also be selected at random from different bulls, ensuring independence from trial to trial. The physiologist is interested in the statistic y, in this experiment the number of female calves born.

Step 2. State the Alternative Hypothesis. In this experiment, the alternative hypothesis is $H_a: \pi \neq 0.5$. Since the physiologist does not know ahead of time what effect the mild acid will have on the sex of newborn calves, this is a two-sided test, or a *two-tailed test*. He will reject the null hypothesis if the outcome is an extreme case in either tail of the binomial distribution.

Step 3. Establish α, the Level of Rejection, and the Region of Rejection. Looking at the binomial distribution $b(y; 20, 0.50)$, he wants to set a rejection level as close to 0.05 as possible (because this is a traditional level used). Since this is a two-tailed test, he wants to reject the null hypothesis if he obtains an outcome with a probability of less than 0.025 at either side of the distribution. He notes from Table 3.2 that

$$P(0 \text{ or } 1 \text{ or } 2 \text{ or } 3 \text{ or } 4 \text{ or } 5)$$
$$= P(0) + P(1) + P(2) + P(3) + P(4) + P(5)$$
$$= 0.000 + 0.000 + 0.000 + 0.001 + 0.005 + 0.015$$
$$= 0.021$$

and that

$$P(15 \text{ or } 16 \text{ or } 17 \text{ or } 18 \text{ or } 19 \text{ or } 20)$$
$$= P(15) + P(16) + P(17) + P(18) + P(19) + P(20)$$
$$= 0.015 + 0.005 + 0.001 + 0.000 + 0.000 + 0.000$$
$$= 0.021$$

so the actual α is 0.042. The region of rejection is all y such that $0 \leq y \leq 5$ or $15 \leq y \leq 20$, and y is called the *test statistic*. Including any more values in the region of rejection would have made α further from 0.05. The symbol y here stands for the number of female calves born (alternatively, y could stand for the number of male calves born).

Step 4. Perform the Experiment and Observe the Outcome. The experiment is now performed, and suppose 6 males and 14 females are born. If the null hypothesis is true, the expected number of female calves would be $E(y) = n\pi = 20 \times 0.5 = 10$. Since the number of female calves observed in the experiment is $y = 14$, the physiologist cannot be especially encouraged by a deviation of only 4 from the number expected by chance alone. However, in the statistical procedure, decisions are based on probability, and the probability of a deviation of this magnitude (or greater) when the treatment is ineffective is needed.

Step 5. Draw Conclusions. The α level and the region of rejection merely specify, prior to the experiment, those outcomes that can be considered plausible and those that would be unusual when the null hypothesis is true. In this experiment, outcomes of less than 6 or more than 14 occur only 0.042 of the time if the null hypothesis is true. Since $y = 14$ is not in the region of rejection, the physiologist does not reject the null hypothesis.

The outcome of 14 deviates by 4 from the expected value of 10 under the null hypothesis $[n\pi = 20(0.5) = 10]$. The probability of a chance deviation this great or greater is

$$P(0) + P(1) + P(2) + P(3) + P(4) + P(5) + P(6) = 0.058$$

plus

$$P(14) + P(15) + P(16) + P(17) + P(18) + P(19) + P(20) = 0.058$$

So the P value is

$$P = 0.058 + 0.058 = 0.116$$

Thus the probability of obtaining a chance deviation of this magnitude (or greater) from the expected 1-to-1 sex ratio is 0.116. This probability is greater than the $\alpha = 0.05$ chosen by the physiologist, hence too large to claim that the experimental sex ratio of 14 to 6 is a significant altering of the proportion of females from $\pi = 0.5$.

Once again, let us remember that it is not known for sure whether or not the addition of a mild acid to bull semen will alter the sex ratio of calves. An experiment based on more than 20 births might verify the change observed in the experiment. However, for this experiment, the physiologist must decide that the experimental outcome is not improbable $(P > \alpha)$ under the null hypothesis and chance alone.

This process of setting up the null hypothesis may still seem rather round-about since the null hypothesis is usually the opposite of the decision the scientist is hoping to make. However, since there is no information about the probability associated with the experimental hypothesis, the null hypothesis must be set up so that known probabilities can be used.

Not all tests of hypotheses are two tailed. Sometimes the experimenter is looking for evidence in a particular direction. The following example will illustrate a *one-tailed test* of hypothesis.

Example 3.2.   Testing a Hypothesis Using a Binomial Distribution

The staff of a reading clinic is interested in determining the sex ratio of children who have a certain reading problem. The children reverse the letter sequences in words; for example, they read "saw" for "was." Someone has claimed that more than 70% of the children with this disorder are boys. The staff decides to look at a random sample of 20 children who have this reading problem. The null hypothesis is H_0: $\pi = 0.7$ and H_a: $\pi > 0.7$ because they are looking for evidence to substantiate the claim. Assuming the null hypothesis is true, they use the binomial distribution $b(y; 20, 0.70)$ as the theoretical model. The number of boys in the random sample of children with this disorder is represented by y.

The level of rejection in this survey is chosen to be as close to 0.05 as possible. Looking at Table 3.2 in Section 3.1, the actual α is seen to be 0.036 and the region of rejection is 18, 19, 20.

Assume the survey reveals that 18 out of the 20 afflicted children are boys. Whether one uses the fact that the test statistic, $y = 18$, is in the region of rejection or that the P value of 0.036 is less than α, the null hypothesis is rejected and it is concluded that there is evidence that more than 70% of the children with this disorder are boys.

We have noted that with this type of test there is no way to be certain whether the null hypothesis is true or false. Although the null hypothesis was rejected in the example above, it is of course possible that it is actually true and a very unlikely outcome just happened to occur. To reject a true null hypothesis is called a *Type I error*. The probability of committing a Type I error in the survey above is 0.036 because $\alpha = 0.036$, that is, there is a 3.6% chance that the null hypothesis is true and sample results lead to rejection of it. The probability of a Type I error is always α, the level of rejection, and is chosen by the experimenter.

If the results had been different, the null hypothesis might not have been rejected. For example, the survey might have shown that 15 out of 20 children displaying reading reversals were boys. Since 15 is not in the region of rejection (and $P = 0.417$), the null hypothesis would not have been rejected, and it could be concluded that among the children with reading reversals 70% or fewer may be boys. In this case, it is possible that the null hypothesis is false, but it has not been rejected. To fail to reject a null hypothesis when it is false is called a *Type II error*.

It is more difficult to determine the probability of a Type II error than of a Type I error. The probability of a Type I error, rejecting a true null hypothesis, is α. The probability of a Type II error is, in this case, the probability that y is not in the region of rejection of the null hypothesis if π is not 0.70. This cannot be determined in this form because there is no specific value for π; $\pi \neq 0.70$ is an infinity of values.

To determine the probability of a Type II error:

1. Choose a reasonable *specific* alternative value of the parameter, $\pi = \pi_a$, that is of clinical importance.
2. Find β, the cumulative frequency in $b(y; n, \pi_a)$ for y in the acceptance region of H_0; that is, $\beta = P(y$ is in the region of acceptance of H_0 if $\pi = \pi_a)$.

The probability β is the probability of failing to reject the null hypothesis when it is false by a specific amount. In more positive terms, the *power* of the experiment or survey, that is, the probability of detecting the specific alternative hypothesis, is $1 - \beta$. Thus power is related to β, and depending on which is easier to compute, we find one from the other by

$$\text{Power} = 1 - \beta \text{ or } \beta = 1 - \text{Power}$$

In the example above, in which 15 out of 20 children with reading reversals were boys, the null hypothesis was not rejected. What is the probability that a false null hypothesis may have been accepted? From knowledge of reading problems, the staff might agree that a reasonable alternative value is $\pi_a = 0.75$. Power depends on the "degree of falseness" of the null hypothesis, so they specify the smallest degree of falseness of practical interest. This means that if in fact 75% of the cases of reading reversals occur in boys the clinic would examine boys very carefully for this problem, but if fewer than 75% were boys they would not examine boys more closely than girls. Referring to the table in the previous section under $b(y; 20, 0.75)$, we find that the probability that $0 \leq y \leq 17$ is $\beta = 0.909$. This means that there is a 90.9% chance of *failing* to reject the null hypothesis if in fact 75% of the children with reading reversals are boys! The chance of detecting the difference is only $1 - 0.909 = 0.091$; the power of this survey is very low.

A *powerful* experiment generally means a power of 0.70 or greater, so the survey above is very poor. This illustrates the need to design an experiment in such a way that there is a reasonable chance of detecting a clinically important difference if it exists. To increase the power in this survey, a much larger sample size is necessary. Another way to increase the power (decrease β) is to increase α.

In practice, many times we do not have enough information to choose a reasonable specific alternative, and thus we are not able to compute β. Fortunately, the power of an experiment usually increases with the size of the sample, so we work with samples that seem large enough to make the experiment powerful. If we can specify the alternative value of the parameter, it is possible to use a repetitive process (likely with the aid of a computer) to determine how large the sample size must be in order to have a specified power. In the reading-reversal example, it is necessary to use a sample size of $n = 501$ to achieve a power of 0.80 in detecting $\pi_a = 0.75$ when the null hypothesis is H_0: $\pi = 0.70$ and $\alpha = 0.05$ (Buckalew, 1974, p. 61). This large size is required because a relatively small difference is specified.

We usually try to achieve a balance between the α level and the power. We want a moderately low α level (as 0.05) and try to get the power as high as possible, usually by taking relatively large samples.

Which type of error is worse depends on the situation. For example, imagine that a medical microbiologist is testing a new antibiotic for effectiveness against a particular bacterium. Currently used antibiotics are known to have a cure rate of $\pi = 0.75$. The two types of error could occur under the following circumstances:

Type I. The microbiologist is testing H_0: $\pi = 0.75$ against H_a: $\pi > 0.75$. The new antibiotic actually has a cure rate of 0.75, but the results of the experiment lead her to conclude that it is better than the antibiotics currently used. If the new one is equal to the others in all other respects, such as price and side effects, then this Type I error is not serious. If, however, the price is higher or the side effects are more severe, then the Type I error is serious.

Type II. The microbiologist is again testing H_0: $\pi = 0.75$ against H_a: $\pi > 0.75$. Now, however, let us assume that the new antibiotic is actually better but she fails to detect this from the results of the experiment. The Type II error here means that a more effective medication will not be used. The seriousness of the error depends on the seriousness of the illness and how much better the new medicine would be. If π is actually 0.78, this would not be much of an improvement so the error is not as serious as if π were 0.98 and a very effective medication were not being used.

The diagram in Figure 3.2 summarizes the various possibilities that occur when testing hypotheses. The specific probabilities listed refer to the reading-reversal study (Example 3.2) used in this section.

Note that the probabilities in the columns of this diagram sum to 1. Also, once the decision is made, only one type of error is possible. If the null hypothesis is rejected, there is then no possibility of a Type II error. Similarly, if we fail to reject the null hypothesis, we no longer need to worry about a Type I error.

| | | H_0 is really | |
		TRUE	FALSE
Decision About H_0 Based on Test of Significance	ACCEPT	*No error:* $1 - \alpha = 0.964$	*Type II error:* P (Type II error) = β $\beta = 0.909$
	REJECT	*Type I error:* P (Type I error) = α $\alpha = 0.036$	*No error:* Power = $1 - \beta = 0.091$

FIGURE 3.2. Type I and Type II errors.

In the discussion of hypothesis testing and errors in this section, we have used only examples that fit the small table of binomial distributions given in Section 3.1. Two similar but larger tables are found in Table A.4a for samples of size $n = 20$ and Table A.4b for samples of size $n = 25$. These tables are used in the same manner as the smaller table in Section 3.1.

If $\alpha = 0.10$ and the test is two tailed, the horizontal lines indicate the regions of rejection and acceptance. If $\alpha = 0.05$ and the test is one tailed, the line in the appropriate tail may be used to indicate the region of rejection. Other α levels can be used, but then the regions must be determined by the user of the table. The probability of a Type II error can also be found from these larger tables; the method is the one just described in this section.

Many other tables are readily available in statistics books and in reference books. If the particular table needed is not available, it can be computed using the Bernoulli formula possibly with the assistance of a computer (see the computer usage sections on the text's Internet site). Approximation methods are also possible; these are discussed in Chapter 7.

A brief summary of this section follows.

Procedure. Test of Hypotheses for a Binomial Parameter π

Region of Rejection Method

H_0: $\pi = \pi_0$
H_a: $\pi \neq \pi_0$ or $\pi > \pi_0$ or $\pi < \pi_0$

Significance level: α
Test statistic: y, the number of successes out of n trials
Using a table for the binomial distribution with probability function $b(y; n, \pi_0)$, determine the region of rejection.
For H_a: $\pi \neq \pi_0$, the region of rejection is $0 \leq y \leq c_L$ and $c_U \leq y \leq n$ such that

$$\sum_0^{c_L} b(y; n, \pi_0) \text{ and } \sum_{c_U}^{n} b(y; n, \pi_0)$$

are each as close as possible to $\alpha/2$.
For H_a: $\pi > \pi_0$, the region of rejection is $c_U \leq y \leq n$ such that

$$\sum_{c_U}^{n} b(y; n, \pi_0)$$

is as close as possible to α.
For H_a: $\pi < \pi_0$, the region of rejection is $0 \leq y \leq c_L$ such that

$$\sum_0^{c_L} b(y; n, \pi_0)$$

is as close as possible to α.
Reject H_0 if y is in the region of rejection.

P-Value Method

For H_a: $\pi \neq \pi_0$, compute $P = P(|y - n\pi_0| \geq |\text{test statistic} - n\pi_0|)$.

For H_a: $\pi > \pi_0$, compute $P = P(y \geq \text{test statistic})$.

For H_a: $\pi < \pi_0$, compute $P = P(y \leq \text{test statistic})$.

Reject H_0 if $P \leq \alpha$.

Error

$P(\text{Type I error}) = \alpha$

$P(\text{Type II error if } \pi = \pi_a) = P(y$ is in the region of acceptance of H_0 if $\pi = \pi_a)$

EXERCISES

3.2.1. Use Tables A.4a and A.4b in the Appendix to find the following:

   a. $P(4 < y < 8)$ when $n = 20$, $\pi = 0.8$

   b. $P(y \leq 2)$ when $n = 25$, $\pi = 0.6$

   c. $P(y \geq 4)$ when $n = 25$, $\pi = 0.25$

   d. $P(y > 15)$ when $n = 20$, $\pi = 0.70$

   e. $P(y < 19)$ when $n = 20$, $\pi = 0.55$

   f. $P(6 \leq y \leq 9)$ when $n = 25$, $\pi = 0.35$

3.2.2. A teacher gives a student a make-up test consisting of 20 true-false questions. The intent of the test is to determine whether the student answers the questions correctly through knowledge of the material or merely by making lucky guesses. Assume the correct answers are a random sequence of "true" and "false" and that the student's guesses are also random.

   a. State a null hypothesis based on the probability of guessing the correct answer to a question.

   b. State a one-tailed alternative hypothesis based on the probability of arriving at the correct answer through knowledge.

   c. Find the region of rejection when α is set as close to 0.05 as possible. (Remember that the null hypothesis will be rejected only if an extreme value occurs on one side of the distribution.)

   d. If the student correctly answers 16 of the 20 questions:

   i. What is the P value?

   ii. What should the teacher conclude?

3.2.3. A carnival operator wants a game that can be won about 30% of the time. If the game is won more frequently, it will not be economical for the operator; if winning is less frequent, potential players will be reluctant to risk their money. He devises a dart-tossing game that he thinks will suit his criterion and tests it on 20 random players.

   a. State a null hypothesis based on his criterion.

   b. State a two-tailed alternative hypothesis.

   c. If the region of rejection is set at $0 \leq y \leq 2$ and $11 \leq y \leq 20$, what is the α level?

   d. What conclusion should the operator draw about the game if there are 9 winners among the first 20 players? What must be assumed about the players in order to accept this conclusion?

3.2.4. A campus parking lot contains 20 spaces, all reserved for faculty members. The administration decides that students may park their cars in the lot after 4:00 PM if faculty usage then drops to less than 70%. A random weekday afternoon is chosen to sample the faculty usage after 4:00 PM.

   a. State the null hypothesis.

   b. State a one-tailed alternative hypothesis that would lead to student usage of the lot.

   c. Find the region of rejection for α as close to 0.05 as possible.

   d. If there are 18 faculty cars in the lot at the time of the survey:

       i. What is the P value?

       ii. What decision should be made about student parking?

   e. Do you see any difficulties in the design of this survey? Suggest a better design.

3.2.5. In the experiment concerning the altering of the sex ratio in newborn calves (Example 3.1), the null hypothesis is H_0: $\pi = 0.5$ and H_a: $\pi \neq 0.5$. There are 20 trials and the region of rejection is $0 \leq y \leq 5$ and $15 \leq y \leq 20$.

   a. The physiologist would consider the experiment a success if the proportion of female calves is 0.70. How likely is it that a change of this magnitude will be detected by the statistical procedure described?

   b. What would you suggest to the physiologist if he does not think that this experimental design is powerful enough to detect this useful change?

3.2.6. In an effort to control mosquitoes without having to use dangerous insecticides, entomologists have taken advantage of two factors in the biological nature of mosquitoes: Male mosquitoes are not bloodsuckers and nearly all female mosquitoes mate but once. Thus the entomologists release massive numbers of sterilized male mosquitoes to reduce the probability of a female mating with a fertile male and consequently producing more mosquitoes. After such a release, the entomologists hypothesize that the probability of a female mating with a fertile male is H_0: $\pi = 0.30$. If 20 females are captured and examined for fertile eggs:

   a. Find the region of rejection if the alternative hypothesis is H_a: $\pi > 0.30$.

   b. What is the power of the experiment if $\pi_a = 0.50$?

   c. What is the power if $\pi_a = 0.70$?

3.2.7. A large corporation is going to purchase 150 company cars for its salesmen and executives. The corporation has already eliminated many makes and models and now must choose between two specific types of cars, A and B, which are comparable in size, purchase price, and maintenance cost. The corporation will base its final decision on the gasoline mileage of these two types. It is known that 70% of the cars of type A average more than 20 miles per gallon, and it is strongly believed that car B has a better record. If B is proved, better they will buy B; otherwise they will buy A.

   a. State the two outcomes that should be considered for a random sample of cars of type B.

   b. State the null hypothesis in terms of cars of type B.

   c. State the one-tailed alternative hypothesis for car B.

   d. Which type of error should be kept to a minimum in this experiment? How can this be accomplished?

3.2.8. A behavioral scientist feels that right-handed people have a tendency to make right-hand turns when they have no other basis for choosing the direction in which they should turn. To conduct a statistical test, she draws a random sample of 20 right-handed individuals from a large group of volunteers. To keep the subjects unaware of the nature of the experiment, she pretends to be conducting a survey of family dietary habits. She has the subjects brought into her office one at a time, questions them about the eating habits of their families, and then directs them out by a different way from the one by which they entered. They are told to go down a hall and out either door at the end. The experimenter watches each subject leave and records whether the subject chooses the door to the right or left as he or she exits.

a. State a null hypothesis which specifies that only chance leads to the choice of the door to the right.

b. For a two-tailed alternative hypothesis, the region of rejection could be $0 \leq y \leq 5$ and $15 \leq y \leq 20$. What is the α level?

c. For a one-tailed alternative hypothesis, the region of rejection could be $14 \leq y \leq 20$. What is the α level?

d. For the specific alternative $\pi_a = 0.70$, which is more powerful, the one-tailed or the two-tailed test?

e. Comment on the deception involved in this experiment.

3.2.9. For a binomial experiment in which $n = 20$ and H_0: $\pi = 0.30$:

a. Find the region of rejection with an α as near 0.05 as possible when H_a: $\pi \neq 0.30$.

b. Find the region of rejection with an α as near 0.05 as possible when H_a: $\pi > 0.30$.

c. For the specific alternative $\pi_a = 0.50$, how much more powerful is the one-tailed test than the two-tailed test?

d. Which of the following statements is true?

   i. The one-tailed test is more powerful because it has a greater α level.

   ii. The one-tailed test is more powerful because it has a greater β.

   iii. The one-tailed test is more powerful because there are more possible y values in its region of rejection.

   iv. The one-tailed test is more powerful because the sum of the probabilities associated with the region of rejection is greater for the specified alternative $b(y; 20, 0.50)$.

3.2.10. After a flood or storm, insurance companies buy damaged goods from stores that carry their policies. To recover some of the loss, they sell the damaged goods to salvage companies. Suppose 30,000 flood-damaged highway safety flares are offered for sale by an insurance company with the claim that 25% of them are too damaged to ignite.

a. State a null hypothesis that would test the insurance company's claim.

b. State the alternative hypothesis of greatest concern to the insurance company.

c. State the alternative hypothesis of greatest concern to a salvage company.

d. Suppose the insurance company's statement about the 30,000 flares is correct. Determine how likely it is that a random sample of 20 flares will have:

   i. Exactly 10 flares that fail to ignite

   ii. At least 10 (that is, $10 \leq y \leq 20$) that fail to ignite

 e. Suppose the insurance company's statement is incorrect and actually 40% are too damaged to ignite.

  i. What is the probability that exactly 10 will fail to ignite?

  ii. What is the probability that at least 10 will fail to ignite?

 f. Suppose H_0: $\pi = 0.25$ is being tested, what is the power of the test when α is as near as 0.05 as possible and π is really 0.40?

3.2.11. Describe how a Type I or Type II error could occur in the following situations and give some of the factors that would determine the seriousness of the errors.

 a. A bookstore is trying to determine what proportion of the students buying a certain textbook will also buy an optional student guide. In the past, 40% of the students buying the text have also bought the guide. The bookstore wants to test H_0: $\pi = 0.40$ against H_a: $\pi > 0.40$.

 b. A seed company wants to claim on a certain seed package that at least 90% of the seeds will germinate. The company decides to check this before the packages are printed and test H_0: $\pi = 0.90$ against H_a: $\pi < 0.90$.

 c. A recreation specialist is planning campsite facilities for a state forest and wants to include several rustic tent-only campsites that will be inaccessible to campers on wheels. He thinks that only 20% of the people camping in the area would desire such facilities. He tests H_0: $\pi = 0.20$ against H_a: $\pi \neq 0.20$.

3.2.12. Archaeologists use pelvic bones to determine whether a skeleton is that of a man or woman. Primitive cultures often buried their outstanding members (rulers, warriors, athletes, and so on) with greater ceremony than ordinary members. Using this fact, much can be learned about the status of women in an early culture by observing the frequency of skeletons of females in ceremonial graves. Suppose that an archaeologist discovers 20 graves that can be assumed to be a random sample of the ceremonial graves of a Stone Age culture in Wiltshire, England.

 a. What is the most logical statistical hypothesis to be tested?

 b. Suppose the region of rejection is: The number of skeletons of females is less than 8. What is the value of α?

 c. Suppose $\pi_a = 0.30$; what is the numerical value of β?

 d. What assumption is necessary to use this test procedure?

3.2.13. A certain dental condition which can be corrected if detected early enough occurs in the population with a frequency of $\pi = 0.20$. An orthodontist believes that this condition occurs more frequently in children who were born with cleft palates and that parents of such children should be warned to watch for early evidence of the dental condition. To test his hypothesis, she follows the dental development of a random sample of 25 children born with cleft palates.

 a. What is the most logical null hypothesis for the orthodontist to check? What alternative hypothesis should she use?

 b. Suppose she wants α to be as close to 0.05 as possible; what region of rejection should she set for y, the number of children in the sample who develop this dental condition?

 c. Suppose 8 of the children in her sample develop the condition. What is the P value? Should she reject the null hypothesis? Why, or why not? What conclusion should she draw?

3.2.14. Sickle-cell disease is a potentially lethal genetic disease in the Black race. It is estimated that 30% of African-Americans in a certain Gulf Coast region have the disease or carry the trait for it. This figure seems too large to a physician in the region, so he takes a random sample of 25 of his African-American patients and examines blood smears.

    a. State the physician's most logical null and alternative hypotheses.

    b. What region of rejection would you suggest he use? What is the α level for this region?

    c. If the percentage in question is really 15%, what is the power of his test?

    d. Which type of error is more serious in his study, Type I or Type II? Why?

    e. Suppose 12 patients of his sample have the condition or seem to be genetic carriers. Should he reject his null hypothesis or not? Why? What is the P value? What conclusion should he draw about the proportion of sickle-cell disease in the Black population?

3.2.15. Cryobiologists have been experimenting for many years with methods of freezing human corneas so that, when thawed, the membranes can be safely used in "eye transplants." If corneas are suspended in ethylene glycol, 70% of membranes survive freezing and thawing. Unfortunately the chemical compound is toxic, and therefore a cornea soaked in it is unsafe for transplant. Suppose a cryobiologist finds a nontoxic chemical that has similar protective properties. He wants to compare its effectiveness with ethylene glycol in the freezing-thawing process.

    a. State the null and alternative hypotheses.

    b. If 20 corneas are to be used in his experiment, give the region of rejection for $\alpha = 0.10$.

    c. Suppose $y = 10$ is the number that survive; should the experimenter feel encouraged or discouraged by the results? Give a reason for your answer.

3.2.16. Vegetable farmers try to avoid the use of insecticides because of expense and health hazards. However, if crops become too heavily infested, it becomes necessary to spray them. Suppose a farmer decides that she will spray her cabbages if their infestation with moth larvae is significantly greater than 20%.

    a. If the farmer samples the crop to determine the percentage of infested cabbages, what is the null hypothesis?

    b. What is the most logical choice for the alternative hypothesis? Why?

    c. For $n = 20$ and α as close to 0.05 as possible, choose the region of rejection that is consistent with the alternative hypothesis.

3.2.17. In times of stress, some people hyperventilate to the point of dizziness and fainting. To determine whether this behavior is equally likely in men and women, a researcher takes a random sample of 25 cases from a hospital emergency room's file on those treated for hyperventilation.

    a. What hypothesis should be tested about the percentage of males among those treated?

    b. What should the region of rejection be if α is to be as near 0.01 as possible?

    c. If 16 of the 25 persons in the sample are men, should the researcher conclude that men are more likely to hyperventilate than women? Why or why not?

3.3. ESTIMATION

So far, our discussion of statistical methods has dealt with only one of the general problems of statistics, decisions about hypotheses. Tests of hypotheses are possible only when we have quite a bit of information about the experimental situations. For example, to analyze the results of the experiment on the sex ratio of calves, the experimenter had to know the sex ratio of newborn calves in an untreated population. In the early stages of experimentation, when less information is available, the scientist often uses *estimation* (Figure 3.3).

Estimation will answer questions like "What proportion of ex-prisoners who have gone through a certain group therapy program will be arrested again within the first two years after release?" If we consider the entire population of prisoners who have gone through or will go through the program during their incarceration and we use as the variable of interest whether or not they are arrested again within two years after release, what is the appropriate value of π, the proportion arrested again?

Since we cannot observe the entire population, we will instead examine a random sample from it and count the number of subsequent arrests in the sample. Recall that this count, based on the results of sampling, is called a *statistic*. Then, using the binomial distribution as a model for this study, we will use the statistic to make a statement about the unknown parameter π, the true proportion of ex-prisoners who will be arrested again (Figure 3.4).

In trying to estimate the unknown parameter, two types of estimates are possible.

1. A *point estimate*—a statistic based on a sample.
2. An *interval estimate*—an inference based on a statistic.

The natural point estimator of a proportion π is

$$\hat{\pi} = \frac{y}{n}$$

in which y is the number of successes in a sample size of n. The estimator $\hat{\pi}$ is read "π hat." In general, placing a caret, or "hat," on a Greek letter indicates an estimator of the parameter.

The estimator $\hat{\pi}$ is not only the natural point estimator but also the best estimator because it has three desirable properties of an estimator:

1. $\hat{\pi}$ *is a maximum-likelihood estimator*. That is, the estimate of π that we get using this estimator makes the outcome that we obtained the one most likely to occur. We can see this by using Table 3.2, where the value of y with the greatest probability, gives the best estimate $\hat{\pi} = y/n$ of the binomial parameter π. In the distribution with probability function $b(y; 20, 0.30)$, $y = 6$ is the most probable outcome and $6/20 = 0.30$; in $b(y; 20, 0.50)$, $y = 10$ is the most probable outcome and $10/20 = 0.50$; in $b(y; 20, 0.70)$, $y = 14$, is the most probable outcome and $14/20 = 0.70$ (see Figure 3.5).

FIGURE 3.3. Types of inference.

FIGURE 3.4. The inferential process.

$\hat{\pi}$ *is unbiased*. That is, if we were to repeat the estimation process, the average of all possible estimates would be the true parameter π.

π *has a minimum variance*. That is, the possible estimates are clustered closer to π than for any other unbiased estimator.

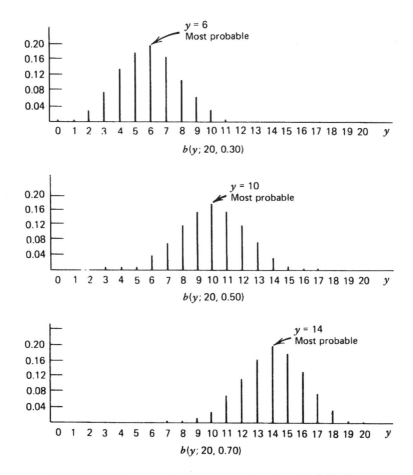

FIGURE 3.5. The most probable outcome in three binomial distributions.

Thus, if we observe a random sample of 20 prisoners who had gone through the therapy program and we find that 6 of them have been arrested again, then the best point estimate of the proportion of subsequent arrests is

$$\hat{\pi} = \frac{6}{20} = 0.30$$

Because of the properties of this estimator, we can be confident that this is likely to be close to the true value. Unfortunately, it will usually not be exactly the true value. A repetition of the survey might yield

$$\hat{\pi} = \frac{8}{20} = 0.40$$

Although we know that both of these estimates are close, we also know that probably neither of them is exactly correct.

One way to avoid this difficulty is to use an *interval estimate*, an inference that the parameter is between certain bounds. The confidence interval is obtained by asking "For which values of π is $\hat{\pi}$ a common or frequent estimate?"

We use the following steps to find an interval estimate.

Procedure. Central Confidence Intervals for π

1. Specify an α level.
2. Take a sample of size n.
3. Find y, the number of successes.
4. Give the interval of all values of π for which y would fall in the region of acceptance for a two-sided α-level test.[†]

For example, if $\alpha = 0.10$, $n = 20$, and $y = 8$, we use Table A.4a in the Appendix; 8 is in the region of acceptance for π between 0.25 and 0.55. Thus $\hat{\pi} = 8/20 = 0.40$ is among the 90% most common estimates of all π values between 0.25 and 0.55. Since $\alpha = 0.10$, when we use this procedure about 90% of the intervals obtained, will include the actual parameter being estimated. The interval is written

$$\text{CI}_{0.90}: 0.25 \leq \pi \leq 0.55$$

and is called the 90% *confidence interval* for π. This method yields a *central* confidence interval since two-sided regions of acceptance are employed. Note that the best point estimate, $\hat{\pi} = 8/20 = 0.40$, is within this interval.

For any given sample size, the method we just outlined gives the narrowest $\text{CI}_{1-\alpha}$. The confidence interval in this example is quite wide; this is because the sample size $n = 20$ is small. If a larger sample is used (and α remains constant), the same statistic $\hat{\pi} = y/n$ will yield a smaller confidence interval. To see this, Tables A.5a through A.5e in the Appendix can be used. These tables list the confidence intervals for various sample sizes and various α levels. (Instructions for reading these tables precede the group.) To see the effect of increased sample size, let $\alpha = 0.10$, $n = 100$, $y = 40$; then $\hat{\pi} = 40/100 = 0.40$ (as in the previous example), and from Table A.5c

$$\text{CI}_{0.90}: 0.318 \leq \pi \leq 0.487$$

which is a smaller interval than the one found for $n = 20$.

[†]The authors are indebted to H. C. Fryer for the graphic determination of confidence intervals in this section and in Tables A.4a and A.4b.

FIGURE 3.6. Linear interpolation yields conservative confidence intervals.

Tables A.5a and A.4b give slightly different 90% confidence intervals for sample size $n = 25$. This difference occurs because Tables A.5a through A.5e were calculated by a different procedure than Tables A.4a and A.4b. The method for finding confidence intervals used in Tables A.4a and A.4b is very instructive but lengthy to compute. The alternative shorter method used for Tables A.5a through A.5e will not be explained here; it is an approximate method and is known to produce reliable confidence intervals.

We can find one-sided confidence intervals as well as central confidence intervals. The method is the same except that the region of acceptance for a one-sided α-level test is used in step 4 of the Procedure given above. If Tables A.5a through A.5e were used, we refer to the α column that is twice as large as the desired α level and use only one of the values L or U that are given. (Example 3.3 demonstrates a one-sided procedure.)

Linear interpolation can be used to obtain confidence intervals for sample sizes between those listed in the tables or it can be used for statistics that fall between values listed in the tables. This method of interpolation of confidence intervals is a conservative estimate because the confidence intervals actually decrease along curves within the straight lines along which interpolation occurs. Since the interpolated values are outside the actual curves, they more than preserve the α level of the tables (Figure 3.6).

As mentioned before, by using an interval estimate, we avoid the almost certain error of a point estimate. If an interval estimate includes the true proportion, then it is correct. It is possible for two different interval estimates to be correct. For example, two polls on the proportion of the American population that approves of the president's economic policy could yield point estimates $\hat{\pi}_1$ and $\hat{\pi}_2$ and interval estimates as in Figure 3.7. If π is the true proportion, both point estimates are wrong. However, both interval estimates are correct. In this particular case, neither interval contains both point estimates but both intervals are still correct.

The question of Type I or Type II errors does not apply to the inference of confidence intervals since no decisions concerning hypotheses are being made. However, the reliability of the estimate made by the confidence interval is expressed in the percentage of confidence. A level of confidence of 95% means that 95% of the intervals that could be determined by this method contain the true population parameter.

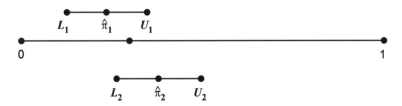

FIGURE 3.7. Confidence intervals for the same parameter obtained from different samples.

Although Tables A.5a through A.5e list confidence intervals, they may also be used to test hypotheses. This is demonstrated in the following example.

Example 3.3.   Using Confidence Intervals to Test Hypotheses

It is generally felt that those opposed to the issuance of a new school bond are more likely to go to the polls to vote than those who favor the bond. Thus a local school board feels that a bond issue must be favored by more than 70% of the registered voters to have a chance of being approved in the bond election.

Since the school board is concerned about detecting whether enough people are in favor of the bond issue, it wants to determine a one-sided confidence interval on π that makes a statement about the smallest possible value that π might be.

Suppose a random sample of $n = 250$ registered voters is surveyed by the school board and $y = 190$ favor the bond issue while $n - y = 60$ oppose it. Using Table A.5d and $y/n = 190/250 = 0.76$, the table is entered at $1 - 0.76 = 0.24$ and the lower bound is $1 - 0.289 = 0.711$. The 95% confidence interval that puts a lower bound on π is

$$\text{one-sided CI}_{0.95}: 0.711 \leq \pi \leq 1.00$$

(The 0.10 column is used because only the lower bound is needed.) This interval shows that the school board can schedule an election and feel confident that the bond issue will pass.

If the board preferred to phrase its investigation in terms of a test of hypothesis, it would test

$$H_0: \pi = 0.70 \quad \text{(bond issue may not pass)}$$

against

$$H_a: \pi > 0.70 \quad \text{(bond issue will pass)}$$

The board would find the one-sided confidence interval for the lowest value of π and conclude that the null hypothesis should be rejected at the 5% significance level because $\pi = 0.70$ is not in the interval.

Similar approaches can be used for two-sided alternatives and one-sided less-than alternatives. The correspondence between confidence intervals and tests of hypotheses is summarized in the following procedure.

Procedure. Testing Hypotheses Using Confidence Intervals

Confidence Interval	Test
Central $\text{CI}_{1-\alpha}: L \leq \pi \leq U$	$H_0: \pi = \pi_0$ $H_a: \pi \neq \pi_0$ α level of rejection Reject H_0 if π_0 is *not* in the confidence interval, that is, $\quad\quad\quad \pi_0 < L \text{ or } \pi_0 > U$
Upper bound	$H_0: \pi = \pi_0$

Confidence Interval	Test
One-sided $CI_{1-\alpha}: 0 \leq \pi \leq U$	$H_a: \pi < \pi_0$ α level of rejection Reject H_0 if π_0 is *not* in the confidence interval, that is, $\pi_0 > U$
Lower bound One-sided $CI_{1-\alpha}: L \leq \pi \leq 1.00$	$H_0: \pi = \pi_0$ $H_a: \pi > \pi_0$ α level of rejection Reject H_0 if π_0 is *not* in the confidence interval, that is, $\pi_0 < L$

EXERCISES

3.3.1. In each case below, the sample size n, the statistic y, the level of confidence $1 - \alpha$, the lower confidence limit L, or the upper confidence limit U are given. Use tables for placing a confidence interval on the binomial parameter π to fill in the missing values in each case.

Case	n	y	$1 - \alpha$	L	U
1	50	20	0.99	—	—
2	—	—	0.95	0.300	0.423
3	250	80	0.95	—	—
4	500	430	0.99	—	—
5	50	16	0.99	—	—
6	—	—	0.95	0.102	0.258
7	500	31	0.90	—	—
8	100	—	—	0.216	0.374
9	—	30	—	0.036	0.093
10	20	—	0.90	0.250	—

3.3.2. In a random sample of 250 inmates of federal prisons, 175 are found to have committed nonviolent crimes.

    a. What is the best estimate of the proportion of such federal offenders?

    b. Place a 95% confidence interval on the proportion of all federal prisoners convicted of nonviolent crimes.

    c. Can you deduce from this that the majority of inmates of all federal prisons have been convicted of nonviolent crimes?

3.3.3. A random sample of 25 precocious readers is drawn and their family backgrounds carefully studied. In 40% of the cases, the child's father is at least 15 years older than the mother. Place a 90% confidence interval on the proportion of such age disparities between the parents of precocious readers.

    a. Using Table A.4b

    b. Using Table A.5a

3.3.4. A random sample of 100 persons suffering from mental depression reveals that 75 of them cannot properly evaluate their job skills.

    a. Give the maximum-likelihood estimate of the binomial parameter.

    b. Set up a 95% confidence interval for this parameter.

3.3.5. In a random sample of 50 kindergarten children, there are 7 who hold crayons in their left hands while coloring a picture.

    a. Give the best point estimate of the proportion of left-handed kindergarten children.

    b. Explain what "best" means in this exercise.

3.3.6. Selected at random, 125 schoolchildren are given their choice of candy made with either light or dark chocolate, but otherwise the candy is the same. Only 30% of them choose the dark chocolate. If a candymaker wants no more than a 1 in 100 chance of being misled by sampling variability, what is the estimate of the proportion of children who prefer dark chocolate?

3.3.7. Selected at random, 250 married couples are given sample ballots containing the names of all candidates for contested offices in the coming election. Husband and wife mark their ballots independently, and their ballots are compared; 130 couples are in perfect agreement in their voting.

    a. What is the estimated numerical value of the binomial parameter for the distribution that models this situation?

    b. Set up a 95% confidence interval for the binomial parameter.

3.3.8. In a random sample of 200 apples from an orchard that had not been sprayed with insecticide, 162 apples bear evidence of insect damage.

    a. What is the best estimate of the proportion of damaged fruit in the orchard?

    b. In what range would you say the "true" proportion lies if you want to have only a 1-in-100 chance of being wrong?

3.3.9. In a random sample of 500 voters from a northern county in West Virginia, 265 of the voters indicate that they will vote for the Democratic candidate for governor.

    a. Set a 99% confidence interval for the proportion of voters in the county who will vote for the Democratic candidate.

    b. The Republican candidate claims that he will win the county by 1% of the votes.

        i. State a null hypothesis for his claim.

        ii. Does the confidence interval in part a lead to acceptance or rejection of this null hypothesis? Why?

        iii. With what α level was the hypothesis tested?

3.3.10. Francis Galton thought everything could be measured and tried to measure everything. He was interested in hot-air ballooning and routinely measured barometric pressure along with the direction and velocity of the wind. As a result, his first major scientific contribution was in meteorology. In his measurements, he noted that the flow of air around a high-pressure area was not counterclockwise as it is around one of low pressure. Because he found it always to be clockwise instead, Galton called the phenomenon an "anticyclone," the term still in use. His conclusion was based on the fact that the number of times there was counterclockwise flow around the n high pressure areas measured by him was $y = 0$. Had confidence intervals been available when he drew his conclusion:

a. Why should he use a two-side interval when all recorded flows were clockwise?

b. Give the $CI_{0.95}$ if his number of observations had been $n = 20, 25, 50, 100$ and π is the proportion of counterclockwise flows in high pressure areas.

c. Instead of using a confidence interval, why is it not possible to test either of the following hypotheses?

   i. H_0: $\pi = 0$ with H_a: $\pi > 0$?

   ii. H_0: $\pi > 0$ with H_a: $\pi = 0$?

3.4. NONPARAMETRIC STATISTICS: MEDIAN TEST

By changing the scale of measurement, we can also use the binomial distribution to analyze data originally recorded on the numerical scale. This is known as a *nonparametric* statistical procedure because inference is made, not about the parameter (or parameters) of the original data, but about the parameter for the new scale of measurement. Disadvantages can result from reducing the scale of measurement, but nonparametric tests are often quick, convenient, and useful statistical tools which need to be examined.

The *one-sample median test* is a nonparametric test in which numerical data are reduced to the nominal scale and analyzed by means of the binomial distribution. The *median (M)* of a distribution is the value which will divide the distribution into halves. Thus the probability is $1/2$ that the median will be exceeded by a random variable u from the distribution, that is, $P(u > M) = 1/2$.

If a random sample of n observations is drawn from a numerical distribution with a known median and we record only y equal to the number of values in the sample which exceed the median, y is a binomial random variable with a $b(y;n,1/2)$ distribution. If the median is not known, we can state a hypothesized value and then use the binomial distribution to test whether approximately half the sample values are greater than the hypothesized median. The procedure will be demonstrated in the following example.

Example 3.4.   The One-Sample Median Test

An oncologist has been studying cervical cancer and has learned that this disease is diagnosed at a median age of 49.5 years ($M = 49.5$). He begins a new study of uterine cancer and soon speculates that this is a disease of older women. To test this belief, he hypothesizes that the median age for victims of uterine cancer is the same as that for those with cervical cancer, and the alternative hypothesis is that uterine cancer victims are older:

$$H_0: P(u > 49.5) = \pi = 0.50$$
$$H_a: \pi > 0.50$$

He then obtains a random sample of 20 women with uterine cancer and finds that $y = 17$ were older than 49.5 years when their condition was diagnosed. This is in the region of rejection for a test with the conventional $\alpha = 0.05$, so he rejects the null hypothesis and concludes that the median age at diagnosis for victims of uterine cancer is greater than it is for those with cervical cancer. In other words, the kind of cancer a woman may have will depend, in part, on her age.

EXERCISES

3.4.1. In a certain large suburban housing development, all the houses were built at approximately the same time, with the same size and initial cost of construction. The median resale price of houses in the development has been established, but a real-estate agent wants to determine if multiple ownership affects the resale price of a house. From records at the county courthouse, she obtains a sample of the resale prices of 25 houses which have had more than one owner. In the sample, 15 were sold below the median price for houses in the area and 10 were sold above the median price.

    a. Give the null and alternative hypotheses.

    b. What is the value of P?

    c. What conclusion should the agent make about the effect of multiple ownership on the resale value of a house in the area?

    d. What factors could affect the validity of the conclusion?

3.4.2. The National Center for Health Statistics has recently reported that the median life expectancy of U.S. white males is 74 years (rounded to an integer value). A physician in the U.S. protectorate of Guam want to see if the same life expectancy holds true for U.S. white males on that island. He obtains a random sample of 20 recent death certificates of U.S. white males, and the ages u of the deceased were

18	59	42	61	38	41	71	40	14	47
73	93	55	51	74	88	60	71	89	63

    a. What hypothesis does the physician want to test?

    b. Why might he want to use a two-sided alternative?

    c. If the null hypothesis is true, what is the expected number of ages greater than $M = 74$? What is the observed number of ages greater than 74?

    d. Compute the P value and compare it to an α of 0.05.

3.4.3. An airline is experiencing a median delay in arrival of 27 minutes and introduces new measures in an effort to make improvements. After the measures have been in effect for a month, a random sample will be taken of arrival times and the median test used to evaluate the effectiveness of the changes.

    a. Give the null and alternative hypotheses which will be used.

    b. For an α as near 0.05 as possible, what will be the region of rejection if the number of flights in the random sample is $n = 25$? $n = 50$? $n = 100$?

REVIEW EXERCISES

Decide whether each of the following statements is true or false. If the statement is false, explain why.

3.1. In a binomial experiment, the outcomes fall into two mutually exclusive classes.

3.2. In a binomial experiment with n trials, y can take on any of n values.

3.3. Binomial distributions are not symmetrical, except when $\pi = 1 - \pi$.

3.4. Because the binomial is a discrete distribution, the expected value will be an integer value.

3.5. If the binomial parameter π is 0.60, the probability of exactly 60 successes out of 120 trials is greater than the probability of 72 successes out of 120 trials.

3.6. If A and B are mutually exclusive events, then $P(A \text{ or } B) = P(A) \times P(B)$.

3.7. The variance for discrete distributions can be computed by using the formula $V(y) = n\pi (1 - \pi)$.

3.8. The addition rule of probability applies only to mutually exclusive events.

3.9. The binomial distribution is an example of a continuous probability distribution.

3.10. To calculate the probabilities in a binomial distribution, the number of trials n and the binomial parameter π must be known.

3.11. The null hypothesis may be H_0: $\pi = 0.05$ and $y/n = 0.05$, but the null hypothesis may still be false.

3.12. A Type I error is defined as "the probability of rejecting the null hypothesis when it is true."

3.13. When the null hypothesis is true, the probability of making a Type I error is equal to α.

3.14. It is impossible to make a Type I error when the null hypothesis is false.

3.15. The symbol β represents the probability of rejecting H_0 when H_0 is false.

3.16. The power of a test of hypothesis is $1 - \alpha$.

3.17. It is impossible to make a Type II error when the null hypothesis is rejected.

3.18. If large sample sizes are used, there is less likelihood of a Type I error and a Type II error.

3.19. If an experiment is well designed and both α and β are small, it should be a good experiment.

3.20. Even when a correct statistical procedure is used, it is possible to accept the null hypothesis when it is false.

3.21. The greater the region of rejection, the more powerful the experiment.

3.22. The probability $P(y$ is in region of rejection$) = \alpha$ whether the null hypothesis is true or false.

3.23. The best point estimate $\hat{\pi} = y/n$ of the parameter π will lie exactly in the middle of the 95% confidence interval for π.

3.24. If the degree of certainty is increased from 0.95 to 0.99, the confidence interval becomes narrower.

3.25. Two methods of estimation are confidence intervals and tests of hypotheses.

3.26. Confidence intervals that are based on large samples are more likely to include the population parameter than those based on smaller samples.

3.27. Other things remaining the same, the larger the value of $\hat{\pi}$, the wider the confidence interval.

3.28. Other things being equal, the greater the level of confidence desired, the wider will be the confidence interval.

3.29. Repeated samples of the same size from the same population will always produce 99% confidence intervals of the same width on the binomial parameter π.

3.30. If the confidence interval does not contain some hypothesized value π_0 of the binomial parameter, the hypothesis can be rejected.

SELECTED READINGS

Angus, J. E., and R. E. Schafer (1984). Improved confidence statements for the binomial parameter. *American Statistician*, **38**, 189–191.

Anderson, T. W., and H. Burstein (1967). Approximating the upper binomial confidence limit. *Journal of the American Statistical Association*, **62**, 857–861.

Anderson, T. W., and H. Burstein (1968). Approximating the lower binomial confidence limit. *Journal of the American Statistical Association*, **63**, 1413–1415; correction, **64** (1969), 669.

Buckalew, I., Jr. (1974). A comparison of the efficiency of the normal approximation to the binomial with the binomial distribution. Master's report, West Virginia University.

Burke, C. J. (1954a). A brief note on one-tailed tests. *Psychological Bulletin*, **50**, 384–387.

Burke, C. J. (1954b). Further remarks on one-tailed tests. *Psychological Bulletin*, **51**, 587–590.

Clopper, C. J., and E. S. Pearson (1934). The use of confidence or fiducial limits illustrated in the case of the binomial. *Biometrika*, **26**, 404–413.

Crow, E. L. (1956). Confidence intervals for a proportion. *Biometrika*, **43**, 423–435.

Feinberg, W. E. (1971). Teaching the Type I and Type II errors: The judicial process. *American Statistician*, **25** (June), 30–32.

Fryer, H. C. (1968). *Concepts and Methods of Experimental Statistics*. Allyn & Bacon, Boston.

Jones, L. V. (1952). Tests of hypotheses: One-sided versus two-sided alternatives. *Psychological Bulletin*, **49**, 43–46.

Jones, L. V. (1954). A rejoinder on one-tailed tests. *Psychological Bulletin*, **51**, 585–586.

Natrella, M. G. (1960). The relation between confidence intervals and tests of significance. *American Statistician*, **14** (February), 20–22, 38.

4 Poisson Distributions

In this chapter we look at a second family of probability distributions, Poisson distributions. Poisson distributions are the appropriate probability model for certain types of experiments. There is an interesting relationship between binomial distributions and Poisson distributions, and this relationship provides a way to approximate some binomial probabilities that are very difficult to compute directly.

4.1. THE NATURE OF POISSON DISTRIBUTIONS

Many scientific experiments involve the random sampling of one or more fixed time intervals, lengths, areas, volumes, or other *sampling units*, and then observing the number of discrete events per sampling unit. For example, a forester might count the number of white-oak trees damaged by deer within sampling quadrants (square areas); an epidemiologist might count the number of new cases of hepatitis in a certain county in one month; a quality control manager might count the number of defects in 25-ft lengths of wire; an ecologist might count the number of parasites per host. In each case the event of interest (damaged white oak, incidence of disease, defect, parasite) is counted for a certain sampling unit (a quadrant, a month, 25 ft, per host).

The outcomes in experiments of this type often have the characteristics of a Poisson process. This process is named after Siméon-Denis Poisson (1781 to 1840), a French mathematician who first studied variables of this type in 1837.

A *Poisson process* consists of discrete events that occur per unit (such as time, length, area, volume, or on an object) and for which:

1. The probability of a single occurrence of the event is directly proportional to the size of the interval, or sampling unit.
2. If the sampling unit is sufficiently small, the probability of two or more occurrences of the event is negligible.
3. The occurrences of the event in nonoverlapping intervals or units are independent, that is, what happens in one sampling unit has no effect on what happens in another nonoverlapping unit.

If an experiment generates a Poisson process and the units are randomly and independently obtained, then the appropriate probability model for the *number of occurrences* of the event in the specified sampling unit is a *Poisson distribution*. The Poisson distribution is a discrete

Statistics for Research, *Third Edition*,   Edited by Shirley Dowdy, Stanley Weardon, and Daniel Chilko. ISBN 0-471-26735-X   © 2004 John Wiley & Sons, Inc.

probability distribution with probability function

$$p(y; \lambda) = \frac{e^{-\lambda}(\lambda)^y}{y!}$$

for $y = 0, 1, \ldots$. In this probability function y is the value of the random variable, $y!$ has the usual meaning of y factorial, e is the constant which is the base of the natural logarithms[†] (equal to 2.7183 if rounded to four decimal places), and λ (the Greek letter "lambda") is the expected number of occurrences in the specified interval. Table A.6 in the Appendix of Useful Tables gives values of $e^{-\lambda}$ for selected values of λ.

To draw statistical inference from data modeled by a Poisson process, the appropriate Poisson probability distribution is needed. As with binomial data, we will rely primarily on the Poisson probability distributions given in the tables in this text. However, it is important to see how these tables can be constructed through the application of mathematical procedures to the probability distribution function.

Note that this probability distribution is completely determined by the parameter λ. If we know λ, we can compute the distribution, as in the following example.

Example 4.1.   A Poisson Probability Distribution

Suppose a certain city has a variable number of suicides per month but the mean is 3 suicides per month. A mental health scientist wants to study this phenomenon and decides to use a Poisson distribution to model the distribution of suicide data. The sampling unit is one month; y is the number of suicides in that month, and $E(y) = \lambda = 3.0$. Then, to compute the probabilities of different numbers of suicides in any specific month, the mental health scientist will use the formula

$$p(y; 3) = \frac{e^{-3}(3)^y}{y!}$$

for $y = 0, 1, 2, \ldots$.

For example, the probability that there will be 0 suicides in a randomly chosen month is

$$P(y = 0) = p(0; 3) = \frac{e^{-3}(3)^0}{0!}$$

Since both $(3)^0$ and $0!$ are each equal to 1, $p(0; 3) = e^{-3}$, which can be found in Table A.6 as 0.0498. Similarly, the probability of exactly one suicide in a randomly chosen month is

$$P(y = 1) = p(1; 3) = \frac{e^{-3}(3)^1}{1!} = e^{-3}(3)$$

Further computations for the distribution are simplified if it is noted that $p(1; 3) = p(0; 3)(3/1)$, $p(2; 3) = p(1; 3)(3/2)$, and in general the probability of any value y can be computed easily from the previous value, $y - 1$,

$$p(y; \lambda) = p(y - 1; \lambda)\left(\frac{\lambda}{y}\right)$$

[†]The irrational number e can also be defined as the limit of the series $(1 + 1/n)^n$, that is, $(1 + 1/1)^1 = 2.0000$, $(1 + 1/2)^2 = 2.5000$, $(1 + 1/3)^3 = 2.3704, \ldots$.

The following table is computed in this manner:

y		$p(y;3)$	
0	$e^{-3} \cdot 3^0/0!$	$= e^{-3}$	$= 0.0498$
1	$e^{-3} \cdot 3^1/1!$	$= p(0)(3/1)$	$= 0.1494$
2	$e^{-3} \cdot 3^2/2!$	$= p(1)(3/2)$	$= 0.2240$
3	$e^{-3} \cdot 3^3/3!$	$= p(2)(3/3)$	$= 0.2240$
4	$e^{-3} \cdot 3^4/4!$	$= p(3)(3/4)$	$= 0.1680$
5	$e^{-3} \cdot 3^5/5!$	$= p(4)(3/5)$	$= 0.1008$
6	$e^{-3} \cdot 3^6/6!$	$= p(5)(3/6)$	$= 0.0504$
7	$e^{-3} \cdot 3^7/7!$	$= p(6)(3/7)$	$= 0.0216$
8	$e^{-3} \cdot 3^8/8!$	$= p(7)(3/8)$	$= 0.0081$
9	$e^{-3} \cdot 3^9/9!$	$= p(8)(3/9)$	$= 0.0027$
10	$e^{-3} \cdot 3^{10}/10!$	$= p(9)(3/10)$	$= 0.0008$
11	$e^{-3} \cdot 3^{11}/11!$	$= p(10)(3/11)$	$= 0.0002$
12	$e^{-3} \cdot 3^{12}/12!$	$= p(11)(3/12)$	$= 0.0001$
13	$e^{-3} \cdot 3^{13}/13!$	$= p(12)(3/13)$	$= 0.0000$

and $p(y) = 0.0000$ (rounded to four decimal places) for $y > 13$.

Poisson probability distributions have some interesting properties. The expected value of y is equal to λ and the variance of y is also λ, that is, $E(y) = V(y) = \lambda$. Also, the sum of two Poisson random variables is a Poisson random variable; thus, if y_1 and y_2 are Poisson random variables with parameters λ_1 and λ_2, respectively, then $y_1 + y_2$ is a Poisson random variable with expected value $\lambda_1 + \lambda_2$. Thus, if we make the sampling unit larger than one month and if we can assume that the number of suicides in one month will be independent from those in another, we can find the expected number of suicides in 2 months as $E(y_1) + E(y_2) = 3 + 3 = 6$, and the expected number during the 3-month summer period (again making the assumption of independence) will be $3(3) = 9$. Similarly, if the sampling unit is made smaller, reducing it by half, for example, we can say that the expected number of suicides in the first half of the month will be $E(y/2) = E(y)/2 = 3/2 = 1.5$. These relationships are important because we usually have a sample of more than just one Poisson random variable.

EXERCISES

4.1.1. The expected number of water mites found on a host, the chironomid fly, is 2.5 and this is a Poisson process.

    a. Are the sampling units water mites, or chironomid flies? Explain.

    b. What is the probability that exactly 1 mite will be found on a fly?

4.1.2. If the accident rate at a certain factory is 7.0 per year and this is a Poisson process:

    a. Find the probability that fewer than 3 accidents will occur in a year.

    b. Find the probability that 3 or more accidents will occur in a year.

4.1.3. The expected number of flaws in 20-ft intervals of wire is 5.0.

    a. What is the number of discrete events, feet or flaws?

   b. What is the expected number in a random 10-ft interval?

   c. What is the probability that there will be 4 flaws in a random 10-ft interval?

4.1.4. In Example 4.1 in this section, involving the number of suicides per month:

   a. What is the probability that no suicides will occur in a month?

   b. What is the probability that more than 6 suicides will occur?

   c. What percentage of months will have at least 1 suicide but not more than 6 suicides?

4.1.5. Additives such as trace minerals, antibiotics, vermifuges, and insecticides are incorporated into animal feeds in parts per million (ppm). For effective mixing, the additives may be compressed into pellets the size of the ground grain in the feed and then colored with vegetable dye for easy identification. Quality control for thoroughness of mixing can be maintained by scooping out a known volume of the mixed feed and counting the number of colored pellets of additives. If properly mixed feed yields a Poisson process with $\lambda = 2.5$ per scoop, find:

   a. The probability that a scoop will contain no pellets of additive

   b. The probability that a scoop will contain exactly 1 such pellet

   c. The probability that a scoop will contain at least 1 pellet

   d. The outcomes that are most likely to occur approximately 80% of the time

4.1.6. In the feed-mixing problem described in Exercise 4.1.5, suppose customary quality control procedures require 10 independently drawn scoops from each batch of mixture. In 10 scoops of properly mixed feed, find:

   a. The expected total number of colored pellets

   b. The probability that there will be no such pellets

4.1.7. a. Compute the Poisson distribution for each of the following values of λ: 0.25, 0.50, 1.00, and 10.00. Round the probabilities to four decimal places.

   b. Graph the Poisson distributions of part a.

   c. Describe the behavior of the graphs of part b.

4.1.8. a. Use the probabilities in Exercise 4.1.7a for $\lambda = 0.25$ to find the expected value of that Poisson distribution. Why is this value slightly different from $E(y) = \lambda = 0.25$?

   b. Use the probabilities computed in Exercise 4.1.7a and $E(y) = 0.25$ to find $V(y)$ for that Poisson distribution. Why is this value slightly different from $V(y) = \lambda = 0.25$?

4.1.9. If y_1 and y_2 are independent Poisson random variables with $\lambda = 0.25$, then $y_1 + y_2$ is a Poisson random variable with $\lambda = 0.50$. Use Exercise 4.1.7 to show that this is true for $y_1 + y_2 = 3$. [*Hint*: Remember that $y_1 + y_2 = 3$ when y_1 and y_2 are respectively (0 & 3), (3 & 0), (1 & 2), or (2 & 1).]

4.2. TESTING HYPOTHESES

Using Table A.7 in the Appendix, which contains the Poisson distributions for selected values of λ, we can test hypotheses with a procedure similar to the one we used for the binomial distribution.

Example 4.2.   Test of Hypothesis for a Poisson Parameter

A biologist studying yeast cells believes that after a certain treatment the cells will be present at a rate of 0.55 per square of a hemacytometer (a microscopic plate usually used to count blood cells). He finds 13 yeast cells in 20 squares and wonders if $13/20 = 0.65$ indicates that a rate of 0.55 is incorrect. To determine whether 13 cells in the 20 squares are likely to occur if his conjectured rate is correct, he uses the Poisson distribution.

The null and alternative hypotheses are

$$H_0: \lambda = 0.55$$
$$H_a: \lambda \neq 0.55$$

Since the sum of two Poisson random variables is also a Poisson random variable, if $\lambda = 0.55$ for one square, then $\lambda = 20(0.55) = 11$ for 20 squares. Using Table A.7, the biologist finds that for α as close to 0.10 as possible the region of rejection is

$$y = 0, 1, 2, 3, 4, 5, 17, 18, 19, \ldots$$

if the test statistic is the number of yeast cells per 20 squares. The actual α level is 0.0933. The count is 13 yeast cells in 20 squares after this treatment, and since 13 does not lie in the region of rejection, the biologist concludes that after the treatment the mean number of yeast cells per square may be 0.55.

Statistical computer programs more often provide a P value rather than a region of rejection, so it may be useful to see again how this probability is obtained and how it is used to make a decision about the null hypothesis. In Example 4.2, $E(y) = 20(0.55) = 11$ yeast cells in 20 squares, and the observed value was $y = 13$, which is 2 yeast cells different from the number expected under the null hypothesis. Because the alternative hypothesis is two sided, the P value measures the probability of a difference from $E(y)$ of 2 or more in either direction, so

$$P = P(y \leq 9) + P(y \geq 13) = 0.3405 + 0.3113 = 0.6518$$

A P value of 0.6518 is very large; hence a difference of this magnitude or even greater could occur easily by chance when the null hypothesis is true. The P value would have to be equal to or less than $\alpha = 0.10$ before we would decide the null hypothesis is false.

For small values of λ the Poisson distributions have relatively large probabilities in the lower tail, so it may be impossible to designate a small α level for a two-tailed alternative or for a one-tailed less-than alternative hypothesis. The technique of using several units—such as the 20 squares in the above example—helps overcome this difficulty.

Table A.7 lists a limited number of values of λ, and the necessary one may not be there. If λ is not too large, the necessary probability distribution can be calculated. For large λ's approximation methods are available; these are discussed in Chapter 7.

Procedure.   Test of Hypotheses for a Poisson Parameter

Region of Rejection Method
$H_0: \lambda = \lambda_0$ (λ = expected number of occurrences in a specified interval)

H_a: $\lambda \neq \lambda_0$, $\lambda < \lambda_0$, or $\lambda > \lambda_0$
Significance level: α

Test statistic: y, the number of occurrences of the phenomenon of concern in a multiple of k specified sampling units.

Using a table for the Poisson distribution with probability function $p(y; \lambda_0 k)$, determine the region of rejection.
For H_a: $\lambda \neq \lambda_0$, the region of rejection is $0 \leq y \leq c_L$ and $c_U \leq y \leq \infty$ such that $\sum_0^{c_L} p(y; \lambda_0 k)$ and $\sum_{c_U}^{\infty} p(y; \lambda_0 k)$ are each as close as possible to $\alpha/2$.
For H_a: $\lambda < \lambda_0$, the region of rejection is $0 \leq y \leq c_L$ such that $\sum_0^{\infty} p(y; \lambda_0 k)$ is as close as possible to α.
For H_a: $\lambda > \lambda_0$, the region of rejection is $c_U \leq y \leq \infty$ such that $\sum_{c_U}^{\infty} p(y; \lambda_0 k)$ is as close as possible to α.

P-Value Method
For H_a: $\lambda \neq \lambda_0$, compute $P = P(|y - \lambda_0 k| \geq |$ test statistic $- \lambda_0 k|)$.
For H_a: $\lambda > \lambda_0$, compute $P = P(y \geq$ test statistic$)$.
For H_a: $\lambda < \lambda_0$, compute $P = P(y \leq$ test statistic$)$.
Reject H_0 if $P \leq \alpha$.

EXERCISES

4.2.1. A physicist wants to verify whether a radioactive substance has a level of radioactivity equal to 4 radioactive particles emitted per millisecond. He measures the radioactivity with a Geiger counter, and it records 18 particles in 3 msec.

   a. What is the expected number of radioactive particles per 3 msec?

   b. Compute the P value for an observed value this far or even farther from the number expected in 3 msec.

   c. Using an α of 0.05, make a test of hypothesis to determine if the radioactivity level is significantly greater than expected.

4.2.2. A certain area of the United States has a rate of 4.5 tornadoes per year. A local religious cult claims that its rituals can reduce this rate. The cult members conduct their rituals and that year 2 tornadoes hit. Use a test of hypothesis with α as close to 0.10 as possible to determine if the rate is significantly less than 4.5 per year. What assumptions are you making as you perform this test?

4.2.3. A hospital emergency center handled victims of automobile accidents at the rate of 10 per week when the local highway had a speed limit of 70 miles per hour. After the speed limit was reduced to 55 miles per hour, 4 highway accident victims were admitted in a randomly selected week. Does this indicate a reduction in emergency admissions for automobile accidents? Could you conclude that lowering the speed limit has reduced highway accidents? Why or why not?

4.2.4. Grain sorghum is a naturally tall-growing plant, but dwarf varieties have been developed so that the crop can be harvested with conventional farm equipment. However, back mutation occurs frequently and tall offspring reappear in a field with an expected value of 1.5 tall plants per 200 ft^2. With each development of a new grain sorghum hybrid, plant breeders must satisfy the farmer that the amount of back mutation has not increased. A hybrid seed company has many experimental hybrids under consideration at a time, and it decides to allot only three 200-ft^2 plots per hybrid. Set up a test of hypothesis for the amount of back mutation.

a. Give the null hypothesis for 3 plots.

b. Give the alternative hypothesis.

c. Give the region of rejection for α as close to 0.05 as possible.

d. Suppose that for a particular hybrid the back mutation doubles to $\lambda_a = 3.0$ per 200 ft^2; what is the power of the test for 3 plots?

e. What is the power for $\lambda_a = 3.0$ if only 1 plot is used? Is it advisable to use more than 1 plot?

4.2.5. The rarest white blood cell is the basophil, which constitutes only 1% of the total white blood cells. Students who are learning to perform white blood cell counts are inclined to mistake other cells for basophils until they have seen them often enough to recognize them. Thus a student's proficiency in performing differential white blood cell counts can be tested by checking whether too many cells have been recorded as basophils. This can be thought of as a Poisson process in which the interval is a count of 100 white blood cells.

a. State a null hypothesis indicating that the student can accurately identify the different kinds of white blood cells.

b. State an alternative hypothesis indicating that the student mistakes other cells for basophils.

c. The instructor decides that any student who records 4 or more basophils per 100 cells counted cannot yet distinguish these cells properly. How likely is it that a student will record cells correctly but have an unusual random sample of cells?

d. The frequency of basophils increases after surgery. Suppose the student is counting white blood cells from a blood smear taken under such conditions and $\lambda = 2.4$ per 100 cells. How likely is it that fewer than 4 basophils are among the 100 cells counted? Should the instructor take precautions that the students are not using blood smears from postoperative patients?

4.2.6. A new synthetic surface has been placed on a university football field, and the team's physician wants to decide whether it has had any effect on the number of knee injuries suffered in a game. Since he has been with the team, it has experienced a mean of $\lambda = 0.7$ knee injuries per game.

a. If the new surface has no effect, what is the expected number of knee injuries in the first 5 games on the new surface?

b. State a null and alternative hypothesis.

c. Suppose that there are a total of $y = 7$ knee injuries in the first 5 games, how likely is a deviation from expected of this magnitude or greater to occur by chance?

d. If the team's physician sets $\alpha = 0.10$, what should he conclude about the effect of the new surface on knee injuries?

e. What caveats about the design should be taken into account when the conclusion is being drawn?

4.3. ESTIMATION

The best point estimate of the Poisson parameter λ is y, the number of occurrences of the event of interest in a randomly selected sampling unit. If several units are sampled, the total number of occurrences is the best estimate for the combined units. Central and one-sided confidence intervals can be found in a manner similar to finding confidence intervals for the binomial parameter π. Table A.7 in the Appendix is used to find the confidence intervals for the Poisson parameter. Because of the relatively large probabilities for low values of y, the horizontal lines in Table A.7 are drawn so that α is as close to 0.20 as possible; thus these lines correspond to approximate 80% central confidence intervals.

Example 4.3.   A Central Confidence Interval for a Poisson Parameter

Foresters are concerned about the number of young trees destroyed by deer. Suppose a forester chooses 4 quarter-acre quadrants at random and finds that in the four plots 8 young trees have been destroyed by deer. She wants to estimate the damage rate per acre by an approximate 80% confidence interval.

Using Table A.7, she finds that 8 is in the region of acceptance for $\lambda = 5.0$ to $\lambda = 12.0$, so the confidence interval is

$$\text{CI}_{0.80}: \quad 5.0 \leq \lambda \leq 12.0$$

in which λ is the damage rate per acre.

The upper and lower bounds on the confidence interval are limited to column entries in Table A.7 so, as was done with the binomial distribution, another table, Table A.8, is given for obtaining more precise upper and lower limits for the confidence interval. Using the same data above, the forester would enter Table A.8 with row entry $y = 8$ and column entry $1 - \alpha = 0.80$; she would find $L = 4.6561$ and $U = 12.9947$, and she obtains the confidence interval

$$\text{CI}_{0.80}: \quad 4.7 < \lambda < 13.0$$

This confidence interval expresses the expected number of damaged trees on a per-acre basis; if she wishes to return it to a per- (quarter-acre) quadrant basis, she divides the upper and lower limits by $k = 4$ and obtains

$$\text{CI}_{0.80}: \quad 1.2 < \lambda < 3.2$$

The greatest row entry for Table A.8 is $y = 20$, and this may not be sufficiently large for some estimates of λ. However, this problem will be addressed in Chapter 7, where it will be seen that when λ is large another distribution can be used to approximate the Poisson distribution.

One-sided confidence intervals can also be determined.

Example 4.4.   A One-Sided Confidence Interval for a Poisson Parameter

The architect for a new hospital in a small city needs to know the maximum number of emergency cases that can be expected in a half-hour period in order to plan adequate facilities. He examines the records at the existing city hospital, which is being replaced; a random selection of 10 half-hour periods gives a total of 6 emergency cases. He can use Table A.7 to find an approximate 90% one-sided confidence interval:

$$\text{One-sided CI}_{0.90}: \quad \lambda \leq 9.0$$

if λ is for a 5-hour period because 9.0 is the largest value of λ for which 6 would be in the region of acceptance. Or he could write

$$\text{One-sided CI}_{0.90}: \quad \lambda \leq 0.90$$

if λ is for a half-hour period.

The one-sided confidence interval indicates that the largest expected value of the Poisson distribution that is likely is 0.90; that is, the largest mean number of cases in a 30-minute period is 0.90. Since 0.90 is the mean, some of the 30-minute periods will have more cases and others less. Since the number of cases in a 30-minute period will usually be within two standard deviations of the expected value λ and in a Poisson distribution $\lambda = V(y)$, the architect can prepare for the worst situation,

$$\lambda = V(y) = 0.90$$

$$\text{sd}(y) = \sqrt{\lambda} = 0.95$$

and the largest number of cases is not likely to be more than

$$\lambda + 2\text{sd}(y) = 0.90 + 2(0.95) = 2.80$$

To be safe, he plans to be able to accommodate 3 cases each half hour.

Procedure. Confidence Intervals for λ

Central

1. Specify α.
2. Take a sample of k sampling units.
3. Observe y, the number of occurrences of the phenomenon of interest in the k units.
4. Give the interval of all values of λ for which y would fall in the region of acceptance for a two-sided α-level test from Table A.7 (or use Table A.8 to get the interval directly).
5. Divide the confidence limits by k to determine the central confidence interval for the rate λ for intervals of the specified unit.

One-Sided, Upper Confidence Limit

Proceed as for a central confidence interval, but in step 4 use the region of acceptance for a one-tailed less-than test of hypothesis in Table A.7 (or double α and use only the upper limit in Table A.8).

One-Sided, Lower Confidence Limit

Proceed as for a central confidence interval, but in step 4 use the region of acceptance for a one-tailed greater-than test of hypothesis in Table A.7 (or double α and use only the lower limit in Table A.8).

EXERCISES

4.3.1. If 3 noxious weeds are found in a 0.25-oz random sample of grass seed, use the Poisson probability distribution to find an 80% confidence interval for the expected number of weeds per 0.25 oz of seed. (Note that using the Poisson model here avoids the necessity of counting all the seeds, a tedious task.) Compare the intervals obtained from Tables A.7 and A.8.

4.3.2. If 8 defects are found in a production process during a random 5-minute interval, find with 90% of confidence the largest mean number of defects that could be expected to occur in a 5-minute period. Compare the intervals obtained from Table A.7 and A.8.

4.3.3. It is found that there are 6 fatal accidents in an underground coal mine for a sample of 20,000,000 employee hours of exposure. Place an approximate 80% confidence interval on the Poisson parameter if the interval is 100,000 employee hours.

4.3.4. In the quality control process described in Exercise 4.1.5, place an approximate 90% confidence interval on the smallest mean number of pellets expected in 1 scoop if 7 pellets are found in 4 random scoops.

4.3.5. Sir Francis Galton (1822 to 1911), one of the early developers of experimental statistics, believed everything could be measured, even boredom. His measure of boredom was a Poisson statistic, the number of signs of unrest that an individual would show per minute. Suppose a student wants to measure how boring a classmate finds the statistics class, so he counts the number of times she yawns, fidgets, looks at her watch, and so on, during 16 half-minute intervals of observation, and the total is 10.

 a. With regard to this survey:

   i. Why must the friend be unaware that her behavior is being observed?

   ii. Why can the time of observation not be for 8 consecutive minutes?

   iii. Is it valid to assume that $E(\lambda)$ remains constant throughout the class period?

 b. Place an 80% confidence interval on the number of signs of boredom she shows per minute.

 c. Do you think a survey of this nature is valid? Ethical?

4.3.6. Suppose the data on trees destroyed by deer in Example 4.3 had been obtained by sampling a 100-acre forest.

 a. What is the estimated number of young trees destroyed by deer in the entire forest?

 b. Set an upper 90% confidence limit for this estimate to get an upper bound for the total number of trees destroyed in the entire forest.

4.4. POISSON DISTRIBUTIONS AND BINOMIAL DISTRIBUTIONS

Besides being useful in its own right, the Poisson distribution is often used as an approximation of the binomial distribution if the number of trials n is large and the probability of success on a single trial π is small. The approximation is possible because it can be shown mathematically that, if π becomes very small while n becomes very large and the product $n\pi$

remains constant, then the binomial distribution will be approximately a Poisson distribution with $\lambda = n\pi$ and the Poisson sampling unit the set of n trials.

Example 4.5. Using a Poisson Distribution to Approximate a Binomial Distribution

A geneticist believes that in a certain experiment the mutation rate is 4 in 1,000,000. She would like to find the probability that in a random sample of 25,000 she will observe no more than one mutation. This experimental situation is appropriately modeled by the binomial distribution $b(y; 25{,}000, 0.000004)$ and she wants to compute

$$P(y \leq 1) = b(0; 25{,}000, 0.000004) + b(1; 25{,}000, 0.000004)$$

$$= \binom{25{,}000}{0}(0.000004)^0(0.999996)^{25{,}000}$$

$$+ \binom{25{,}000}{1}(0.000004)^1(0.999996)^{24{,}999}$$

This computation is not feasible directly, and logarithms or a calculator with a y^x function would have to be used to compute an approximate answer.

Instead, the geneticist could approximate this probability by using a Poisson distribution. The Poisson parameter would be $\lambda = n\pi = 25{,}000(0.000004) = 0.100000$; that is, the expected number of mutations per 25,000 trials is 0.1. For the Poisson distribution

$$P(y \leq 1) = p(0; 0.1) + p(1; 0.1)$$

$$= \frac{e^{-0.1}(0.1)^0}{0!} + \frac{e^{-0.1}(0.1)^1}{1!}$$

$$= 0.904837 + 0.904837(0.1)$$

$$= 0.995321$$

Using this very simple computation, the geneticist can be relatively certain that in a random sample of size 25,000 she will observe no more than one mutation.

This approximation of the binomial distribution by the Poisson distribution is good only for small π and large n. Some statisticians suggest as a rule of thumb that $\lambda = n\pi$ should be less than 7.

Procedure. Poisson Approximation of a Binomial Distribution

For $n\pi < 7$, a binomial distribution may be approximated by a Poisson distribution: $b(y; n, \pi)$ is approximated by $p(y; n\pi)$.

It is important that we recognize the difference between a Poisson distribution and a binomial distribution so that we use the proper one to model an experiment and so that we know when it is appropriate to approximate a binomial by a Poisson. The following summary may be helpful:

Binomial	Poisson
1. Random variable: y = number of successes in n trials	1. Random variable: y = number of successes in a specified sampling unit
2. Number of trials: n, a finite number	2. Number of trials: infinite, since we count discrete events (successes) in a unit
3. Two parameters: π = probability of success for a single trial n = number of trials	3. One parameter: λ = mean number of successes per sampling unit
4. $E(y) = n\pi$ $V(y) = n\pi(1 - \pi)$	4. $E(y) = V(y) = \lambda$

EXERCISES

4.4.1. If it is known that the probability of having a bad reaction to a certain injection is 0.001, what is the probability that more than 1 person in 100 will have a bad reaction?

4.4.2. If the rate of accidental drownings per year is 0.000003 (i.e., 3 per 1,000,000 population), what is the probability that there will be more than 2 drownings in a city with a population of 400,000?

4.4.3. A manufacturer of TV sets initiates an inspection system to reduce the number of defective sets leaving the plant. Prior to this system the proportion of defective sets was 1 in 80. After the new system is in effect, in a random sample of 320 sets there are 2 defective sets. Use a test of hypothesis to decide if the proportion of defects has been reduced.

4.4.4. Suppose routine blood typing for 400 army recruits reveals that 6 of them have AB-negative blood.

a. What assumptions would you have to make for this to be considered a random sample of army personnel? Of the entire country?

b. Place an approximate 80% confidence interval on the proportion with AB-negative blood among army recruits.

c. Assuming it can be justified, place an approximate 80% confidence interval on the proportion of those with AB-negative blood in the entire country.

4.4.5. Fish and game commissions measure the hunting pressure on large game in their states by taking random samples of hunters and recording their successes during the hunting season. The following data record the number of white-tailed deer taken by a random sample of 50 Texas deer hunters:

Number of Deer Killed	Hunters
0	45
1	4
2	1

Because the fish and game commission wishes to protect against overhunting, place an approximate 90% of confidence interval on the largest mean number of deer taken per 50 hunters in the state.

REVIEW EXERCISES

Decide whether each of the following statements is true or false. If the statement is false, explain why.

4.1. In a Poisson distribution, $E(y) = n\pi$ and $V(y) = n\pi(1 - \pi)$.

4.2. Poisson data consist of discrete, countable observations.

4.3. Because $E(y)$ is usually small for a Poisson distribution, a relatively large number of sampling units is needed to estimate λ effectively.

4.4. A unique characteristic of Poisson distributions is that for any specified distribution the expected value will be numerically greater than the variance.

4.5. The Poisson distribution is sometimes called the "distribution of rare events" and hence is seldom encountered in experimentation.

4.6. The shape of a Poisson frequency distribution is symmetrical around its expected value.

4.7. In testing a hypothesis about the Poisson parameter, the alternative hypothesis may be one tailed or two tailed.

4.8. Confidence intervals for a Poisson parameter are symmetrical around the point estimate y.

4.9. There is a separate Poisson distribution for every value of λ and n.

4.10. The Poisson distribution can always be used to approximate the probabilities of a binomial distribution.

4.11. Because λ is usually small, small values of y are much more probable than large values when sampling from a Poisson distribution.

4.12. The power of a test of hypothesis for the Poisson parameter is increased as the number of units sampled is increased.

4.13. Because the random variable y can be an integer value between 0 and infinity, the Poisson distribution is a continuous probability distribution.

4.14. A characteristic of the Poisson distribution is the relationship $p(y; \lambda) = p(y - 1; \lambda) (\lambda/y)$.

4.15. The mean and standard deviation of the Poisson distribution are both λ.

4.16. If certain conditions are met, arithmetic can be simplified by using the binomial distribution to approximate the Poisson.

4.17. If there is only one sample unit, y is the best point estimate of the Poisson parameter.

4.18. The Poisson parameter must be a positive value.

4.19. One may have a countable number of discrete events which occur in a specified sampling unit but still not have a Poisson process.

4.20. $p(0; \lambda) = e^{-\lambda}$.

SELECTED READINGS

Haight, F. (1967). *Handbook of the Poisson Distribution*. Wiley, New York.

Hoaglin, D. C. (1980). A Poissonness plot. *American Statistician*, **34**, 146–149.

Sheu, S. S. (1984). The Poisson approximation to the binomial distribution. *American Statistician*, **38**, 206–207.

"Student" [William Sealy Gosset] (1906). On the error of counting with a haemacytometer. *Biometrika*, **5**, 351–360.

5 Chi-Square Distributions

In this chapter we study some uses of a continuous probability distribution called the *chi-square distribution*. Although this theoretical probability distribution is usually not a direct model of a population distribution, it has many uses when we are trying to answer questions about populations. For example, the chi-square distribution can be used to decide whether or not a set of data fits a specified theoretical probability model—a "goodness-of-fit" test. It can also be used to decide whether or not several samples came from the same population even when the model of the population is unspecified—a chi-square test of homogeneity. It is possible to make these and other decisions about populations because the chi-square distribution is often a model for the distribution of some statistic obtained by sampling from the population.

5.1. THE NATURE OF CHI-SQUARE DISTRIBUTIONS

In 1876, Frederick R. Helmert did some of the early work on the theoretical chi-square distributions. We can get some feeling for the nature of these distributions from the graphs of their probability density functions (Figure 5.1). The symbol usually used for the chi-square random variable is the compound symbol χ^2 (the exponent should not be confused with the squaring operation).

If χ^2 is a random variable with a chi-square distribution:

1. χ^2 is a positive real number.
2. The density function $f(\chi^2)$ for χ^2 depends on only one parameter, v (pronounced "nu"), called the *degrees of freedom*.
3. The expected value of χ^2 is equal to the degrees of freedom, that is, $E(\chi^2) = v$.
4. The variance of χ^2 is two times the degrees of freedom, that is, $V(\chi^2) = 2v$.
5. The maximum value of $f(\chi^2)$ is at $\chi^2 = v - 2$ if $v > 2$.
6. The graph of $f(\chi^2)$ is not symmetrical but approaches symmetry as the degrees of freedom increase.

Table A.9 in the Appendix of Useful Tables gives selected critical values for some of the chi-square distributions. The degrees of freedom are listed at the left; thus each row is from a different chi-square distribution. The headings at the top of the columns give α, the area to the right of the chi-square values listed in the tables. For example, if χ^2 has a chi-square distribution with 4 degrees of freedom, then a vertical line at $\chi^2 = 0.484$ divides the chi-square distribution so that $\alpha = 0.975$ of the area under the curve is to the right of 0.484 and $1 - \alpha = 0.025$ of the area is to the left (see Figure 5.2). We write $\chi^2_{0.975,4} = 0.484$. Critical

Statistics for Research, Third Edition, Edited by Shirley Dowdy, Stanley Weardon, and Daniel Chilko. ISBN 0-471-26735-X © 2004 John Wiley & Sons, Inc.

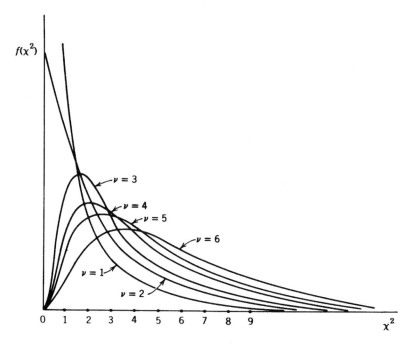

FIGURE 5.1. Chi-square distributions with v degrees of freedom. (Adapted from P. G. Hoel, *Elementary Statistics*, 4th ed., Wiley, New York, 1979, p. 249.)

values are used to determine regions of rejection because for continuous random variables areas correspond to probabilities. The probability that a chi-square random variable with 4 degrees of freedom has a value greater than 0.484 is equal to 0.975.

Another example is given in Figure 5.3. If χ^2 is a chi-square random variable with 15 degrees of freedom, then 5% of the area is to the right of a vertical line at $\chi^2 = 24.996$ and

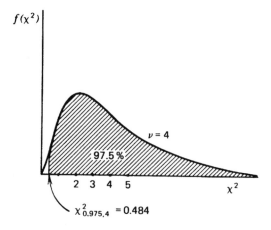

FIGURE 5.2. Meaning of values in the chi-square table.

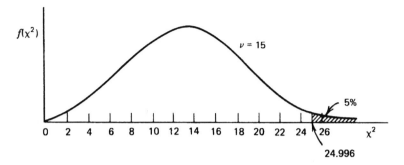

FIGURE 5.3. A chi-square distribution.

95% of the area is to the left of this line, or $\chi^2_{0.05,15} = 24.996$. This distribution has a mean of 15, a variance of 30, and the graph has a maximum at 13.

Helmert studied these theoretical distributions with apparently no idea that they could be used for a test of significance. In 1900 Karl Pearson was able to use Helmert's chi-square distributions to test hypotheses about multinomial experiments. A multinomial experiment is a generalization of a binomial experiment.

A *multinomial experiment* is an experiment in which:

1. There are k possible outcomes and the probability of the ith outcome is π_i with $\sum_{i=1}^{k} \pi_i = 1$.
2. The experiment is repeated n times, that is, there are n trials.
3. The π_i's are constant from trial to trial.
4. The trials are independent.
5. We are interested in o_i, the number of times the ith outcome occurs; $\sum_{i=1}^{k} o_i = n$.

Note that a binomial experiment is a multinomial experiment with $\pi_1 = \pi$, $\pi_2 = 1 - \pi$ in which π is the probability of success on a single trial, and $o_1 = y$, $o_2 = n - y$ in which y is the number of successes in n trials. Like the binomial distribution, the expected number of occurrences of the ith outcome is $n\pi_i$.

Example 5.1.   A Multinomial Experiment

If palomino horses are bred to other palominos, they produce progeny in the ratio of 1 dark-colored colt to 2 palominos to 1 light-colored colt. An experiment involving a random sample of 96 colts of palominos would be a multinomial experiment.

1. There are $k = 3$ outcomes: dark, palomino, light.
   $P(\text{dark}) = 1/4 = \pi_1$; $P(\text{palomino}) = 1/2 = \pi_2$; $P(\text{light}) = 1/4 = \pi_3$;
   $1/4 + 1/2 + 1/4 = 1$.
2. $n = 96$.
3. The π_i's are constant from trial to trial.
4. Since this is a random sample, the trials are independent.
5. We are interested in the number of colts of each color type.

If a geneticist questioned whether the ratios specified above were correct, he could use Pearson's approach to resolve the question. Pearson was looking for a simple statistic, a value that could be easily computed and that would indicate whether the results of an experiment deviated from expected results. He proposed the following statistic:

$$w = \sum_{i=1}^{k} \frac{(o_i - e_i)^2}{e_i}$$

in which $e_i = n\pi_i$, the expected value of o_i. A small value of w would indicate close agreement of the experimental results with the theory and a large value would indicate disagreement with the theory.

Pearson's statistic is a discrete random variable since it is composed of arithmetic operations on the discrete random variables $o_1, o_2, \ldots, o_k$. The probability distribution of w can be shown to be approximately Helmert's chi-square distribution with $k - 1$ degrees of freedom. Since the probabilities have been tabulated for the theoretical chi-square distribution, it is possible to use Pearson's statistic in a more precise way than just as a descriptive statistic; we can do a statistical test of hypothesis. Since Pearson's statistic is approximately a chi-square random variable, many people write

$$\chi^2 = \sum_{i=1}^{k} \frac{(o_i - e_i)^2}{e_i}$$

We also write χ^2 instead of w. It should be remembered, however, that the theoretical chi-square distribution studied by Helmert is a continuous probability distribution, whereas Pearson's statistic, which arises from multinomial experiments, is a discrete random variable. A test of hypothesis to check that specified probabilities in a multinomial experiment are correct is called the *multinomial chi-square test.*

Example 5.2.   A Multinomial Chi-Square Test

The geneticist mentioned above found that in the random sample of 96 colts of palominos there are 21 dark-colored colts, 52 palomino colts, and 23 light-colored colts. He wants to check whether $\pi_1 = 1/4$, $\pi_2 = 1/2$, and $\pi_3 = 1/4$ are correct parameters for a probability model. Thus he decides to test

$$H_0: \pi_1 = \frac{1}{4}, \ \pi_2 = \frac{1}{2}, \ \pi_3 = \frac{1}{4}$$

against

$$H_a: \pi_1 \neq \frac{1}{4} \ \text{or} \ \pi_2 \neq \frac{1}{2} \ \text{or} \ \pi_3 \neq \frac{1}{4}$$

that is, at least one inequality. He will reject the null hypothesis if the experimental results are unusual when the null hypothesis is true, that is, if they occur by chance alone less than $\alpha = 0.05$ of the time.

The expected number in each category is

$$e_1 = n\pi_1 = 96\left(\frac{1}{4}\right) = 24$$

$$e_2 = n\pi_2 = 96\left(\frac{1}{2}\right) = 48$$

$$e_3 = n\pi_3 = 96\left(\frac{1}{4}\right) = 24$$

He then uses the following table to organize his computations.

Category	Observed o_i	Expected e_i	$o_i - e_i$	$(o_i - e_i)^2$	$(o_i - e_i)^2/e_i$
Dark	21	24	-3	9	0.375
Palomino	52	48	4	16	0.333
Light	23	24	-1	1	0.042
					$\chi^2 = 0.750$

Since there are $k = 3$ categories, this statistic is distributed approximately as the chi-square random variable with $v = 3 - 1 = 2$ degrees of freedom. Referring to Table A.9 and recalling that large deviations from the expected values will give a large chi-square statistic, the geneticist finds that for $v = 2$ the theoretical chi-square value of 5.991 divides the lower 95% of the distribution from the upper 5%. He will reject the null hypothesis if the chi-square statistic is greater than or equal to 5.991. Since this is not the case, he concludes that there is no evidence that the theory is incorrect and that the specified ratios may be correct.

If the geneticist in this example wanted to find the P value associated with this test, P would equal $P(\chi^2 > 0.750)$. It is not possible to find the specific value of this probability from Table A.9. Using the second row, for $v = 2$, the most that can be said is that $P > 0.05$.

Since binomial experiments are a special case of multinomial experiments, the multinomial chi-square test can be used to test the correctness of a binomial parameter. There will be two categories, success and failure, and thus one degree of freedom. This procedure has an advantage over the test given in Chapter 3; it is independent of sample size and the specified binomial parameter, so a multitude of binomial tables is unnecessary—Table A.9 is sufficient. If the experimenter had to rely on available binomial tables, he might be tempted to tailor the experiment to fit the table. He might pick a sample size that appears in the table even if it is not the best sample size; or he might discard data if he cannot control the sample size (as in many genetics experiments) so that it fits the tables. Needless to say, these are not ideal scientific procedures. The multinomial chi-square test helps to avoid these pitfalls.

There are two disadvantages, however, to using a multinomial chi-square test when testing a binomial parameter. First, because of the nature of the chi-square statistic, one-tailed alternatives are more involved than we will discuss here. Thus, if a one-tailed alternative is desired, the exact binomial distribution should be used (in the case of large sample sizes, the approximation procedure that will be explained in Chapter 7 may be used). The second disadvantage is that the approximation of the discrete sampling chi-square distribution by the continuous theoretical chi-square distribution is not very good for 1 degree of freedom with small sample sizes. For $n \leq 25$, a continuity correction should be made in the chi-square statistic:

$$\text{corrected } \chi^2 = \sum_{i=1}^{k} \frac{(|o_i - e_i| - 0.5)^2}{e_i}$$

For degrees of freedom other than 1, there is no appropriate continuity correction. However, except for very small samples, the approximation of the discrete chi-square distribution by the continuous one is good. Some statisticians recommend that all expected values should be at least 5 in order to have an acceptable approximation. Others feel this is too conservative and indicate that no expected value should be less than 1, and not more than 20% of the expected values should be less than 5. We suggest these latter guidelines. If these conditions are not met, it is sometimes possible to combine categories to raise the expected value. Care should be taken, however, that the experimental question can still be answered when the categories are combined.

Besides being convenient, the chi-square test has another property to recommend it. In many situations the chi-square test is the most powerful one available—that is, it is the test that is most likely to detect a deviation from the null hypothesis if one exists.

Procedure. Multinomial Chi-Square Test

H_0: $\pi_1 = \pi_{1_0}, \pi_2 = \pi_{2_0}, \ldots, \pi_k = \pi_{k_0}$
H_a: *At least one inequality*

Significance level: α

Test statistic:

$$\chi^2 = \sum_{i=1}^{k} \frac{(o_i - e_i)^2}{e_i}$$

o_i = observed number of outcomes in ith category

$$e_i = n\pi_{i_0} \text{ with } n = \sum_{i=1}^{k} o_i$$

Region of rejection: $\chi^2 \geq \chi^2_{\alpha,k-1}$

EXERCISES

5.1.1.   Use Table A.9 in the Appendix of Useful Tables to find the following:

   a. $\chi^2_{0.01,7}$

   b. $\chi^2_{0.995,10}$

   c. $\chi^2_{0.025,70}$

    d. $P(\chi^2 > 31.410)$ if χ^2 is a chi-square random variable with 20 degrees of freedom

    e. $P(\chi^2 < 27.488)$ if χ^2 is a chi-square random variable with 15 degrees of freedom

    f. b if $P(\chi^2 > b) = 0.05$ and χ^2 is a chi-square random variable with 10 degrees of freedom

    g. b if $P(\chi^2 \le b) = 0.995$ and χ^2 is a chi-square random variable with 22 degrees of freedom

    h. the degrees of freedom if $P(\chi^2 < 0.831) = 0.025$ and χ^2 is a chi-square random variable

5.1.2. Computer programs for producing tables of random digits are often called pseudo-random-number generators because there is no way to prove that the digits are in random order. However, some properties of randomness can be tested. As an exercise, suppose that the 50 digits in row 1 of Table A.1 in the Appendix are a random sample.

    a. State a null hypothesis about the proportion of even digits if the table is random.

    b. State an alternative hypothesis that would indicate a lack of randomness.

    c. Use a multinomial chi-square test with $\alpha = 0.05$ to test the above null hypothesis.

5.1.3. Assume the first three rows of Table A.1 are a random sample of size 150 and test that each of the digits $0, 1, \ldots, 9$ is equally frequent in the whole table by means of a multinomial chi-square test ($\alpha = 0.05$). What is the P value associated with this test?

5.1.4. Within some populations the proportion of those who are carriers of the sickle-cell trait is estimated to be 30%. A public health officer on a Caribbean island wonders whether this estimate is correct for the citizens of that island. Assuming that it will be a random sample, he requests that the next 150 blood tests performed in a certain clinic also include a microscopic examination for the sickling phenomenon. Given that there are 57 cases of sickling in the sample, perform a multinomial chi-square test to determine whether this proportion is correct. Use $\alpha = 0.05$. State the final conclusion.

5.1.5. When a certain red-flowering plant is self-fertilized, genetic theory indicates that the plants developed from the resulting seed should be in the ratio of 3 red-flowering plants to 1 white-flowering plant. If a random sample of 100 such seeds is collected and 68 produce red-flowering plants, 29 produce white-flowering plants, and 3 do not germinate, do these results agree with the theory? Use a multinomial chi-square test with $\alpha = 0.01$. What assumption must be made about the nongerminating seeds for this to be a valid test?

5.1.6. Analyze the data in part d of Exercise 3.2.3 by means of a multinomial chi-square test at $\alpha = 0.05$. Since the sample size is below 25 and there is only 1 degree of freedom, use the continuity correction. Does your conclusion agree with the conclusion you reached in Exercise 3.2.3?

5.1.7. A congressional representative circulates a questionnaire to all constituents to determine which national issue should be given the highest priority. A random sample of 500 gives the following:

Issue	Number Who Felt This Issue Deserves Highest Priority
Pollution	40
Economy	97
Energy	31

Issue	Number Who Felt This Issue Deserves Highest Priority
Medical care	85
Foreign policy	53
Defense	71
Questionnaire not returned	123

The representative wants to know if there is a preference for one of the issues. Test the hypothesis that all of the issues are equally preferred against the hypothesis that some preference exists. What is the P value? What conclusion should the representative draw from this study? What assumption must be made about those who did not return the questionnaire in order for this analysis to be valid?

5.1.8. On the basis of size, blue crabs are categorized by marine biologists as young, juvenile, mature. In a healthy crab population that is being acceptably harvested by commercial fishermen, the percentage of each type is

50% young   30% juvenile   20% mature

Deviations from these percentages usually indicate an unhealthy or overfished population. Fish and game biologists can dredge the bottom of a bay or estuary with nets to obtain a sample of crabs in an area close to commercial crabbing to determine if there is an unacceptable distribution of ages. Suppose that a small bay is dredged and the following categories of crab are netted:

58 young   33 juvenile   39 mature

a. Give the most logical null and alternative hypotheses for this study.

b. For this study, which is more serious, a Type I or Type II error? Why?

c. Perform a test of significance at $\alpha = 0.05$.

d. What is the experimental conclusion?

e. Suppose it is known that fishermen keep all mature and some juvenile crabs they net; all others are released unharmed. It is also known that young crabs are most susceptible to pollution, with juveniles the second most susceptible. Based on this information and the test of significance, which of the following is the appropriate action?

  i. Allow continued harvesting of crabs in the bay.

  ii. Close the bay to commercial crabbing because of overfishing.

  iii. Close the bay due to possible pollution.

  iv. Close the bay because of both overfishing and possible pollution.

5.1.9. In studying the genetic association between hair and eye color in human beings, a geneticist might hypothesize that the genes for hair color and eye color are located on the *same* chromosome. If a large group of dark-haired and brown-eyed people were to intermarry with another large group of light-haired and blue-eyed people, Mendel's law could be used to predict the characteristics of the second generation if the genes for hair color and eye color were on *different* chromosomes. The ratio of dark-haired and brown-eyed people to dark-haired and blue-eyed people to light-haired and brown-eyed people to light-haired and blue-eyed people would be 9:3:3:1. If the genes are on the same chromosome, this ratio does not appear.

a. What are the null and alternative hypotheses that should be used for this experiment?

b. Assume 1317 offspring of this type are located and classified with the following results:

Dark hair, brown eyes	782
Dark hair, blue eyes	234
Light hair, brown eyes	241
Light hair, blue eyes	60

What should the geneticist conclude?

5.1.10. In a certain state the distribution of the population by age is as follows:

Age (years)	Population (thousands)
Under 15	475
15–24	304
25–34	182
35–44	190
45–54	208
55–64	170
65–74	111
Over 74	72

a. Find the proportion of the population in each age group.

b. A certain planned city in this state claims that its inhabitants have the same proportion of people in each age group as the state as a whole. What null and alternative hypotheses should be used to test its claim?

c. If the city has a population of 12,500, compute the expected values for each age category if the null hypothesis is true.

d. If the city has the following distribution of ages, complete the test at the 5% significance level and state the conclusion.

Age (years)	Population (thousands)
Under 15	3016
15–24	2438
25–34	2037
35–44	2031
45–54	1253
55–64	977
65–74	585
Over 74	163

5.2. GOODNESS-OF-FIT TESTS

The multinomial chi-square test discussed in Section 5.1 is one type of *goodness-of-fit test*. It can be used to determine if the outcomes from a multinomial experiment fit a distribution with specified proportions of responses in certain categories.

A similar procedure can be used to determine whether a response variable for some population can be modeled by some other probability distribution. For the case in which the parameters of the probability distribution are known, the test is very similar to the multinomial chi-square test. If the parameters are unknown and must be estimated, an adjustment in the degrees of freedom is necessary.

Example 5.3.   Goodness-of-Fit Test with a Specified Parameter

Each day a salesperson calls on 5 prospective customers and she records whether or not the visit results in a sale. For a period of 100 days her record is as follows:

Number of sales:	0	1	2	3	4	5
Frequency:	15	21	40	14	6	4

A marketing researcher feels that a call results in a sale about 35% of the time, so he wants to see if this sampling of the salesperson's efforts fits a theoretical binomial distribution for 5 trials with 0.35 probability of success, $b(y; 5, 0.35)$. This binomial distribution has the following probabilities and leads to the following expected values for 100 days of records:

y	$p(y)$	$e = 100p(y)$
0	0.1160	11.60
1	0.3124	31.24
2	0.3364	33.64
3	0.1812	18.12
4	0.0487	4.87
5	0.0053	0.53

Since the last category has an expected value of less than 1, he combines the last two categories to perform the goodness-of-fit test.

Category A_i	Observed Frequency o_i	$P(A_i)$	Expected Frequency e_i	$o_i - e_i$	$(o_i - e_i)^2$	$(o_i - e_i)^2 / e_i$
0	15	0.1160	11.60	3.40	11.5600	0.9966
1	21	0.3124	31.24	-10.24	104.8576	3.3565
2	40	0.3364	33.64	6.36	40.4496	1.2024
3	14	0.1812	18.12	-4.12	16.9744	0.9368
4 or 5	10	0.0540	5.40	4.60	21.1600	3.9185
					$\chi^2 =$	10.4108

In this goodness-of-fit test the hypotheses are:

H_0: *This sample is from* $b(y; 5, 0.35)$

H_a: *This sample is not from* $b(y; 5, 0.35)$

The degrees of freedom are $v = k - 1 = 5 - 1 = 4$. The critical value is $\chi^2_{0.05,4} = 9.488$. The null hypothesis is rejected if this value is exceeded. Thus the marketing researcher rejects the null hypothesis. The sales do not follow the pattern of this binomial distribution.

If the salesperson has no idea of the proportion of the times she is successful, she could estimate π by dividing the total number of sales by the total number of visits, $187/500 = 0.374$. She could then test to see if her sales fit $b(y; 5, 0.374)$. The procedure is similar to the above, except now the degrees of freedom are $k - 2 = 5 - 2 = 3$. One additional degree of freedom is lost because of the estimated parameter. In general, $v = k - 1 - r$, where r is the number of parameters that are estimated.

A goodness-of-fit test for a Poisson distribution can be done in a similar manner.

Example 5.4. Goodness-of-Fit Test with an Unspecified Parameter

If the same typesetter sets all the copy for a book, the error rate should be approximately the same throughout the book. With this assumption, the number of misprints per page may be a Poisson random variable. To check whether the Poisson model is correct, an efficiency expert collects the following data from a random sample of 100 pages:

Number of mistakes per page:	0	1	2	3	4	5	6
Observed frequency o_i:	13	24	31	18	11	2	1

He wants to test

H_0: *This sample is from a Poisson distribution*

against

H_a: *This sample is not from a Poisson distribution*

To estimate λ, the average number of errors per page, he computes the total number of errors and divides by the number of pages, $200/100 = 2.00$. Thus 2.00 is an estimate of λ in the Poisson distribution. Looking at the Poisson distribution with $\lambda = 2.00$, he finds

Y	Probability
0	0.1353
1	0.2707
2	0.2707
3	0.1804
4	0.0902
5	0.0361
6	0.0120
Over 6	0.0045

If these 8 categories are used for a goodness-of-fit test, the expected values for the last 3 categories will all be less than 5. Since $3/8 = 0.375$, too many expected values are under 5. To take care of this, he can combine the last three categories and compute the chi-square statistic as follows:

Category A_i	Observed o_i	$P(A_i)$	Expected e_i
0	13	0.1353	13.53
1	24	0.2707	27.07
2	31	0.2707	27.07
3	18	0.1804	18.04
4	11	0.0902	9.02
Over 4	$\underline{3}$	0.0526	5.26
	100		

and

$$\chi^2 = \sum_{i=1}^{k} \frac{(o_i - e_i)^2}{e_i} = 2.345$$

The null hypothesis will be rejected if this computed chi-square value is greater than or equal to $\chi^2_{0.05,4} = 9.488$. There are 4 degrees of freedom because $v = k - 1 - 1 = 6 - 2 = 4$; the additional degree of freedom is lost because of the estimation of λ. The efficiency expert does not reject the null hypothesis in this study, and he concludes that the errors per page may be modeled by a Poisson distribution.

Both of the examples used in this section concern discrete probability distributions. It is also possible to do a chi-square goodness-of-fit test for continuous probability distributions. An example is given in Exercise 7.1.7.

Procedure. Chi-Square Goodness-of-Fit Test

H_0: *This sample is from distribution A*

H_a: *This sample is not from distribution A*

Significance level: α

Test statistic:

$$\chi^2 = \sum_{i=1}^{k} \frac{(o_i - e_i)^2}{e_i}$$

$o_i =$ observed number of outcomes in category A_i

$$e_i = nP(A_i) \quad n = \sum_{i=1}^{k} o_i$$

Region of rejection:

$$\chi^2 \geq \chi^2_{\alpha,\nu}$$

$$\nu = k - 1 - r$$

r = number of parameters in distribution A estimated from the sample

EXERCISES

5.2.1. Sixty sample groups of 4 persons in each group contain the following distribution for the number of persons with type O blood:

Number with type O:	0	1	2	3	4
Frequency:	8	18	21	8	5

Are these sample groups of four from the binomial distribution $b(y; 4, 0.40)$? What is the P values?

5.2.2. Assume the number of defects in a hundred 20-ft sections of wire are

Number of defects:	0	1	2	3	4
Frequency:	88	10	1	0	1

Does this fit a Poisson distribution with $\lambda = 0.10$?

5.2.3. A campground has 5 rustic campsites not accessible to campers on wheels. Some nights, some of these campsites are unoccupied because of the small number of campers with equipment for such campsites. The ranger keeps track of the number of unoccupied sites for 50 nights.

Number unoccupied:	0	1	2	3	4	5
Frequency:	22	20	7	1	0	0

Do these data fit a binomial distribution?

5.2.4. If the number of parasites found on 80 hosts are

Number of parasites:	0	1	2	3	4	5
Number of hosts:	20	28	19	9	3	1

does this fit a Poisson distribution?

5.2.5. It seems that the history of the Supreme Court with respect to the occurrence of appointments within a year might be an example of a Poisson distribution (Kinney,

1973; Wallis, 1936). Test the following data for Poissonness using a chi-square goodness-of-fit test at the 0.05 significance level:

Number of Appointments per Year	Number of Years (1790–1972)
0	108
1	55
2	19
3	1
4 or more	0

5.3. CONTINGENCY TABLE ANALYSIS

With goodness-of-fit tests, we can determine whether a single sample comes from a population that has a certain probability model. Sometimes we want to know whether or not *several samples* all come from the same population and perhaps we do not even know the appropriate model for the population. A *chi-square test of homogeneity* can often be used in this case.

For example, a speech pathologist might want to know whether the proportion of males among stammerers and the proportion of males among lispers are the same. Her null and alternative hypotheses are

$$H_0: \pi_S = \pi_L$$
$$H_a: \pi_S \neq \pi_L$$

in which π_S is the proportion of stammerers who are male and π_L is the proportion of lispers who are male. Note that the values of π_S and π_L are not specified in the null hypothesis. (The proportions for females could also be included in the null hypothesis, but this is unnecessary since there are only two classes, male and female, and the proportions must sum to 1.)

The speech pathologist collects information from two random samples, one of stammerers and the other of lispers (that is, a stratified random sample), and arranges the data in the form of a two-way table called a *contingency table*. (The following data are simplified in order to keep the arithmetic simple in this first example.)

	SAMPLES	
	Stammer	Lisp
Male	32	28
Female	18	22
Total	50	50

The proportion of males in the sample of stammerers is 32/50 and the proportion of males in the sample of lispers is 28/50. Are these sample proportions so different that they indicate that the population proportions are not equal, $\pi_S \neq \pi_L$? To answer this, the speech pathologist computes the total number of males and females in the samples and uses these totals to find the expected value for each of the cells in the two-way layout if the null hypothesis is true.

	OBSERVED					EXPECTED		
	Stammer	Lisp	Total			Stammer	Lisp	Total
Male	32	28	60		Male	30	30	60
Female	18	22	40		Female	20	20	40
Total	50	50	100		Total	50	50	100

The expected number of male stammerers is 30 because if the two populations are the same, $60/100 = 0.60$ of the people with speech problems are males and $0.60(50) = 30$, that is, there are 50 stammerers and 30 of them on the average should be males. There are two ways that the rest of the cells can be filled with expected values. Each expected value can be computed similarly to the one for the male stammerers; however, since the totals are known, the remaining cells can be filled by subtraction. For example, the expected number of male lispers is $60 - 30 = 30$.

To find the expected value for a cell directly from the totals, we divide the product of the two corresponding marginal totals by the grand total. For the male stammerers this is $(50)(60)/100 = 30$. We can summarize this procedure by using the following symbols in which i identifies the row and j the column.

	OBSERVED		Total		EXPECTED	
	o_{11}	o_{12}	$o_{1.}$		e_{11}	e_{12}
	o_{21}	o_{22}	$o_{2.}$		e_{21}	e_{22}
Total	$o_{.1}$	$o_{.2}$	$o_{..}$			

$$e_{ij} = \frac{(o_{i.})(o_{.j})}{o_{..}}$$

Once we have found the expected value, the χ^2 statistic is computed in the usual way.

Class	o_{ij}	e_{ij}	$o_{ij} - e_{ij}$	$(o_{ij} - e_{ij})^2$	$(o_{ij} - e_{ij})^2/e_{ij}$
Male, stammer	32	30	+2	4	0.133
Female, stammer	18	20	−2	4	0.200
Male, lisp	28	30	−2	4	0.133
Female, lisp	22	20	+2	4	0.200
					$\chi^2 = 0.666$

In a chi-square test of homogeneity, the degrees of freedom are $v = (r - 1)(c - 1)$ in which r is the number of rows and c is the number of columns. In this illustration $v = 1$. This corresponds to the fact that once we have computed one expected value from the totals in the two-by-two layout, all of the other values are determined.

The critical chi-square value for 1 degree of freedom is $\chi^2_{0.05,1} = 3.841$, and the null hypothesis is rejected if the chi-square statistic is greater than or equal to this value. The speech pathologist notes that the computed chi-square value is less than the critical value, and she decides that the proportion of males among stammerers may be the same as the proportion of males among lispers. She concludes that when males are tested for speech problems they should not be tested for a specific problem such as stammering but should be given a general test that would identify both stammerers and lispers.

A chi-square test of homogeneity is used to determine whether two or more samples are from the same multinomial population. In the example just completed, the decision concerned two samples from binomial populations. In the next example three multinomial samples will be examined.

Example 5.5. Chi-Square Test of Homogeneity

A political scientist is interested in determining how important the promise of no tax increase is for voters of different political affiliations. Using voter registration lists, she chooses random samples of 100 from each of the groups, Democrats, Republicans, and Independents, and she asks the subjects to rate the importance of no tax increase on a scale from 1 to 4. The results are as follows:

	Very Important 1	2	3	Not Important 4	Total
Democrats	42	26	19	13	100
Republicans	55	21	14	10	100
Independents	38	30	22	10	100
Total	135	77	55	33	300

In words, the hypotheses are

H_0: *Members of the three parties agree on the importance of no tax increase*
   (homogeneity)
H_a: *Members of the three parties do not agree on the importance of no tax increase*
   (lack of homogeneity)

Note that in this example the three samples are in the rows, whereas in the previous example about speech defects, the samples were in the columns.

Using the totals and the formula

$$e_{ij} = \frac{(o_{i.})(o_{.j})}{o_{..}}$$

the expected values are

	1	2	3	4	Total
Democrats	45.0	25.7	18.3	11.0	$100 = o_{1.}$
Republicans	45.0	25.7	18.3	11.0	$100 = o_{2.}$
Independents	45.0	25.7	18.3	11.0	$100 = o_{3.}$
Total	$135 = o_{.1}$	$77 = o_{.2}$	$55 = o_{.3}$	$33 = o_{.4}$	$300 = o_{..}$

The χ^2 statistic is computed.

Class	o_{ij}	e_{ij}	$(o_{ij} - e_{ij})^2/e_{ij}$
Democrats			
1	42	45.0	0.200
2	26	25.7	0.004
3	19	18.3	0.027
4	13	11.0	0.364
Republicans			
1	55	45.0	2.222
2	21	25.7	0.860
3	14	18.3	1.010
4	10	11.0	0.091
Independents			
1	38	45.0	1.089
2	30	25.7	0.719
3	22	18.3	0.748
4	10	11.0	0.091
			$\chi^2 = \overline{7.425}$

Since there are 3 rows and 4 columns in the contingency table,

$$v = (r - 1)(c - 1) = (3 - 1)(4 - 1) = 6$$

At the 0.05 level of rejection, the null hypothesis is rejected if the computed chi-square value is greater than or equal to

$$\chi^2_{0.05,6} = 12.592$$

Since this is not the case in this study, the null hypothesis is accepted and the political scientist concludes that there is no evidence to indicate that the three samples are different with respect to their opinions on the importance of no tax increase.

The chi-square test of homogeneity is applied to two or more samples when the samples have been classified by *one characteristic*. There is a similar chi-square test that can be used to analyze data from a *single sample* when the data have been classified by *two characteristics*. For example, in a state in which party affiliation is not declared at voter registration, a single sample of 300 registered voters could be selected at random and asked for their opinion on the importance of no tax increase and also for their party preference. The contingency table would look similar to the table in Example 5.5 except that it is not likely that there would be exactly 100 from each party. The political scientist would be trying to determine whether party affiliation is related to opinion about taxes, and the test procedure is called a *chi-square test of independence*.

> H_0: *Party reference is independent of opinion about the importance of no tax increase*
>
> H_a: *Party reference is related to opinion about the importance of no tax increase*

The test statistic and region of rejection are determined as in a test for homogeneity; the difference is in how the sample was chosen. The test of homogeneity involves a stratified sample. The test of independence involves a simple random sample.

A worked-out example follows.

Example 5.6.   A Chi-Square Test of Independence

Football coaches feel that a football team has an advantage when it is playing a home game in its own stadium. The enthusiasm of the crowd, familiarity with the field, and the lack of fatigue from travel all seem to contribute to this assumed advantage. A coach wants to test this theory at his school. If the theory is wrong, whether a game is won or lost is independent of whether the game is played at home or away. The hypotheses are

> H_0: *Winning is independent of where the game is played*
>
> H_a: *Winning depends on where the game is played*

The coach examines the records at his school over the past 31 years, a single sample. He classifies the results as follows (ties and bowl games are omitted):

OBSERVED

	Home	Away	Total
Won	97	69	166
Lost	42	83	125
Total	139	152	291

Intuitively the data seem to confirm the coach's theory. Using the marginal totals, he computes the following expected values:

EXPECTED

	Home	Away
Won	79.3	86.7
Lost	59.7	65.3

He then computes the chi-square statistic:

Class	o_{ij}	e_{ij}	$o_{ij} - e_{ij}$	$(o_{ij} - e_{ij})^2$	$(o_{ij} - e_{ij})^2/e_{ij}$
Won/home	97	79.3	17.7	316.3	3.99
Lost/home	42	59.7	−17.7	316.3	5.30
Won/away	69	86.7	−17.7	316.3	3.65
Lost/away	83	65.3	17.7	316.3	4.84
					$\chi^2 = \overline{17.78}$

Since $\chi^2_{0.05,1} = 3.841$, the null hypothesis is rejected and the coach concludes that if these 31 years are a random sample of this school's games, there is evidence that the probability of winning depends on where the game is played.

To interpret the dependence, he would note that the predictor classification is the location of the game (the column categories) and the predicted classification is the outcome of the game (the row categories). He would then examine the proportions in the columns, the predictor classifications. He finds that $97/139 = 0.697$ of the games at home are won while only $42/139 = 0.302$ of the home games are lost. Also, only $69/152 = 0.454$ of the away games are won, while $83/152 = 0.546$ of the away games are lost. From this he would conclude that playing at home increases the probability of winning. There is evidence of a home team advantage. Odds can also be used to summarize the data (see Section 5.4).

Since 2×2 contingency tables have 1 degree of freedom, the continuity correction should be used to improve the approximation of the discrete sampling distribution by the continuous theoretical chi-square distribution if $n < 25$.

As in goodness-of-fit tests, contingency table tests do not work well for small expected values (below 5). In the 2×2 case, another test can be used when the expected values are small, *Fisher's exact test*. References to this test are given at the end of this chapter (Finney, 1948; Fisher, 1973; Latscha, 1955).

Procedure. Contingency Table Analysis

Chi-Square Test of Homogeneity

H_0: *The populations sampled are the same with respect to the categorization*

H_a: *The populations sampled are different with respect to the categorization*

Chi-Square Test of Independence

H_0: *The row categories are independent of the column categories*

H_a: *The row categories and the column categories are dependent*

Significance level: α

Test statistic:

$$\chi^2 = \sum_i \sum_j \frac{(o_{ij} - e_{ij})^2}{e_{ij}}$$

o_{ij} = number of occurrences in the ijth cell

$$e_{ij} = \frac{(o_{i.})(o_{.j})}{o_{..}}$$

$$o_{i.} = \sum_j o_{ij}$$

$$o_{.j} = \sum_i o_{ij}$$

$$o_{..} = \sum_i \sum_j o_{ij}$$

Region of rejection:

$$\chi^2 \geq \chi^2_{\alpha,v} \quad v = (r - 1)(c - 1)$$

r = number of rows

c = number of columns

EXERCISES

5.3.1. A serum thought to be effective in preventing colds is given to 300 persons. Their records for one year are compared with those of 200 untreated persons with the following results:

	No Colds	One Cold	More Than One Cold
Treated	145	80	75
Untreated	80	70	50

Use a chi-square test of homogeneity to analyze these data.

5.3.2. A social scientist wants to determine if the feelings that parents have toward young people "living together" are affected by the age of their youngest child.

	Parents' Feelings	
Age of Youngest Child	Approve	Disapprove
Over 26	50	10
18–26	10	40
Under 18	60	30

a. State the null hypothesis verbally in terms of independence.

b. Perform a chi-square test of independence at the 0.05 level of significance.

c. Which classification is the predictor? Which is the predicted?

d. Use the proportions of the predictor classifications to state a specific conclusion about the dependency.

5.3.3. It is reported that offspring of users of a certain recreational drug may have a higher incidence of birth defects than the general population. To obtain information about a possible relationship between this drug and birth defects, 100 offspring of female rats fed the drug and 100 offspring from untreated female rats are examined. The results are given below:

	Progeny	
Females	Birth Defects	Normal
Treated	30	70
Untreated	20	80

Analyze these data. What do you conclude from the study? Is this a test of homogeneity or independence?

5.3.4. A consumer's union would like to compare three brands of flashlight batteries. Its testers randomly select 100 batteries of each brand and classify them into 3 groups depending on lifetimes:

Brand	Less than 5 Hours	5 to 10 Hours	Over 10 Hours	Total
X	30	60	10	100
Y	15	60	25	100
Z	30	30	40	100

a. State the null and alternative hypotheses to be tested.

b. Compute the chi-square statistic.

c. What are the statistical decision and the experimental conclusion?

5.3.5. An entomologist is interested in determining whether certain insecticides have a differential effect on black flies. The results of his experiment are

Insecticide	Dead	Alive
A	165	35
B	172	28
C	173	27

a. What null hypothesis can be tested with these data?

b. If the entomologist sets the rejection level at 1%, how large must the chi-square statistic be in order for him to reject the null hypothesis?

c. Compute the statistic.

    d. How likely is it that a sample as unusual as this will be obtained when the null hypothesis is true?

    e. What decision should the entomologist make about the null hypothesis? What conclusion should be drawn?

5.3.6. A study is conducted on adult male cancer patients to determine whether there is any association between the kinds of work they perform and the kinds of cancer they have. The data are classified by the two categories as below:

Site of Malignancy

Occupation	Skin	Stomach	Prostate
Professional	25	58	37
Managerial	34	90	36
Laborer	41	52	27

    a. State the null hypothesis verbally.

    b. Give the critical value of the test statistic for $\alpha = 0.05$.

    c. Compute the expected value for the category laborer and stomach.

    d. The computed value of χ^2 is 10.49. Which of the following statements are appropriate to this survey?

        i. The type of work one does causes certain kinds of cancer.

        ii. The location of a cancer is independent of occupation.

        iii. There is a significant association between occupation and kind of cancer.

    e. Specify the predictor and predicted classification.

    f. What specific conclusion can be drawn about the kind of cancer associated with each of the occupations in the study?

5.3.7. Feminine beauty was another variable Francis Galton measured. He even tried to draw a "beauty map" of Britain patterned after the weather maps he had already created. Being a proper Victorian English gentleman, however, he wanted to observe and record without being observed observing and recording. So he would tear a piece of paper in the shape of a cross and put it in his jacket pocket along with a tailor's straight pin. Then upon seeing a woman in an area he had not yet mapped, he would use the pin to put a hole in the top of the cross if she was attractive, in the arms of the cross if she was of medium attractiveness, and in the bottom of the cross if she was unattractive. Later, he would record the number of pin holes and their locations. He reported that he found women in London more attractive than those in Aberdeen. Suppose that conclusion was based on the following data:

City	Attractive	Medium	Unattractive	Total
Aberdeen	55	100	45	200
London	75	100	25	200
Total	130	200	70	400

    a. Give the null and alternative hypotheses.

    b. Perform the test of significance and draw conclusions.

    c. What are the odds Galton would encounter an attractive woman in London?

    d. How could you compare the odds of encountering an attractive woman in each of the two cities?

5.4. RELATIVE RISKS AND ODDS RATIOS

The contingency table analysis for 2×2 tables described in Section 5.3 tests the hypothesis that $\pi_1 - \pi_2$ is equal to zero. There are situations where the difference between the two proportions might not be the best way to interpret the data. If π_1 is the probability of an unfavorable outcome for a treatment group and π_2 is the probability of an unfavorable outcome for a placebo group, then a difference of 0.1 when $\pi_1 = 0.1$ and $\pi_2 = 0.2$ might be more important than a difference of 0.1 when $\pi_1 = 0.4$ and $\pi_2 = 0.5$.

    Consider the following two examples.

1. The risk for heart attacks is relatively low for adults whose cholesterol is less than that 200 mg/dL. However the American Heart Association estimates that about 50% of adult Americans have cholesterol greater than 200 mg/dL. Suppose a study shows that a program of modest physical activity without any other lifestyle changes can reduce the percentage of adults with high cholesterol to 40%.

2. The National Center for Chronic Disease Prevention and Health Promotion estimates that 20% of American children and adolescents are overweight. Again suppose a study shows that a program of modest physical activity can reduce the percentage of overweight children and adolescents to 10%.

While the improvement is 10% for both populations, the 10% change for the overweight children represents an improvement for 1 out of every 2 while the 10% change for the adults with high cholesterol represents an improvement for only 1 out of every 5.

    Many of the above situations also can be generalized as follows. There are two categorical variables. One variable can be designated as the *explanatory variable* and the other as the *response variable*. The explanatory variable has two categories and the response has two categories. The numbers of individuals with each combination of the two categories are counted. The counts are displayed in the 4 cells of a 2×2 table. By convention, the rows (the side of the table) are assigned to the explanatory variable and the columns (the top of the table) are assigned to the response.

    The response variable is sometimes called the *outcome* variable. One category of the outcome variable is called the *primary outcome*. For example, in a study of the effects of smoking, the category lung cancer might be the primary outcome. No lung cancer would be the other category. One of the categories of the explanatory variable is called *a risk factor*. Smoker could be that category. Non-smoker could be the other category.

    Many medical studies focus on the effectiveness of intervention procedures. For example, a study might focus on the use of aspirin for preventing coronary heart disease. In such studies one of the categories of the explanatory variable is the use of some drug or procedure as prevention or treatment and the other category is a placebo. The risk factor is the placebo. The primary outcome is a disease such as coronary heart disease.

    The goal of these studies is to determine if the risk factor is related to the primary outcome. The studies can be broadly classified as e*xperimental* or *observational*. In experiments, explanatory factors are assigned to samples of subjects. In observational studies (surveys), subjects from a target population are selected and the explanatory factors that are present are simply observed in each subject. The presence of one or the other of the outcomes is determined for each subject.

There are two types of observational studies, *prospective* and *retrospective*. In each, two random samples are selected for comparison. The primary difference has to do with whether the samples were selected on the basis of the explanatory variable or on the basis of the response variable.

In prospective studies, one of the random samples consists of subjects who have the risk factor and the other random sample consists of subjects who do not. After a period of time the subjects in both samples are examined to determine which have the primary outcome.

In retrospective studies, one of the random samples consists of subjects who have shown the primary outcome (often called the *cases*) and the other random sample consists of the subjects who have not shown the primary outcome (called the *controls*). The subjects are examined to determine how many in each sample have the risk factor. The degree of usefulness of retrospective studies is related to the selection of the random sample of subjects not exhibiting the primary outcome. An attempt should be made to match the controls to the cases as much as possible. If there is a difference in the proportion of subjects with the primary outcome, there should be no uncertainty that the difference can be attributed to the risk factor.

Both prospective and retrospective studies have important roles in research. A prospective study that follows random samples of smokers and nonsmokers might be useful, but it could take a long time to complete because it could not be accomplished without following the subjects through their entire lives. Prospective studies can be very expensive because very large samples are required to get enough positive primary outcomes to allow for statistical inference. With the current proactive attitude toward smoking cessation, such an experiment could be viewed as unethical.

Example 5.7.   A Retrospective Study on Relative Risk and Odds Ratio

A physician at a clinic in southern Appalachia is concerned about the number of underweight newborns he sees in his practice. He gives health surveys to the mothers and observes that many of the mothers with serious gum disease have underweight babies. He summarizes the data in the following table:

	Underweight Baby		
Gum Disease	Yes	No	Total
Yes	17	83	100
No	117	783	900
Total	134	866	1000

Are there more underweight babies born to mothers with gum disease? Unless there are an equal number of babies born to mothers with gum disease and without gum disease, it is difficult to make useful comparisons directly from the table. The question of interest is whether the proportion of underweight babies is the same for each group of mothers. He can calculate conditional proportions of underweight babies for each group. For the mothers with gum disease, the proportion of underweight babies is $17/100 = 0.17$. For the mothers without gum disease, the proportion of underweight babies is $117/900 = 0.13$. If the proportions are multiplied by 100%, they are percentages. The number 0.17 might also be viewed as the probability that a randomly selected mother with gum disease has an underweight baby.

Because underweight babies are susceptible to more disease and developmental problems, the proportions also are referred to as the *risks* of an underweight baby.

The *relative risk* of an outcome for two categories of an explanatory variable is the ratio of the risk for each category. For the table above, the explanatory variable is gum disease or no gum disease and the relative risk is $0.17/0.13 = 1.31$. It is usually expressed as a multiple. A relative risk of 1.31 means the risk of an underweight baby for a mother with gum disease is 1.31 times the risk of an underweight baby for a mother without gum disease. A relative risk of 1 means the risk is the same for both categories.

Sometimes the increase in risk is presented as a percentage instead of a multiple:

$$\% \text{ increased risk} = \frac{\text{change in risk}}{\text{original risk}} \times 100\%$$

or

$$\% \text{ increased risk} = (\text{relative risk} - 1) \times 100\%$$
$$= (1.31 - 1) \times 100\% = 31\%$$

Mothers with gum disease have a 31% increased risk for underweight babies compared to mothers without gum disease.

Odds are an alternative way to express that a randomly selected individual will fall into a particular group for a categorical variable. The *odds* of an underweight baby is the number of babies who are underweight divided by the number of babies who are not underweight. Again, we can calculate the odds for each group of mothers. The odds for an underweight baby for mothers with gum disease is $17/83 = 0.205$. The odds for an underweight baby for mothers without gum disease is $117/783 = 0.149$. The *odds ratio* for an outcome for two categories of an explanatory variable is the ratio of the odds for each category. For the table above, the odds ratio is $0.205/0.149 = 1.38$.

Notice that risks and odds are two ways of looking at the same problem. If we know that the risk of an underweight child for a mother with gum disease is $17/100$, then the odds are $17/(100 - 17) = 17/83$. Likewise, if we know that the odds are $17/83$, then the risk is $17/(17 + 83) = 17/100$. In addition, the relative risk and the odds ratio are about the same if the risks are small for both groups. Note that in the example the relative risk is 1.31 and the odds ratio is 1.38.

While the relative risk might be easier to understand, the odds ratio gives researchers a wider range of statistical methods for binary data. The odds ratio is the only parameter that describes the binary outcomes for the explanatory categories that can be estimated from retrospective studies. Notice that the proportion of underweight babies in mothers with serious gum disease provides no information about the proportion of mothers with gum disease among mothers of underweight babies. Similarly, a retrospective study of smoking and lung cancer cannot be used to estimate the individual proportions of smokers and nonsmokers or their difference among those who get lung cancer.

The odds ratio is the same regardless of which variable is considered to be the response. Consider the underweight baby example above. The odds ratio is the same regardless of which variable, underweight baby or mother with gum disease, is considered as the response. The odds of underweight babies among women with gum disease is 1.38 times the odds of underweight babies among women without gum disease. The odds of gum disease among mothers of underweight babies is 1.38 time the odds of no gum disease among mothers of underweight babies.

Procedure. Relative Risk and Odds Ratio

For 2×2 contingency tables of the form

	Response Variable	
Explanatory Variable	Yes	No
Yes	o_{11}	o_{12}
No	o_{21}	o_{22}

$$\text{Relative risk} = \frac{o_{11}/(o_{11} + o_{12})}{o_{21}/(o_{21} + 2_{12})} = \frac{o_{11}(o_{21} + o_{12})}{o_{21}(o_{11} + o_{12})}$$

$$\text{Odds ratio} = \frac{(o_{11})/(o_{12})}{(o_{21})/(o_{22})} = \frac{(o_{11})(o_{22})}{(o_{21})(o_{12})}$$

EXERCISES

5.4.1. A serum thought to be effective in preventing colds is given to 300 persons. Their records for one year are compared with those of 200 untreated persons with the following results (see Exercise 5.3.1):

	No Colds	Colds
Treated	145	155
Untreated	80	120

a. Is this a prospective or a retrospective study?

b. What is the relative risk for cold for the untreated?

c. What is the odds ratio?

5.4.2. It is reported that offspring produced by users of a certain drug may have a higher incidence of birth defects than the general population. To obtain information about a possible relationship between this drug and birth defects, 100 offspring of female rats fed the drug and 100 offspring from untreated female rats are examined. The results are given below (see Exercise 5.3.3):

	Progeny	
Females	Birth Defects	Normal
Treated	30	70
Untreated	20	80

   a. Is this an experimental or an observational study?
   b. What is the relative risk of birth defects for treated rats?
   c. What is the odds ratio of birth defects for treated rats?

5.4.3. An aortic aneurysm is a marked dilation of the aorta either in its thoracic or abdominal portion. A group of physicians has collected information from new patients for several years. One item is the initial aneurysm size determined by radiology. Another item is whether it ruptured. Their data can be summarized in the following table:

| | Rupture | |
Aneurysm Size	Yes	No
≥ 5 cm	10	128
< 5 cm	3	163

   a. Is this an experimental or an observational study?
   b. What is the relative risk of ruptures for the larger aneurysms?
   c. What is the odds ratio for ruptures for the larger aneurysms?

5.4.4. For a one-year period the magistrate court in a certain city randomly assigned some of the drivers found guilty of vehicular injury to a 4-week defensive driving course in addition to the usual penalties. Drivers who appeared in court were identified as repeat offenders and as participants of the course. A summary of this study is given in the following table.

| | Second Accident | |
Defensive Driving Course	Yes	No
Yes	18	30
No	22	30

   a. Is this an experimental or an observational study?
   b. What is the relative risk of a second accident for the non-participants of the defensive driving course?
   c. What is the odds ratio of a second accident for the non-participants of the defensive driving course?
   d. Comment on the utility of the defensive driving course.

5.5. NONPARAMETRIC STATISTICS: MEDIAN TEST FOR SEVERAL SAMPLES

Contingency chi-square procedures can also be used for a nonparametric test that several populations all have the same median. Numerical data from several samples are reduced to the nominal scale by recording only whether or not each value is greater than the median. Then, the contingency chi-square procedure is used to determine whether there are any significant differences, from sample to sample, in the proportions above and below the median.

Example 5.8.   Two-Sample Median Test

A cancer research team has two random samples, each of 20 women with cervical cancer. The difference between the two groups is the kind of cancer cells involved, LCNK or SM. It is of interest to know if there are differences between the two groups—to know whether younger women tend to have one type of cancer cell and older women have the other.

The median age for the 40 women was found to be $M = 48$ years. Among the 20 women with LCNK cancer cells, there were 10 who were older than 48, 9 who were younger, and 1 who was 48. Among those with SM cells, there were 9 older than 48, 10 younger, and 1 who was 48. Because the data are to be reduced to the nominal scale of "above" or "below" median age, it is customary to discard any values which fall on the median. When this is done, the following table is obtained:

	Cell Type		
	LCNK	SM	Total
Above median	10	9	19
Below median	9	10	19
Total	19	19	38

The hypothesis is that the probability that a cancer victim will be above median age, $P(u > M) = \pi$, will be the same irrespective of which group she is in. The alternative hypothesis is that there is an association between cell type and the probability she will be above median age:

$$H_0: \pi_1 = \pi_2 = 0.50$$
$$H_a: \pi_1 \neq \pi_2$$

The usual contingency chi-square analysis yields $\chi^2 = 0.1053$ with one degree of freedom, which is clearly nonsignificant at any conventional α level. Thus there is no evidence of an association between age and type of cancer cell.

Example 5.8 involved only two groups; hence it would be called a two-sample median test. For any number of samples, the analysis is called a k-sample median test, but the procedure remains essentially the same.

Procedure. Median Test

1. The median, or middle value, is found for all the observations irrespective of group.
2. Each numerical observation, u, is compared to the median and recorded on the nominal scale as being "above" or "below" the median. All $u = M$ are discarded.
3. The data which have been transformed to the nominal scale are then summarized in a $2 \times k$ table.
4. A contingency chi-square analysis is conducted.

EXERCISES

1. A Peace Corps volunteer wants to see which of four species of fast-growing tropical trees will do best in a reforestation program in Haiti. She plants enough trees to obtain 2-year growth data from a random sample of 30 trees of each species. Lacking computing equipment for an analysis of data at the numerical scale of measurement, she decides to perform a median test on the following transformed data:

	Species			
Growth	A	B	C	D
Above median	16	10	11	23
Below median	14	20	19	7

   a. What null hypothesis can be tested with these data?
   b. Give the alternative hypothesis.
   c. What is the critical value of the test statistic for $\alpha = 0.05$?
   d. Perform the test of significance and draw a conclusion.

2. The president of a nationwide accounting firm asks the personnel office to examine the firm's records to see whether inadvertent sexual discrimination has taken place with regard to promotion. Among other data which are gathered, there are random samples of 25 men and women respectively who were originally employed eight years earlier and who still work for the firm. There is a record of the number of months each employee worked for the firm before promotion to senior level. The data are given below, ordered within sex for convenience:

Women					Men				
21	25	26	26	31	8	8	16	20	23
31	37	40	43	43	25	26	27	28	28
51	54	56	61	62	29	30	31	36	37
62	66	68	71	71	38	38	41	44	47
72	76	80	84	85	48	50	53	70	82

   a. The median for an even number of observations is usually given as the value half-way between the two middle observations, or in this example the value half-way between the ordered 25th and 26th observations. Show how that value is found to be 40.5 months.
   b. What percentage of the women in the sample were promoted to senior level within their first 40.5 months of employment? What percentage of men? Are the two percentages significantly different at the 0.05 level?

3. Although lacking any satisfactory numerical scale of measurement, behavioral biologists can rank the members of a group according to behavioral attributes such as aggressiveness and greediness. Wanting to determine whether there is any association between these two attributes, a biologist is able to observe the behavior of a tribe of 64 adult tamarins (small South American primates) living under nearly natural conditions at a modern zoo. She learns to identify each of the animals at sight and is able to give

each a rank according to aggressiveness and a second rank according to greediness. She wants to see whether those above median rank with respect to aggressiveness are also above median rank with respect to greediness. The results are given below:

	Aggressiveness	
Greediness	Below Median	Above Median
Above median	12	20
Below median	20	12

a. State the null hypothesis in terms of independence.

b. Why is the expected value equal to $(1/4)n$ for all cells?

c. Perform the test of significance and then draw conclusions about the relationship between these two behavioral characteristics.

REVIEW EXERCISES

Decide whether each of the following statements is true or false. If a statement is false, explain why.

5.1. There is only one chi-square distribution.

5.2. The chi-square statistic does not have a continuous distribution, but the continuous distribution attributed to Helmert provides reliable probability statements.

5.3. If the computed value of χ^2 is greater than the critical value, the null hypothesis is false.

5.4. H_0: $\pi = 0.7$ with H_a: $\pi \neq 0.7$ can be tested with either the binomial distribution or the chi-square distribution; if the sample size is large, the conclusion should be the same for the two tests.

5.5. If women are twice as likely as men to suffer spousal abuse, then the odds ratio is 2.0.

5.6. To say that a computed chi-square value is "significant" indicates that it is numerically smaller than the critical value against which it is compared.

5.7. In a multinomial experiment to test H_0: $\pi_1 = 0.25$, $\pi_2 = 0.50$, $\pi_3 = 0.25$, 3 degrees of freedom should be used.

5.8. If the sample size is less than 25, a correction for continuity should be made when testing a 1:2:1 ratio.

5.9. As the degrees of freedom for the chi-square distribution increase, the probability of rejecting a true null hypothesis decreases.

5.10. With random sampling, a computed chi-square value greater than the critical value can be obtained, even when the null hypothesis is true.

5.11. If there is close agreement between the observed and expected frequencies, the chi-square statistic should be relatively large.

5.12. The critical value at $\alpha = 0.05$ for a multinomial chi-square test about a 27:9:9:9:3:3:3:1 genetic ratio is 14.067.

5.13. To test whether a set of samples can be modeled by a Poisson distribution, the experimenter must specify the Poisson parameter before sampling.

5.14. If the null hypothesis for a goodness-of-fit test is not rejected, it can be concluded that the data are from a population with the specified probability distribution.

5.15. A chi-square contingency table analysis is not appropriate if it is suspected that the row and column categories are not independent.

5.16. To reject the null hypothesis in a chi-square test of independence is to decide that the categories in the rows are independent of those in the columns.

5.17. The chi-square test of homogeneity can be used if hypothetical ratios are unknown but may be equal for all populations sampled.

5.18. A chi-square test of independence for a $k \times 2$ table has $k - 1$ degrees of freedom associated with it.

5.19. A chi-square test of homogeneity can be used to test the equality of the parameters in two binomial distributions.

5.20. The expected value and the variance of a given chi-square distribution are equal.

SELECTED READINGS

Chapman, D. G., and R. C. Meng (1966). The power of chi-square tests for contingency tables. *Journal of the American Statistical Association*, **61**, 965–975.

Chase, G. R. (1972). On the chi-square test when the parameters are estimated independently of the sample. *Journal of the American Statistical Association*, **67**, 609–611.

Cochran, W. G. (1952). The chi-square test of goodness of fit. *Annals of Mathematical Statistics*, **23**, 315–345.

Cochran, W. G. (1954). Some methods for strengthening the common chi-square tests. *Biometrics*, **10**, 417–451.

Conover, W. J. (1974). Some reasons for not using the Yates correction on 2×2 contingency tables. *Journal of the American Statistical Association*, **69**, 374–382.

Cox, C. P. (1982). An alternative way of calculating the χ^2 independence or association test statistic for a $2 \times k$ contingency table. *American Statistician*, **36**, 133.

Davis, L. J. (1986). Exact tests for 2×2 contingency tables. *American Statistician*, **40**, 139–141.

Finney, D. J. (1948). The Fisher-Yates test of significance in 2×2 contingency tables. *Biometrika*, **35**, 145–156.

Fisher, R. A. (1935). The logic of inductive inference. *Journal of the Royal Statistical Society*, Section A, **98**, 39–54.

Fisher, R. A. (1973). *Statistical Methods for Research Workers*, 14th ed. Hafner, New York.

Good, I. J. (1973). What are degrees of freedom? *American Statistician*, **27**, 227–228.

Grizzle, J. E. (1967). Continuity correction in the chi-square test for 2×2 tables. *American Statistician*, **21** (Oct.), 28–32.

Guenther, W. C. (1977). Power and sample size for approximate chi-square tests. *American Statistician*, **31**, 83–85.

Hoel, P. G. (1938). On the chi-square distribution for small samples. *Annals of Mathematical Statistics*, **9**, 158–165.

Kinney, J. (1973). Poisson updated (letter to the editor). *American Statistician*, **27**, 195.

Lackritz, J. (1983). Exact P-values for chi-squared tests. *Proceeding of the Section on Statistical Education, American Statistical Association*, 130–132.

Latscha, R. (1955). Tests of significance in a 2×2 contingency table: Extension of Finney's table. *Biometrika*, **40**, 74–86.

Liddell, F. D. K. (1972). Correcting the correction in the chi-square test in 2×2 tables. *Biometrics*, **28**, 268–269.

Plackett, R. L. (1964). The continuity correction in 2×2 tables. *Biometrika*, **51**, 327–337.

Roscoe, J. T., and J. A. Byars (1971). An investigation of the restraints with respect to sample size commonly imposed on the use of the chi-square statistic. *Journal of the American Statistical Association*, **66**, 755–759.

Shapiro, S. H. (1982). Collapsing contingency tables—a geometric approach. *American Statistician*, **36**, 43–46.

Upton, G. J. G. (1982). A comparison of alternative tests for the 2 × 2 comparative trials. *Journal of the Royal Statistical Society*, Series A, **145**, Part 1, 86–105.

Wallis, W. A. (1936). The Poisson distribution and the Supreme Court. *Journal of the American Statistical Association*, **31**, 376–380.

Williams, C. A., Jr. (1950). On the choice of the number and width of classes for the chi-square test of goodness of fit. *Journal of the American Statistical Association*, **45**, 77–86.

Yarnold, J. K. (1970). The minimum expectation in chi-square goodness of fit tests and the accuracy of approximations for the null distribution. *Journal of the American Statistical Association*, **65**, 864–886.

Yates, F. (1934). Contingency tables involving small numbers and chi-square tests. *Journal of the Royal Statistical Society Supplement*, **1**, Series B, 217–235.

6 Sampling Distribution of Averages

In Chapters 3 through 5 we discussed techniques for analyzing certain types of data that are collected on the nominal scale or were reduced to that scale. All of the procedures in those chapters dealt with data that are in the form of counts. This chapter is a transition to data that are collected on a numerical scale. The remainder of this book will deal mainly with data that arise from measurements rather than frequency counts.

6.1. POPULATION MEAN AND SAMPLE AVERAGE

As in the case of count data, researchers use statistical analysis of measurement data to make statements about populations that are not totally accessible from information obtained from properly chosen samples.

One of the parameters of a population that is often of interest is the *population mean*, because it is one way to describe the population's center or location. If the population were totally accessible, its mean would be computed by the formula

$$\mu = \frac{\sum y}{N}$$

in which μ (the lower-case Greek letter mu) is the symbol for the population mean, $\sum y$ is the sum of all of the values of the variable of interest for the whole population, and N is the number of elements in the population. We rarely have an opportunity to use this formula since most of the populations we study are not totally accessible; they either are too large, perhaps even infinite, or would be destroyed in the process of measurement.

Example 6.1.   Computing a Population Mean

Historians often use the frequency of certain grammatical constructions to help identify the writings of a historical person. For example, a historian might determine the number of occurrences of a parallel series of adjectives such as "the worker was *tired* and *weary*" in 3000-word sections of a person's known writings. Imagine that the population of all of the known writings of the person can be arranged into 10 sections of 3000 words each, and the number of occurrences are

19   21   18   24   19   21   22   19   22   22

Statistics for Research, *Third Edition*,   Edited by Shirley Dowdy, Stanley Weardon, and Daniel Chilko.
ISBN 0-471-26735-X   © 2004 John Wiley & Sons, Inc.

To find the population mean, the historian finds the sum of these data and divides by the number of observations:

$$\mu = \frac{\sum y}{N}$$

$$= \frac{19 + 21 + 18 + 24 + 19 + 21 + 22 + 19 + 22 + 22}{10}$$

$$= 20.7$$

That is, the mean number of parallel adjectives per 3000 words used by this author is 20.7.

If the population data are arranged in the form of a frequency distribution in which y is the value of the variable of interest and f is the number of occurrences, then the population mean can be computed by the formula

$$\mu = \frac{\sum yf}{N}$$

in which the summation is over the different values of y. To use this formula, a third column is added to the frequency table and the sum is found:

y	f	yf
18	1	18
19	3	57
21	2	42
22	3	66
24	1	24
	$N = 10$	$207 = \sum yf$

and then

$$\mu = \frac{\sum yf}{N}$$

$$= \frac{207}{10}$$

$$= 20.7$$

If relative frequencies are given in the population table where

$$\text{relative frequency} = \mathbf{f} = f/N$$

then the computation of the population mean is simplified to

$$\mu = \sum y\mathbf{f}$$

Thus

y	**f**	y**f**
18	0.1	1.8
19	0.3	5.7
21	0.2	4.2
22	0.3	6.6
24	0.1	2.4

$$\mu = \sum y\mathbf{f} = 20.7$$

We could represent the population by a graph (Fig. 6.1), and then the mean μ can be interpreted as the balancing point of the distribution (Fig. 6.2).

Since it is often impossible to obtain the population mean, statistical inference is used to estimate μ or to test a hypothesis concerning μ. The basic tool for these inferences (as in the case of count data) is a probability distribution that is a model of the population. We are already familiar with the concept of the expected value $E(y)$ of a probability distribution (see Section 2.5). If a certain probability distribution is the appropriate model for a population, then $E(y)$ will coincide with the population mean μ. Because of this, the expected value of a probability distribution is often called its mean, and we write $\mu = E(y)$. We should recall at this point that the expected value of a discrete probability distribution can be computed by the formula

$$E(y) = \sum yp(y)$$

This is analogous to the formula for a population mean if the values are arranged in a relative frequency distribution:

$$\mu = \sum y\mathbf{f}$$

Statistical inference about a population mean requires, in addition to a probability

FIGURE 6.1. A population distribution.

FIGURE 6.2. The population mean as the balancing point.

distribution to model the population, some information obtained from a sample of the population. A reasonable statistic to use is the *sample average*. The sample average is analogous to a population mean. If $\bar{y}$ is used as the symbol for a sample average,[†] then

$$\bar{y} = \frac{\sum y}{n}$$

in which y is the value of the variable of interest for each of the members in the sample, the sum is over those values, and n is the number of observations in the sample. (The symbol $\bar{y}$ is read "y bar.") As in the case of population means, this formula can be modified for data arranged in a frequency table; then

$$\bar{y} = \frac{\sum yf}{n}$$

If the data are in a relative frequency table, then

$$\bar{y} = \sum y\mathbf{f}$$

Example 6.2.   Computing a Sample Average

A random sample of 100 high-school students is taken prior to their senior year and the number of books they read that summer is recorded:

y	f
0	0.15
1	0.20
2	0.30
3	0.15
4	0.10
5	0.05
6	0.02
7	0.02
8	0.00
9	0.00
10	0.01

[†]To avoid confusion, the expression "average" will be used for a sample and "mean" for a population.

The sample average is computed by adding a third column to the relative frequency table and summing:

y	$\mathbf{f}$	$y\mathbf{f}$
0	0.15	0.00
1	0.20	0.20
2	0.30	0.60
3	0.15	0.45
4	0.10	0.40
5	0.05	0.25
6	0.02	0.12
7	0.02	0.14
8	0.00	0.00
9	0.00	0.00
10	0.01	0.10

$$\bar{y} = \sum y\mathbf{f} = 2.26 \text{ books}$$

A sample average $\bar{y}$ is used as an estimator of the population mean μ. We write $\bar{y} = \hat{\mu}$ (which is read "mu hat") when we want to indicate that the sample average is an estimator of the population mean. The sample average is usually a maximum-likelihood estimator. It is usually also unbiased and has a minimum variance among unbiased estimators (see Section 3.3).

Procedure. Measures of Location

	Ungrouped Data	Grouped Data	
		Frequency Distribution	Relative Frequency Distribution
Population Mean	$\mu = \dfrac{\sum y}{N}$	$\mu = \dfrac{\sum yf}{N}$	$\mu = \sum y\mathbf{f}$
	N = population size	N = population size f = frequency	$\mathbf{f}$ = relative frequency
Sample Average	$\bar{y} = \dfrac{\sum y}{n}$	$\bar{y} = \dfrac{\sum yf}{n}$	$\bar{y} = \sum y\mathbf{f}$
		n = sample size f = frequency	$\mathbf{f}$ = relative frequency
Expected Value of a Discrete Probability Distribution		$E(y) = \sum yp(y)$	

EXERCISES

6.1.1. Find the population mean for the heights of the 50 male students given in Exercise 2.2.4.

6.1.2. Use the data in Exercise 2.2.4 for the following:

    a. Arrange the heights into a population frequency distribution.

    b. Compute the population mean from the population frequency distribution.

    c. Find the population relative frequency distribution.

    d. Compute the population mean from the relative frequency distribution.

6.1.3. The following data from a random sample of 5-year-old children in the United States represent the number of cavities in their teeth:

$$4 \quad 0 \quad 1 \quad 0 \quad 3 \quad 2 \quad 1 \quad 0 \quad 4 \quad 3 \quad 2 \quad 3 \quad 4 \quad 2 \quad 2 \quad 3 \quad 2 \quad 1 \quad 1 \quad 2$$

    a. Find the sample average from this ungrouped data.

    b. Arrange the data into a frequency table.

    c. Find the sample average from the frequency table.

    d. Estimate the mean number of cavities for the population of all 5-year-old children in the United States.

6.1.4. At a certain university a total census is made of all graduating seniors to determine how many courses they have failed during their undergraduate education. The population is as follows:

y:	0	1	2	3	4	5
f:	0.870	0.071	0.031	0.012	0.011	0.005

Find the population mean.

6.2. POPULATION VARIANCE AND SAMPLE VARIANCE

A second population parameter that is often of interest is σ^2, the *population variance*. Variance is a measure of the spread of the population. Suppose we want to choose between two investment plans and are told that both have mean earnings of 10% per annum; we might conclude that they were equally good. However, suppose we learn that plan A has a variance twice as large as plan B. This gives us additional information on which to base a choice. If we want to be relatively certain that our earnings are close to 10%, we would select plan B. If we are willing to gamble that our earnings might be considerably in excess of 10% (or possibly considerably below 10%), we would choose plan A.

    A population variance can be computed from ungrouped data or from data that are grouped into a frequency or relative frequency distribution if the population is of the accessible variety.

    For ungrouped data, a population variance is defined to be

$$\sigma^2 = \frac{\sum (y - \mu)^2}{N}$$

in which σ^2 is read "sigma squared" and represents the population variance. In practice, it is more convenient to use an equivalent computational form of this formula, especially when using a hand-held calculator or electronic spreadsheet—hence called the "machine equation":

$$\sigma^2 = \frac{\sum y^2 - \left(\sum y\right)^2 / N}{N}$$

Example 6.3.   Computing a Population Variance from Ungrouped Data

Consider again the small population of sections of all known writings of a historical person. The number of usages of parallel adjectives per 3000-word sections are

$$19 \quad 21 \quad 18 \quad 24 \quad 19 \quad 21 \quad 22 \quad 19 \quad 22 \quad 22$$

and the mean usage is $\mu = 20.7$. The population variance is the average squared deviation from the mean. In tabular form, the computations are as follows:

y	$y - \mu$	$(y - \mu)^2$
19	$19 - 20.7 = -1.7$	2.89
21	$21 - 20.7 = 0.3$	0.09
18	$18 - 20.7 = -2.7$	7.29
24	$24 - 20.7 = 3.3$	10.89
19	$19 - 20.7 = -1.7$	2.89
21	$21 - 20.7 = 0.3$	0.09
22	$22 - 20.7 = 1.3$	1.69
19	$19 - 20.7 = -1.7$	2.89
22	$22 - 20.7 = 1.3$	1.69
22	$22 - 20.7 = 1.3$	1.69
		$\sum (y - \mu)^2 = 32.10$

and

$$\sigma^2 = \frac{\sum (y - \mu)^2}{N} = \frac{32.10}{10} = 3.210$$

This process can be shortened by using the machine equation, the equivalent computational formula that is more adaptable to a calculating device:

$$\sigma^2 = \frac{\sum y^2 - \left(\sum y\right)^2 / N}{N}$$

$$\sum y = 207 \quad \sum y^2 = 4317 \quad N = 10$$

so

$$\sigma^2 = \frac{4317 - (207)^2 / 10}{10}$$

$$= 3.210$$

Sometimes population data are grouped into frequency or relative frequency tables. In these cases the formulas can be adapted. For a frequency table,

$$\sigma^2 = \frac{\sum (y - \mu)^2 f}{N} = \frac{\sum y^2 f - \left(\sum yf\right)^2 / N}{N}$$

and for relative frequency tables,

$$\sigma^2 = \sum (y - \mu)^2 \mathbf{f} = \sum y^2 \mathbf{f} - \left(\sum y\mathbf{f}\right)^2$$

This last formula is analogous to the computation of the variance of a discrete probability distribution:

$$V(y) = \sum [y - E(y)]^2 p(y)$$
$$= \sum y^2 p(y) - \left[\sum yp(y)\right]^2$$

If a probability distribution is used to represent a population and a certain probability distribution is an appropriate model, then σ^2, the variance of the population, will be the same as $V(y)$, the variance of the probability distribution. Because of this, σ^2 is often used when speaking of the variance of a probability distribution.

Usually we will be estimating the population variance by using a statistic from a random sample of the population. The statistic that is an estimator of the population variance is the *sample variance*, or s^2:

$$s^2 = \frac{\sum (y - \bar{y})^2}{n - 1} = \frac{\sum y^2 - \left(\sum y\right)^2 / n}{n - 1}$$

Note that the denominator of s^2 is $n - 1$, an unusual way to "average" the squared deviations from the sample average. This modification is necessary so that the sample variance will be an unbiased estimator of the population variance. We write

$$s^2 = \hat{\sigma}^2$$

to indicate that the sample variance is an estimator of the population variance.

The formula for sample variance can be modified for data that are grouped into a frequency table:

$$s^2 = \frac{\sum (y - \bar{y})^2 f}{n - 1} = \frac{\sum y^2 f - \left(\sum yf\right)^2 / n}{n - 1}$$

Example 6.4.   Computing a Sample Variance from Grouped Data

In the high-school reading study (Example 6.2) of Section 6.1, the frequency table can be expanded to find $\sum yf$ in the third column and $\sum y^2f$ in the fourth column:

y	f	yf	y^2f
0	15	0	0
1	20	20	20
2	30	60	120
3	15	45	135
4	10	40	160
5	5	25	125
6	2	12	72
7	2	14	98
8	0	0	0
9	0	0	0
10	1	10	100
	$n = 100$	$\sum yf = 226$	$\sum y^2f = 830$

Thus

$$s^2 = \frac{\sum y^2f - \left(\sum yf\right)^2 / n}{n-1}$$

$$= \frac{830 - (226)^2 / 100}{99}$$

$$= 3.22$$

A summary of the computational procedures for variances follows.

Procedure.   Measures of Spread

	Grouped Data	
Ungrouped Data	Frequency Distribution	Relative Frequency Distribution
Population Variance		
$\sigma^2 = \dfrac{\sum (y-\mu)^2}{N}$	$\sigma^2 = \dfrac{\sum (y-\mu)^2 f}{N}$	$\sigma^2 = \sum (y-\mu)^2 \mathbf{f}$
$= \dfrac{\sum y^2 \left(\sum y\right)^2 / N}{N}$	$= \dfrac{\sum y^2f - \left(\sum yf\right)^2 / N}{N}$	$= \sum y^2 \mathbf{f} - \left(\sum y\mathbf{f}\right)^2$
	$N = \sum f$	
N = population size	f = frequency	$\mathbf{f}$ = relative frequency

Ungrouped Data	Grouped Data	
	Frequency Distribution	Relative Frequency Distribution

Sample Variance

$$s^2 = \frac{\sum(y-\mu)^2}{n-1}$$

$$s^2 = \frac{\sum(y-\mu)^2 f}{n-1}$$

Convert relative frequencies to frequencies and method to the left user

$$= \frac{\sum y^2 - \left(\sum y\right)^2 / n}{n-1}$$

$$= \frac{\sum y^2 f - \left(\sum yf\right)^2 / n}{n-1}$$

$$n = \sum f$$

n = sample size

f = frequency

Variance of a Discrete Probability Distribution

$$V(y) = \sum [y - E(y)]^2 p(y)$$

$$= E(y^2) - [E(y)]^2$$

$$= \sum y^2 p(y) - \left[\sum yp(y)\right]^2$$

We might wonder at this point about the meaning of the numerical value of population and sample variances. Larger variances indicate a larger spread for the distribution, but can more than this be said? One approach is to use the result worked out by the Russian mathematician P. L. Chebyshev (1821 to 1894).

Chebyshev used the standard deviation, a measure related to the variance. A *population standard deviation* is the positive square root of the population variance:

$$\sigma = \sqrt{\sigma^2}$$

And a sample standard deviation is the positive square root of the sample variance:

$$s = \sqrt{s^2}$$

The standard deviation has the advantage of being in the same units of measurement as the data, whereas the variance is in squared units that often have no intuitive meaning (as "squared books" in Example 6.4).

Chebyshev proved that in any collection of data *at least* three-fourths of the values lie within two standard deviations of the mean (or average) and at least eight-ninths of the values lie within three standard deviations of the mean (or average). In general, the theorem states

TABLE 6.1. Chebyshev's Theorem for Some Values of $k > 1$

At least this proportion of the data:	Lies within this interval: Population	Sample
$1 - 1/2^2 = 3/4$	$\mu \pm 2\sigma$	$\bar{y} \pm 2s$
$1 - 1/3^2 = 8/9$	$\mu \pm 3\sigma$	$\bar{y} \pm 3s$
$1 - 1/4^2 = 15/16$	$\mu \pm 4\sigma$	$\bar{y} \pm 4s$
$1 - 1/k^2$	$\mu \pm k\sigma$	$\bar{y} \pm ks$

TABLE 6.2. The Empirical Rule

Approximately this proportion of the data:	Lies within this interval: Population	Large Sample
0.682	$\mu \pm 1\sigma$	$\bar{y} \pm 1s$
0.954	$\mu \pm 2\sigma$	$\bar{y} \pm 2s$
0.997	$\mu \pm 3\sigma$	$\bar{y} \pm 3s$

that for real numbers k, $k > 1$, at least $1 - 1/k^2$ of the values lie within k standard deviations of the mean (or average). Table 6.1 summarizes this result.

Note that the theorem is true for *any* population or sample. Although this theory gives only a lower bound for the proportion of the data within certain intervals, it is applicable to all data sets regardless of the shape of their distribution and regardless of their size.

If a population or a large sample is symmetrical and mound shaped, an estimate is possible for the proportion of the data within certain intervals. The estimates in Table 6.2 are often called the *empirical rule*. (These proportions are determined from the standard normal distribution; see Section 7.1.)

EXERCISES

6.2.1. Find the population variance for the heights of the 50 males given in Exercise 2.2.5.

6.2.2. Use the height data and the tables found in Exercise 6.1.2 for the following:

a. Compute the population variance from the population frequency distribution.

b. Compute the population variance from the relative frequency distribution.

6.2.3. Use the sample data from Exercise 6.1.3 for the following:

a. Find the sample variance from the ungrouped data.

b. Find the sample variance from the frequency table.

6.2.4. Use the data from the population in Exercise 6.1.4 and find the population variance.

6.2.5. Consider the following three samples:

I	1	2	2	3	3	3	4	4	5		
II	7	8	8	9	9	9	10	10	11		
III	1	1	1	2	2	3	4	4	5	5	5

    a. Graph the frequency distribution for each of the three samples.

    b. Compute the average of each sample.

    c. Compute the variance of each sample.

    d. Compare the average of samples I and II. What characteristic of the two data sets explains the difference in the averages?

    e. Notice that the variances of sets I and II are equal. What geometric property of these two distributions accounts for this equality?

    f. Note that sets I and III have the same average. Why is this possible for two data sets that seem so different?

    g. Compare the shape of distributions I and III. Why would you expect the variance of I to be smaller than the variance of III?

6.2.6. Each mating season, birds of a certain species usually lay a clutch of 6 eggs in their nests. A biologist notices, however, that clutch number deviates from the usual when the birds feed on a certain kind of berry containing a narcotic alkaloid. He examines the nests of 7 such birds and finds the following numbers of eggs:

$$8 \quad 2 \quad 5 \quad 7 \quad 4 \quad 10 \quad 6$$

    a. Is there evidence that the alkaloid causes the birds to lay fewer eggs than usual?

    b. Compute the variance of the sample.

6.2.7. Show that Chebyshev's theorem is true for the population in Exercise 6.1.4 for $k = 2$ and $k = 3$.

6.3. THE MEAN AND VARIANCE OF THE SAMPLING DISTRIBUTION OF AVERAGES

When dealing with binominal data, the useful statistic for inference is the number of occurrences in a certain category. This count summarizes the entire sample. Similarly, when dealing with numerical data, there is a useful statistic which summarizes all of the measurements from the sample; this statistic is $\bar{y}$, the sample average. In many types of inference, we use the summary statistic $\bar{y}$ rather than the actual values obtained from the individuals in the sample. Since we use the sample average, it is necessary to further develop the properties of this statistic.

The first thing we should note is that $\bar{y}$ is a random variable; that is, it has a numerical value that is associated with the outcome of an experiment or survey. The sample average $\bar{y}$ depends upon the particular random sample chosen and varies for different samples, even those from the same distribution.

Because $\bar{y}$ is a random variable, it has a probability distribution. The probability distribution associated with $\bar{y}$ is called the *sampling distribution of sample averages*. This sampling distribution consists of all possible values of $\bar{y}$ for a fixed sample size and the probabilities associated with these values of the random variable.

If the random variable is discrete and has a finite number of values, we can actually display the sampling distribution of averages. For example, if the population consists of the numbers 1, 2, 3, 4 and all of these values are equally likely, then the population can be represented by the following probability distribution:

y	1	2	3	4
$p(y)$:	1/4	1/4	1/4	1/4

This probability distribution could be the model for several different experiments. For example, imagine a lottery device that contains 4 lightweight balls numbered 1, 2, 3, and 4. Air randomly forces one of the balls to be displayed. This probability distribution would be a model of the infinite population of possible outcomes when the variable is the number of the ball displayed. Another experiment modeled by this distribution consists in selecting a card at random with replacement from a deck containing 10 cards of each of 1, 2, 3, and 4 and observing the number on the card. *Sampling with replacement* means that after the card is selected and the number is observed the card is returned to the deck before the next card is selected. Sampling with replacement effectively creates an infinite population from a finite one.

If samples of size 2 are selected at random from an infinite population represented by this probability distribution (or from a finite population with replacement), then the averages of all possible samples of size 2 are given in the body of the following table:

	$\bar{y}$	Observation 2			
		1	2	3	4
	1	1	3/2	2	5/2
Observation 1	2	3/2	2	5/2	3
	3	2	5/2	3	7/2
	4	5/2	3	7/2	4

If the random variable is continuous or has an infinite number of values, we cannot enumerate all of the averages but we can still think about them. To illustrate the properties of sampling distributions of averages, we will use the above small discrete example; however, the same properties are true for all sampling distributions of averages.

Since the sampling distribution of averages of all samples of a fixed size is a probability distribution, it has an expected value (mean) and a variance, and these parameters are related to the mean and variance of the underlying population.

In the discrete example concerning equally likely numbers, the mean of the population is

$$\mu_y = E(y) = \sum yp(y)$$
$$= (1 + 2 + 3 + 4)\left(\frac{1}{4}\right)$$
$$= \frac{5}{2}$$

and the variance of the population is

$$\sigma_y^2 = V(y) = \sum \left(y - \frac{5}{2}\right)^2 p(y)$$
$$= \frac{5}{4}$$

To find the mean and the variance of the sampling distribution of averages of all samples of size $n = 2$, we first give the probability distribution in tabular form:

$\bar{y}$:	1	3/2	2	5/2	3	7/2	4
$p(\bar{y})$:	1/16	2/16	3/16	4/16	3/16	2/16	1/16

The graph of the sampling distribution of averages appears in Figure 6.3. The mean is

$$\mu_{\bar{y}} = E(\bar{y}) = \sum \bar{y}p(\bar{y})$$
$$= 1\left(\frac{1}{16}\right) + \left(\frac{3}{2}\right)\left(\frac{2}{16}\right) + \cdots + 4\left(\frac{1}{16}\right)$$
$$= \frac{5}{2}$$

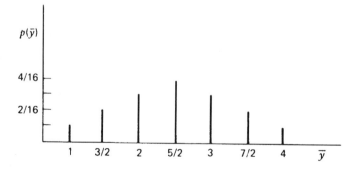

FIGURE 6.3. A sampling distribution of averages.

and the variance is

$$\sigma_{\bar{y}}^2 = V(\bar{y}) = \sum \left(\bar{y} - \frac{5}{2} \right)^2 p(\bar{y})$$

$$= \frac{5}{8}$$

We should note the following about this example of a sampling distribution of averages:

1. The sampling distribution of averages has the same mean as the underlying population.
2. The sampling distribution of averages has a smaller variance than the underlying population.
3. The sampling distribution of averages is symmetric and unimodal.

One particular illustration, of course, does not prove that these properties always hold. However, it can be proved mathematically that for all sampling distributions of averages:

1. $\mu_{\bar{y}} = \mu_y$.
2. $\sigma_{\bar{y}}^2 = \sigma_y^2 / n$.
3. If the sample size n is sufficiently large, then the distribution of $\bar{y}$ is symmetric and unimodal or approximately so.

Another property of sampling distributions of averages is taken up in Chapter 7 after the discussion of normal distributions. In Chapters 7 and 8, the sampling distribution of averages is used for making an inference about the population mean.

In this section, as well as in the rest of this book, unless specified otherwise, we assume that sampling is from an infinite population or from a finite population and the sampling is with replacement. If the sampling is without replacement and from a finite population, we assume that the sample size is 5% or less of the population size. Many of the properties discussed in this text do not hold if sampling is without replacement from a finite population and the sample size is more than 5% of the population size.

EXERCISES

6.3.1. Let y be a discrete random variable with the following distribution:

$$p(y) = \frac{1}{3} \quad \text{for } y = 5, 7, 10$$

$$p(y) = 0 \quad \text{elsewhere}$$

  a. Draw the graph of this probability distribution.
  b. Find $E(y)$ and $V(y)$.
  c. Find the sampling distribution of averages of all samples of size $n = 2$ from a population that is modeled by this distribution. Graph the sampling distribution of averages.

d. Compute $E(\bar{y})$ to show that it is equal to $E(y)$.

e. Compute $V(\bar{y})$ to show that it is equal to $V(y)/n$.

6.3.2. Let x and y be two independent random variables each with the distribution described in Exercise 6.3.1. Show that:

a. $E(x + y) = E(x) + E(y)$

b. $E(x - y) = E(x) - E(y)$

c. $E(3y) = 3E(y)$

d. $V(x + y) = V(x) + V(y)$

e. $V(x - y) = V(x) + V(y)$

f. $V(3y) = 9V(y)$

6.3.3. The properties of expected value and variance illustrated in Exercise 6.3.2 are true in general:

$$E(x + y) = E(x) + E(y)$$

$$E(x - y) = E(x) - E(y)$$

$$E(ay) = aE(y), \text{ for a constant } a$$

$$V(x + y) = V(x) + V(y), \text{ if } x \text{and } y \text{ are independent}$$

$$V(x - y) = V(x) + V(y), \text{ if } x \text{ and } y \text{ are independent}$$

$$V(ay) = a^2 V(y), \text{ for a constant } a$$

Use these properties to show that in general, if $\bar{y} = \sum y/n$ in which the y's are independent, then:

a. $E(\bar{y}) = E(y)$

b. $V(\bar{y}) = V(\bar{y})/n$

6.3.4. For the population of heights given in Exercise 2.2.4:

a. What is $E(\bar{y})$ for all random samples of size 10? (See Exercise 6.1.1).

b. What is $V(\bar{y})$ for all random samples of size 10? (See Exercise 6.2.1).

6.3.5. Six female college students have heights (in inches) as follows: 62, 64, 65, 66, 65, 68. If these 6 students are considered to be a population from which sampling is done with replacement:

a. Draw the frequency distribution of the population.

b. Find the sampling distribution of averages for all samples of size 2 (with replacement) taken from this population. Draw its graph.

c. Find the population mean.

d. Find the mean of the sampling distribution of averages and confirm that it is the same as the population mean.

e. Find the variance of the population.

f. Find the variance of the sampling distribution of averages for samples of size $n = 2$ from the population variance.

6.4. SAMPLING WITHOUT REPLACEMENT

The previous section provided a discussion of sampling distributions for infinite populations or for finite populations when the sampling is with replacement. In sampling with replacement, $p(y)$ for a particular value of y remains constant even though that value may already have been selected. There is another situation called *sampling without replacement* which is frequently encountered in the social sciences.

Consider again a variable y with values 1, 2, 3, 4 in equal frequency. We saw that, when selection is with replacement and the sample is of size $n = 2$, $E(\bar{y}) = 5/2$ and $V(\bar{y}) = 5/8$. This time, however, consider these 4 integers as a finite population, so that once any one of them has been selected for the first member of a sample of size $n = 2$, it is no longer available to be the second number in that sample. We could think of a set of 4 cards each containing one of the numbers 1, 2, 3, or 4. Two cards are to be selected at random, and after the first one is chosen, it is not returned to the set. Hence we call this sampling without replacement. The possible sample means are then

		Observation 2			
$\bar{y}$		1	2	3	4
	1		3/2	2	5/2
Observation 1	2	3/2		5/2	3
	3	2	5/2		7/2
	4	5/2	3	7/2	

We can readily verify that

$$\mu_{\bar{y}} = E(\bar{y}) = \sum \bar{y}p(\bar{y})$$

$$= \left(\frac{3}{2}\right)\left(\frac{2}{12}\right) + 2\left(\frac{2}{12}\right) + \cdots + \left(\frac{7}{2}\right)\left(\frac{2}{12}\right)$$

$$= \frac{5}{2}$$

and the variance is

$$\sigma_{\bar{y}}^2 = V(\bar{y}) = \sum \left(\bar{y} - \frac{5}{2}\right)^2 p(\bar{y}) = \frac{5}{12}$$

We notice that $E(\bar{y})$ remains the same whether or not we sample with replacement, but $V(\bar{y})$ is smaller when we sample from a finite population without replacement. There is a constant relationship between the variances for the two types of sampling; if the variance among sample averages is $\sigma_{\bar{y}}^2$ for sampling with replacement, then the variance for sampling without replacement is

$$\frac{(N-n)}{(N-1)}\sigma_{\bar{y}}^2$$

where N is the size of the population and n is the size of the sample. We can verify the relationship for our demonstration population and compute the variance of the sample means for sampling without replacement as

$$\frac{(N - n)}{(N - 1)} \sigma_{\bar{y}}^2 = \frac{(4 - 2)}{(4 - 1)} \left(\frac{5}{8}\right) = \frac{5}{12}$$

The multiplier $(N - n)/(N - 1)$ is called the *finite population correction factor* and is often written as $(1 - n/N)$ because when N is large $N - 1$ is almost equal to N. Notice that this correction factor is close to 1 if n is small relative to N. If n/N is less than $1/20$, then the correction faction is greater than 0.95, that is, it is almost 1; effectively this means that the finite population correction factor can be dropped from the formula if n/N is less than $1/20$.

EXERCISES

6.4.1. A finite population is of size $N = 8$, with $\mu = 8$ and $\sigma^2 = 5.25$.

    a. What is $V(\bar{y})$ if sampling is with replacement and $n = 1, 3, 5, 8$, respectively?

    b. Use the formula with the finite population correction faction to find $V(\bar{y})$ if sampling is without replacement and $n = 1, 3, 5, 8$.

6.4.2. Chimpanzees have no known numbering system, but they may have a sense of quantity. To test this, a behavioral biologist presents a hungry chimp with 7 bunches of bananas containing, respectively, $y = 1, 2, 3, 4, 5, 6, 7$ bananas. The chimp has been trained to understand that it may choose any 2 bunches of bananas.

    a. How many combinations of 2 bunches are there?

    b. Would this situation constitute sampling with or without replacement?

    c. If it chooses at random, that is, it has no sense of quantity, what is the expected average number of bananas per bunch for the chimp's choice of two bunches? What is $V(\bar{y})$? What outcomes lie within two standard deviations of $E(\bar{y})$?

    d. Suppose the chimp chooses the bunches with six and seven bananas. How many ways can this particular choice be made? What is the probability that this is just a random choice, meaning the chimp has no sense of quantity? Is there evidence that the animal has a sense of quantity?

REVIEW EXERCISES

Decide whether each of the following statements is true or false. If a statement is false, explain why.

6.1. It is appropriate to compute the average of a set of data collected on a nominal scale.

6.2. The sample average is always one of the values in the sample.

6.3. For any sample, $\sum (y - \bar{y}) = 0$.

6.4. If y is measured in inches, the unit of measurement for the standard deviation is squared inches.

6.5. If for each value y in a sample $x = y + 10$, then $\bar{x} + 10 = \bar{y}$.

6.6. If for each value y in a sample $x = y + 10$, then the variance of y is equal to the variance of x.

6.7. If for each value y in a sample $x = ay$, then $\bar{x} = a\bar{y}$ and the variance of x is a^2 times the variance of y.

6.8. If y_1 and y_2 are random variables with the same probability distribution, then $E(y_1 - y_2) = 0$ and $V(y_1 - y_2) = 0$.

6.9. If two populations have the same mean, then they also have the same variance.

6.10. For many random samples the sample average $\bar{y}$ is not equal to the mean μ of the population from which the sample was chosen.

6.11. Because $\bar{y}$ is an unbiased estimator of μ, $\bar{y} = \mu$.

6.12. A sample average is computed in the same manner as a population mean.

6.13. A sample variance is computed in the same manner as a population variance.

6.14. If a population has a mean of 10 and a standard deviation of 2, then the sampling distribution of averages of samples of size $n = 2$ has a mean of 10 and a standard deviation of 1.

6.15. The variance of a sampling distribution of averages is larger than the variance of the underlying population because $\bar{y}$ has more distinct values than y.

6.16. Chebyshev's theorem shows that in all samples most of the data lie within three standard deviations of the average.

6.17. One of the advantages of using a sample average instead of a single observation to estimate the population mean is that the sample average is more likely to be close to the population mean.

6.18. The empirical rule cannot be applied to skewed distributions.

6.19. If the sampling is with replacement, the expected value of the sampling distribution of averages is different from the expected value when the sampling is without replacement.

6.20. A public opinion poll in which no person can be interviewed more than once is an example of sampling without replacement.

7 Normal Distributions

In Chapters 3 and 4 we discussed two types of discrete distributions, binomial and Poisson, that may be appropriate models for some discrete variables encountered in research. In Chapter 5 we discussed a continuous probability distribution, the chi-square distribution, which is not usually a direct model for a population but which can be used in an indirect way to answer questions about populations. In this chapter we discuss a second type of continuous probability distribution, the family of normal distributions. A normal distribution is sometimes the appropriate model for a population with a variable of interest that is continuous.

7.1. THE STANDARD NORMAL DISTRIBUTION

Some continuous variables can be modeled by a bell-shaped theoretical probability distribution called a *normal distribution*, also called a *Gaussian distribution* after Carl Friedrich Gauss (1777 to 1855), who investigated its mathematical properties.

For example, the sample of heights of 100 women measured to the nearest inch, as given in Table 7.1, can be grouped into a relative frequency distribution:

y	f	y	f
60	0.01	67	0.14
61	0.04	68	0.08
62	0.03	69	0.01
63	0.07	70	0.01
64	0.26	71	0.01
65	0.19	72	0.01
66	0.14		

We should like to find a continuous probability distribution that can be used to model the population from which this sample was taken. Looking at the graph of the sample (Figure 7.1), we see that it is not perfectly bell shaped, but the departures are not extreme. A sample of size 100 will resemble the population from which it was taken, but it will not be exactly like the population. It seems possible that the population of heights could be modeled by a theoretical normal distribution (Figure 7.2), with the following density function:

$$f(y) = \frac{1}{\sigma\sqrt{2\pi}} e^{-(y-\mu)^2/2\sigma^2}$$

Statistics for Research, *Third Edition*, Edited by Shirley Dowdy, Stanley Weardon, and Daniel Chilko.
ISBN 0-471-26735-X © 2004 John Wiley & Sons, Inc.

TABLE 7.1. Heights in a Sample of 100 Women

66	65	68	67	68	67	67	64	64	68
65	60	64	64	64	64	63	67	64	65
70	64	64	68	65	64	65	62	65	66
64	65	66	72	66	66	67	64	65	67
65	66	67	66	71	67	67	64	63	65
66	62	68	61	69	63	66	61	65	64
64	65	67	65	64	68	67	64	66	67
68	63	63	67	68	65	64	65	66	62
65	65	63	64	66	61	64	67	64	64
63	66	61	64	65	66	64	64	64	65

The density function $f(y)$ gives the height of the curve above the y axis. In this density function, y is the random variable; y has all real numbers for its values. There are three constants in the density function: 2, π, and e. The constant π is the irrational number equal to approximately 3.14 (this use of π is not related to the binomial parameter), and the irrational e, approximately equal to 2.72, is the base of natural logarithms. There are two independent parameters in the density, μ and σ^2; μ can be any real number and σ^2 can be any nonnegative real number. In any particular normal density function, μ and σ^2 are fixed; thus there is a different normal distribution for each pair μ, σ^2.

The normal density function describes a curve that is

1. unimodal,
2. symmetrical,
3. asymptotic to the y axis, and
4. bell shaped.

The normal distribution has

1. $E(y) = \mu$,
2. $V(y) = \sigma^2$,
3. inflection points at $\mu - \sigma$ and $\mu + \sigma$,
4. total area between the curve and the y axis equal to 1, and
5. more than 99% of the area between $\mu - 3\sigma$ and $\mu + 3\sigma$.

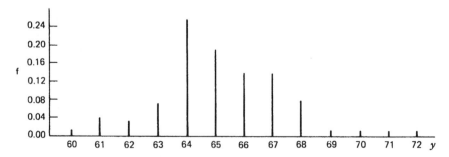

FIGURE 7.1. Heights in a sample of 100 women.

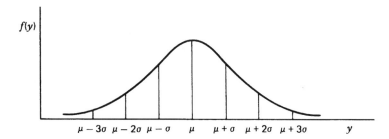

FIGURE 7.2. The normal distribution $N(\mu, \sigma^2)$.

In the sample of women's heights given above, the sample average $\bar{y}$ is 65.2 and the sample variance s^2 is 4.392. Thus, this sample might be from a population that can be modeled by a normal distribution with $E(y) = \mu = 65.2$ and $V(y) = \sigma^2 = 4.392$. We write $N(65.2, 4.392)$ to represent this theoretical distribution. (In Exercise 7.1.7 a goodness-of-fit test is described which can be used to check whether or not this is a good model; it is.)

Probabilities related to continuous random variables are represented by areas. Calculus (in particular, numerical integration) is necessary to find the areas of various sections under the normal curve. Tables, however, have been derived for the normal distribution $N(0, 1)$, called the *standard normal distribution*. These tables can also be used to find the areas of sections under any normal curve by means of a standardization process.

The standard normal random variable is usually represented by z to distinguish it from other random variables. Table A.10 in the Appendix of Useful Tables gives the probabilities that the random variable z is greater than a designated value between 0 and 3.09. For example, if $P(z > 1.36)$ is desired, the table is entered at row 1.30 and column 0.06, and the entry in the body of the table indicates that 0.087 of the area under the curve is to the right of $z = 1.36$ (Figure 7.3). To make this more practical, imagine that we have a freezer with temperatures that follow a standard normal distribution when measured on the Fahrenheit scale (the mean temperature is 0°F and the standard deviation is 1°F); then 8.7% of the time the temperature is above 1.36°F. Or we could say that the probability is 0.087 that the temperature is above 1.36°F. Areas relative to negative z values can be found by using the symmetry of the normal distribution. For example, $P(z < -1.36) = P(z > 1.36) = 0.087$.

If y is normally distributed with a mean of μ and variance σ^2, then y can be *standardized* by the formula

$$z = \frac{y - \mu}{\sigma}$$

FIGURE 7.3. $P(z > 1.36) = 0.087$.

FIGURE 7.4. Standardization preserves area.

Since z is the number of standard deviations y is from μ, z is sometimes called the *standard normal deviate*. If we want to find the probability that y is between 3 and 6 in $N(2, 4)$, we compute

$$z = \frac{3-2}{2} = 0.5 \quad \text{and} \quad z = \frac{6-2}{2} = 2$$

Then

$$P(3 \leq y \leq 6) = P(0.5 \leq z \leq 2)$$
$$= 0.309 - 0.023$$
$$= 0.286 \text{ (Figure 7.4)}$$

Another example follows.

Example 7.1.   Using the Standard Normal Distribution to Find Probabilities

Assume that an ecologist is studying the lungs of wild rabbits for possible contamination from a local power station. He has to build a trap to catch the rabbits, and he wants to make the door wide enough to catch a good percentage of them. Assume he knows that the mean width of rabbits' shoulders is $\mu = 3.80$ in. with a variance of $\sigma^2 = 0.36$ in.^{2} If he makes the door 5 in. wide, what percentage of rabbits will be able to go through the door? That is, what is $P(y < 5)$?
    He finds that the standard normal deviate is

$$z = \frac{y - \mu}{\sigma} = \frac{5.0 - 3.8}{0.6} = 2.00$$

So the door is 2.00 standard deviations wider than the mean width of rabbits' shoulders. Using Table A.10, he finds that $P(z < 2.00) = 1 - 0.023 = 0.977$. This means that the area under the standard normal curve to the left of 2.00 is 0.977. It also means that, in the normal distribution $N(3.80, 0.36)$, 0.977 of the area under the curve is to the left of 5; so 97.7% of the wild rabbits will fit through the door.

EXERCISES

7.1.1. Use Table A.10 to find:
    a. $P(-1 \leq z \leq 2)$
    b. $P(-3.02 < z < 0)$

c. $P(-0.5 < z < 0.5)$

d. $P(z > 2.34)$

e. $P(z > 0)$

f. $P(z \geq -1.58)$

g. $P(0.56 < z \leq 0.98)$

h. $P(-2.44 < z < -0.12)$

i. $P(|z| > 1)$

j. $P(|z| > 2)$

k. $P(|z| > 3)$

7.1.2. Use Table A.10 to find:

a. $P(y < 4)$ if y is distributed as $N(5, 0.64)$

b. $P(10 < y < 13)$ if y is distributed as $N(12, 4)$

c. $P(y > 13)$ if y is distributed as $N(15, 9)$

d. $P(y < 0 \text{ or } y > 3)$ if y is distributed as $N(1, 9)$

7.1.3. In $N(100, 400)$, find:

a. The proportion of the values greater than 70

b. The values of y within the central 90% of the distribution

c. The smallest value of y that exceeds 85% of the distribution

d. The largest value of y that is below 60% of the distribution

7.1.4. Assume that Graduate Record Examination (GRE) scores follow a normal distribution with a mean of 1000 and a standard deviation of 200.

a. What percentage of graduates who take this exam have GRE scores greater than 750?

b. What GRE score separates the upper 30% of graduates from the other 70%?

c. Between what values are the scores of the central 90% of the graduates?

d. How likely is it that a randomly selected graduate will be one who has a GRE score greater than 1000?

e. How likely is it that a random sample of 10 graduates will contain more than 7 who have GRE scores greater than 1000?

f. Suppose that a group of 10 graduates contains 8 who have GRE scores greater than 1000.

   i. Does this appear to be a random sample?

   ii. Why?

7.1.5. The greater the sulfur content of coal, the less desirable it is as a heating fuel. Given that the variability among assays for sulfur in coal from a certain mine is $\sigma = 6$ lb/ton and that they follow a normal distribution, answer the following:

a. Mines that assay 80 lb of sulfur per ton are considered worthless for heating fuel. How likely is it that a mine with mean sulfur content of $\mu = 62$ lb/ton will be placed in the worthless category on the basis of one random 1-ton sample?

b. Some cities will not permit the sale of coal within the city limits if its assay for sulfur is as great as 34 lb/ton. How likely is it that coal with $\mu = 40$ lb/ton will be allowed to be sold within the city limits on the basis of one random 1-ton sample?

7.1.6. A researcher in industrial relations notices that many men who receive high salaries are tall of stature. She decides to investigate the question whether height is related to salary. She wants to classify a man as tall if he is in the upper 10% of the heights of adult males. If adult male heights are normally distributed with a mean of 68 in. and a variance of 1.44 in.^{2}, what is the shortest height (to the nearest inch) that this researcher will classify as tall?

7.1.7. In the sample of women's heights given in this section, the sample average is $\bar{y} = 65.2$ in. and the sample variance is $s^2 = 4.392$, or $s = 2.1$ in. Use these sample values as estimates of μ and σ^2 in the normal distribution and perform a chi-square goodness-of-fit test. Since two parameters are estimated, the degrees of freedom will be $k - 1 - 2$. Use the categories 59.5 to 60.5, 60.5 to 61.5, and so on. Expected values can be computed by finding the probability that a height is in such a section and multiplying by the sample size. If necessary, combine categories to prevent the expected values from becoming too small.

7.1.8. In Francis Galton's time some political candidates included in their campaign material the "total marks" (score) they had received in a grueling (44 hours over 8 days) but prestigious mathematics examination. Galton felt many politicians claimed higher scores than they received. He obtained marks actually given on two successive examinations and found them to compare favorably to a $N(\mu, \sigma^2)$ distribution. His data consisted of the scores received by 800 men, and only 6.7% of them were greater than 1500 marks, which was minimally sufficient to be awarded the title of "wrangler of mathematics."

    a. If the data are from a normal distribution with $\mu = 900$, show how to find $\sigma^2 = 1600$.

    b. The one of the approximately 400 students who receives the greatest number of marks is called "senior wrangler." If scores are normally distributed, what score is likely to qualify for that distinction. *Hint:* What z value will have $1/400$ of the area under the standard normal curve to the right of it?

    c. To address the concern Galton was investigating, suppose 140 candidates have reported scores they claim they received on the examination.

        i. What assumptions must be made in order to use the normal distribution for inference?

        ii. If the assumptions can be made, what is the expected number with scores greater than 1500 marks

        iii. Suppose 24 of the 140 claim they received scores greater than 1500 marks, what would you conclude about the truthfulness of the scores claimed?

7.2. INFERENCE FROM A SINGLE OBSERVATION

Whenever possible, we use samples consisting of several observations in order to make inference about a population. However, there are times when it is necessary to make a judgment about an unknown parameter from a single observation.

One example in which multiple observations are not feasible is a test of a certain type of concrete slab to determine its load-carrying capacity. Since it is expensive and time consuming to construct the slab and since it will be destroyed by the test, it is desirable to draw whatever inferences are possible from a single trial.

Imagine that a civil engineer measured the number of pounds per square inch (psi) required to crack a certain type of slab and found it to be 2500 psi. Is it possible that these slabs crack at values that are from a normal distribution with $\mu = 2300$ and $\sigma^2 = 6400$? To answer this question, he could standardize 2500 as discussed in Section 7.1. Then

$$z = \frac{y - \mu}{\sigma} = \frac{2500 - 2300}{80} = 2.5$$

The standardized value could then be compared with the 95% most common z values which would occur if the distribution is $N(2300, 6400)$. In the standard normal distribution 95% of the area is between -1.96 and 1.96. We write $z_{0.025} = 1.96$ to indicate that 2.5% of the area is to the right of 1.96. Thus $-1.96 = z_{0.975} = -z_{0.025}$ (Figure 7.5).

The value of 2500 corresponds to a z value of 2.5; that is, it is 2.5 standard deviations above the mean. Since this is to the right of 1.96, it would be a very unusual result from a distribution which is $N(2300, 6400)$ and the engineer would conclude that the mean is not 2300 psi. It appears that this concrete slab has a higher load-carrying capacity.

If the population mean is unknown, it is possible to carry out a test of hypothesis from a single observation (we stress, however, that, whenever possible, a larger number of observations should be used).

Example 7.2.   Testing a Hypothesis about a Mean with a Sample of One Observation

Suppose a person showed many of the symptoms of hypothyroidism (an underactive thyroid gland). At one time her physician would have sent her to the hospital for a basal metabolism test. The test was fairly involved and somewhat lengthy and required that the patient be in a fasting condition. Thus the decision whether or not to administer thyroid extract depended on a single observation of the patient's basal metabolism rate.

The mean basal metabolism rate for people with properly functioning glands is 40 calories per square meter per hour; a person suffering from hypothyroidism will have a reduced basal metabolism rate. Thus the null and alternative hypotheses are

$$H_0: \mu = 40 \quad \text{and} \quad H_a: \mu < 40$$

The variability in basal metabolism rate among people with properly functioning thyroids is also known, and for this example it is assumed that the population of such rates is distributed as $N(40, 16)$. If the physician did not want more than 0.05 probability of a

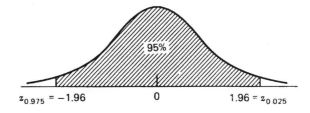

FIGURE 7.5. The standard normal distribution.

misdiagnosis of a person with a properly functioning thyroid ($\alpha = 0.05$), he would compute the test statistic

$$z = \frac{y - \mu_0}{\sigma} = \frac{y - 40}{4}$$

in which μ_0 is the value of μ in the null hypothesis and σ is the known standard deviation. Evidence that the null hypothesis is false would be a large negative value of z since low basal metabolism rates are transformed to the left tail of the standard normal distribution (Figure 7.6). This z statistic is compared with the critical value of $z_{0.95} = -1.64$; if $z \leq -1.64$, H_0 is rejected.

If the physician did not understand how to carry out this test of hypothesis, he might ask a biostatistician to find the basal metabolism rate y that divides the area under the $N(40, 16)$ curve into the lower 5% of the area and the upper 95% of the area. This is done by placing the critical value of z in the equation and solving for y. Thus

$$-1.64 = \frac{y - 40}{4}$$
$$y = 40 - 1.64(4) = 33.44$$

The physician would then make $y = 33.44$ his decision point. If the patient's basal metabolism rate was less than or equal to 33.44 calories, the diagnosis would be hypothyroidism and thyroid extract would be prescribed. In statistical terms, the null hypothesis of normal thyroid function would be rejected. If the patient's basal metabolism rate was greater than this value, the hypothesis would not be rejected, and the physician would investigate something other than the thyroid as the cause of the symptoms.

Procedure. Inference About a Single Observation from a Normal Distribution

Test of Hypothesis
$H_0: \mu = \mu_0$
$H_a: \mu \neq \mu_0$ or $\mu > \mu_0$ or $\mu < \mu_0$
Significance level: α
Test statistic:

$$z = \frac{y - \mu_0}{\sigma}$$

Region of rejection: $|z| \geq z_{\alpha/2}$ or $z \geq z_\alpha$ or $z \leq -z_\alpha$, respectively.

FIGURE 7.6. Low values in $N(40, 16)$ which occur only 5% of the time.

EXERCISES

7.2.1. Use Table A.10 in the Appendix to find:

    a. $z_{0.05}$

    b. $z_{0.95}$

    c. $z_{0.01}$

    d. $z_{0.99}$

    e. $z_{0.005}$

    f. $z_{0.995}$

7.2.2. Assume that the temperatures of healthy infants follow an $N(99, 1)$ distribution when measured on a Fahrenheit scale.

    a. If a particular infant has a temperature of $100.5°F$, should his temperature be considered "normal"? That is, test the hypothesis $H_0: \mu = 99$ against $H_a: \mu \neq 99$ at $\alpha = 0.05$.

    b. Give the P value.

7.2.3. Legend has it that Archimedes made his discovery concerning specific gravity (Archimedes' principle) while trying to determine whether the king's crown was made of pure gold or an alloy. Working with metal samples which he knew to be pure gold or alloys, he found that his device for measuring specific gravity produced a mean determination of $\mu = 19.3$ for pure gold, whereas all alloys tested yielded lower mean specific gravities. For the sake of this problem, suppose Archimedes' measuring device followed an $N(\mu, 0.09)$ distribution.

    a. What would be a suitable null hypothesis for such an experiment?

    b. What would be the most logical alternative hypothesis?

    c. If $\alpha = 0.05$, what should be the region of rejection for this experiment?

    d. How likely is it that a random sample of an alloy with a specific gravity determination of 18.7 would be mistakenly called pure gold in this experiment?

7.2.4. A dairy farmer buys a heifer (female calf) from a Holstein-Friesian herd that is thought to be genetically superior to others in the region. The quantity of milk production among mature cows in the herd is normally distributed with $\mu = 18,000$ lb/year and $\sigma = 2500$ lb/year. Assuming the new owner can provide feed, shelter, and other environmental factors equivalent to those for the herd from which the calf was bought:

    a. Give the numerical value of $E(y)$, the expected milk production of the calf when it reaches maturity.

    b. What is the probability that the calf will produce at a greater rate than the mean of the herd from which it was bought?

    c. What is the probability that it will produce at a rate greater than the breed mean of $\mu = 14,000$?

7.3. THE CENTRAL LIMIT THEOREM

Although normal distributions occur frequently in experiments, many random variables are not normally distributed, and it would be inappropriate to use a normal distribution as the model. In spite of this, if the samples are large enough, a normal distribution can often still be used to find certain probabilities associated with the experiment because of some results that are known from the mathematical theory of statistics. The theory relevant to this use concerns the properties of the sampling distribution of averages.

In Section 6.3 we noted that the sampling distribution of averages has the following properties:

1. $\mu_{\bar{y}} = \mu_y$; that is, the mean of the sampling distribution of averages is the same as the mean of the underlying population.
2. $\sigma_{\bar{y}}^2 = \sigma_y^2/n$; that is, the variance of the sampling distribution of averages is equal to the variance of the underlying population divided by the sample size.
3. If n is sufficiently large, then the sampling distribution of averages is symmetrical and unimodal or approximately so.

The third property can now be made more explicit. If a population is normal, the sampling distribution of averages is normal. If a population is not normal, the sampling distribution of averages is approximately normal for large n.

This last property is known as the *central limit theorem*. It is because of this property that normal distributions come into play in many statistical analyses. With very few exceptions,[†] no matter what form the underlying population distribution takes, as n increases, the sampling distribution of averages approaches a normal distribution; thus the normal distribution can be used to approximate probabilities in cases of reasonably large samples ($n \geq 30$) from nonnormal distributions.

Usually in statistics we observe a sample and use the data collected to make decisions about the population. If we compute the sample average, we have one value from the sampling distribution of averages. Using the three properties just discussed, we can answer probability questions about sample averages. If the underlying population is normally distributed, the sampling distribution of averages is also normally distributed and has the same expected value as the population distribution and a variance that is $1/n$ of the population variance. If the underlying distribution is not normal, the sampling distribution of averages for large n is approximately normal and has the same expected value as the population distribution and a variance of $1/n$ times the population variance.

Example 7.3.   Probabilities Associated with a Sample Average

An educational psychologist is working with a random sample of 5 adults. They are going to take a standardized intelligence (IQ) test with scores that are normally distributed with a mean of 105 and a standard deviation of 15. The psychologist wants to know how likely it is that the average score of the 5 subjects will be greater than 108, that is, $P(\bar{y} > 108)$.

Since she is working with a sample average, she has a single value from the sampling distribution of averages that is normally distributed with a mean of 105 and a variance of $\sigma_{\bar{y}}^2 = \sigma_y^2/n = 15^2/5 = 45$. Thus

$$P(\bar{y} > 108) = P(z > 0.45) = 0.326$$

because

$$z = \frac{\bar{y} - \mu_{\bar{y}}}{\sigma_{\bar{y}}} = \frac{\bar{y} - \mu_y}{\sigma_y/\sqrt{n}} = \frac{108 - 105}{\sqrt{45}} = 0.45$$

The psychologist concludes that the probability is 0.326 that the average scores of her 5 subjects will be above 108.

[†]It is sufficient that the distribution have a finite variance.

EXERCISES

7.3.1. If the basal metabolism rate for people with properly functioning thyroid glands can be modeled by a normal distribution with mean 40 calories per square meter per hour and a standard deviation of 4, find:

   a. The probability that a healthy person chosen at random will have a rate less than 35

   b. The probability that 5 healthy persons chosen at random will all have a rate less than 35

   c. The probability that the average rate of 5 healthy persons chosen at random is less than 35

7.3.2. A certain aptitude test for job trainees follows a normal distribution with a mean of 80 and a standard deviation of 16.

   a. What is the probability that a random sample of 4 trainees will all have scores above 88?

   b. What is the probability that the average score for a random sample of 4 trainees will be above 88?

7.4. INFERENCES ABOUT A POPULATION MEAN AND VARIANCE

Although it is sometimes necessary to make decisions on the basis of a single observation (as in Section 7.2), in general this is not the preferred procedure. Larger samples yield more information on which to base decisions. If we are interested in making a decision about μ or an estimate of μ, then using $\bar{y}$ with $n > 1$ instead of a single observation has the advantage that $\bar{y}$ is less variable than y. A smaller variance increases the probability of obtaining a sample value close to the true population mean. Another advantage of using averages of samples is that, even if the original population does not have a normal distribution, the sampling distribution of averages for large n is approximately normal (central limit theorem).

Tests of hypotheses based on averages are analogous to the procedure for an individual observation. For a single observation, the standardization procedure is

$$z = \frac{y - \mu}{\sigma}$$

For averages of samples of size n, the standardization procedure is

$$z = \frac{\bar{y} - \mu}{\sigma/\sqrt{n}}$$

because the mean of the sampling distribution of averages is the same as the original mean and the standard deviation of the sampling distribution is $\sigma/\sqrt{n}$. (This denominator is sometimes called the *standard error*. "Error" in this context implies, not a mistake, but variability due to sampling.)

Example 7.4. Using the Standard Normal Distribution to Test a Hypothesis about μ

An aneurysm is a weakness in an artery that causes it to balloon and possibly burst. If it is in the blood vessel receiving blood as it is pumped out of the heart (called a TAA for thoracic aortic aneurysm), it is almost always life threatening. Corrective surgery is possible, but it too is risky, so rather than chance an unneeded operation, surgeons prefer to wait until there is evidence that the aorta is in danger of bursting. Fortunately the size of the aneurysm provides a good indication of its danger of bursting. So, to gain useful information, thoracic surgeons at a medical center conduct a study on the sizes of aneurysms at first diagnosis. Suppose they obtain the following TAA information on 30 patients randomly sampled from a nationwide database:

cm	mm
7	0 2 5
6	2 5 6 8
5	1 4 5 5 5 6 8 9
4	0 1 2 3 5 6 7 8 9
3	0 6 6 8 9
2	9

The aneurysm sizes are presented in a *stem-and-leaf plot*, a useful graphic summary of the measures which retains all values as well as shows something about how they are distributed. The first column shows the first digit of a measurement and the second column gives the rest of the measurement. So the first row of data represents three patients with aneurysms 7 cm or greater in diameter. The values of these measures are 7.0, 7.2, and 7.5 cm, respectively. The usual terminology for a stem-and-leaf plot is to call the entry in the first column the *stem*, or *node*, and those in the second column the *leaves*.

The plot shows that the distribution of measures is unimodal, with more data located on the 4.0- and 5.0-cm stems than on any others. It is also somewhat symmetric, but it's best to say only that it resembles a normal distribution. Still, by taking advantage of the central limit theorem, the standard normal distribution can be used to make statistical inference about the mean size of TAA at first diagnosis.

Suppose the standard text on thoracic surgery reports median TAA as 4.7 cm and the surgeons want to test whether that is the mean value of the population from which their sample is drawn. So they would like to test H_0: $\mu = 4.7$ against the alternative H_a: $\mu \neq 4.7$. They will compute a z value as their test statistic, and for a test at the 5% level of significance, they will reject H_0 if $|z| \geq z_{0.025} = 1.96$. But before they can compute z they must obtain the sample average

$$\bar{y} = \frac{\sum y}{n} = \frac{153.0}{30} = 5.1$$

and because the population variance is unknown, it is estimated ($\hat{\sigma}^2$) by the sample variance,

$$\hat{\sigma}^2 = s^2 = \frac{\sum y^2 - \left(\sum y\right)^2 / n}{n-1} = \frac{825.78 - (153.0)^2 / 30}{29} = 1.568$$

Once they have these two sample statistics, they can make the test of hypothesis:

$$z = \frac{\bar{y} - \mu_0}{\hat{\sigma}/\sqrt{n}} = \frac{5.1 - 4.7}{1.25/\sqrt{30}} = 1.12$$

Since $1.12 < 1.96$, the sample average does not deviate significantly from the hypothesized mean. The surgeons do not reject the null hypothesis and conclude that the mean TAA at first diagnosis could indeed be 4.7 cm.

Confidence intervals on μ can also be determined from samples with $n > 1$.

Example 7.5. Using the Standard Normal Distribution to Find a Confidence Interval on μ

Assume that a researcher at an agricultural experiment station knows that the variance in butterfat production for Holstein-Friesian dairy cattle is $\sigma^2 = 6400$ (lb/year)^2. He treats a group of dairy cattle by adding inorganic nitrate to their diet because he knows the bacteria in cows' rumens can metabolize inorganic nitrogen and thereby possibly reduce the cost of having to feed cattle more expensive sources of nitrogen. However, not knowing what effect it may have on production, he wants to know the mean butterfat production for this treatment group, that is, the value of μ. He would perform a test of hypothesis to get some information about μ, the mean for the treatment group. If the null and alternative hypotheses are

$$H_0: \mu = \mu_0$$
$$H_a: \mu \neq \mu_0$$

and $\alpha = 0.05$, he would use the formula

$$z = \frac{\bar{y} - \mu_0}{\sigma/\sqrt{n}}$$

He would not reject the null hypothesis if

$$-1.96 \leq \frac{\bar{y} - \mu_0}{\sigma/\sqrt{n}} \leq 1.96^\dagger$$

or, the equivalent, if

$$\bar{y} - 1.96 \frac{\sigma}{\sqrt{n}} \leq \mu_0 \leq \bar{y} + 1.96 \frac{\sigma}{\sqrt{n}}$$

Thus the 95% confidence interval on μ is

$$CI_{0.95}: \bar{y} \pm 1.96 \frac{\sigma}{\sqrt{n}}$$

---

†Strictly speaking, we do not reject the null hypothesis if $-1.96 < z < 1.96$. Since this is a continuous distribution, however, $P(z = 1.96) = 0$ and the two types of inequalities are equivalent.

and if $\bar{y} = 465$ and $n = 25$, then

$$\text{CI}_{0.95}: 465 - 1.96\left(\frac{80}{5}\right) \leq \mu \leq 465 + 1.96\left(\frac{80}{5}\right)$$

$$433.64 \leq \mu \leq 496.36$$

for the treatment group.

If the population variance σ^2 is unknown (as is commonly the case), it can be estimated by the sample variance

$$s^2 = \frac{\sum(y - \bar{y})^2}{n - 1} = \frac{\sum y^2 - \left(\sum y\right)^2/n}{n - 1}$$

If the sample size is large ($n \geq 30$), s^2 can be used in place of σ^2 in inferences concerning μ.

=====

Procedure. Inferences about a Population Mean

Assumptions: 1. $n < 30$, population normal, and σ known, or
2. $n \geq 30$

Confidence Intervals

$$\text{CI}_{1-\alpha}: \bar{y} - z_{\alpha/2}\frac{\sigma}{\sqrt{n}} \leq \mu \leq \bar{y} + z_{\alpha/2}\frac{\sigma}{\sqrt{n}}$$

if σ is known. If σ is unknown and $n \geq 30$, estimate σ by s.

Test of Hypothesis
$H_0: \mu = \mu_0$
$H_a: \mu \neq \mu_0$ or $\mu > \mu_0$ or $\mu < \mu_0$
Significance level: α
Test statistic:

$$z = \frac{\bar{y} - \mu_0}{\sigma/\sqrt{n}}$$

if σ is known. If σ is unknown and $n \geq 30$, estimate σ by s.
Region of rejection: $|z| \geq z_{\alpha/2}$ or $z \geq z_\alpha$ or $z \leq -z_\alpha$, respectively.

=====

Sometimes the parameter of interest is not the population mean, but rather the population variance. Several examples follow. A teacher is interested in the variability of the grades for a class; a large variance may indicate that although the class as a whole is performing well some individuals may not be performing at an acceptable level. During the manufacturing of drugs, the variance of the potency is of concern and also the variance of the purity level. During the machine filling of boxes or bottles with a product, the variance of the quantity put into the container is of concern. Variability of sentence length has been used to establish authorship. These are only some of the areas in which the investigator needs information about the variance.

It is possible to test hypotheses and determine confidence intervals for a population variance if the population is normal. These procedures make use of the fact that

$$\frac{\sum(y - \bar{y})^2}{\sigma^2} = \frac{(n-1)s^2}{\sigma^2}$$

is distributed as a chi-square distribution with $n-1$ degrees of freedom if y is normally distributed.

Example 7.6.   Inference about the Variance of a Normal Population

In a certain city, the mean electric consumption for residence is 7.2 thousand kWh with a variance of 2.25 thousand kWh^2. Differences in home consumption are due to the energy efficiency of the house and the life-style of the occupants.

In a sample of 101 homes from an area in which all of the residences are of equal size and equal energy efficiency, the sample variance is 1.21 thousand kWh^2. Does this indicate that uniform energy-efficient homes significantly lower the variance of electric consumption?

The null and alternative hypotheses are

$$H_0: \sigma^2 = 2.25 \quad H_a: \sigma^2 < 2.25$$

The test statistic is

$$\chi^2 = \frac{(n-1)s^2}{\sigma_0^2}$$

with $n-1 = 100$ degrees of freedom. At $\alpha = 0.05$ the region of rejection is

$$\chi^2 \le \chi^2_{0.95,100} = 77.929$$

The value of the test statistic is

$$\chi^2 = \frac{100(1.21)}{2.25} = 53.778$$

Thus the null hypothesis is rejected and there is evidence that uniform housing significantly reduces the variability of electric consumption. This result suggests that a program to encourage persons to make their homes more energy efficient might be worthwhile.

If desired, a central confidence interval can be determined for σ^2 for the population of uniform residences of the type sampled:

$$\text{CI}_{0.95}: \frac{(n-1)s^2}{\chi^2_{0.025,n-1}} \le \sigma^2 \le \frac{(n-1)s^2}{\chi^2_{0.975,n-1}}$$

$$\frac{100(1.21)}{129.561} \le \sigma^2 \le \frac{100(1.21)}{74.222}$$

$$0.93 \le \sigma^2 \le 1.63$$

The inferences relative to the variance of a normal population can be summarized as follows.

Procedure. Inferences about a Population Variance

Assumption: Normality

Confidence Intervals

$$CI_{1-\alpha}: \frac{(n-1)s^2}{\chi^2_{\alpha/2,n-1}} \leq \sigma^2 \leq \frac{(n-1)s^2}{\chi^2_{1-\alpha/2,n-1}}$$

Test of Hypothesis

$H_0: \sigma^2 = \sigma_0^2$

$H_a: \sigma^2 \neq \sigma_0^2$, or $\sigma^2 > \sigma_0^2$ or $\sigma^2 < \sigma_0^2$

Significance level: α

Test statistic:

$$\chi^2 = \frac{(n-1)s^2}{\sigma_0^2}$$

Region of rejection: $\chi^2 \leq \chi^2_{1-\alpha/2,n-1}$ or $\chi^2 \geq \chi^2_{\alpha/2,n-1}$, or $\chi^2 \geq \chi^2_{\alpha,n-1}$, or $\chi^2 \leq \chi^2_{1-\alpha,n-1}$, respectively.

EXERCISES

7.4.1. On an IQ test which is distributed as $N(100, 225)$, the average IQ score for a certain second grade in a private school in Victoria, Texas, is $\bar{y} = 106$. If $\alpha = 0.05$, how often might a deviation this large or larger occur by chance in a random sample of 25?

7.4.2. A certain intelligence test has an $N(100, 100)$ distribution. To see whether intelligence is inherited, tests are given to the eldest child of each of a random sample of 16 acclaimed scholars. The average score of the children is 105.

   a. Give the null hypothesis to be tested.

   b. Give the alternative hypothesis.

   c. Perform the test.

   d. How likely is it that data like these represent a sample from a population in which the null hypothesis is true?

7.4.3. A synthetic female hormone (DES) has been used to fatten livestock. If this substance appears in the meat, it affects the sexual maturity of young animals eating the meat. Biological assays can be used to test for the presence of DES in meat. Young female rats are fed the suspected meat, and if they mature earlier than expected, it is probably because of DES in the meat. Suppose for a given strain of rat that time until sexual maturity in the females follows an essentially normal distribution with a mean of 90 days and a variance of 144.

   a. What is the probability that a randomly selected female rat will reach sexual maturity before 90 days? Before 86 days?

   b. What is the probability that the average time until sexual maturity of nine female rats will be less than 90 days? Less than 86 days?

   c. A random sample of nine female rats is fed a diet including meat suspected of containing DES.

      i. What are the most logical null and alternative hypotheses?

      ii. If $\alpha = 0.05$, which values of the sample average will lead to the rejection of the null hypothesis?

      iii. Suppose for female rats on a diet containing DES sexual maturity follows an $N(86, 144)$ distribution; what is the probability of making a Type II error?

7.4.4. A coal research scientist has discovered that West Virginia coal contains an ore rich in aluminum. Although it is present in coal only as a trace mineral, it may be economically practical to recover the ore from the ash left when coal is burned in large boilers of power plants. To estimate the quantity of the ore in coal, the scientist takes a random sample consisting of 100 observations and computes the following:

$$\sum y = 8400 \, \text{ppm}$$

$$\left(\sum y\right)^2 = 70{,}560{,}000 \, \text{ppm}^2$$

$$\sum y^2 = 715{,}500 \, \text{ppm}^2$$

    a. What is the best estimate of the mean content of aluminum ore in West Virginia coal?

    b. Show that the sample standard deviation is 10 ppm.

    c. A coal economist calculates that the recovery of the ore will be profitable if it is present to an extent greater than 82.3 ppm in the coal burned in the boilers. On the basis of these data, would you recommend attempting to recover the ore?

7.4.5. The following stem-and-leaf plot gives the weight in kilograms of 30 stalks of an experimental variety of plantain fruit that has been genetically altered to contain a greater level of protein:

kg	kg/10
9	8
8	2 4 6
7	3 5 7 8
6	0 1 3 7 8
5	1 2 3 4 7 8 8
4	1 2 4 6 7
3	0 1 3 5 7

    a. Compute s^2.

    b. Find a 95% confidence interval for σ^2.

    c. Perform a test of hypothesis at the 5% level of significance to determine whether or not this sample came from a population that has a variance of 3.0.

    d. Find a 95% confidence interval for μ using s^2 to approximate σ^2.

7.4.6. Many organic phosphorous compounds are effective insecticides, but they are also chemically stable and likely to get into the human food chain. They have even been detected in the digestive tracts of recently born infants, but it is not known to what extent this is via mother's milk and to what extent these compounds pass through the placental membrane prior to birth. To get answers to these questions, a medical research team draws samples of amniotic fluid from the wombs of 64 pregnant women and performs chemical analyses for a certain organic phosphorous insecticide. The following data are obtained:

$$\sum y = 320.00 \, \text{ppm}$$

$$\sum y^2 = 1761.28 \, \text{ppm}^2$$

a. Estimate the mean ppm of the compound found in amniotic fluid.

b. Show that the sample variance is 2.56 ppm^2.

c. Place a 95% confidence interval on the mean.

d. Place a 95% confidence interval on the variance.

7.4.7. It can be illustrated that $s^2 = \sum (y - \bar{y})^2/(n - 1)$ is an unbiased estimator of σ^2 by the following special case. Let the population be an equally likely distribution of 1, 2, 3, 4. This population was discussed in Section 6.3.

a. List all possible samples (with replacement) of size 2.

b. Compute the sample variance of each sample.

c. Find the relative frequency of each different sample variance found in part b.

d. Find $E(s^2)$ and show that $E(s^2) = \sigma^2$.

7.5. USING A NORMAL DISTRIBUTION TO APPROXIMATE OTHER DISTRIBUTIONS

A normal distribution can sometimes be used to approximate the probabilities associated with response variables that follow a binomial or a Poisson distribution.

In the case of a binomial distribution, the central limit theorem implies that if n is fairly large ($n \geq 25$) and π is fairly close to 0.5 ($0.2 \leq \pi \leq 0.8$), then the binomial random variable y can be transformed into a random variable that is distributed approximately as the standard normal random variable

$$z \cong \frac{y - n\pi}{\sqrt{n\pi(1 - \pi)}}$$

Note that $n\pi = \mu$ is the mean of the binomial distribution and $\sqrt{n\pi(1 - \pi)}$ is the standard deviation.

Example 7.7.   Using a Normal Distribution to Approximate Probabilities for a Binomial Random Variable

A sociologist studying families headed by a single parent would like to know the probability of finding 40 or more such families in a random sample of 100 families if 30% of families are of this type.

Since $E(y) = n\pi = 100(0.30) = 30$ and $V(y) = n\pi(1 - \pi) = 100(0.30)(0.70) = 21$, then

$$P(y \geq 40) \cong P\left(z \geq \frac{40 - 30}{\sqrt{21}}\right)$$

$$= P(z \geq 2.18)$$
$$= 0.015$$

Thus, if the sociologist needs at least 40 cases for a study, a sample of 100 families will probably not be sufficient.

Since the binomial distribution is discrete and the normal distribution is continuous, the approximation will be poor in the case of small sample sizes. To compensate for this, a continuity correction of 0.5 is often made. If we represent the binomial probabilities by bars of unit width so that the area of the bar centered over y is the probability of y and we represent the normal distribution by a smooth curve, we can see (Figure 7.7) that using 40 as the cutoff point in the above example does not take into consideration half of the bar below 40. Thus, instead of finding $P(y \geq 40)$, we should find $P(y \geq 39.5)$. The sociologist above would then find

$$P(y \geq 39.5) \cong P\left(z \geq \frac{39.5 - 30}{\sqrt{21}}\right)$$

$$= P(z \geq 2.07)$$
$$= 0.019$$

The additional accuracy may be important in some experiments.

A test of hypothesis can also be done about the binomial parameter, making use of the fact that $(y - n\pi)/\sqrt{n\pi(1 - \pi)}$ is approximately standard normal. This procedure is especially helpful for large sample sizes since exact binomial tables may not be available.

Example 7.8.   Using a Normal Distribution to Test a Hypothesis About π

Most people have a dominant eye which looks directly ahead while the other eye adjusts to it in order to bring a viewed object into focus. A reading specialist wants to determine whether there is any tendency for one eye to be dominant in children with a certain reading problem. She takes a random sample of 225 children with the reading problem and determines the dominant eye for each of them. Suppose she finds that for 144 of the children the right eye is

This half of the bar is missing →

FIGURE 7.7. Approximating a binomial distribution by a normal distribution.

dominant. The null and alternative hypotheses are

$$H_0: \pi = 0.5 \quad \text{and} \quad H_a: \pi \neq 0.5$$

The test statistic is

$$z \cong \frac{y - n\pi_0}{\sqrt{n\pi_0(1 - \pi_0)}}$$

$$= \frac{144 - 225(0.5)}{\sqrt{225(0.5)(0.5)}}$$

$$= 4.2$$

At $\alpha = 0.05$, she will reject the null hypothesis if $|z| \geq 1.96$. Since $|4.2| > 1.96$, she rejects the null hypothesis and concludes that more than half the children with this reading problem have a dominant right eye.

If the specialist in the above example would like to find a confidence interval for π, she could make use of the fact that

$$z \cong \frac{y - n\pi}{\sqrt{n\pi(1 - \pi)}} = \frac{y/n - \pi}{\sqrt{\dfrac{\pi(1 - \pi)}{n}}}$$

and that y/n is the best point estimate of π. Analogous to confidence intervals on μ, the confidence interval on π would be

$$\mathrm{CI}_{1-\alpha}: y/n \pm z_{\alpha/2}\sqrt{\frac{\pi(1 - \pi)}{n}}$$

However, since π is unknown, it must be estimated in the standard error by y/n, giving

$$\mathrm{CI}_{1-\alpha}: y/n \pm z_{\alpha/2}\sqrt{\frac{(y/n)(1 - y/n)}{n}}$$

In the sample, since $y = 144$, she would find

$$\mathrm{CI}_{0.95}: \frac{144}{225} \pm 1.96\sqrt{\frac{(144/225)(1 - 144/225)}{225}}$$

$$0.640 \pm 1.96(0.0320)$$

$$0.640 \pm 0.0627$$

$$0.577 \leq \pi \leq 0.703$$

If desired, the statistic

$$z \cong \frac{y/n - \pi_0}{\sqrt{\dfrac{\pi_0(1 - \pi_0)}{n}}}$$

can be used for tests of hypothesis. This is equivalent to the method illustrated in the example.

Procedure. Normal Approximation of a Binomial Distribution

Assumptions: $n \geq 25$ and $0.2 \leq \pi \leq 0.8$

Confidence Intervals

$$\text{CI}_{1-\alpha}: y/n - z_{\alpha/2}\sqrt{\frac{(y/n)(1 - y/n)}{n}} \leq \pi \leq y/n + z_{\alpha/2}\sqrt{\frac{(y/n)(1 - y/n)}{n}}$$

Tests of Hypotheses

$H_0: \pi = \pi_0$
$H_a: \pi \neq \pi_0$ or $\pi > \pi_0$ or $\pi < \pi_0$
Significance level: α
Test statistic:

$$z \cong \frac{y - n\pi_0}{\sqrt{n\pi_0(1 - \pi_0)}} = \frac{y/n - \pi_0}{\sqrt{\pi_0(1 - \pi_0)/n}}$$

Region of rejection: $|z| \geq z_{\alpha/2}$ or $z \geq z_\alpha$ or $z \leq -z_\alpha$, respectively.

The normal distribution can also be used to approximate probabilities related to variables that follow a Poisson distribution. This approximation arises from the central limit theorem. If y is a Poisson random variable and λ is large, y can be transformed into a random variable that is distributed approximately as the standard normal random variable

$$z \cong \frac{y - \lambda}{\sqrt{\lambda}}$$

Note that λ is the mean and $\sqrt{\lambda}$ the standard deviation of the Poisson distribution.

Example 7.9.   Using a Normal Distribution to Approximate Probabilities for a Poisson Random Variable

A traffic control specialist wants to know the probability that more than 30 vehicles will pass a given intersection in a 3-minute period at 3:00 PM if the expected number of vehicles to pass that intersection in 3 minutes at that time is 25:

$$P(y > 30) \cong P\left(z > \frac{30.5 - 25}{\sqrt{25}}\right)$$

$$= P(z > 1.1)$$

$$= 0.136$$

This computation is much simpler than working with the exact Poisson distribution. Note that a continuity correction is used because the discrete Poisson distribution is being approximated by the continuous normal distribution.

Tests of hypotheses about λ can also be done with a z statistic using the fact that $(y - \lambda)/\sqrt{\lambda}$ is approximately standard normal for large λ.

Procedure. Normal Approximation of a Poisson Distribution

Test of Hypothesis
H_0: $\lambda = \lambda_0$
H_a: $\lambda \neq \lambda_0$ or $\lambda > \lambda_0$ or $\lambda < \lambda_0$
Significance level: α
Test statistic:

$$z = \frac{y - \lambda_0}{\sqrt{\lambda_0}}$$

Region of rejection: $|z| \geq z_{\alpha/2}$ or $z \geq z_\alpha$ or $z \leq -z_\alpha$, respectively.

When two populations have proportions π_1 and π_2 with corresponding odds ω_1 and ω_2, a useful alternative to comparing the difference in proportions $(\pi_2 - \pi_1)$ is the odds ratio ϕ:

$$\phi = \frac{\omega_2}{\omega_1} = \frac{\pi_2/(1 - \pi_2)}{\pi_1/(1 - \pi_1)}$$

We can estimate the odds from randomly sampled data summarized in a 2×2 contingency tables of the form

	Response variable		
Explanatory variable	Yes	No	Sample sizes
Yes	o_{11}	o_{12}	n_1
No	o_{21}	o_{22}	n_2

The estimated odds ratio is

$$\hat{\phi} = \frac{\hat{\omega}_2}{\hat{\omega}_1} = \frac{\hat{\pi}_2/(1 - \hat{\pi}_2)}{\hat{\pi}_1/(1 - \hat{\pi}_1)} = \frac{o_{11}/o_{12}}{o_{21}/o_{22}} = \frac{(o_{11})(o_{22})}{(o_{21})(o_{12})}$$

The estimated odds ratio is not normally distributed; however, the sampling distribution of the natural log of the estimated odds ratio is approximately normally distributed if the sample sizes n_1 and n_2 are large. The mean and variance of the natural log[†] of the estimated odds ratio

---

[†]The natural log ($\log_e$) has e as its base rather than the more common log ($\log_{10}$) which has 10 as its base. The relationship is $\log_e(y) = 2.3026 \log_{10}(y)$. Table A.17 provides values of $\log_{10}(y)$.

are

$$E(\log_e \hat{\phi}) = \log_e \phi$$

$$V(\log_e \hat{\phi}) = \frac{1}{n_1 \pi_1(1 - \pi_1)} + \frac{1}{n_2 \pi_2(1 - \pi_2)}$$

The variance of the distribution of the log odds ratio depends on π_1 and π_2, which are unknown. For confidence intervals, the proportions π_1 and π_2 will be replaced by their individual sample estimates, and the standard error of estimate is

$$\text{s.e.}(\log_e \hat{\phi}) = \sqrt{\frac{1}{n_1 \hat{\pi}_1(1 - \hat{\pi}_1)} + \frac{1}{n_2 \hat{\pi}_2(1 - \hat{\pi}_2)}}$$

For testing hypothesis about the equality of the odds in two populations, each proportion will be replaced by the estimate of the common proportion

$$\hat{\pi}_c = \frac{o_{11} + o_{21}}{n_1 + n_2}$$

and the standard error of estimate is

$$\text{s.e.}(\log_e \hat{\phi}) = \sqrt{\frac{1}{n_1 \hat{\pi}_c(1 - \hat{\pi}_c)} + \frac{1}{n_2 \hat{\pi}_c(1 - \hat{\pi}_c)}}$$

We will perform statistical inference for the log odds ratio by using a normal approximation and then restate the results for the odds ratio.

Example 7.10.   Using the Normal Distribution for Inference about an Odds Ratio

The results of Dr. Jonas Salk's experiment of his polio vaccine were as follows:

	Proportion with Paralytic Polio	Number in Study
Inoculated group	0.00016	200,745
Control group	0.00057	201,229

To test the hypothesis that the odds ratio for Dr. Salk's vaccine is greater than 1:

$$H_0: \log_e \phi = 0 \quad \text{i.e., } \phi = 1$$
$$H_a: \log_e \phi > 0 \quad \text{i.e., } \phi > 1$$

The test statistic is

$$z = \frac{\log_e \hat{\phi}}{\sqrt{(1/n_1 \hat{\pi}_c(1 - \hat{\pi}_c)) + (1/n_2 \hat{\pi}_c(1 - \hat{\pi}_c))}} = \frac{1.27}{0.164} = 7.74$$

where

$$\hat{\phi} = \frac{\hat{\pi}_2/(1 - \hat{\pi}_2)}{\hat{\pi}_1/(1 - \hat{\pi}_1)} = \frac{0.00057/(1 - 0.00057)}{0.00016/(1 - 0.00016)} = 3.56$$

$$\log_e \hat{\phi} = \log_e 3.56 = 2.3026 \log_{10}(3.56) = 2.3026(0.5514) = 1.27$$

$$\hat{\pi}_c = \frac{o_{11} + o_{12}}{n_1 + n_2} = \frac{32 + 115}{200{,}745 + 201{,}229} = 0.00037$$

$$\text{s.e.}(\log_e \hat{\phi}) = \sqrt{\frac{1}{n_1 \hat{\pi}_c(1 - \hat{\pi}_c)} + \frac{1}{n_2 \hat{\pi}_c(1 - \hat{\pi}_c)}} = \sqrt{\frac{1}{74.4} + \frac{1}{74.2}} = 0.164$$

$$z = \frac{1.27}{0.164} = 7.74$$

With $\alpha = 0.05$, we will reject the null hypothesis if $z > 1.645$. Since $z > 1.645$, we reject the null hypothesis and conclude that the odds of paralytic polio is greater for the control group than for the inoculated group.

In Dr. Salk's experiment the odds for members of the unvaccinated group was $\hat{\phi} = 3.56$ times greater than those for those receiving the vaccine. However, this is a point estimate, and for inference an interval estimate is preferred. The formula for a confidence interval for the log odds ratio is

$$\text{CI}_{1-\alpha}: \log_e \hat{\phi} \pm z_{\alpha/2} \sqrt{\frac{1}{n_1 \hat{\pi}_1(1 - \hat{\pi}_1)} + \frac{1}{n_2 \hat{\pi}_2(1 - \hat{\pi}_2)}}$$

and the formula for a confidence interval for the odds ratio is

$$\text{CI}_{1-\alpha}: \hat{\phi} \pm e^{z_{\alpha/2} \sqrt{\frac{1}{n_1 \hat{\pi}_1(1 - \hat{\pi}_1)} + \frac{1}{n_2 \hat{\pi}_2(1 - \hat{\pi}_2)}}}$$

For Dr. Salk's data the 95% confidence interval is

$$3.56 \pm e^{1.96\sqrt{\frac{1}{32.1} + \frac{1}{114.6}}} = 3.56 \pm 1.48$$

$$2.08 \leq \phi \leq 5.04$$

With 95% confidence it could be concluded that people who have not been vaccinated are 2.08 to 5.04 times more likely to contract paralytic polio than are those who received the vaccine.

Procedure. Normal Approximation for the Log Odds Ratio

Confidence Intervals:

$$\text{CI}_{1-\alpha}: \hat{\phi} \pm e^{z_{\alpha/2} \sqrt{\frac{1}{n_1 \hat{\pi}_1(1 - \hat{\pi}_1)} + \frac{1}{n_2 \hat{\pi}_2(1 - \hat{\pi}_2)}}}$$

Test of Hypotheses
H_0: $\phi = 1$
H_a: $\phi > 1$
Significance level: α
Test statistic:

$$z = \frac{\log_e \hat{\phi}}{\sqrt{\dfrac{1}{n_1 \hat{\pi}_c(1 - \hat{\pi}_c)} + \dfrac{1}{n_2 \hat{\pi}_c(1 - \hat{\pi}_c)}}}$$

where

$$\hat{\phi} = \frac{\hat{\pi}_2/(1 - \hat{\pi}_2)}{\hat{\pi}_1/(1 - \hat{\pi}_1)} \quad \text{and} \quad \hat{\pi}_c = \frac{o_{11} + o_{12}}{n_1 + n_2}$$

Region of rejection: $z > z_\alpha$

EXERCISES

7.5.1. A physical education professor claims that 35% of third-grade children can do a handstand. If this claim is true:

  a. Find the probability that 10 or more third-grade children out of a random sample of 25 can do a handstand.

   i. Use the exact binomial distribution.

   ii. Use the normal distribution without a continuity correction.

   iii. Use the normal distribution with a continuity correction.

  b. Find the probability that 40 or more third-grade children out of a random sample of 100 can do a handstand.

   i. Use the normal distribution without a continuity correction.

   ii. Use the normal distribution with a continuity correction.

  c. Based on the results of parts a and b, is the correction for continuity more important in large or in small samples?

7.5.2. A customer relations bureau located in a large eastern city claimed that 80% of the complaints registered with it were settled to the satisfaction of the customers. The local newspaper, doubting whether the percentage was really that large, takes a random sample of 40 complainants and asks them whether they had received satisfaction. Only 12 indicate that they had. Use the normal approximation to make a test of significance at $\alpha = 0.01$.

7.5.3. In a certain Midwestern community, 25% of the population consists of third-generation descendants of one Finnish immigrant family. Within the community there is a remittent nervous disorder that may be transmitted genetically. There are 75 cases of the disorder on which to base studies.

  a. If the disorder is *not* genetic or in any way associated with ethnic origin, what percentage of those with the disorder are likely to be third-generation descendants of that family?

b. What are the most logical null and alternative hypotheses to test whether the disorder is genetically controlled?

c. If 28 of the 75 cases are third-generation descendants of the Finnish family, carry out the test at the 0.05 level of significance.

7.5.4. A random sample of 100 high-school dropouts in Pittsburgh aged 17 to 19 revealed that 20% of them were unemployed.

a. Place a 95% confidence interval on the percentage of all similar people in that area who are unemployed.

b. The average unemployment rate for the entire work force in Pittsburgh is 7.0%. Is the unemployment rate among high-school dropouts significantly higher than for the entire work force? Justify your answer.

7.5.5. Many people claim they can distinguish the difference in taste between fish that has been frozen and fish that is prepared fresh. In an experiment, a random sample of 100 consumers is presented with two portions of cooked fish, one of each kind. Of these consumers, 64 can correctly distinguish between the fresh and the frozen fish.

a. Use a point estimate to estimate the proportion of people in the population who can make this distinction.

b. The answer to part a is an estimate and thus subject to variability. What is the estimated variance of this estimate?

c. Use the normal approximation to the binomial distribution in order to place a 95% confidence interval on the proportion.

d. Is there statistically significant evidence that some people can distinguish fresh fish and are not just guessing? Explain.

7.5.6. The theory of radioactive decay predicts that a certain material is expected to emit 40 radioactive particles in 10 msec.

a. What is the probability that at least 35 particles will be emitted in 10 msec?

b. What is the probability that between 30 and 35 particles (inclusive) will be emitted?

7.5.7. A nuclear physicist suspects that a counter is missing some radioactive particles because it has a certain "dead" period as it counts; that is, if two particles are emitted very close together, the counter misses the second one. Assume that the theory correctly states that the expected number of radioactive particles emitted in 10 msec from a certain material is 40. If a counter counts 26 particles in 10 msec, does the physicist have evidence that the counter is giving undercounts?

7.5.8. A serum thought to be effective in preventing colds is given to 300 persons. Their records for one year are compared with those of 200 untreated persons with the following results:

	No Colds	Colds
Treated	145	155
Untreated	80	120

Construct a 95% confidence interval for the odds ratio for colds in the untreated group compared to the treated group.

7.5.9. It is reported that offspring of users of a certain recreational drug may have a higher incidence of birth defects than the general population. To obtain information about a possible relationship between this drug and birth defects, 100 offspring of female rats fed the drug and 100 offspring from untreated female rats are examined. The results are given below:

	Progeny	
Females	Birth Defects	Normal
Treated	30	70
Untreated	20	80

Using a 0.05 level of significance, is there statistical evidence to support the experimental hypothesis that the odds ratio for birth defects in the treated group compared to the untreated group is greater than 1?

7.5.10. In Exercise 7.1.8, the proportion of scores on a mathematics examination that are high enough to achieve prestigious recognition is $\pi = 0.067$, but 24 of 140 politicians claim they received such scores. What is the probability of so many of them in a random sample of 140 people?

7.6. NONPARAMETRIC STATISTICS: A TEST BASED ON RANKS

There are situations in which data are not normally distributed but the mean and variance of the distribution are known. An especially useful distribution of this sort is the distribution of the N consecutive ranks from 1 to N. This is a discrete uniform distribution with $\mu = (N + 1)/2$ and $\sigma^2 = (N^2 - 1)/12$. (The denominator 12 is a constant which arises in the computation of σ^2 and is not related to the number of ranks involved.)

If we are concerned about the average rank $\bar{r}$ in a random sample without replacement of n of the N consecutive ranks, the expected value and variance of the average rank in the sample will be

$$E(\bar{r}) = \mu = \frac{N + 1}{2}$$

and

$$V(\bar{r}) = \frac{\sigma^2}{n} = \frac{(N - n)(N + 1)}{12n}$$

With this knowledge and a sample sufficiently large for the central limit theorem, we can compute the probability of obtaining a given average rank in a random sample from N consecutive ranks with

$$z = \frac{\bar{r} - (N + 1)/2}{\sqrt{(N - n)(N + 1)/12n}}$$

Example 7.11. Applying the Central Limit Theorem to Rank Data

There is strong consumer preference for clear fruit juices, so food chemists often evaluate different methods of clarifying the juices and nectars of fruits. Suppose a chemist is comparing the effectiveness of filtration with and without prior enzyme treatment. He takes a large volume of apple juice as it comes through the company's presses, divides it into subsamples, and applies the methods of clarification using 20 vials of juice per method.

When he attempts to obtain quantitative measures of the optical density (or clarity), he discovers that his optical density reader is producing faulty results and requires repair. The experiment will need to be repeated, but to salvage whatever results possible, he holds each vial of juice to the light and discovers that he can satisfactorily rank the 40 vials from clearest to cloudiest. Ranks 1 through 40 are assigned to the vials according to their clarity and the data below are obtained:

Treatment					Rank						Average
Enzyme	1	3	5	6	7	8	9	10	13	14	
	15	16	19	21	22	28	29	31	32	36	16.25
Control	2	4	11	12	17	18	20	23	24	25	
	26	27	30	33	34	35	37	38	39	40	24.75

It appears that the vials containing juice without enzyme treatment have greater ranks (greater cloudiness) than the other, but a statistical test is still desired for the probability statement it provides.

Under the null hypothesis, the vials of juice treated with enzyme are simply a random sample of 20 of the ranks from 1 through 40, and hence the expected average rank is

$$E(\bar{r}) = \frac{N+1}{2}$$

$$= \frac{40+1}{2}$$

$$= 20.5$$

and it can be shown that the variance is

$$V(\bar{r}) = \frac{(N-n)(N+1)}{12n}$$

$$= \frac{20(40+1)}{12(20)}$$

$$= 3.42$$

If the conditions for the central limit theorem hold, the hypothesis

$$H_0: E(\bar{r}) = 20.5$$

versus

$$H_a: E(\bar{r}) \neq 20.5$$

can be tested using the normal variate z as the test statistic,

$$z = \frac{\bar{r} - E(\bar{r})}{\sqrt{V(\bar{r})}} = \frac{16.25 - 20.5}{\sqrt{3.42}}$$

$$= -\frac{4.25}{1.85}$$

$$= -2.30$$

The P value $= P(|z| > 2.30) = 2(0.011) = 0.022$ is less than the conventional $\alpha = 0.05$; hence the null hypothesis can be rejected, and it can be concluded that apple juice which is not treated with the enzyme prior to filtration has a significantly greater rank for cloudiness than does that which receives the enzyme treatment.

The example above is a variation of the Mann–Whitney–Wilcoxon test, and the procedure is the basis of the group of nonparametric procedures known as *rank tests*. Even when data are recorded on the continuous numerical scale, they can be transformed by replacing them with their ranks and a hypothesis tested about average rank. It is generally advised that at least one of the samples be 20 or larger before the central limit theorem applies. For both samples less than 20, it has been suggested that the continuity correction be used,

$$z = \frac{\bar{r} - 1/2 - E(\bar{r})}{\sqrt{V(\bar{r})}}$$

Also, there are tables for the exact distribution of a related statistic when both samples are less than 20 [see Conover (1998) or Daniel (1990)].

Procedure. Rank Test for Sample of n of Integers 1 to N

H_0: $E(\bar{r}) = (N + 1)/2$ (*This is a random sample of n of the integers 1 to N.*)
H_a: $E(\bar{r}) \neq (N + 1)/2$ (*The ranks in the sample tend to be lower or higher than a random sample.*)
Significance level: α
Test statistic:

$$z = \frac{\bar{r} - (N + 1)/2}{\sqrt{(N - n)(N + 1)/12n}}$$

Region of rejection: $|z| \geq z_{\alpha/2}$ or $z \geq z_\alpha$ or $z \leq -z_\alpha$, respectively.

EXERCISES

7.6.1. The consecutive ranks from 1 to $N = 50$ are randomly sampled.

    a. What is the numerical value of $E(\bar{r})$ when $n = 10, 20, 30, 40$, respectively?

    b. What is the numerical value of $V(\bar{r})$ when $n = 10, 20, 30, 40$, respectively?

7.6.2. Odor is used in the identification of certain organic chemical compounds, and because women are thought to have a keener sense of smell than men, they may have a natural advantage in being able to identify these chemicals. To test this, all of the organic chemistry graduate students in a large department are given the same dilution of an aromatic organic compound to smell. They are asked to tell their professor the name of the compound as soon as they think they have identified the odor. The order in which female (F) and male (M) students correctly identified the compound is given below, from first to last:

(First)	F	F	M	F	F	F	M	F	M	M	
	M	M	M	M	M	M	F	F	F	F	
	M	M	M	M	M	M	M	M	M	M	(Last)

   a. What is the highest scale of measurement available here: nominal, ordinal, or numerical?

   b. If there is no difference between men and women with respect to keenness of smell, what is the expected average rank of the 10 women in the study?

   c. What is the variance of a random sample of 10 of the consecutive integers from 1 through 30?

   d. What null and alternative hypotheses would be appropriate?

   e. Using $\alpha = 0.05$, make the test of significance and draw conclusions.

7.6.3. Given below are particulate data from samples of the flumes of two coal-burning generators. The two are adjacent, using coal from the same mine, and otherwise identical, except that a scrubber has been installed on one in an effort to reduce particulate emission.

With Scrubber					Without Scrubber					
0.40	0.50	0.65	1.41		1.87	2.10	3.55	3.57	3.82	3.94
2.32	2.45	2.46	2.73		4.27	4.32	4.53	4.65	4.70	4.73
3.19	3.20	4.75	4.77		5.06	6.33	6.51	7.09	7.57	9.63

Rank the data and make a 0.05 test of the effectiveness of the scrubber in reducing particulate level.

REVIEW EXERCISES

Decide whether each of the following statements is true or false. If a statement is false, explain why.

7.1. Neither of the parameters of a normal distribution can be negative.

7.2. All bell-shaped distributions are normal distributions.

7.3. In a normal distribution, if μ has a large numerical value, then σ^2 will also tend to be large.

7.4. In a normal distribution, about 95% of the values lie within -2 to $+2$.

7.5. If the variance of a population that follows a normal distribution is known, then, if necessary, a test of hypothesis concerning the mean can be performed from a sample of size $n = 1$.

7.6. If possible, samples of size larger than 1 should be used for purposes of inference.

7.7. According to the central limit theorem, if n is large, the sampling distribution of averages is closely approximated by a normal distribution.

7.8. The central limit theorem can only be applied to symmetrical distributions.

7.9. A test of hypothesis involving the z statistic is frequently used because most experimental populations follow normal distributions with known variances.

7.10. If a population has variance $\sigma^2 = 12$, then the variance among the averages of all samples of size 3 drawn at random with replacement from the population will be $\sigma_{\bar{y}}^2 = 4$.

7.11. For a test of hypothesis using a z statistic, the region of rejection is uniquely determined by the alternative hypothesis and the sample size.

7.12. The danger in misusing a one-tailed test when a two-tailed test should be used is that it makes α larger than for the proper test.

7.13. The danger in misusing a two-tailed test when a one-tailed test should be used is that it makes β larger than for the proper test.

7.14. Other things being equal, in a test of hypothesis, the larger the sample size, the smaller the α level.

7.15. Other things being equal, in a confidence interval, the larger the sample size, the narrower the interval.

7.16. If a population distributed as $N(\mu, \sigma^2)$ is randomly sampled and $(\bar{y} - \mu)/(s/\sqrt{n})$ is used to compute a z statistic, the probabilities will be reliable only if n is large.

7.17. If the $1 - \alpha$ central confidence interval on μ does not contain the value of μ in the null hypothesis, then a two-tailed test would lead to rejection of the null hypothesis at the α level of significance.

7.18. If the variance of a normal distribution is unknown and is estimated by s^2, then two separate random samples of the same size could produce two confidence intervals of different widths.

7.19. Hypotheses about the binomial parameter π tested by the exact binomial distribution and by the normal approximation give exactly the same probabilities.

7.20. When n is large and π is near 0.5, the binomial distribution is approximately a normal distribution.

SELECTED READINGS

Adams, W. J. (1974). *The Life and Times of the Central Limit Theorem*. Kaedman, New York.

Conover, W. J. (1998). *Practical Nonparametric Statistics*, 3rd ed. Wiley, New York.

Daniel, W. W. (1990). *Applied Nonparametric Statistics*, 2nd ed. Duxbury Press, Pacific Grove, California.

Griffin, M. P., and J. T. Smith (1982). Deriving the normal and exponential densities using EDA techniques, *The American Statistician*, **36**, 373–377.

Pearson, K. (1924). Historical note on the origin of the normal curve of errors. *Biometrika*, **16**, 402–404.

Plane, D. R., and K. R. Gordon (1982). A simple proof of the nonapplicability of the central limit theorem to finite populations. *American Statistician*, **36**, 175–176.

Tate, R. F., and G. W. Klett (1959). Optimal confidence intervals for the variance of a normal distribution. *Journal of the American Statistical Association*, **54**, 674–682.

8 Student's t Distribution

In most experimental situations, the population variance is unknown. In Chapter 7 we noted that if a population variance is unknown and the sample size is 30 or more, the population variance can be estimated by the sample variance and then the standard normal distribution can be used for inference. If the sample size is below 30, this procedure will not give reliable probabilities. We discuss the appropriate procedure for such situations in this chapter.

8.1. THE NATURE OF t DISTRIBUTIONS

At the beginning of the twentieth century, William Sealy Gosset (1896 to 1937) was an employee of the Guinness brewery in Dublin, where he interpreted data and planned barley experiments. In 1906 and 1907 he was sent to University College, London, to study statistics with Karl Pearson. In 1908 he published a paper in which he noted that if random samples of size less than 30 are taken from a normal distribution and the samples used to estimate the variance, then the statistic

$$\frac{\bar{y} - \mu}{s/\sqrt{n}}$$

is not normally distributed. The probabilities in the tails of this distribution are greater than for the standard normal distribution (Figure 8.1).

This is reasonable since

$$z = \frac{\bar{y} - \mu}{\sigma/\sqrt{n}}$$

contains only one random variable $\bar{y}$, while

$$\frac{\bar{y} - \mu}{s/\sqrt{n}}$$

contains two random variables $\bar{y}$ and s. Gosset also noticed that as n increases this new distribution approaches the standard normal distribution.

Gosset published his findings under the pseudonym "Student" because of the Guinness company's restrictive policy on publication by its employees. The sampling distributions he studied are called *Student's t distributions*, and we write

$$t = \frac{\bar{y} - \mu}{s/\sqrt{n}}$$

Statistics for Research, *Third Edition*, Edited by Shirley Dowdy, Stanley Weardon, and Daniel Chilko. ISBN 0-471-26735-X © 2004 John Wiley & Sons, Inc.

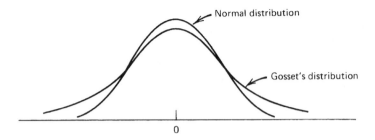

FIGURE 8.1. Comparison of the standard normal distribution and a *t* distribution.

The density functions for Student's *t* distributions are known, and a description of the curve may be helpful (see Figure 8.2).

Student's *t* distributions are

1. unimodal;
2. asymptotic to the horizontal axis;
3. symmetrical about zero, $E(t)$;
4. dependent on v, the degrees of freedom (for the statistic under discussion, $v = n - 1$);
5. more variable than the standard normal distribution, $V(t) = v/(v - 2)$ for $n > 2$;
6. approximately standard normal if v is large.

Table A.11 in the Appendix of Useful Tables gives many of the critical values of the *t* distributions needed for inference. The *t* distributions are listed by degrees of freedom. In the table, α corresponds to the probability that *t* exceeds the tabular value; thus $P(t > 1.721$ if $v = 21) = 0.05$. We write $t_{0.05, 21} = 1.721$.

Since the *t* distribution is symmetrical, critical values for the lower tail can be obtained from the upper tail, $t_{1-\alpha,v} = -t_{\alpha,v}$

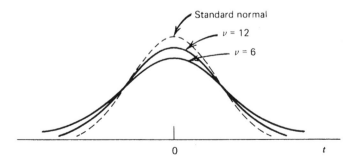

FIGURE 8.2. Student's *t* distributions.

Thus

$$t_{0.95,16} = -t_{0.05,16} = -1.746$$

It should be emphasized that the t statistic arises only when we are sampling from a population with a normal distribution and when σ^2 is estimated by s^2. Whether the sample size is large or small,

$$\frac{\bar{y} - \mu}{s/\sqrt{n}}$$

has a t distribution. However, since the t distribution is quite close to the standard normal for $n \geq 30$, it is common to approximate the probabilities in the t distribution by the standard normal for large sample sizes. If more accuracy is desired and the appropriate table or computer program is available, the t distribution can be used.

It is permissible to use the t distribution to estimate probabilities when we are sampling from a distribution that is not normal if the distribution is at least symmetrical, unimodal, and with a variance that is not inordinately large. In this case, the t distribution is a good estimate of the actual sampling distribution.

EXERCISES

8.1.1. Use Table A.11 to find:

    a. $t_{0.01,\ 10}$

    b. $t_{0.99,\ 10}$

    c. $t_{0.025,\ 7}$

    d. $t_{0.975,\ 7}$

    e. $t_{0.005,\ 23}$

    f. $t_{0.995,\ 23}$

8.1.2. Use Table A.11 to find:

    a. $P(t > 2.145$ if $v = 14)$

    b. $P(t < 2.518$ if $v = 21)$

    c. $P(t < -1.782$ if $v = 12)$

    d. $P(t > -1.363$ if $v = 11)$

    e. $P(-2.120 \leq t \leq 2.120$ if $v = 16)$

    f. $P(|t| \geq 2.831$ if $v = 21)$

8.1.3. A random sample is taken of 16 women who are the sole support of their families, and information is obtained about their annual income (in dollars):

$$\sum y = 128{,}000$$

$$\sum y^2 = 1{,}177{,}600{,}000$$

Assume that the distribution of incomes is normal.

a. Find the best point estimate of the mean income of all women who are the sole support of their families.

b. Estimate the population variance.

c. If μ is actually \$6400, compute

$$t = \frac{\bar{y} - \mu}{s/\sqrt{n}}$$

d. How likely is it that a *t* statistic of this magnitude or larger will arise when choosing random samples of size 16 from this population?

8.2. INFERENCE ABOUT A SINGLE MEAN

Under the following conditions, *t* distributions may be used for inference about μ:

1. The population distribution is normal (or at least symmetrical and unimodal).
2. The population variance is unknown and estimated by the sample variance.
3. The sample is random.

Tests of hypothesis about a population mean μ and confidence intervals for μ using t distributions are analogous to using the standard normal distribution.

Example 8.1.   Using a *t* Distribution to Find a Confidence Interval for μ

After running about 17 miles, marathon runners encounter a form of physiological stress which they call "hitting the wall." To better pinpoint where in a race to expect this phenomenon, a sports physiologist has 12 male marathon runners race until each feels this stress. The variable of interest is the number of miles run until the stress occurs. These are

$$15.8 \quad 16.5 \quad 15.3 \quad 16.2 \quad 17.1 \quad 16.4$$
$$17.5 \quad 17.3 \quad 16.9 \quad 16.6 \quad 17.0 \quad 17.7$$

The physiologist would like to use a *t* distribution to find a 95% confidence interval on the mean distance a marathon runner covers before "hitting the wall." He finds that $\sum y = 200.4$ miles and $\sum y^2 = 3{,}352.08$. He computes a point estimate for the mean,

$$\bar{y} = \frac{200.4}{12} = 16.70$$

and the sample variance is

$$s^2 = \frac{\sum y^2 - \left(\sum y\right)^2 / n}{n - 1} = \frac{3352.08 - (200.4)^2 / 12}{11} = 0.4909$$

The sample standard deviation is $s = 0.70$ and the standard error of the mean is $s/\sqrt{n} = 0.70/\sqrt{12} = 0.20$. Since there are 12 subjects, the degrees of freedom are $n - 1 = 12 - 1 = 11$. Thus

$$\text{CI}_{0.95}: \bar{y} \pm t_{0.025,11} \frac{s}{\sqrt{n}}$$

$$16.70 \pm 2.201(0.20)$$

$$16.70 \pm 0.44$$

$$16.26 \le \mu \le 17.14$$

For this to be valid, the physiologist must be able to assume that the variable of interest is normally distributed, or at least approximately so. Perhaps he might be able to base the assumption on some theoretical knowledge of the physiological changes that occur during running, but more likely he will need empirical evidence. If he has been observing this phenomenon for some time in the course of his other investigations of marathon runners, he may have accumulated enough rough measurements to draw a graph and check on the symmetry and unimodality. Two graphical representations of data are often included in statistical packages to provide some visual evidence about the assumption. For the 12 observations in the sample, these are shown in Figure 8.3, where the experimenter would find the familiar histogram along with another graphic.

The histogram would show him that there is only one mode, but it might cause him to be concerned about symmetry, and the second schematic is provided for visual examination of the validity of that assumption. Above the histogram is a *box-and-whisker plot*, often simply called a box plot. Using the same horizontal scale as the histogram, the vertical line in the middle of the rectangle gives the location of the *median*, and the edges of the rectangle locate

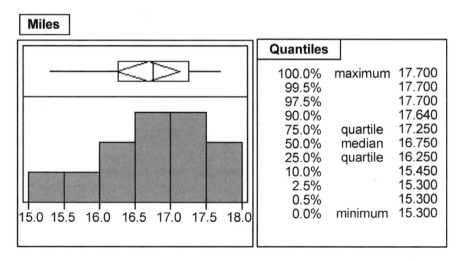

FIGURE 8.3. Graphics used for examining distribution of data.

the *upper and lower quartiles*. Thus the n observations in the sample are divided, as nearly as possible, into $n/4$ equal portions so that approximately half of the sample data lie within the range of the box, one-fourth lie to the left of the rectangle, and the remaining one-fourth to the right. The lines extending from the right and left of the box are called *whiskers*, and they extend, respectively, to the largest and smallest numerical values in the sample. Consequently, if the data were perfectly symmetrical, the physiologist would see a "mirror-image" diagram centered at the median. Although there is some evidence of lack of symmetry, the visual evidence from the two graphics[†] should lead him to feel his sample satisfies the assumption. If he is unable to justify the assumption, he will have to be cautious about how much faith he has in the accuracy of the interval.

Another condition for the validity of this confidence interval (as well as for other inferences) is that the subjects are a random sample from the population of interest. To obtain a completely random sample of 12 runners from the population of all male marathon runners in this country is not feasible. Often the investigator must rely on local volunteers. It would be better if he could find a list or runners from across the country and try to obtain a sample of distance runners from this group. If only local runners are feasible, the generalization to all runners is not as credible. There could be some local condition that affects the variable of interest, for example, altitude.

At a later state in the experimentation, the physiologist may want to test a hypothesis about the distance until stress occurs. For example, he might decide to extend his investigation to female runners. An immediate question would be whether the distance until stress for women is also 17 miles.

Example 8.2.   Using a t Distribution to Test a Hypothesis about μ

The sports physiologist would like to test H_0: $\mu = 17$ against H_a: $\mu \neq 17$ for female marathon runners. In a random sample of 8 female runners, he finds

$$\bar{y} = 18.2 \quad \text{and} \quad s^2 = 0.65$$

Since $n = 8$, the degrees of freedom are $v = 7$, and at $\alpha = 0.05$ the null hypothesis will be rejected if $|t| \geq t_{0.025,7} = 2.365$. The test statistic is

$$t = \frac{\bar{y} - \mu_0}{s/\sqrt{n}} = \frac{18.2 - 17}{\sqrt{0.65/8}} = 4.21$$

Thus he rejects the null hypothesis and concludes that for women the distance until stress is more than 17 miles.

A two-tailed test was used in the above example. If the physiologist had some previous information that stress occurs later, if at all, for women, then a one-tailed test in the upper tail would have been appropriate. Using H_a: $\mu > 17$, at $\alpha = 0.05$ the region of rejection is $t \geq t_{0.05,\ 7} = 1.895$.

It is possible to make inference about another type of mean, the mean of the *difference between two matched groups*. For example, the mean difference between pretest scores and

---

[†]Both graphics are needed because data can be symmetric but not unimodal.

post-test scores for a certain course or the mean difference in reaction time when the same subjects have received a certain drug or have not received the drug might be desired. In such situations, the experimenter will have two sets of sample data (in the examples just given, pretest/post-test or received/did not receive); however, both sets are obtained from the same subjects. Sometimes the matching is done in other ways, but the object is always to remove extraneous variability from the experiment. For example, identical twins might be used to control for genetically caused variability or two types of seeds are planted in identical plots of soil under identical conditions to control for the effect of environment on plant growth.

If the experimenter is dealing with two matched groups, the two sets of sample data contain corresponding members—thus he has, essentially, one set consisting of pairs of data. Inference about the mean difference between these two dependent groups can be made by working with the differences within the pairs and using a t distribution with $n - 1$ degrees of freedom in which n is the number of pairs.

Example 8.3. Matched-Pair t Test

Two types of calculators are compared to determine if there is a difference in the time required to perform a certain common statistical calculation. Twelve students chosen at random are given drills with both calculators so that they are familiar with the operation of each type. Then the time they take to complete the calculation on each device is measured in seconds (which calculator they are to use first is determined by some random procedure to control for any additional learning during the first calculation). The data are as follows:

Student	Calculator A	Calculator B	Difference y_d	(Difference)^2 y_d^2
1	23	19	4	16
2	18	18	0	0
3	29	24	5	25
4	22	23	-1	1
5	33	31	2	4
6	20	22	-2	4
7	17	16	1	1
8	25	23	2	4
9	27	24	3	9
10	30	26	4	16
11	25	24	1	1
12	27	28	-1	1
			$\sum y_d = 18$	$\sum y_d^2 = 82$

The null hypothesis is H_0: $\mu_d = 0$ and H_a: $\mu_d \neq 0$ in which μ_d is the population mean for the difference in time on the two devices. Thus

$$\bar{y}_d = \frac{\sum y_d}{n} = \frac{18}{12} = 1.5$$

$$s_d^2 = \frac{\sum y_d^2 - \left(\sum y_d\right)^2 / n}{n - 1} = \frac{82 - (18)^2 / 12}{11} = 5$$

The test statistic is

$$t = \frac{\bar{y}_d - \mu_{d_0}}{s_d/\sqrt{n}} = \frac{1.5 - 0}{\sqrt{5/12}} = 2.325$$

Using $\alpha = 0.05$ and $v = 12 - 1 = 11$, $t_{0.025,11} = 2.201$, and since $t > 2.201$, the test is significant and the two calculators differ in the time necessary to perform the calculation. Looking at the data, since $\bar{y}_d$ is positive, the experimenter concludes that the calculation is faster on machine B.

In the above example, the experimenter was interested in whether there is a difference in time required on the two calculators; thus $\mu_d = 0$ was tested. The population mean specified in the null hypothesis need not be zero; it could be some other specified amount. For example, in an experiment about the reaction time the experimenter might hypothesize that after taking a certain drug reaction times are slower by 2 seconds; then H_0: $\mu_d = 2$ would be tested, with $y_d = y_{\text{after}} - y_{\text{before}}$. The alternative hypothesis may be one-tailed or two-tailed, as appropriate for the experimental question.

Using a matched-pair design is a way to control extraneous variability. If the study of the two calculators involved a random sample of 12 students who used calculator A and another random sample of 12 students who used calculator B, additional variability would be introduced because the two groups are made up of different people. Even if they were to use the same calculator, the means of the two groups would probably be different. If the differences among people are large, they interfere with our ability to detect any difference due to the calculators. If possible, a design involving two dependent samples that can be analyzed by a matched-pair t test is preferable to two independent samples. The analysis proper for two independent samples is discussed in Section 8.3.

If confidence intervals are desired for the mean of the difference between two dependent samples, they can also be computed:

$$\text{CI}_{1-\alpha}\colon \bar{y}_d \pm t_{\alpha/2,n-1}\frac{s_d}{\sqrt{n}}$$

Procedure. Inference About a Mean Using a t Distribution

Assumptions: normality, or at least symmetry and unimodality; unknown population variance

Confidence Intervals

$$\text{CI}_{1-\alpha}\colon \bar{y} - t_{\alpha/2,n-1}\frac{s}{\sqrt{n}} \leq \mu \leq \bar{y} + t_{\alpha/2,n-1}\frac{s}{\sqrt{n}}$$

Test of Hypothesis

H_0: $\mu = \mu_0$
H_a: $\mu \neq \mu_0$ or $\mu > \mu_0$ or $\mu < \mu_0$
Significance level: α
Test statistic:

$$t = \frac{\bar{y} - \mu_0}{s/\sqrt{n}}$$

Region of rejection: $|t| \geq t_{\alpha/2,n-1}$ or $t \geq t_{\alpha,\,n-1}$ or $t \leq -t_{\alpha,\,n-1}$, respectively.

EXERCISES

8.2.1. From a random sample of 16 applicants for certain graduate fellowships, the following statistics are obtained about their GRE scores:

$$\sum y = 16{,}000$$

$$\left(\sum y\right)^2 = 256{,}000{,}000$$

$$\sum y^2 = 18{,}400{,}000$$

a. Give the best point estimate of the population mean.

b. Estimate the standard error of this estimate.

c. Place a 95% confidence interval on this population mean.

8.2.2. The mean pulse rate for active males of college age is 72 beats per minute, but it is thought to be greater for less active men of the same age. A physician at a student health center questions her male patients on whether they participate in leisure-time sports and measures the pulse rates of a random sample of 12 who do not. The following pulse rates, in stem-and-leaf format, are obtained:

Tens	Units
9	1
8	1 3 6
7	2 4 5 5 6 8
6	6 7

a. Criticize the sample on the basis of the population it may represent.

b. Assuming some valid inference can be made, prepare for a test of hypothesis by giving:

i. The most logical null and alternative hypotheses

ii. The critical region of the test statistic for $\alpha = 0.05$

c. Conduct the test of significance by computing:

i. The sample average and variance

ii. The value of the test statistic

d. Assume the inference is valid; what would you conclude from this study?

8.2.3. Distance runners are known to have lower pulse rates than their contemporaries. Suppose pulse rates are measured on a random sample of 25 runners 5 minutes after they have completed a 10-kilometer run. The data yield $\bar{y} = 58.2$ beats per minute and $s^2 = 72.25$.

a. Compute the standard error of the average.

b. Use the standard error to set a 95% confidence interval for the mean pulse rate of distance runners.

8.2.4. Fruit flies (*Drosophila melanogaster*) are attracted to light. This phenomenon is called positive phototaxis, and it may be an inherited behavior. Suppose a geneticist measures the phototactic response of all flies for one generation and finds a mean response time of 80 seconds. He then mates the male and female that showed the fastest response times. The following data are obtained on the phototactic response times of their offspring:

$$n = 30$$

$$\sum y = 2136 \text{ seconds}$$

$$\sum y^2 = 155{,}225.2$$

a. If phototactic behavior is inherited, should the offspring of the male and female that showed the most rapid response have an average response time greater or less than that of the previous generation?

b. Use the answer to part a to set up the most logical null and alternative hypotheses.

c. Perform the test of significance and state the conclusion.

8.2.5. Organic phosphorous insecticides are very stable chemically and are known to collect in the soil and water and eventually to enter the food chain of human beings. In a study made in an agricultural region in the Orient, the milk of 40 nursing mothers was examined and found to have an average of 4.2 ppm of organic phosphorous insecticides. The sample standard deviation was 1.2 ppm.

a. Place a two-sided 99% confidence interval on the mean level of these compounds in mothers' milk in the region.

b. Place a one-sided 99% confidence limit on the worst the mean contamination might be.

8.2.6. The mean score on the Graduate Record Exam is 1000 for all students who take the exam. No extensive study has been made to determine whether higher or lower mean scores are attained by students 30 years of age or older. A pilot study is done, and the following data are obtained:

$$n = 18$$

$$\sum y = 18{,}972$$

$$\sum (y - \bar{y})^2 = 435{,}200$$

a. Prepare for a test of significance by giving:

    i. The most logical null and alternative hypotheses

    ii. The critical value for the test statistic for $\alpha = 0.05$

b. Compute the average and variance.

c. Conduct the test of significance and state the conclusion.

8.2.7. At a certain university, an English proficiency test must be passed before undergraduates can receive their degrees. Some students have been known to take the test twice before passing it. A random sample of 25 such students was taken, and the number of "comma errors" was counted on the first and second tests. The average difference on the two tests was a decrease of 2.4 errors. The standard deviation was 6.0.

    a. If a college administrator wants to test to show that there was no improvement, what are the null and alternative hypotheses?

    b. Perform the test.

8.2.8. One side of the brain is dominant over the other. A psychologist wishes to determine whether the reaction time for voluntary movement is more rapid for the hand controlled by the dominant side of the brain. Fifteen random subjects are given five instructions for each hand in random order and the difference in total reaction time for each hand is recorded for each subject.

    a. Give the most logical null and alternative hypotheses.

    b. What is the test statistic?

    c. Give the degrees of freedom and the critical value at $\alpha = 0.05$.

8.2.9. Agronomists have identified 7 different geographical areas with respect to raising corn in West Virginia and have managed to obtain an experimental farm in each area. To see if a single variety of corn can be recommended for the entire state, the two leading varieties are compared for yield at all 7 localities. The following yields in bushels per acre are obtained:

Variety	Geographical Area						
	1	2	3	4	5	6	7
A	45	41	58	60	42	32	57
B	47	44	62	63	46	35	59
$(B - A)$	2	3	4	3	4	3	2
$(B - A)^2$	4	9	16	9	16	9	4

    a. Why is it a good design to compare the two varieties at each location?

    b. What is the average difference in the yields?

    c. Show that the estimated standard error of this difference is 0.309.

    d. The seed company that sells variety B claims it will exceed variety A in yield by more than 2 bushels per acre. Test this claim at $\alpha = 0.05$.

    e. What is your conclusion about the seed company's claim?

    f. Find a 95% central confidence interval on the mean difference in yield of the two types of seed. How is this confidence interval related to the test in part d?

8.2.10. An industrial psychologist devises a 50-point questionnaire to measure a worker's attitude toward his job; the higher the score, the more favorably the worker views it. The industrial psychologist is concerned that attitude may be affected by the relationship of the day questioned to payday, with a worker responding more favorably if he has been recently paid. To evaluate the effect of payday, she draws a random sample of 16 workers and gives them all the same questionnaire the day before (with score y_1) and the day after (with score y_2) they are paid. The difference in each worker's two scores ($y_d = y_1 - y_2$) is the variable analyzed.

    a. Give the most logical null and alternative hypotheses.

b. Use the sample data

$$\sum y_1 = 512$$

$$\sum y_2 = 608$$

$$\sum (y_d - \bar{y}_d)^2 = 1500$$

and $\alpha = 0.05$ to give the critical value of the test statistic. Make the test of significance.

c. Is there a payday effect?

8.2.11. Listed below are the gains in pounds of a random sample of pairs of twin lambs in which one member of each pair is treated with an antibiotic and the other remains untreated (control).

Pair:	1	2	3	4	5	6	7
Treated:	33.5	29.0	29.0	20.0	30.0	33.0	15.0
Control:	30.0	34.0	18.0	16.5	25.0	19.5	15.0
y_d:	3.5	−5.0	11.0	3.5	5.0	13.5	0.0

Pair:	8	9	10	11	12	13	14
Treated:	15.0	21.0	31.0	20.5	22.0	22.0	29.0
Control:	18.0	23.0	24.0	28.0	18.0	26.0	20.0
y_d:	−3.0	−2.0	7.0	−7.5	4.0	−4.0	9.0

Pair:	15	16	17	18	Total
Treated:	26.0	22.0	38.0	25.0	461.0
Control:	18.0	32.0	32.0	16.0	413.0
y_d:	8.0	−10.0	6.0	9.0	48.0

a. If $\sum y_d^2 = 890.0$, compute s_d^2.

b. If you had no knowledge before this experiment of the effect of antibiotics on weight gain, give the most logical null and alternative hypotheses.

c. Conduct the test at $\alpha = 0.05$, stating your decision about the null hypothesis and your experimental conclusion.

d. Place a 95% confidence interval on the mean difference in weight gain and explain how this confidence interval could be used to test the null hypothesis.

8.3. INFERENCE ABOUT TWO MEANS

At the end of Section 8.2 we discussed a matched-pair t procedure for two dependent samples. In this section we discuss the appropriate procedure for two *independent* random samples that meet the following conditions:

1. The experimenter is interested in the difference of two population means, $\mu_1 - \mu_2$.
2. The two samples, one from each population, are independent.

3. Both populations are normal, or at least approximately so.

4. The population variances are unknown but are the same for both populations, $\sigma_1^2 = \sigma_2^2 = \sigma^2$.

Example 8.4.   Group Comparison t Test

Chemical compounds that are carcinogenic to mammals also commonly cause genetic mutations in lower organisms. Thus preliminary screening of possible cancer-producing compounds can be performed by testing whether these compounds increase the mutation rate of microorganisms.

Suppose an experimenter uses this procedure as the first safety screening of an aromatic hydrocarbon that could be used as an industrial solvent. He adds the compound to a medium of an *Ascomycetes* fungus in several petri dishes and compares the mutation rate of this group (the treatment group) with the control group (untreated group).

The variable measured is the number of mutant colonies per petri dish. The experimenter realizes that this discrete random variable probably is not normally distributed but rather has a Poisson distribution. Since he would like to use a t test to make the comparison, he first transforms his counts, x, by letting $y = \log_{10} x$. [If there are any zero counts, he will use $y = \log_{10}(x + 1)$.] Experience has shown him that in this situation his transformation will yield distributions that, although discrete, are approximately normal. After the transformation, his data are summarized as follows:

Sample Data	Control Group				Treatment Group			
	2.13	1.59	1.14	1.77	1.42	1.73	1.57	1.49
	1.36	1.46	1.19		2.52	1.83	1.35	1.53

From his previous work he believes that the variances of the two populations, although unknown, are in fact equal. The closeness of the sample variances seems to confirm this. (If he were in doubt, he could apply the test to be described in Section 8.4 to the sample variances in order to test the hypothesis $\sigma_1^2 = \sigma_2^2$.) Since he believes the two variances are equal, the best point estimate of this common variance will be an average of the two sample variances weighted by the degrees of freedom. This weighted average is called the *pooled sample variance* and is computed as follows:

$$s_p^2 = \frac{\sum (y_1 - \bar{y}_1)^2 + \sum (y_2 - \bar{y}_2)^2}{(n_1 - 1) + (n_2 - 1)}$$

$$= \frac{(n_1 - 1)s_1^2 + (n_2 - 1)s_2^2}{n_1 + n_2 - 2}$$

In this experiment,

$$s_p^2 = \frac{6(0.12) + 7(0.14)}{7 + 8 - 2} = 0.131$$

He would like to test

$$H_0: \mu_1 - \mu_2 = 0 \quad \text{against} \quad H_a: \mu_1 - \mu_2 < 0$$

In other words,

$$H_0\text{: } \mu_1 = \mu_2 \quad \text{against} \quad H_a\text{: } \mu_1 < \mu_2$$

The test statistic has $v = n_1 + n_2 - 2 = 13$ degrees of freedom, corresponding to the denominator of the pooled sample variance, and

$$t = \frac{(\bar{y}_1 - \bar{y}_2) - (\mu_1 - \mu_2)_0}{\sqrt{\dfrac{s_p^2}{n_1} + \dfrac{s_p^2}{n_2}}} = \frac{(1.52 - 1.68) - 0}{\sqrt{\dfrac{0.131}{7} + \dfrac{0.131}{8}}} = -0.85$$

The critical value at $\alpha = 0.05$ is $t_{0.95,13} = -1.771$. Thus the null hypothesis is not rejected, and the experimenter concludes that there is no evidence that this aromatic hydrocarbon increases the mutation rate of the fungus.

Note that the t statistic, although different from the statistic used for one-sample or matched-pair tests, is still of the same form:

$$t = \frac{(\text{Estimate of the parameter}) - (\text{Hypothesized value of the parameter})}{(\text{Estimated standard error of the estimator})}$$

The estimator of $\mu_1 - \mu_2$ is $\bar{y}_1 - \bar{y}_2$. Since the variances of the two groups are equal ($\sigma_1^2 = \sigma_2^2 = \sigma^2$) and the samples are independent,

$$V(\bar{y}_1 - \bar{y}_2) = V(\bar{y}_1) + V(\bar{y}_2)$$

$$= \frac{\sigma^2}{n_1} + \frac{\sigma^2}{n_2}$$

This is estimated by

$$\frac{s_p^2}{n_1} + \frac{s_p^2}{n_2}$$

and the standard error of the estimator is estimated by

$$\sqrt{\frac{s_p^2}{n_1} + \frac{s_p^2}{n_2}}$$

A caution about this procedure: The test is not reliable if the variances of the two groups are unequal. If there is doubt, this should be checked by the method to be described in the next section. If the variances prove to be unequal and the sample sizes are small ($n_1 < 30$ or $n_2 < 30$), then there is no exact test available and an approximation procedure such as the one in the next section should be used.

The test in this section is the appropriate one for two *independent* samples. Two independent samples should *not* be analyzed by means of a matched-pair procedure, for the degrees of freedom will be lower, increasing the magnitude of the critical value and reducing the power of the test.

If the combined sample size is large ($n_1 + n_2 \geq 30$), the critical value may be estimated by a z value for convenience. If both samples are large ($n_1 \geq 30$ and $n_2 \geq 30$), the test statistic may be replaced by

$$z = \frac{(\bar{y}_1 - \bar{y}_2) - (\mu_1 - \mu_2)_0}{\sqrt{\dfrac{s_1^2}{n_1} + \dfrac{s_2^2}{n_2}}}$$

eliminating the need to pool the sample variances. Whether or not the population variances are equal, this z statistic is valid for two large samples. If the actual population variances are known, then

$$z = \frac{(\bar{y}_1 - \bar{y}_2) - (\mu_1 - \mu_2)_0}{\sqrt{\dfrac{\sigma_1^2}{n_1} + \dfrac{\sigma_2^2}{n_2}}}$$

is the appropriate statistic for all sample sizes.

Confidence intervals for $\mu_1 - \mu_2$ may also be computed. For $n_1 < 30$ or $n_2 < 30$ with $\sigma_1^2 = \sigma_2^2$ and σ_1^2, σ_2^2 unknown, use

$$\text{CI}_{1-\alpha}: \bar{y}_1 - \bar{y}_2 \pm t_{\alpha/2, n_1+n_2-2} \sqrt{\frac{s_p^2}{n_1} + \frac{s_p^2}{n_2}}$$

For $n_1 \geq 30$ and $n_2 \geq 30$ with σ_1^2, σ_2^2 unknown, use

$$\text{CI}_{1-\alpha}: \bar{y}_1 - \bar{y}_2 \pm z_{\alpha/2} \sqrt{\frac{s_1^2}{n_1} + \frac{s_2^2}{n_2}}$$

If σ_1^2 and σ_2^2 are known, use

$$\text{CI}_{1-\alpha}: \bar{y}_1 - \bar{y}_2 \pm z_{\alpha/2} \sqrt{\frac{\sigma_1^2}{n_1} + \frac{\sigma_2^2}{n_2}}$$

regardless of sample size.

Procedure. Inference About Two Independent Samples

Assumptions: normality or at least symmetry and unimodality

$$\sigma_1^2, \sigma_2^2 \text{ unknown}, \quad \sigma_1^2 = \sigma_2^2, \quad \text{and} \quad n_1 \text{ or } n_2 < 30$$

Confidence Interval on $\mu_1 - \mu_2$

$$\text{CI}_{1-\alpha}: \bar{y}_1 - \bar{y}_2 \pm t_{\alpha/2, n_1+n_2-2} \sqrt{\frac{s_p^2}{n_1} + \frac{s_p^2}{n_2}}$$

with

$$s_p^2 = \frac{(n_1 - 1)s_1^2 + (n_2 - 1)s_2^2}{n_1 + n_2 - 2}$$

Test of Hypothesis

H_0: $\mu_1 - \mu_2 = (\mu_1 - \mu_2)_0$

H_a: $\mu_1 - \mu_2 \neq (\mu_1 - \mu_2)_0$ or $\mu_1 - \mu_2 > (\mu_1 - \mu_2)_0$

    or $\mu_1 - \mu_2 < (\mu_1 - \mu_2)_0$

Significance level: α

Test statistic:

$$t = \frac{\bar{y}_1 - \bar{y}_2 - (\mu_1 - \mu_2)_0}{\sqrt{\dfrac{s_p^2}{n_1} + \dfrac{s_p^2}{n_2}}} \quad \text{with } s_p^2 \text{ as above}$$

Region of rejection: $|t| \geq t_{\alpha/2, n_1 + n_2 - 2}$ or $t \geq t_{\alpha, n_1 + n_2 - 2}$ or $t \leq -t_{\alpha, n_1 + n_2 - 2}$, respectively.

Assumptions: n_1 and $n_2 \geq 30$

Confidence Interval on $\mu_1 - \mu_2$

$$\text{CI}_{1-\alpha}: \bar{y}_1 - \bar{y}_2 \pm z_{\alpha/2} \sqrt{\frac{\sigma_1^2}{n_1} + \frac{\sigma_2^2}{n_2}}$$

Use s_1^2 and s_2^2 to estimate σ_1^2 and σ_2^2 if the population values are unknown.

Test of Hypothesis

H_0: $\mu_1 - \mu_2 = (\mu_1 - \mu_2)_0$

H_a: $\mu_1 - \mu_2 \neq (\mu_1 - \mu_2)_0$ or $\mu_1 - \mu_2 > (\mu_1 - \mu_2)_0$

    or $\mu_1 - \mu_2 < (\mu_1 - \mu_2)_0$

Significance level: α

Test statistic:

$$z = \frac{\bar{y}_1 - \bar{y}_2 - (\mu_1 - \mu_2)_0}{\sqrt{\dfrac{\sigma_1^2}{n_1} + \dfrac{\sigma_2^2}{n_2}}}$$

Use s_1^2 and s_2^2 to estimate σ_1^2 and σ_2^2 if the population values are unknown.

Region of rejection: $|z| \geq z_{\alpha/2}$ or $z \geq z_\alpha$ or $z \leq -z_\alpha$, respectively.

EXERCISES

8.3.1. After an extended dry period, measurements are taken on atmospheric pollution in urban and rural locations. The data are summarized as follows:

	Urban	Rural
n	7	5
$\bar{y}$	26.0 ppm	12.2 ppm
s^2	91	126

a. Compute the pooled variance.

b. What are the null and alternative hypotheses if the experimenter is looking for evidence of higher pollution in the urban locations?

c. Perform the test of significance at $\alpha = 0.05$ assuming that the variables meet the assumptions for a group comparison t test.

d. Place a 95% confidence interval on the maximum difference between the two means.

8.3.2. A study is done on insecticide residues on fruit. Normal spraying practices are followed in an apple orchard. After the fruit is picked, a random sample of 16 apples is washed individually by hand. A second sample of 16 is washed mechanically. The experimenter is unsure which method would be more effective in removing insecticide residues. The level of insecticide present on each fruit is determined chemically, yielding the following data:

By Hand	By Machine
$\bar{y} = 3.5\,\text{ppm}$	$\sum y = 48\,\text{ppm}$
$\sum y^2 = 200.5$	$\sum (y - \bar{y})^2 = 5.1$

Test for a significant difference of insecticide residue at the 0.01 level of significance.

8.3.3. A certain industrial solvent absorbs atmospheric moisture very rapidly. The absorbed moisture dilutes the solvent and lessens its usefulness. Two types of containers are used in an effort to find a method of storage that will retard moisture absorption. After two months of storage, 10 containers are chosen at random from each kind and are examined for moisture content:

	Container A	Container B
$\sum y$	100	120
$\sum y^2$	1012	1450.5

Place a 99% central confidence interval on the difference in the moisture content of the two types of containers.

8.3.4. In a study of the effect of protein quality in the diet, two groups of juvenile female rats are fed diets of the same caloric content, but they differ in the quality of the protein. The experimenter believes that by the end of the experiment the rats on a high-quality protein diet will gain on the average more than 5 grams more than those on a low-quality diet. The experiment begins with equal numbers of rats on each diet, but some are mistakenly assigned to another experiment and have to be eliminated from the protein experiment. Data on the weight gain (in grams) of the remaining rats are collected and summarized:

	High Quality	Low Quality
Sample size	12	7
Sample average	119.7	101.2
Sample standard deviation	21.4	20.6

a. Give the most appropriate null and alternative hypotheses for this experiment.

b. What assumptions are necessary in order to apply a *t* test for two independent groups?

c. Assuming the two populations have the same variance, test the null hypothesis.

d. What do you conclude about the diets?

8.3.5. At a certain university, Graduate Record Exam scores are compared for doctoral students who completed their PhD work within 7 years of their bachelor's degree and those who did not complete their work within that time. Random sampling provides the following results:

	Completed Work	Did Not Complete Work
Sample size	25	25
Average score	1056	912
Standard deviation	295	270

Is there any evidence that those who finish their PhD work within 7 years score higher on the GRE than those who do not finish within that time? Do you believe that lower GRE scores can be used to predict those who will have difficulty completing their doctoral work on time? Why or why not?

8.3.6. An environmental chemist is performing a study of iron in atmospheric particulate measured downwind from a steel mill. She is concerned that wind velocity at the time of measurement may affect the readings, so she decides to obtain observations on 30 randomly chosen days during the period of peak operation of the mill and compare measurements taken on days when the wind is calm (velocity ≤ 5 knots) with measurements taken on windy days (velocity >5 knots). The data and some summary information are presented below:

	Calm Days			Windy Days						
y	0.68 0.74	0.88		0.25	0.29	0.30 0.43	0.45	0.50	0.60	
	0.89 0.97	1.00		0.65	0.69	0.74 0.80	0.87	0.87	0.89	
	1.17 1.25	1.27		0.91	0.92	0.93 0.95	1.01	1.03	1.16	
$\sum y$	8.85					15.24				
$\sum (y - \bar{y})^2$	0.3592					1.4347				

a. What hypothesis can be tested about the effect of wind velocity on the measurement of iron in atmospheric particulate?

b. What assumptions must be made in order to perform a *t* test on these data?

c. Find the pooled sample variance.

d. Perform the *t* test and draw a conclusion.

8.3.7. Two experimental methods of controlling acid drainage from coal mines are compared. The data are as follows, with greater numerical values indicating the more effective method:

	Method A	Method B
Average	5.60	6.70
Variance	0.98	0.85
Sample size	6	9

a. Place a 95% confidence interval on the difference between the means for the two methods.

b. Using the confidence interval, what decision would you make about the equality of the means for the two methods?

8.3.8. An educator thinks that engineers, although known to be equal to physical scientists in quantitative skills, have less verbal ability. To test this, GRE verbal scores are compared for large random samples of engineering and physical-science seniors.

	Engineering	Physical Science
Average	414	422
Standard deviation	30	40
Sample size	100	100

a. State the most logical null and alternative hypotheses.

b. Take advantage of the large sample sizes and perform the appropriate z test.

c. What conclusion should be drawn from this study?

8.4. INFERENCE ABOUT TWO VARIANCES

In Section 8.3 we described procedures for analyzing data from two populations having equal variances. There are situations, of course, in which the variances of the two populations under consideration are different. The variability in the weights of elephants is certainly different from the variability in the weights of mice, and in many experiments, even though we do not have these extremes, the treatments may affect the variances as well as the means.

The null hypothesis $H_0: \sigma_1^2 = \sigma_2^2$ is tested by using a statistic that is in the form of a ratio rather than a difference; the statistic is s_1^2/s_2^2. Intuitively, if the variances are equal, this ratio should be approximately equal to 1, so values that differ greatly from 1 indicate inequality.

It has been found that the statistic s_1^2/s_2^2 from two normal populations with equal variances follows a theoretical distribution known as an F distribution. The density functions for F distributions are known, and we can get some understanding of their nature by listing some of their properties. Let us call a random variable that follows an F distribution F; then the following properties exist:

1. $F > 0$.

2. The density function of F is not symmetrical.

3. F depends on an ordered pair of degrees of freedom v_1 and v_2; that is, there is a different F distribution for each ordered pair v_1, v_2. (v_1 corresponds to the degrees of freedom of the numerator of s_1^2/s_2^2 and v_2 corresponds to the denominator.)

4. If α is the area under the density curve to the right of the value F_{α,v_1,v_2}, then

$$F_{\alpha,v_1,v_2} = 1/F_{1-\alpha,v_2,v_1}$$

5. The F distribution is related to the t distribution:

$$F_{\alpha,1,v_2} = (t_{\alpha/2,v_2})^2$$

Table A.12 in the Appendix gives upper critical values for F if $\alpha = 0.050, 0.025, 0.010, 0.005, 0.001$. Lower-tail values can be found using property 4 above.

Example 8.5.   Testing for the Equality of Two Variances

Both rats and mice carry ectoparasites that can transmit disease organisms to humans. To determine which of the two rodents presents the greater health hazard in a certain area, a public health officer traps (presumably at random) both and counts the number of ecto-parasites each carries. The data are presented first in side-by-side stem-and-leaf plots and then as side-by-side box-and-whisker plots:

Mice		Rats	
Tens	Units	Tens	Units
3		3	0 4
2		2	0 0 0 1 2 3 3
1	0 1 2 2 6 8	1	3 3 5 5 5 5 5 6 6 6 7 7 7 8 8
0	7 8 9	0	3 6 7 8 8 8

	n	s^2	$\bar{y}$
Rats	31	43.4	16.3
Mice	9	13.0	11.4

He wants to test for the equality of means with a group comparison t test. He assumes that these discrete counts are approximately normally distributed, but because he is studying animals of different species, sizes, and body surface areas, he has some doubts about the equality of the variances in the two populations, and the box plots seem to support that concern. Thus he first must test

$$H_0: \sigma_1^2 = \sigma_2^2 \quad \text{against} \quad H_a: \sigma_1^2 \neq \sigma_2^2$$

with the test statistic $F = s_1^2/s_2^2 = 43.4/13.0 = 3.34$. Since $n_1 = 31$ and $n_2 = 9$, the degrees of freedom for the numerator are $v_1 = n_1 - 1 = 30$ and for the denominator $v_2 = n_2 - 1 = 8$. In Table A.12 he finds

$$F_{0.05,30,8} = 3.079 \quad \text{and} \quad F_{0.05,8,30} = 2.266$$

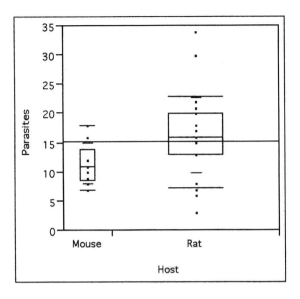

Means and Std Deviations

Level	Number	Mean	Std Dev	Std Err Mean
Mouse	9	11.4444	3.60940	1.2031
Rat	31	16.2581	6.58770	1.1832

thus the region of rejection (Figure 8.4) at $\alpha = 0.10$ is

$$F \geq F_{0.05,\,30,8} = 3.079 \quad \text{and} \quad F \leq F_{0.95,30,8} = \frac{1}{F_{0.05,8,30}} = \frac{1}{2.266} = 0.441$$

Since the computed F equals 3.34, the null hypothesis is rejected, and the public health officer concludes that the variances are unequal. Since one of the sample sizes is small, he may not perform the usual t test for two independent samples.

One-tailed tests of hypotheses involving the F distribution can also be performed, if desired, by putting the entire probability of a Type I error in the appropriate tail.

Central confidence intervals on σ_1^2/σ_2^2 are found as follows:

$$\text{CI}_{1-\alpha}: \quad \frac{s_1^2}{s_2^2} \frac{1}{F_{\alpha/2,v_1,v_2}} \leq \frac{\sigma_1^2}{\sigma_2^2} \leq \frac{s_1^2}{s_2^2} F_{\alpha/2,v_2,v_1}$$

Although the public health officer cannot perform the usual t test for two independent samples because of the unequal variances and the small sample size, there are approximation

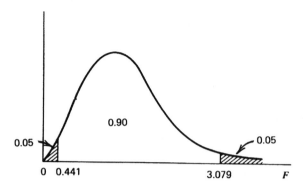

FIGURE 8.4. Regions of rejection in an *F* distribution.

methods available. One such test is called the Behrens–Fisher, or the *t'* test for two independent samples and using adjusted degrees of freedom.

Example 8.6.   Testing $\mu_1 - \mu_2$ if $\sigma_1^2 \neq \sigma_2^2$

To test H_0: $\mu_1 = \mu_2$ against H_a: $\mu_1 \neq \mu_2$ at $\alpha = 0.05$, the health officer uses the test statistic

$$t' = \frac{(\bar{y}_1 - \bar{y}_2) - (\mu_1 - \mu_2)_0}{\sqrt{\dfrac{s_1^2}{n_1} + \dfrac{s_2^2}{n_2}}} = \frac{(16.3 - 11.4) - 0}{\sqrt{\dfrac{43.4}{31} + \dfrac{13.0}{9}}} = 2.90$$

with adjusted degrees of freedom

$$\nu \cong \frac{\left(\dfrac{s_1^2}{n_1} + \dfrac{s_2^2}{n_2}\right)^2}{\dfrac{\left(\dfrac{s_1^2}{n_1}\right)^2}{n_1 - 1} + \dfrac{\left(\dfrac{s_2^2}{n_2}\right)^2}{n_2 - 1}} = \frac{\left(\dfrac{43.4}{31} + \dfrac{13.0}{9}\right)^2}{\dfrac{\left(\dfrac{43.4}{31}\right)^2}{30} + \dfrac{\left(\dfrac{13.0}{9}\right)^2}{8}} = 24.93$$

With $\nu = 25$ H_0 will be rejected if $|t'| \geq t_{0.025,25} = 2.306$. Since $|t'| = 2.90 > 2.060$, the null hypothesis is rejected, and the public health officer concludes that on the average there are more ectoparasites on rats than on mice.

If not an integer value, as in the example, the adjusted degrees of freedom may be rounded to the closest integer or interpolation may be used in the *t* table for a more accurate critical value. Since this *t'* test is only an approximate procedure and is usually very conservative (rejection is difficult), it should be avoided if possible. Instead, larger sample sizes should be obtained when feasible.

Survey sampling texts, for instance (Lohr and Schaeffer *et al.*, listed in the Selected Readings of Chapter 2) deal with optimum allocation of sample size when variances are unequal. When population sizes are very large compared to sample sizes and costs per observation are about the same for each group, sampling theory advises that larger samples are needed from more variable populations. This is also intuitive, for we seem to know that if a population is not too variable, the average of even a small sample will be quite reliable. For example, we need count the number of intact ears of only a few maras (large South American rodents) to know that, along with other mammals, $\bar{y} = 2$ is a reliable estimate of the mean number of ears for the species. Similarly, we know that when the variable of interest has a large variance we must have a large sample in order to obtain a satisfactory estimate of μ. Thus, if we wish to estimate the mean weight of *Equus caballas*, the horse species, we must plan for a very large sample that will measure weights from those of dog-sized ponies to huge dray horses.

When the assumption of equal variances can be made, a t-test with $n_1 = n_2$ will have the smallest standard error. However, when variances are unequal, the smallest standard error is obtained when the sample size for each group is proportional to its variance,

$$\frac{n_1}{n_2} = \frac{\sigma_1^2}{\sigma_2^2}$$

Experience and simulation studies have also shown that the t-test is reasonably *robust* when this condition is met. A statistically robust t-test is one that gives fairly reliable P values even when certain of the assumptions of the test are not met. Because the t'-test is so very conservative, when sample sizes are proportional to variances, a better test might be the t-test with s_p^2 replaced with s_1^2 and s_2^2, respectively. However, when variances are unequal, it is always best to have large samples from each group as well as being proportional to group variances.

A summary of several test statistics in the form of a flowchart for making a decision about the appropriate procedure is given in Figure 8.5. Degrees of freedom involved in the t, F, and χ^2 procedures are indicated by subscripts; for example, t_{n-1} means that the test has $n - 1$ degrees of freedom. Since a matched-pair t test is essentially a one-sample procedure (the set of differences is a single sample), this test does not appear explicitly in the flowchart.

EXERCISES

8.4.1. Use Table A.12 to find:
   a. $F_{0.01,11,7}$
   b. $F_{0.01,\ 7,11}$
   c. $F_{0.05,\ 20,\ 15}$
   d. $F_{0.95,15,\ 20}$
   e. $F_{0.99,\ 8,\ 3}$

8.4.2. The writings of different authors can be partially characterized by the variability in the lengths of their sentences. Two manuscripts, A and B, are found by a historian and she wants to know whether they have the same author. Several sentences from each are chosen at random, and word counts are taken; the variable of interest y is the number of words per sentence.

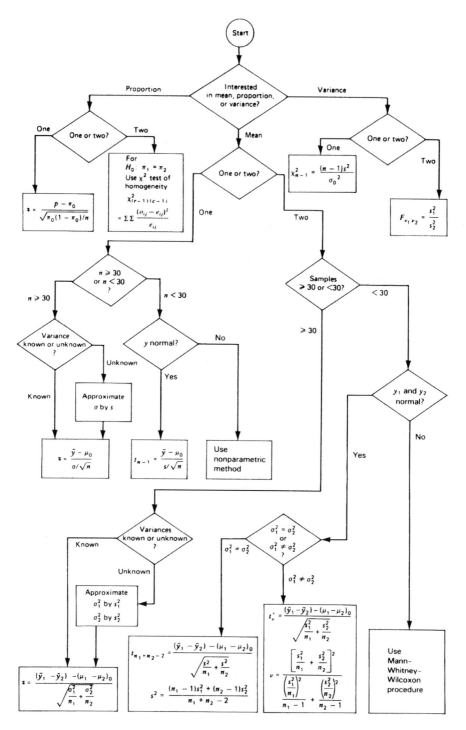

FIGURE 8.5. Flowchart of test statistics.

	Manuscript A	Manuscript B
n	15	15
$\sum y$	141	210
$\sum y^2$	1327	2942

Is there evidence of different authorship at the 0.02 level of significance?

8.4.3. A highway engineer wishes to compare the resin content of asphalt from a Caribbean source with those from a North American source. The following statistics are obtained:

	Sample Value		
	Average	Variance	Size
Caribbean	21.4	0.44	10
North American	22.0	0.11	8

Given only this information, perform the appropriate test of hypothesis to determine if there is a difference in the mean resin content from the two sources (use $\alpha = 0.10$).

8.4.4. A nutritionist wishes to study vitamin B production by bacteria in the caecum (a portion of the digestive tract) and wishes to use either mice or meadow voles, whichever have the larger mean caecum volume. The sample data on which he must make his decision are:

	Mice	Voles
Number of observations	16	11
Average caecum volume	6.5	8.9
Variance	4.6	13.1

a. Should he use a t test or a t' test? (Use $\alpha = 0.10$.)

b. Test to see if there is a significant difference in the average caecum volumes. (Use $\alpha = 0.10$.)

c. What would you suggest to the nutritionist?

8.4.5. The following values were computed from the length of life of two brands of light bulbs (in hours):

	Brand A	Brand B
n	9	16
$\bar{y}$	1560	1573
$\sum (y - \bar{y})^2$	440	1860

a. Is there a difference in the variability of lifetimes for the two brands of bulbs? (Use $\alpha = 0.02$.)

b. Find a 98% confidence interval on the ratio of the two variabilities.

8.5. NONPARAMETRIC STATISTICS: MATCHED-PAIR AND TWO-SAMPLE RANK TESTS

Two of the most commonly used rank tests are the nonparametric counterparts of the matched-pair and two-sample *t* tests. As we have seen before, data may be recorded on the ordinal scale of measurement or data on the numerical scale may be reduced to the ordinal scale by replacing observations with their ranks. Whether the ranks are obtained as the original scale of measurement or as transformations from the numerical scale, statistical inference is based on whether or not the ranks seem to be randomly distributed among the experimental groups. This is the null hypothesis for rank tests; the alternative hypothesis is that observations in one group tend to rank higher than those in another.

There are many conveniences to rank tests. The computations are relatively simple and straightforward, especially when sample sizes are not too large and there are few observations that tie for the same rank. The mean and variance of the original data need not be known. With the transformation to the ranks from 1 to N, the value of $E(\bar{r})$ and $V(\bar{r})$ under the null hypothesis are known rather than estimated. The original data need not have a normal distribution. The rank tests are almost as powerful as the corresponding z or t test when the original data are normally distributed, and they have been shown to be even more powerful for certain non-normal data. Consequently, rank tests are useful analytical tools for research workers.

The *Wilcoxon signed-rank test* is the counterpart in rank statistics to the matched-pair procedure covered earlier in this chapter. It tests the hypothesis that plus and minus signs are randomly assigned to the integers 1 through N. When the null hypothesis is true, the difference between the members of pairs are just random and the difference $y_d = B - A$ will be positive or negative by chance alone. It would be as though we recorded the absolute difference between the members of all pairs and then tossed a coin and assigned a plus sign in front of the difference if the coin showed a head or a minus sign if the coin showed a tail. Under these conditions, $E(y_d) = 0$. In the Wilcoxon test we simply replace the $|y_d|$ with their ranks, reattach the observed plus or minus signs, and then test to determine whether the average rank is significantly different from zero.

Using this technique, when the null hypothesis is true,

$$E(\bar{r}) = \mu = 0$$

and it has been shown that

$$V(\bar{r}) = (N + 1)(2N + 1)/6N$$

Consequently, when the sample size is large enough to meet the conditions of the central limit theorem, we can use the normal distribution to test the null hypothesis

$$H_0: \mu = 0$$

against either a one- or two-sided alternative. The test statistic will be

$$z = \frac{\bar{r} - \mu_0}{\sqrt{V(\bar{r})}} = \frac{\bar{r} - 0}{\sqrt{(N + 1)(2N + 1)/6N}}$$

Example 8.7.   Wilcoxon Signed-Rank Test

Suppose that a college dean is interested in whether there is any predictable change in the academic performance of international students from the first to the second semester of their first year at a U.S. university. She selects a random sample of 20 such students and obtains their first- and second-semester grade point averages, GPAs.

Student	First	Second	Sign	$\|y_d\| = \|F - S\| =$ |Difference|	Rank	Signed Rank
A	1.53	3.67	−	2.14	19	− 19
B	2.00	2.74	−	0.74	11	− 11
C	1.93	3.50	−	1.57	17	− 17
D	3.90	3.27	+	0.63	8	+ 8
E	2.14	1.97	+	0.17	3	+ 3
F	1.52	1.54	−	0.02	1	− 1
G	0.91	3.42	−	2.51	20	− 20
H	1.95	1.04	+	0.91	13	+ 13
I	3.00	2.45	+	0.55	6	+ 6
J	1.67	2.09	−	0.42	5	− 5
K	2.78	2.00	+	0.78	12	+ 12
L	1.21	3.00	−	1.79	18	− 18
M	1.66	1.78	−	0.12	2	− 2
N	1.75	2.31	−	0.56	7	− 7
O	2.96	2.25	+	0.71	10	+ 10
P	1.50	2.20	−	0.70	9	− 9
Q	2.25	0.91	+	1.34	16	+ 16
R	2.66	1.52	+	1.14	15	+ 15
S	1.87	1.61	+	0.26	4	+ 4
T	3.50	2.56	+	0.94	14	+ 14
					Sum	− 8
					Average	− 0.40

The signed-rank value for student A is obtained by first finding the difference between the GPA for the first semester and that for the second semester, $y_d = F - S = 1.53 - 3.67 = -2.14$. The negative sign is recorded in the column for signs and the absolute difference of 2.14 is recorded in the next column. After all the absolute values are entered, they are ranked and student A has the 19th greatest difference. In the last column the negative sign is reattached, giving − 19 as the signed rank for student A. The same procedure is followed for each student.

The null hypothesis that there is no difference between the first- and second-semester GPA is

$$H_0: \mu = 0$$

and because there is no prior information about whether the second-semester GPA should be greater or smaller than that for the first semester, the alternative hypothesis is

$$H_a: \mu \neq 0$$

The test statistic is computed as

$$z = \frac{-0.40 - 0}{\sqrt{(20 + 1)(40 + 1)/6(20)}} = \frac{-0.40}{\sqrt{7.175}} = \frac{-0.40}{2.679} = -0.15$$

The *P* value for a two-sided alternative hypothesis is $P(|z| > 0.15) = 2(0.440) = 0.880$, indicating that results such as these could easily be attributed to chance. Hence there is no statistical basis for rejecting the null hypothesis, and the dean concludes that there is no difference between the first- and second-semester GPA of international students during their first year of study in the United States.

In all the rank tests which are examined, we use data which are recorded on the ordinal scale or which have been transformed from the numerical scale to rank data. Under these circumstances, we are dealing with the integers 1 to *N*, and the expected value and variance are mathematically known for a statistic, such as $\bar{r}$, which is derived from a random grouping of these consecutive integers. If the null hypothesis is true, the grouping of ranks with plus or minus signs is truly random, so we commonly use the expression "under the null hypothesis" when we talk about the values of μ and σ which are used in the *z* test.

To use the normal distribution in a rank test, *N* must be large enough for the central limit theorem to hold true. For Wilcoxon's signed-rank test, it is generally recommended that *N* be at least 20; however, it is suggested that fairly reliable *P* values can be obtained when *N* is smaller if the continuity correction is used:

$$z = \frac{\bar{r} - 1/2 - \mu_0}{\sqrt{V(\bar{r})}} = \frac{\bar{r} - 1/2 - 0}{\sqrt{(N + 1)(2N + 1)/6N}}$$

Also, for small values of *N*, tables are available for the exact distribution of a small sample test statistic.

When data are measured on the continuous numerical scale, strictly speaking, there will be no ties, but the same recorded value does occur in experimental data because these are rounded values. Thus it is important to know how to handle tied observations in rank tests. In the Wilcoxon test, there are two types of ties to consider:

1. Both members of a pair are the same.
2. There are tied differences between pairs.

When both members of a pair are the same, the difference $y_d = 0$, and since zero is neither positive nor negative, it has no sign. Therefore differences of zero must be discarded and the value of *N* reduced accordingly.

When differences are tied, they should received the same ranks, and it is customary to give them the average of the ranks they occupy as a group. In the example above, students *O* and *P* have very nearly the same absolute difference between the first- and second-semester GPA. Had the absolute differences been exactly the same, say $|y_d| = 0.70$ for both students, then they would be tied for ranks 9 and 10, and the average rank of 9.5 would be entered for each student in the column of ranks.

Ties of this nature cause the variance to become smaller. The reduction in the size of the variance depends on the number of ties and the number of members in a tie. The computation of the variance can be found in textbooks on nonparametric statistics [see Conover (1998) or

Daniel (1990)]. However, the presence of tied observations usually causes little change in the computed value of z, and in practice the reduction in the size of the variance due to ties is unimportant unless there are a great number of ties or unless z is very near the critical value before the reduction is applied.

Procedure. Rank Test for Matched Pairs

To obtain the average signed-rank of the difference between pairs:

1. Find the difference between pairs.
2. Record the sign of the difference in one column and the absolute value of the difference in another.
3. Rank the absolute differences from smallest to largest.
4. Reattach signs of differences to their respective ranks to obtain signed ranks, which are then averaged to obtain $\bar{r}$.

Test of Hypothesis

$H_0: E(\bar{r}) = \mu = 0$

$H_a: \mu \neq 0$ or $\mu > 0$ or $\mu < 0$

Significance level: α
Test statistic:

$$z = \frac{\bar{r} - \mu_0}{\sqrt{V(\bar{r})}} = \frac{\bar{r} - 0}{\sqrt{(N+1)(2N+1)/6N}}$$

for $N \geq 20$ or

$$z = \frac{\bar{r} - 1/2 - \mu_0}{\sqrt{V(\bar{r})}} = \frac{\bar{r} - 1/2 - 0}{\sqrt{(N+1)(2N+1)/6N}}$$

for $N < 20$.
Region of rejection: $|z| \geq z_{\alpha/2}$ or z/z_α or $z < -z_\alpha$, respectively.

The rank test counterpart for testing the difference between means of two groups has already been discussed in Section 7.6. However, even though there are two groups, we need to compute $\bar{r}$ for only one group and test whether it is significantly different from $E(\bar{r})$. This is because the transformed data consist of the ranks 1 through N, and if $\bar{r}$ is known for one of the groups, then we could always find the corresponding average for the other group.

More precisely, if the two groups have sample sizes n_1 and n_2 and their averages are $\bar{r}_1$ and $\bar{r}_2$, respectively, then

$$n_1\bar{r}_1 + n_2\bar{r}_2 = \frac{N(N+1)}{2}$$

where $n_1 + n_2 = N$, because $N(N+1)/2$ is the sum of the consecutive integers from 1 to N. So generally we compute whichever average seems easier and then perform the z test,

$$z = \frac{\bar{r}_1 - (N+1)/2}{\sqrt{(N-n_1)(N+1)/12n_1}}$$

or

$$z = \frac{\bar{r}_2 - (N+1)/2}{\sqrt{(N - n_2)(N+1)/12n_2}}$$

as is appropriate.

EXERCISES

8.5.1. One of the side effects of cancer chemotherapy is that the treatment may interfere with nerve action. An oncologist is evaluating the effect of a heavy metal compound as a treatment for cervical cancer, and on each patient a measurement is taken on ulnar sensory nerve amplitude (in microamperes) before treatment begins and after the patient has been on treatment for 6 months. A significant decrease in nerve amplitude would indicate that the treatment has a potentially harmful side effect.

Patient:	1	2	3	4	5	6	7	8	9
Before:	6.7	7.0	7.1	9.0	9.8	10.0	10.1	10.9	11.0
After:	7.6	3.3	9.1	9.3	10.7	7.2	12.3	6.7	9.5
y_d:	−0.9	3.7	−2.0	−0.3	−0.9	2.8	−2.2	4.2	1.5

Patient:	10	11	12	13	14	15	16	17	18
Before:	11.3	11.5	11.7	11.9	12.4	12.5	12.6	12.8	14.0
After:	7.9	11.3	14.2	11.0	5.0	10.3	9.4	8.8	14.0
y_d:	3.4	0.2	−2.5	0.9	7.4	2.2	3.2	4.0	0.0

Patient:	19	20	21	22	23	24	25	26
Before:	14.2	14.6	14.8	15.0	15.0	15.6	16.6	18.1
After:	8.5	12.5	11.7	16.0	12.6	14.4	15.8	14.4
y_d:	5.7	2.1	3.1	−1.0	2.4	1.2	0.8	3.7

a. Why would the Wilcoxon signed-rank test be appropriate for analyzing these data?

b. What would be the most appropriate null and alternative hypotheses?

c. Show that $\bar{r} = 8.48$.

d. Perform the test of significance and draw conclusions about whether or not the treatment has a harmful side effect on nerve activity.

8.5.2. Use a nonparametric test to analyze the data in Exercise 8.2.11.

8.5.3. Use a nonparametric test to analyze the data in Exercise 8.3.6.

8.5.4. In Exercise 2.3.5 the following fictitious data were presented as supporting Galton's idea that skills are inherited and hence young children of skilled laborers should show greater manual dexterity than those of unskilled laborers:

Frequencies of Dexterity Skill Scores

Father:	x	g	f	e	d	c	b	a	A	B	C	D	E	F	G	X
Skilled:	0	0	0	1	0	0	1	1	0	1	1	1	0	1	2	1
Not:	1	1	1	0	2	1	0	0	1	1	0	0	1	1	0	0

On this scale lowercase x is the lowest possible measurement and an uppercase X the highest.

   a. Why can rank order statistics be used for a nonparametric test to compare the skills of the two groups of children?

   b. What assumptions of that test should be of concern for a statistical analysis of these data?

   c. Give the null hypothesis for the nonparametric test and the alternative that agrees with Galton's experimental hypothesis

   d. Test the null hypothesis and draw conclusions about the skills of the two groups of children.

REVIEW EXERCISES

Decide whether each of the following statements is true or false. If a statement is false, explain why.

8.1. The t distribution is appropriate for small sample sizes irrespective of whether or not the variance is known.

8.2. For each positive-integer degree of freedom, there is a different t distribution.

8.3. Gosset discovered that when n is small s^2 tends to overestimate σ^2.

8.4. For a one-sample t test, the region of rejection is uniquely determined by the alternative hypothesis and sample size.

8.5. For a fixed α level, as the degrees of freedom increase in a t test, the absolute value of the critical value increases.

8.6. $CI_{0.95}: \bar{y} \pm t_{0.025}s/\sqrt{n}$ contains 95% of all population means.

8.7. $\bar{y} \pm t_{\alpha/2,v}s/\sqrt{n}$ is narrower than the corresponding interval based on the standard normal distribution $\bar{y} \pm z_{\alpha/2}s/\sqrt{n}$.

8.8. If two samples consist of pairs of data, the experimenter may choose between the matched-pair t test or the t test for two independent samples.

8.9. In the matched-pair t test, the parameter in the null hypothesis must equal zero.

8.10. In a paired comparison t test involving 20 pairs of twins, there are 38 degrees of freedom.

8.11. A paired comparison t test should always be used when $\sigma_1^2 = \sigma_2^2$.

8.12. If a t test determines that the difference between two sample averages is significant, then the experimenter should conclude that two different populations were sampled.

8.13. If in a two-sample t test $\mu_1 = \mu_2$, then the computed value of t will be exactly zero.

8.14. If for two populations $\sigma_1^2 = \sigma_2^2$ the best estimate of the common variance is $(s_1^2 + s_2^2)/2$ irrespective of other considerations.

8.15. If H_0: $\mu_1 = \mu_2$ is true, then for the group comparison t test the t statistic should be close to 0.

8.16. If $\sigma_1^2 = \sigma_2^2$ is true, then the F statistic should be close to 0.

8.17. When σ_1^2 and σ_2^2 are unequal and unknown and the samples are small, there is no exact test for a hypothesis of equality of means from the two populations.

8.18. There are many F distributions, one for each ordered pair of degrees of freedom.

8.19. In a box-and-whisker plot, the "box" is constructed so that 50% of the observations lie within it.

8.20. $1/F_{0.005,6,8} = F_{0.995,8,6}$.

SELECTED READINGS

Boland, P. J. (1984). A biographical glimpse of William Sealy Gosset. *American Statistician*, **38**, 179–183.

Boneau, C. A. (1960). The effects of violations of assumptions underlying the t test. *Psychological Bulletin*, **57**, 49–64.

Box, J. F. (1981). Gosset, Fisher, and the t distribution. *American Statistician*, **35**, 61–66.

Conover, W. J. (1998). *Practical Nonparametric Statistics*, 3rd ed. Wiley, New York.

Daniel, W. W. (1990). *Applied Nonparametric Statistics*, 2nd ed. PWS-KENT, Boston.

Eisenhart, C. (1979). On the transition from "Student's" z to "Student's" t, *American Statistician*, **33**, 6–10.

Gayen, A. K. (1949). The distribution of "Student's" t in random samples of any size drawn from non-normal universes. *Biometrika*, **36**, 353–369.

Geary, R. C. (1947). Testing for normality. *Biometrika*, **34**, 209–242.

Grunow, D. G. C. (1951). Test for the significance of the difference between means in two normal populations having unequal variance. *Biometrika*, **38**, 252–256.

Guenter, W. C. (1981). Sample size formulas for normal theory t tests. *American Statistician*, **35**, 243–244.

Lackritz, J. R. (1984). Exact p values for F and t tests. *American Statistician*, **38**, 312–314.

Neyman, J. (1938). Mr. W. S. Gosset. *Journal of the American Statistical Association*, **33**, 226–228.

Owen, D. B. (1965). The power of Student's t test. *Journal of the American Statistical Association*, **60**, 320–333.

Scheffé, H. (1943). On solutions of the Behrens–Fisher problem based on the t-distribution. *Annals of Mathematical Statistics*, **14**, 35–44.

Scheffé, H. (1944). A note on the Behrens–Fisher problem. *Annals of Mathematical Statistics*, **15**, 430–432.

"Student" [William Sealy Gosset] (1908). The probable error of a mean. *Biometrika*, **6**, 1–25.

Walsh, J. E. (1947). On the power efficiency of a t-test formed by pairing sample values. *Annals of Mathematical Statistics*, **18**, 601–604.

Welch, B. L. (1937). The significance of the difference between two means when the population variances are unequal. *Biometrika*, **29**, 350–362.

9 Distributions of Two Variables

Thus far our discussion of inference has focused on the values of a *single* variable of interest obtained from a random sample. We saw in Chapter 2, however, that it is possible to consider more than one variable associated with a given population. For example, *two* variables from the same population that might be considered are age and blood pressure. Other examples are height and weight, caloric intake and weight loss, and hours of study and grade on an exam. In this chapter we consider pairs of variables and possible relationships between these variables. In all of the sections except 9.5 both variables are numerical. In Section 9.5 the variables are nominal. It is also possible to study the relationship among several variables; for example, blood pressure is related to age, weight, and exercise. Relationships among more than two variables are discussed in Chapter 14. Relationships between two variables, one of which is nominal and the other numerical, are also discussed in Chapter 14.

9.1. SIMPLE LINEAR REGRESSION

A question often asked about a pair of variables x and y is, "How do changes in x affect the value of y?" For example, as a man ages five years, how will this affect his blood pressure? Or we might ask a related question, "What is the expected value of y for a certain value of x?" For example, if a man is 30 years old, what is his expected blood pressure?

The x variable age is called the *independent variable* or the *predictor variable*, and the y variable blood pressure is called the *dependent variable* or the *response variable*. If x and y have a relationship with each other, to predict y from x, we have to be able to find a model for the relationship. The simplest model of a relationship is a straight line. If a straight-line model is appropriate, the line is called the *regression line* and we say that we are regressing y on x. This type of regression is called *simple linear regression*; "simple" indicates that there is only one independent variable and "linear" indicates that the model is a straight line.

When dealing with pairs of variables, we have the same difficulty as with a single variable, namely, we usually are unable to measure all possible members of the population. In the single-variable case, we solved this difficulty by using a random sample to make inference about the population. We do the same for pairs of variables. For example, if we are interested in studying a possible linear relationship between age and blood pressure in adult males, we use a random sample of men, obtain sample data about age and blood pressure, and then see if a straight line fits the data.

Statistics for Research, *Third Edition*, Edited by Shirley Dowdy, Stanley Weardon, and Daniel Chilko.
ISBN 0-471-26735-X © 2004 John Wiley & Sons, Inc.

Say a random sample of 10 adult males yields the following data:

Age x:	28	23	52	42	27	29	43	34	40	28
Systolic blood pressure (mm Hg) y:	70	68	90	75	68	80	78	70	80	72

We begin our analysis by plotting the pairs x, y as points (Figure 9.1). This graph is called a *scatter plot*. The points certainly do not fall exactly on a straight line, but there does appear to be a general linear upward trend such that higher ages are associated with higher systolic blood pressure. Regression is used to fit a straight line to such data in a unique way so that the line can be used to predict systolic blood pressure from age.

It is possible, of course, that two variables are related in some other manner than by a straight-line relationship, or perhaps they are not related to each other at all. Thus our discussion of simple linear regression must include a method for determining whether or not a straight line is the appropriate model for a given set of data (Section 9.2).

Since the simplest possible relationship between two variables is a straight line, it is natural to try to use this model before considering more complex models. Sometimes, even if the true relationship is something other than a straight line (as in Figure 9.2), a straight line may be close enough to the true relationship for a preliminary analysis. A straight line is convenient to use because the mathematics involved is relatively simple.

Sometimes the true relationship is definitely not linear and a straight line is a very poor model of the relationship. One example is the relationship between the amount of nitrogen fertilizer used on a field and the yield of the crop. The true relationship is quadratic and would be represented by a parabola. In this example, however, economy limits the amount of fertilizer that the farmer would consider using, and in the economical range the relationship might be approximated by a straight line (Figure 9.3). Unfortunately, not every curvilinear relationship will have such a subset of x values that are the main interest of the investigator. Curvilinear relationships are discussed in Sections 14.6 and 14.7.

To understand how a straight line is fitted to a set of data that consists of pairs of values obtained for two variables, we consider an overly simplified example. Imagine that an efficiency expert is investigating a possible linear relationship between the number of hours of instruction employees receive about a certain assembly procedure in a factory and the number of units they are able to produce per hour. The following data are collected from five employees:

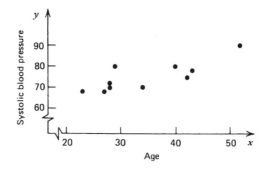

FIGURE 9.1. A scatter plot of age and systolic blood pressure.

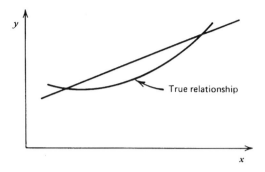

FIGURE 9.2. A relationship that is approximated by a straight line.

Hours of Instruction x	Units per Hour y
1	5
2	4
3	6
4	8
5	7

In a real study the investigator would take a random sample of several employees from the groups of employees with the different levels of instruction. However, to keep this illustration simple, we imagine a random sample of just one employee at each level. The approach is the same for several employees at each level.

The first thing the investigator does is graph the scatter diagram (Figure 9.4). If there are enough points in the scatter diagram, it may indicate the general shape of the curve or line that can possibly be used as a model for the variables. A generalized random scatter may indicate that there is no relationship between the variables.

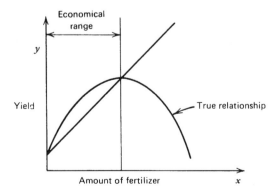

FIGURE 9.3. A relationship that is approximated by a straight line in a certain region of the independent variable.

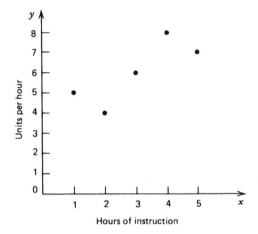

FIGURE 9.4. Scatter diagram for the production study.

Even if the relationship is linear, not all of the points will lie exactly on the line. The model (Figure 9.5) is of the form

$$y = \alpha + \beta x + \varepsilon$$

The regression line is given by the function

$$f(x) = \alpha + \beta x$$

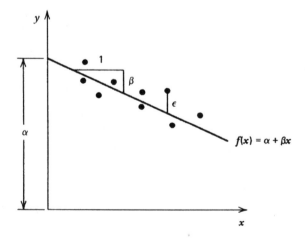

FIGURE 9.5. A regression line.

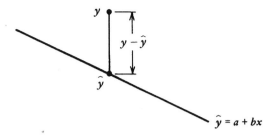

FIGURE 9.6. A vertical deviation from a least-squares line.

in which α is the y intercept and β is the slope[†] (the change in y per unit increase in x). The term ε indicates the vertical deviation of a particular point from the line, that is, the line represents the mean y response at a given x value, but individuals will deviate from the mean response due to random variability.

Returning now to the factory example, if the investigator thinks the relationship is linear, the problem is to specify the line that characterizes the relationship by finding the equation of the line. Since only a sample is available, the parameters α and β must be approximated. One approach is simply to draw a line that seems to fit the data; however, this would not be a unique solution. Another approach is to draw a line that has an equal number of points above and below; this is not unique either. Or the line might be drawn such that the vertical deviations would sum to zero; but again, this is not unique.

The problem of approximating the true regression line is solved by using the *least-squares trend line*, also called the *sample regression line*. The least-squares trend line is that unique line for which the sum of the squares of the vertical distances of the sample points from the line is as small as possible (Figure 9.6). Assume that the least-squares line is of the form

$$\hat{y} = a + bx$$

in which a is the y intercept and b is the slope. We minimize the function

$$f(a, b) = \sum (y - \hat{y})^2$$

in which y is an observed value and $\hat{y}$ is the value predicted by the line for the corresponding x. That is, we find a and b such that this sum is as small as possible. This is done using calculus and leads to two simultaneous equations called the *normal equations*:

$$an + b \sum x = \sum y$$

$$a \sum x + b \sum x^2 = \sum xy$$

Solving these two equations simultaneously, the slope is

$$b = \frac{\sum xy - \left(\sum x\right)\left(\sum y\right)/n}{\sum x^2 - \left(\sum x\right)^2/n}$$

[†]Note that this use of α and β is entirely different from the use of these symbols in connection with Type I and Type II error.

and

$$a = \bar{y} - b\bar{x}$$

The denominator of the slope should be familiar; it is similar to the computational form for the sum of squared deviations that appears in a sample variance,

$$\sum (x - \bar{x})^2 = \sum x^2 - \left(\sum x\right)^2 / n$$

The numerator of the slope can be shown to be a sum of products:

$$\sum (x - \bar{x})(y - \bar{y}) = \sum xy - \left(\sum x\right)\left(\sum y\right) / n$$

Because expressions of this type are used so frequently in regression, it is convenient to use some brief symbols to represent them. We use

$$S_{xx} = \sum (x - \bar{x})^2 = \sum x^2 - \left(\sum x\right)^2 / n$$

and

$$S_{xy} = \sum (x - \bar{x})(y - \bar{y}) = \sum xy - \left(\sum x\right)\left(\sum y\right) / n$$

for the sum of the squared x deviations and for the sum of the products of deviations. Then the estimated slope is

$$b = \frac{S_{xy}}{S_{xx}}$$

The least-squares line has the property of containing the point $(\bar{x}, \bar{y})$, in which $\bar{x}$ is the sample average of the x values and $\bar{y}$ is the sample average of the y values. This point may or may not be one of the sample points; in this example it happens to be a data point (Figure 9.7). Since one of the points on the line is known, $(\bar{x}, \bar{y})$, the line can be determined once we know its slope. The slope is given by the formula

$$b = \frac{S_{xy}}{S_{xx}} = \frac{\sum xy - \left(\sum x\right)\left(\sum y\right) / n}{\sum x^2 - \left(\sum x\right)^2 / n}$$

so it can be computed as follows:

x	y	x^2	xy
1	5	1	5
2	4	4	8
3	6	9	18
4	8	16	32
5	7	25	35
15	30	55	98

$$b = \frac{98 - (15)(30)/5}{55 - (15)^2/5} = \frac{8}{10} = 0.8$$

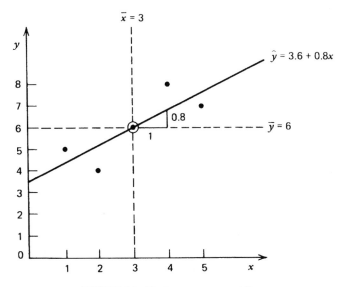

FIGURE 9.7. The least-squares trend line.

The slope indicates that as x increases one unit y increases 0.8 units. An additional hour of instruction increases mean productivity by 0.8 units per hour. Using the slope and starting at $(\bar{x}, \bar{y}) = (3, 6)$, we move one unit to the right and 0.8 unit up to locate a second point on the line (if the slope had been negative, we would move down). Since two points determine a unique straight line, the least-squares trend line can now be drawn.

The y intercept can be found from the formula

$$a = \bar{y} - b\bar{x}$$
$$= 6 - 0.8(3)$$
$$= 3.6$$

Thus the equation of the line is

$$\hat{y} = 3.6 + 0.8x$$

This is the sample regression line, and assuming that it is the proper model for the investigation, it is used to predict y for a given x; that is, it can predict the number of units per hour that would be produced if an employee had a certain number of hours training. Only values between 1 and 5 may be specified for the independent variable x, since data were collected only for that range. Extrapolation outside the range of the x variable is not reliable since the relationship may not be linear in other regions.

Remember that a sample regression line may be used for prediction only if the model is appropriate. It is always possible to compute the least-squares line; its usefulness for prediction is a different question, which will be dealt with in the next section.

The slope of the least-squares line gives us some information about the nature of the relationship. If b is close to zero, it may be approximating a true slope of $\beta = 0$. A slope of $\beta = 0$ indicates that there is no relationship between x and y, or that the y means have a constant value, or it could indicate a nonlinear relationship (however, not all nonlinear relationships have $\beta = 0$). If x and y are linearly related and increase together, then b approximates $\beta > 0$. If y decreases as x increases, then b approximates $\beta < 0$ (Figure 9.8).

Note that the slope of the least-squares line is not a pure number, but it is expressed in certain units of measurement. For example, if the variables are x, height in inches, and y, weight in pounds, then b is expressed in

$$\frac{(\text{inches})(\text{pounds})}{(\text{inches})^2} = \frac{\text{pounds}}{\text{inch}}$$

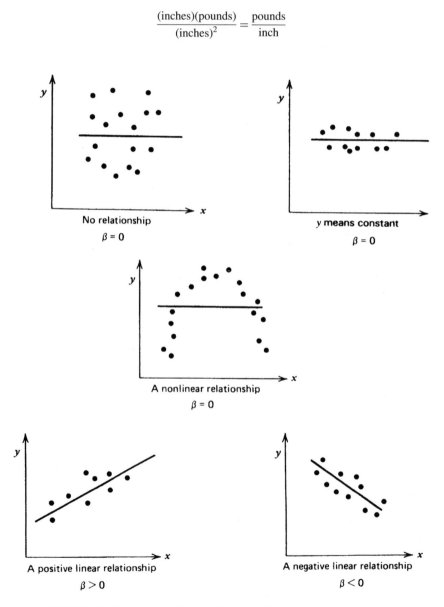

FIGURE 9.8. Various types of scatter diagrams with population regression lines.

that is, in pounds per inch. If the same subjects were measured in centimeters and kilograms, b would have a different value because it would be in different units of measurement. Because of this, the magnitude of the slope cannot be used as a measure of the strength of the linear relationship. A measurement used to express the degree of association between x and y is the *correlation coefficient.* This is discussed in Section 9.4.

Further, we should note that the equation

$$\hat{y} = a + bx$$

is the sample regression line for the *regression of y on x.* The *regression of x on y* is usually a different line. Thus, if x is hours of sleep per night and y is pounds overweight, we might regress pounds overweight on hours of sleep; that is, we would want to predict pounds overweight from hours of sleep (if in fact there was a linear relationship). On the other hand, we might be interested in the regression of hours of sleep on pounds overweight; that is, we would want to predict hours of sleep from pounds overweight. In most studies, the two lines would be different.

Procedure. The Least-Squares Trend Line

Given n pairs of observations x, y, the least-squares trend line or sample regression line for the regression of y on x is

$$\hat{y} = a + bx$$

To find this line, compute

$$\sum x, \quad \sum x^2, \quad \sum y, \quad \text{and} \quad \sum xy$$

and then compute

$$\bar{x} = \sum x / n$$

$$\bar{y} = \sum y / n$$

$$S_{xx} = \sum x^2 - \left(\sum x\right)^2 / n$$

$$S_{xy} = \sum xy - \left(\sum x\right)\left(\sum y\right) / n$$

The slope is

$$b = \frac{S_{xy}}{S_{xx}}$$

and the y intercept is

$$a = \bar{y} - b\bar{x}$$

EXERCISES

9.1.1. Which of the following completes the statement correctly? In the equation $\hat{y} = a + bx$, the value of a:

a. Can never be negative

b. Determines the slope of the trend line

c. Determines the point at which the trend line intersects the y axis

d. Determines the point at which the trend line intersects the x axis

9.1.2. Draw a scatter diagram and find the least-squares trend line for the following sample data.

Number of hours of study x:	4	5	6	7	8	9	10	11	12
Grade on exam y:	55	60	50	70	70	70	80	90	85

9.1.3. If x is measured in pounds and y is measured in days, what are the units of measurement for the slope of the least-squares trend line?

9.1.4. In each case below, use the information given to obtain the numerical value of the slope of the least-squares trend line.

a. $\hat{y} = 5$ if $x = 10$, and $\hat{y} = 10$ if $x = 20$.

b. $\sum (x - \bar{x})(y - \bar{y}) = 30$, $\sum (y - \bar{y})^2 = 10$, and $\sum (x - \bar{x})^2 = 5$.

c. $\hat{y} = -3 + 15x$.

d. $\bar{y} = 10$, $\bar{x} = 13$, and $\hat{y} = 15$ if $x = 15$.

9.1.5. A botanist studying *Arabadopsis thaliana* notes a relationship between the number of branches on the plant and the number of seed pods it produces. A preliminary analysis yields the following data:

Branches x:	14	15	16	17	18
Seed pods y:	50	60	70	100	120

a. Find $\sum (x - \bar{x})(y - \bar{y})$.

b. Compute the slope of the trend line.

c. Give the equation of the trend line.

d. What is the predicted number of seed pods on a plant with 16 branches?

9.1.6. Obesity in mice is inherited. For every gram above mean mature weight that a female mouse is in her generation, the mean of her daughters' mature weights is $2/5$ g above the mean weight in their generation.

a. What is the slope of the regression line?

b. Predict the mature weight of a daughter if her mother's weight is 28 g, the mean for the mother's generation is 23 g, and the mean for the daughter's generation is 20 g.

c. Predict the mature weight of a daughter if her mother's weight is 23 g, the mean for the mother's generation is 20 g, and the mean for the daughter's generation is 22 g.

9.1.7.   A study of nursing activities is conducted in a 100-bed hospital in Kansas. The nursing staff remains constant through the study, but the patient load varies, so it is possible to observe how nurses allocate their duty time with different patient loads. One of the nursing activities observed and measured is patient care and another is the time spent on records and reports. A separate study is made for each hospital ward, and the data below represent the minutes per staff duty hour spent on these activities by the nurses in the surgery ward under varying patient loads:

Patient load:	2	3	4	6	7	8
Patient care:	44.7	53.0	71.7	111.3	129.4	159.9
Records and reports:	15.8	16.0	13.3	10.4	7.2	9.3

  a. Examine the relationship between patient load and time spent in patient care.

     i. What sort of linear relationship seems logical, positive or negative?

     ii. Do the data tend to support the experimental hypothesis?

     iii. Compute the slope of the least-squares trend line that shows how an increase in patient load affects staff time allocated to patient care.

     iv. What are the units of measurement for the slope of the trend line?

     v. Find the equation that would allow surgery-ward nurses to predict the amount of time they have to allocate per staff duty hour for a given number of patients in their ward.

     vi. Use the equation to estimate the amount of time required for patient care if there were only one patient in the ward. (Since one patient is outside the range of the data collected, this may be a poor estimate.) Use it to estimate the time required for 5 patients.

  b. Examine the relationship between patient load and time spent on records and reports.

     i. Does the linear relationship appear to be positive or negative?

     ii. Does such a relationship seem intuitively logical prior to the survey or is the relationship one that can be rationalized after the data are collected?

     iii. Compute the least-squares trend line that shows how an increase in patient load affects the staff time allocated to records and reports.

     iv. Suppose that a minimum of 5 minutes per staff duty-hour is required for necessary records and reports. Assume that the trend can be extrapolated and estimate the point at which patient load becomes so heavy that the surgical nursing staff no longer has adequate time for record keeping.

9.1.8.   When a straight line is fitted to data that follow a binomial distribution, a special procedure known as *probit analysis* is employed. This procedure takes into account such conditions as the relationship between the mean and the variance of the binomial distribution and the fact that the trend is rarely linear over the full range of π. However, the first step in probit analysis is to fit a "provisional" line to the data, and this can be done by employing the least-squares procedure developed in this section. Suppose an advertising firm wants to determine the relationship between the number of times a commercial is shown on national television and the percentage of viewers who have seen the commercial.

Number of times commercial shown x:	10	15	20	25	30
Percentage of viewers y:	13	32	35	53	67

a. Use least-squares procedures to find the slope of the trend line.
b. Give the equation of the "provisional" line.
c. Use the equation to estimate how many times a television commercial must be shown before 50% of the viewers have seen it. (This is called the *50% effective dose*, or *ED_50*, in probit analysis.)

9.1.9. Francis Galton extended least-squares techniques by employing them in a study of the relationship between mature heights of fathers and their sons. He collected hundreds of observations, plotted them on graph paper, and noted a straight-line relationship among average heights. Some of his data in inches might be as follows:

Fathers' height:	65	66	67	68	69	70	71
Average height of sons:	66.9	67.8	68.0	67.9	69.6	69.2	70.1

a. What is the average height of the fathers' generation?
b. What is the average height of the sons' generation?
c. If a group of fathers are each 1 in. above average height for their generation, what is the expected average deviation of their sons from the average height of their respective generation?

9.1.10. A study is made to determine the rate of disappearance from the environment of radioactive chemicals after a nuclear accident. Strontium 85 is released in an alfalfa field in a simulated accident. Twenty goats are allowed to graze the field, and at 30-day intervals the level of strontium 85 is measured in dried samples of alfalfa as well as in the goats' milk. The alfalfa data are given below:

Days after release x:	30	60	90	120	150
Strontium level in dried alfalfa y, ppm:	1.85	1.43	1.21	1.19	1.37

a. Compute the least-squares trend line.
b. What are the units of measure for the slope? For the y intercept?
c. The measured level of strontium 85 in alfalfa on day 150 seems somewhat contrary to the trend shown in the other data. Compute the predicted level for $x = 150$. Compute the deviation of the observed value from this point on the trend line.

9.1.11. Fit a straight line to the age and blood pressure data given in this section.

9.2. MODEL TESTING

The least-squares line can always be computed for any set of two or more points with different x values. It may not be appropriate, however, to predict from this line. For prediction, two conditions are necessary:

1. The straight-line model fits the data.
2. The straight line being estimated is not horizontal ($\beta \neq 0$); that is, the regression line is a better predictor of y than $\bar{y}$.

In this section we discuss each of these conditions in turn.

First we need to be more precise as we speak of a regression line being a model for a certain research situation. Two variables x, y (Figure 9.9) meet the conditions for the regression of y on x if:

1. The x values are fixed by the experimenter and are measured with negligible error.[†]
2. For each x value there is a normal distribution of y values. (This assumption is necessary for inference.)

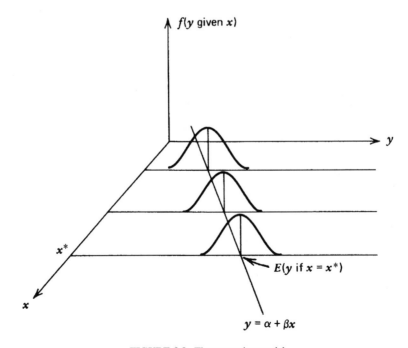

FIGURE 9.9. The regression model.

[†]Regression analysis is also possible in cases where x is a random variable (see Section 9.4).

3. The distribution of y for each x has the same variance, symbolized as $\sigma_{y \cdot x}^2$ and read as the "variance of y independent of x" to indicate that the variance around the trend line is the same irrespective of the value of x.
4. The expected values of y for each x lie on a straight line.

Another way to express these conditions is to say that the variables satisfy the model

$$y = \alpha + \beta x + \varepsilon$$

in which the ε's are normally distributed with a mean of zero and a variance of $\sigma_{y \cdot x}^2$ and the ε's are independent of the x's and independent of each other.

One way to test for violations of these assumptions is by an examination of the residuals $y - \hat{y} = e$ that result from fitting the least-squares line to the sample data. In the small example about employee training used for illustration purposes in Section 9.1, the residuals could be computed as follows:

x	y	$\hat{y}$	$y - \hat{y} = e$
1	5	$3.6 + 0.8(1) = 4.4$	0.6
2	4	$3.6 + 0.8(2) = 5.2$	-1.2
3	6	$3.6 + 0.8(3) = 6.0$	0.0
4	8	$3.6 + 0.8(4) = 6.8$	1.2
5	7	$3.6 + 0.8(5) = 7.6$	-0.6

Since the e's estimate the ε's in the model, to check for normality, an overall plot of the residuals can be drawn as a dot diagram (Figure 9.10). In this unrealistically small example it is difficult to check for departures from normality because of the small number of points. Some patterns that appear with larger samples are illustrated in Figure 9.11.

Linearity can be checked by plotting the residuals e against the predicted values $\hat{y}$ (Figure 9.12). A linear relationship is reflected in a random scatter about a horizontal line at $e = 0$. If the relationship is nonlinear, it usually results in a systematic plot that has some pattern. A systematic pattern could also indicate that another independent variable is affecting y.

Equality of variances can be checked by plotting the residuals e against the predicted values $\hat{y}$ or the independent variable x (Figure 9.13). Equal variances result in a horizontal band of points, whereas variances that depend on the magnitude of x will result in a fan-shaped distribution. In situations where the variance of y is proportional to the magnitude of x and the trend line passes through the origin, the trend line is usually estimated by the ratio of the two means, $\bar{y}/\bar{x}$ (see Section 9.7).

The regression model assumes independence of the ε's. This means that the random error in one observation does not affect the random error in another observation. This assumption is sometimes violated. If the observations have a natural sequence in time or space, the lack of independence is called *autocorrelation*.

FIGURE 9.10. An overall plot of residuals.

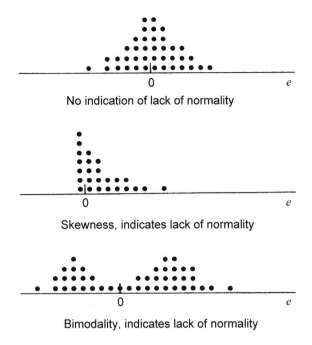

FIGURE 9.11. Checking overall plots of residuals for violations of normality.

Autocorrelation may occur for several reasons: The dependent variable may follow economic trends; an instrument may be drifting out of calibration; batch processes in a reactor system may leave some of the product to be carried over to the next batch; observations may be from adjacent experimental plots that have similar conditions. These are only some examples. Diagnosis is difficult, but this type of dependence can sometimes be detected by plotting the residuals against the time order or the spatial order of the observations (Figure 9.14).

The visual inspection of the original scatter diagram of the data and the various types of residual plots is an important first step in any regression analysis and should not be omitted. Statistical programs on computers make it possible to inspect these diagrams with little labor. If the diagrams reveal any departures from the assumptions required for regression, a different model may be necessary, or perhaps a transformation can be used on the data before the regression analysis (Sections 14.6 and 14.7). If the visual inspection does not turn up any departures from assumptions, we have not proved that the model is correct, but at least there is no overwhelming evidence that it is wrong.

Besides these visual checks of the assumptions, there is a statistical test that can be performed to see if there is a significant lack of fit with a straight line. Repeated observations are necessary at each x value to carry out such a test (see Draper and Smith 1998). This test for lack of fit is found in some statistical computer packages such as SAS and JMP.

If we decide that a straight line seems to be a reasonable model, then we need to determine that the line is not horizontal. A horizontal line indicates that x does not make a significant

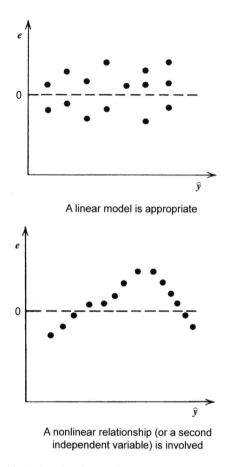

A linear model is appropriate

A nonlinear relationship (or a second
independent variable) is involved

FIGURE 9.12. Residuals plotted against predicted values to check for a linear relationship.

contribution to the prediction of y; that is, there is no linear relationship. To test whether the line is horizontal, we test

$$H_0: \beta = 0$$

in which β is the slope of the population regression line. Rejection of this hypothesis is evidence that the line explains a significant portion of the variability in y. Acceptance of this hypothesis means that there is no advantage to considering the values of x as we attempt to predict y. We could do just as well by using the model $\hat{y} = \bar{y}$.

The test statistic is a t statistic in which b is the estimator of the parameter β. To estimate the standard error of the estimator b for the denominator of the t test, we first must consider the variance of the y values about the sample regression line. We use the residuals and compute the sum of the squared residuals, and then we divide this sum by the degrees of freedom that are $n - 2$ for simple linear regression (thus a minimum of 3 points is required for this test).

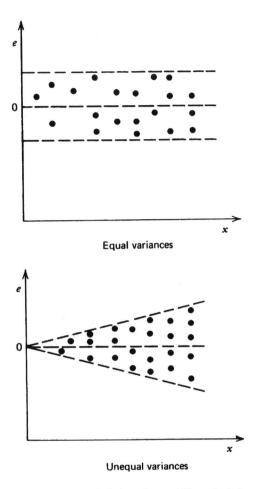

Equal variances

Unequal variances

FIGURE 9.13. Residuals plotted against the independent variable to check for equality of variances.

It may be helpful to explain why the degrees of freedom in the denominator for the variance around the sample trend line are $n - 2$ rather than the $n - 1$ we use when computing the variance around the sample mean. The explanation begins by remembering that the sample trend line is

$$\hat{y} = a + bx$$

so the sum of squared deviations around the trend line is

$$\sum (y - \hat{y})^2 = \sum (y - a - bx)^2$$

Since a and b, respectively, are estimates of α and β, the two parameters of the straight line, we simply continue the practice we first began in Section 5.2 of subtracting a degree of freedom for each parameter we estimate.

Independence of ϵ's with each other

Autocorrelation

FIGURE 9.14. Residuals plotted against the order in which they were observed.

For example, in the employee training example, the variance of the observations about the least-squares line is computed as follows:

$y - \hat{y}$	$(y - \hat{y})^2$
0.6	0.36
-1.2	1.44
0.0	0.00
1.2	1.44
-0.6	0.36
	3.60

and

$$s^2_{y \cdot x} = \frac{\sum (y - \hat{y})^2}{n - 2} = \frac{3.60}{5 - 2} = 1.2$$

in which n is the number of pairs of data. Variance about the trend line is the variance in y when we have removed the effect of the x variable. In the employee training example, *before* we have removed the effect of the x variable, the variance in y is

$$s_y^2 = \frac{\sum(y - \bar{y})^2}{n - 1} = \frac{10}{4} = 2.5$$

This represents the variance of the data points about $\bar{y}$. In contrast, $s_{y \cdot x}^2$ is the variance about the trend line and is the variance in y independent of x. Note that 2.5 is reduced to 1.2 when the effect of x is removed (Figure 9.15).

In practice, it is usually easier to use the short computational formula

$$\sum(y - \hat{y})^2 = \sum(y - \bar{y})^2 - \left[\sum(x - \bar{x})(y - \bar{y})\right]^2 / \sum(x - \bar{x})^2$$

$$= S_{yy} - \frac{S_{xy}^2}{S_{xx}}$$

$$= S_{yy} - bS_{xy}$$

in which

$$S_{yy} = \sum(y - \bar{y})^2 = \sum y^2 - \left(\sum y\right)^2 / n$$

Using $s_{y \cdot x}^2$, the standard error of b can be shown to be

$$\frac{s_{y \cdot x}}{\sqrt{S_{xx}}}$$

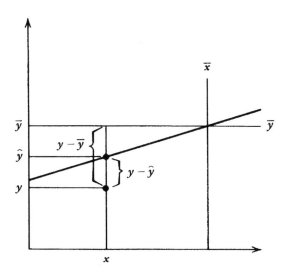

FIGURE 9.15. Deviation of an observed y value from the average y value and from a predicted y value.

and the t statistic for a test of H_0: $\beta = 0$ is

$$t = \frac{b - \beta_0}{s_{y \cdot x} / \sqrt{S_{xx}}}$$

with $n - 2$ degrees of freedom.

In the training example, to test H_0: $\beta = 0$ against H_a: $\beta > 0$ at $\alpha = 0.05$, we would reject the null hypothesis if $t \geq t_{0.05,3} = 2.353$. A one-tailed test is used because additional training is expected to increase productivity if it is of any effect at all. Then

$$t = \frac{0.8 - 0}{\sqrt{1.2} / \sqrt{10}} = 2.31$$

and the null hypothesis is not rejected. Thus the line seems to be horizontal and the equation of the trend line should not be used for prediction. Note that the t statistic of 2.31 is very close to the critical value, so it is possible that a larger sample size might provide evidence that the line does contribute significant information about y. We repeat again that the small sample size here is unrealistic and is used only to keep the computations to a minimum.

If it is possible to reject $\beta = 0$, then prediction from the least-squares line is appropriate. Prediction may be done only for values of x within the range of the collected data. Extrapolation outside of that range is not reliable.

Values other than zero may be used in the null hypothesis when testing the slope parameter if this is reasonable for the experiment. The test procedure is analogous.

Procedure. Testing the Slope Parameter

Assumption: $y = \alpha + \beta x + \varepsilon$ with the ε's independently normally distributed with a mean of 0 and a variance $\sigma_{y \cdot x}^2$

Test of Hypothesis
H_0: $\beta = \beta_0$
H_a: $\beta \neq \beta_0$ or $\beta > \beta_0$ or $\beta < \beta_0$

Significance level: α
Test statistic:

$$t = \frac{b - \beta_0}{s_{y \cdot x} / \sqrt{S_{xx}}}$$

with

$$b = \frac{S_{xy}}{S_{xx}} \quad \text{and} \quad s_{y \cdot x}^2 = \frac{S_{yy} - b S_{xy}}{(n - 2}$$

Region of rejection: $|t| \geq t_{\alpha/2}^{n-2}$, or $t > t_{\alpha, n-2}$ or $t < -t_{\alpha, n-2}$, respectively.

EXERCISES

9.2.1. For the data in Exercise 9.1.2:
   a. Carry out a residual analysis.
   b. Show that $s_{y \cdot x}^2 = 33.57$.
   c. To test the significance of the least-squares line:
      i. Give the most logical null and alternative hypotheses.
      ii. Give the critical value.
      iii. Compute the test statistic and state the conclusion.

9.2.2. Explain the difference between y and $\hat{y}$.

9.2.3. If y is the number of fish caught in x hours of fishing, give the units of measurement for:
   a. The slope of the trend line
   b. A predicted y value
   c. The point in which the trend line meets the y axis

9.2.4. Some species of tropical fish bear their young alive rather than lay eggs. An aquarium keeper wants to determine whether the number of young increases with each parity (time when young are produced). The following data are available for study:

Order of parity:	1	2	3	4	5
Number of young:	7	11	9	13	15

   a. Find the slope of the sample regression line.
   b. Compute the sample variance about the trend line.
   c. What are the most logical null and alternative hypotheses about the slope of the regression line?
   d. Why is a two-sided alternative inappropriate?
   e. To perform the test:
      i. What assumptions must be made about the distributions of x and y?
      ii. If the assumptions are valid, what conclusion should be drawn?

9.2.5. Review Exercise 9.1.7 of this chapter, in which there is a discussion of the effect of patient load on nursing activities in a hospital.
   a. Conduct a test of hypothesis to see if patient load can be used to predict the time spent on patient care.
      i. Give the null hypothesis in symbols and in a complete sentence.
      ii. Why should the alternative hypothesis be one-sided?
      iii. Give the critical value of the test statistic for $\alpha = 0.05$.
      iv. Perform the test of significance.
   b. Conduct a test of hypothesis about patient load as a predictor of the time available for records and reports.
      i. Give the null hypothesis.
      ii. Why should the alternative hypothesis be two-sided?
      iii. Perform the test of significance at $\alpha = 0.01$.

9.2.6. When experimentation with lysergic acid diethylamide (LSD) first began, the hallucinogenic effect was noted as so similar to the symptoms of schizophrenia that medical scientists thought they had discovered a chemical cause of the mental disorder. Because an increase in the level of copper in the blood is frequently (but not always) associated with schizophrenia, a study was made to see whether the level of blood copper increased with the administration of increasing dosages of LSD.

a. What null hypothesis would be used in an analysis of this experiment?

b. What would be the alternative hypothesis?

c. Dosages were calibrated according to the percentage of those receiving the dosage who hallucinate. The level of blood copper was measured at each dosage. The data obtained were as follows:

Effective dosage (%):	0	25	50	75	100
Level of blood copper (mg/liter):	0.87	0.98	0.70	0.90	1.05

i. Compute the slope of the least-squares trend line.

ii. Test the $\beta = 0$ at $\alpha = 0.5$.

iii. Draw conclusions, answering the following questions: Do increasing dosages of LSD cause significant increases in blood copper level? Because increased blood copper is a common condition in schizophrenia, is there significant evidence that LSD may be a chemical cause of schizophrenia?

9.2.7. Review Exercise 9.1.10 in which a nuclear accident is simulated by releasing strontium 85 in an alfalfa field.

a. Compute $\sum (y - \hat{y})^2$ by using the short computational formula.

b. Compute $\sum (y - \hat{y})^2$ by finding the expected value on the trend line for each value of x and subtracting it from the observed value.

c. In performing a test of significance of the least-squares trend line:

    i. What is the null hypothesis?

    ii. Why is the alternative H_a: $\beta < 0$?

    iii. What is the critical value of the test statistic for $\alpha = 0.05$?

    iv. What is the decision about the null hypothesis? What should be concluded?

9.2.8. In Exercise 9.1.9 involving the relationship between fathers' and sons' heights:

a. Compute the expected height of sons y of fathers of each height x given in the experiment.

b. Compare observed height y with expected height $\bar{y}$ and compute:

    i. The sum of the deviations from the trend line, $\sum (y - \hat{y})$

    ii. The sum of the squared deviations from the trend line, $\sum (y - \hat{y})^2$

c. Compare observed height y and expected height $\hat{y}$ in terms of how they deviate from the average; compute:

    i. The sums of the deviations from the average, $\sum (y - \bar{y})$ and $\sum (\hat{y} - \bar{y})$

    ii. The sums of the squared deviations from the average, $\sum (y - \bar{y})^2$ and $\sum (\hat{y} - \bar{y})^2$

d. Use the above computations to empirically verify the following mathematical identities:

  i. The sum of squares from the average equals the sum of squares due to the linear trend plus the sum of squares from the trend line:

$$\sum (y - \bar{y})^2 = \sum (\hat{y} - \bar{y})^2 + \sum (y - \hat{y})^2$$

  ii. The sum of squares due to the linear trend is

$$\sum (\hat{y} - \bar{y})^2 = \frac{\left[\sum (y - \bar{y})(x - \bar{x}) \right]^2}{\sum (x - \bar{x})^2} = \frac{S_{xy}^2}{S_{xx}} = bS_{xy}$$

  iii. The sum of squares from the trend line is

$$\sum (y - \hat{y})^2 = \sum (y - \bar{y})^2 - \frac{\left[\sum (y - \bar{y})(x - \bar{x}) \right]^2}{\sum (x - \bar{x})^2} = S_{yy} - bS_{xy}$$

9.3. INFERENCES RELATED TO REGRESSION

The term "*regression*" originated with the work of Francis Galton. The studies of inheritance inspired by Darwin's work led Galton to believe that everything could be studied quantitatively. One of his studies involved the linear trend between the heights of fathers and their sons. The slope of the trend line in this particular study was positive but less than 1, so Galton called the relationship a "regression toward the mean." The term "regression" was then applied to any linear trend. It was an unfortunate term, however, because the slope of a least-squares trend line need not be less than 1.

TABLE 9.1.   Inferences Related to Regression

Parameter	Test Statistic $\nu = n - 2$	$1 - \alpha$ Central Confidence Interval
α	$t = \dfrac{a - \alpha_0}{s_{y \cdot x} \sqrt{1/n + \bar{x}^2/S_{xx}}}$	$a \pm t_{\alpha/2, n-2} s_{y \cdot x} \sqrt{\dfrac{1}{n} + \dfrac{\bar{x}^2}{S_{xx}}}$
β	$t = \dfrac{b - \beta_0}{s_{y \cdot x}/\sqrt{S_{xx}}}$	$b \pm \dfrac{t_{\alpha/2, n-2} s_{y \cdot x}}{\sqrt{S_{xx}}}$
$\mu_y^* = E(y \text{ if } x = x^*)$	$t = \dfrac{\hat{y} - (\mu_y^*)_0}{s_{y \cdot x} \sqrt{1/n + (x^* - \bar{x})^2/S_{xx}}}$	$\hat{y} \pm t_{\alpha/2, n-2} s_{y \cdot x} \sqrt{\dfrac{1}{n} + \dfrac{(x^* - \bar{x})^2}{S_{xx}}}$

Several types of inference are possible in relation to the regression line. Confidence intervals and tests of hypotheses are possible for parameters α and β and for $\mu_y^* = E(y$ if $x = x^*)$, the expected value of y for a specific value x^* of x. These procedures are summarized in Table 9.1.

The following example will illustrate the use of some of these procedures.

Example 9.1. Inferences Related to Regression

If the efficiency expert in Section 9.1 had obtained the following data instead of that previously given,

x:	1	1	2	4	4	5	6	6	7
y:	3	6	4	3	6	5	9	10	8

he could organize the regression analysis as follows:

$$n = 9 \qquad \sum x = 36 \qquad \sum x^2 = 184$$
$$\sum xy = 248 \qquad \sum y = 54 \qquad \sum y^2 = 376$$
$$\bar{x} = \sum x/n = 36/9 = 4.0$$
$$\bar{y} = \sum y/n = 54/9 = 6.0$$
$$S_{xx} = \sum (x - \bar{x})^2 = \sum x^2 - \left(\sum x\right)^2 / n = 184 - (36)^2/9 = 40$$
$$S_{yy} = \sum (y - \bar{y})^2 = \sum y^2 - \left(\sum y\right)^2 / n = 376 - (54)^2/9 = 52$$
$$S_{xy} = \sum (x - \bar{x})(y - \bar{y}) = \sum xy - \left(\sum x\right)\left(\sum y\right)/n = 248 - (36)(54)/9 = 32$$

The estimated slope is

$$b = \frac{\sum (x - \bar{x})(y - \bar{y})}{\sum (x - \bar{x})^2} = \frac{S_{xy}}{S_{yy}} = \frac{32}{40} = 0.80$$

The y intercept is

$$a = \bar{y} - b\bar{x} = 6 - 0.8(4.0) = 2.8$$

The least-squares trend line is

$$\hat{y} = 2.8 + 0.8x$$

Assuming that a residual analysis uncovers no deviations from the assumptions, it is valid to predict from this line because, testing

$$H_0: \beta = 0 \quad \text{against} \quad H_a: \beta > 0$$

at $\alpha = 0.05$, we find

$$s_{y \cdot x}^2 = \frac{\sum (y - \hat{y})^2}{n - 2} = \frac{S_{yy} - S_{xy}^2/S_{xx}}{n - 2}$$

$$= \frac{52 - (32)^2/40}{7} = 3.78$$

$$s_{y \cdot x} = \sqrt{3.78} = 1.95$$

and

$$t = \frac{b - 0}{s_{y \cdot x}/\sqrt{S_{xx}}} = \frac{0.8}{1.95/\sqrt{40}} = 2.595$$

with

$$t_{0.05,7} = 1.895$$

The 95% central confidence interval on β is

$$\text{CI}_{0.95}: b \pm t_{0.025,7} s_{y \cdot x}/\sqrt{S_{xx}}$$

$$0.8 \pm 2.365(1.95)/\sqrt{40}$$

$$0.8 \pm 0.73$$

If the researcher wants to find the average productivity with 3.5 hours of instruction, he finds

$$\hat{y} = 2.8 + 0.8x = 2.8 + 0.8(3.5) = 5.6$$

This is the estimate of the average productivity for 3.5 hours of instruction, $E(y$ if $x = 3.5)$. The 95% central confidence interval on this parameter is

$$\text{CI}_{0.95}: \hat{y} \pm t_{0.025,7} s_{y \cdot x} \sqrt{\frac{1}{n} + \frac{(x^* - \bar{x})^2}{S_{xx}}}$$

$$5.6 \pm 2.365(1.95)\sqrt{\frac{1}{9} + \frac{(3.5 - 4)^2}{40}}$$

$$5.6 \pm 1.58$$

If an experimenter is interested in predicting the next y observation at a given level x^* of x, the point estimate is the same as for the expected y value at that level:

$$\hat{y} = a + bx^*$$

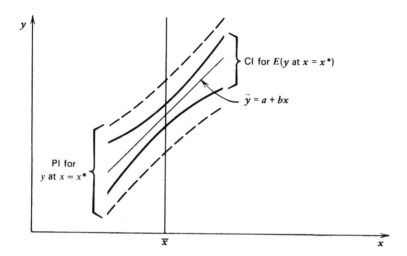

FIGURE 9.16. Prediction intervals and confidence intervals.

However, the formula for the prediction interval on the next observation is slightly different than the formula for the confidence interval on the expected value:

$$PI_{1-\alpha}: \hat{y} \pm t_{\alpha/2,n-2}s_{y\cdot x}\sqrt{1 + \frac{1}{n} + \frac{(x^* - \bar{x})^2}{S_{xx}}}$$

These prediction intervals are wider than the corresponding confidence intervals, and this seems logical because we are trying to predict a single value rather than the population mean for all values of y with a common x^*. Both types of intervals are narrowest at $x^* = \bar{x}$ (Figure 9.16).

EXERCISES

9.3.1. The linear relationship between weight y (in grams) and age x (in days) has been studied in a strain of inbred guinea pigs. The following values have been computed. The guinea pigs ranged from 8 to 14 days of age.

$$n = 16, \quad b = 5.0, \quad \bar{x} = 11, \quad \bar{y} = 87$$
$$\sum (x - \bar{x})(y - \bar{y}) = 200, \quad \sum (y - \bar{y})^2 = 1{,}126$$

a. Find $\sum (x - \bar{x})^2$.
b. Compute the variance about the least-squares trend line.
c. Place a 95% confidence interval on the mean weight of 8-day-old guinea pigs.

9.3.2. A random sample of 27 college men yields the following data in a study of the relationship between arm length x (in inches) and leg length y (in inches):

$$\sum x = 675 \quad \sum y = 810 \quad b = 1.2$$

$$\sum (x - \bar{x})^2 = 25 \quad \sum (y - \bar{y})^2 = 136$$

a. Compute the variance around the sample regression line.
b. Make a test of significance of this line against the most logical alternative.
c. Find a 95% confidence interval for β.
d. Predict the leg length of a man with arms 25 in. long.
e. Find the 95% prediction interval for this length.

9.3.3. In an effort to find a method of predicting the dental work required by army recruits, an army dentist studies the dental records of a random sample of 10 recruits completing their service. She computes the relationship between the number of cavities filled in the first two years of service y with the number of cavities filled in the two years before service x.

a. State the null hypothesis that should be used to test for the usefulness of the regression line.
b. Give the alternative hypothesis you would suggest to the dentist and the reason for that alternative.
c. Give the critical value.

9.3.4. Suppose the following statistics are computed for the dental study in Exercise 9.3.3:

$$\sum x = 50, \quad \sum y = 52, \quad \sum xy = 321$$

$$\sum (x - \bar{x})^2 = 68, \quad \sum (y - \bar{y})^2 = 75.6$$

a. Find the estimate of the slope of the trend line.
b. Find the standard error of the estimate of the slope.
c. Find 95% central confidence intervals for:

  i. The slope of the trend line.
  ii. The average number of cavities an enlistee will have filled during his first two years of service.

d. Find the 95% prediction interval for the number of cavities to be filled in the teeth of a new enlistee who in the previous two years had 3 new fillings.

9.3.5. In an experiment involving 12 female mice and their first litters, a study is made of the relationship between the rate of weight gain (gain divided by original weight) of the female during pregnancy x and the birth weight y of her litter. The following statistics are computed:

$$\bar{x} = 0.10, \quad \bar{y} = 20.00, \quad \sum xy = 24.48$$

$$\sum (x - \bar{x})^2 = 0.16, \quad \sum (y - \bar{y})^2 = 15.84$$

    a. Find b.

    b. Find the sample variance about the trend line.

    c. Test the significance of the trend line against the most logical one-sided alternative hypothesis.

    d. Estimate the average birth weight of a litter for a mouse that gained 0.12 during pregnancy.

    e. Place a 95% confidence interval on this estimate.

    f. Find the intersection of the trend line with the y axis.

    g. Place a 90% confidence interval on α.

    h. Comment on the validity of parts d through g.

9.3.6. Refer again to Exercise 9.1.7, which discusses the effect of patient load on nursing activities.

    a. Place a one-sided 95% confidence interval on the lowest value of the slope of the trend line that relates time spent on patient care with patient load.

    b. Place a two-sided 95% confidence interval on the slope of the trend line relating time spent on records and reports with patient load.

9.3.7. For Exercise 9.2.6, which examines the relationship between LSD dosage and blood copper level:

    a. Compute a 90% two-sided confidence interval on the slope.

    b. Compute a 90% central confidence interval on the y intercept.

    c. Compute a 90% confidence interval for the lowest mean copper level of those receiving a 50% dosage.

    d. Find the 90% prediction interval for the lowest copper level of an individual who would receive a 70% dosage.

    e. Is it valid to use these intervals?

9.3.8. For Exercise 9.1.10, which involves a simulated nuclear accident:

    a. Place a 95% central confidence interval on the mean ppm of all alfalfa samples that could be taken on the 150th day.

    b. Place a 95% central prediction interval on the ppm of a single sample that could be taken that day.

    c. How does the observed sample correspond to these intervals?

    d. The data do not record the amount of strontium 85 released and immediately available to the alfalfa at the start of the experiment.

        i. Estimate this from the data available.

        ii. Place a 99% confidence interval on this estimate.

        iii. Would you have any hesitation about using these estimates?

9.4. CORRELATION

The main use of regression is prediction. Suppose our example involving the efficiency expert reflected a practical situation. We would want first to test to determine whether there is a significant linear relationship between the hours of instruction an employee receives and the number of units per hour that employee can produce. Once armed with a significant linear trend, we would then want to choose a sensible number of hours of instruction, x^* (which does not extrapolate beyond the data used in the analysis), and predict the resulting mean hourly production, μ_y^*. However, there are situations in which the x variable is not "fixed" or readily chosen by the experimenter but instead is a random *covariate* to the y variable; that is, x and y vary together. In such situations, we may be more interested in determining the strength of the linear relationship than in prediction, and the sample correlation coefficient r is the statistic employed for this purpose.

In Example 8.3 in the previous chapter, we used the matched-pair t test because we anticipated a strong linear association between the length of time required by a student to perform a calculation using calculator A and the length of time required by the same student to perform a similar calculation using calculator B. In mathematical terminology, length of time for calculation on A (the x variable) and length of time for calculation on B (the y variable) are covariates and are said to have a linear bivariate distribution, simply meaning that we can use a straight line to model the manner in which they vary together. Furthermore, the variance of difference in time, $d = x - y$, is found to be

$$V(d) = V(x - y) = \sigma_x^2 + \sigma_y^2 - 2\rho\sigma_x\sigma_y$$

in which σ_x^2 is the variance of x, σ_y^2 is the variance of y, and ρ is the correlation coefficient. This equation, containing the correlation coefficient ρ as a parameter of the linear bivariate distribution of x and y, shows why the variance of the differences will be small when ρ is large.

In correlation studies, we are interested in the strength of the linear relationship between two variables, so we estimate the correlation coefficient, make statistical inference about it, and see how the variability in the experiment is affected by association between the two variables.

To demonstrate how the sample correlation is computed, we will turn again to the data in Example 8.3 giving the times for each student when similar calculations are performed on different calculators:

Student number:	1	2	3	4	5	6	7	8	9	10	11	12
Calculator A, x:	23	18	29	22	33	20	17	25	27	30	25	27
Calculator B, y:	19	18	24	23	31	22	16	23	24	26	24	28

The same sample statistics are computed as in regression analysis, namely

$$S_{xx} = 262.67 \quad S_{xy} = 199.67 \quad \text{and} \quad S_{yy} = 191.67$$

and with these, we can compute the sample correlation coefficient

$$r = \frac{S_{xy}}{\sqrt{S_{xx}S_{yy}}} = \frac{199.67}{\sqrt{(262.67)(191.67)}} = 0.89$$

Unlike the regression coefficient b, the correlation coefficient has no units of measurement associated with it. Thus, from the magnitude of the absolute value of r, we can get a feeling of the strength of the linear association. In all cases $-1 \leq r \leq +1$. If $r = -1$, there is a perfect negative relationship and all the data points are on a sample regression line with negative slope. If $r = +1$, the relationship is a perfect positive one with all sample points on a regression line with positive slope. As r gets closer to zero, there is less association between the variables. Thus the direction and, to some degree, the strength of association can be judged simply by looking at the sign and magnitude of r. With a sample correlation coefficient $r = 0.89$, we can see that there is a positive and relatively strong linear association between the students' respective computing times using each calculator. Because of this strong correlation, the variance of differences will be small, and hence the matched-pair t test is a very efficient method of analysis.

In the matched-pair t test, we deal with $x - y$, which is a linear combination of the two covariates, and $V(x - y)$ is estimated as the random variation in the experiment. In regression, for the estimate of random variability, we estimate the variance of a different linear combination of x and y, namely $V(y - a - bx)$. In these two situations, and in others to follow in later chapters, we anticipate that there is a linear association between x and y, and if there is, the experimental variance will be smaller after we have explained the variability due to the correlation between x and y.

When we discuss the variability in y which is explained by the linear association between x and y, we frequently use another statistic which is related to the sample correlation coefficient. This is the sample *coefficient of determination* r^2. The coefficient of determination has the following interpretation:

$$\left\{ \begin{array}{c} \text{The proportion of variability} \\ \text{in } y \text{ unexplained by the} \\ \text{linear relationship} \end{array} \right\} = \frac{\sum (y - \hat{y})^2}{\sum (y - \bar{y})^2}$$

$$= \frac{\sum (y - \bar{y})^2 - \left[\sum (x - \bar{x})(y - \bar{y}) \right]^2 / \sum (x - \bar{x})^2}{\sum (y - \bar{y})^2}$$

$$= 1 - \frac{S_{xy}^2}{S_{xx}S_{yy}} = 1 - r^2$$

and so

$r^2 = 1 -$ the proportion of unexplained variability in the population

$=$ the proportion of variability in y which is explained by the linear relationship

Thus r^2 indicates the proportion of the variability in y explained by the linear bivariate association with x. If r^2 is large (close to 1), most of the variability is explained by the relationship, and knowledge of the numerical value of the x variable is almost as efficient as knowledge of y. If r^2 is close to zero, then there is little linear association between the two variables, and information about the size of the x variable provides very little information about the size of the y variable. There are studies in which r^2 is the most meaningful statistic to be computed, and even in regression analysis it is frequently the first statistic which is computed in order for the experimenter to determine whether a regression equation will be useful for predicting y.

In the data from Example 8.3, we found that $r = 0.89$, and hence $r^2 = 0.79$. Thus 79% of the variability among the students' computing times with calculator B can be explained on the basis of the linear relationship between their respective computing times on the other calculator. While we cannot predict perfectly how long it will take a student to perform a calculation on B, there is evidence that anyone who is fast when using calculator A will also be fast when using calculator B, and vice versa.

We have seen that in a regression or correlation analysis, we apportion the sum of squares for experimental variability into two parts:

$$\sum (y - \bar{y})^2 = \sum (y - \hat{y})^2 + \frac{\left[\sum (x - \bar{x})(y - \bar{y})\right]^2}{\sum (x - \bar{x})^2}$$

where

$$\frac{\left[\sum (x - \bar{x})(y - \bar{y})\right]^2}{\sum (x - \bar{x})^2} = \text{sum of squares due to the trend line}$$

and

$$\sum (y - \hat{y})^2 = \text{sum of squares around the trend line.}$$

When there is no correlation between the variables x and y, these two sums of squares can be divided by their degrees of freedom and provide two independent estimates of σ_y^2, the variance of y. We have seen that there are $n - 2$ degrees of freedom associated with $\sum (y - \hat{y})^2$. We have also seen that bS_{xy} is an alternative method of computing the sum of squares due to regression; hence there is one parameter estimated and consequently 1 degree of freedom associated with that sum of squares. Thus we can use an F test to determine whether these two terms are simply independent estimates of the same variance or whether the linear association explains significant variability in the y variable. The F test is

$$F = \frac{(S_{xy}^2/S_{xx})/1}{(S_{yy} - S_{xy}^2/S_{xx})/(n - 2)} = \frac{r^2}{(1 - r^2)/(n - 2)}$$

if both numerator and denominator are divided by S_{yy}. This F test is a routine part of most regression analyses performed on a computer. It will be examined in further detail in Section 9.6 on JMP analysis, and it is an integral part of multiple regression analysis, which is covered in Chapter 14.

Notice that

$$F = \frac{(S_{xy}^2/S_{xx})/1}{(S_{yy} - S_{xy}^2/S_{xx})/(n - 2)} = \frac{(S_{xy}/S_{xx})^2 S_{xx}}{s_{y \cdot x}^2}$$

$$= \left[\frac{b}{s_{y \cdot x}/\sqrt{S_{xx}}}\right]^2 = t^2$$

that is, the F test for the significance of the correlation coefficient is equivalent to the t test for a zero slope.

Care should be taken in the interpretation of regression and correlation. If there is a significant linear relationship, this in itself does not indicate that changes in the x variable *cause* changes in the y variable. In the efficiency example, it is possible that increased instruction causes increased productivity; however, the significance of the regression line alone does not prove this. Causality must be demonstrated by an argument outside the statistical analysis. In many cases there may be no causality involved. If there is a strong linear association between length of upper arm and that of lower arm, it would be difficult to claim that a long upper arm is the cause of a long lower arm. Instead, both variables reflect the growth pattern of the individual. Furthermore, in Example 8.3, there is probably no causality, but instead the calculating times on each calculator are just two different measures of a student's manual dexterity.

The foregoing discussion of correlation and regression indicates that they are different but not mutually exclusive techniques. Roughly, regression is used for prediction, whereas correlation is used to determine the degree of association.

Besides the different functions served by regression and correlation, different assumptions are used to develop the theory behind these procedures (see Table 9.2 and Figure 9.17). As a result of these models, the following guidelines should be used. All regression procedures (Sections 9.1 to 9.3) may be applied to both models.

Also, the computation of the sample correlation coefficient and the coefficient of determination may be applied to both models. However, inference about the population correlation coefficient should only be made if the experimenter believes the variables are bivariate normal (fit the correlation model); for example, the statistic r may be used as an estimate of the population correlation coefficient ρ. If $\rho = 0$ for a bivariate normal distribution, then there is no useful linear relationship and we can also conclude that x and y are *independent* in the statistical sense. (Recall that in regression analysis, if $\beta = 0$, it is still possible that x and y are related by some type of relationship other than a linear one.)

The hypothesis $H_0: \rho = 0$ is tested with a t statistic having $n - 2$ degrees of freedom:

$$t = \frac{r}{\sqrt{\dfrac{1 - r^2}{n - 2}}}$$

TABLE 9.2.   Difference between Regression and Correlation

Regression Model	Correlation Model
1. x is fixed at levels chosen by the experimenter. (Scientists call this an independent variable.) At each fixed x level, subjects are chosen at random and y is measured. (Scientists call y the dependent variable.)	1. Subjects are sampled at random and the x, y measurements are recorded.
2. x is measured without error; that is, there is no sampling variability in x. Only y contains sampling variability.	2. Both x and y contain sampling variability.
3. For each value of x there is a normal distribution of y.	3. For each value of x there is a normal distribution of y, and for each value of y there is a normal distribution of x.
4. Each distribution of y has the same variance.	4. The x distributions have the same variance. The y distributions have the same variance.
5. The expected value of the normal y distributions lie on a straight line.	5. The joint distribution of x and y is the bivariate normal distribution.

The regression model

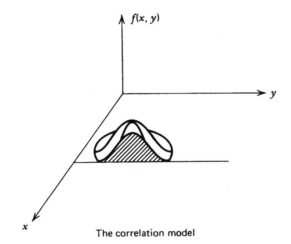

The correlation model

FIGURE 9.17. The different assumptions for regression and correlation.

Example 9.2.   Inference from the Sample Correlation

Some people have life-threatening reactions to vaccines, so an immunologist is looking for a measurement which can be made on a patient before vaccination and which will be highly correlated with the patient's reaction to the vaccine. Suppose that the following (fictional) data are obtained when a small amount of a hepatitis vaccine is used in a skin test on a random sample of patients and then their skin test results are compared to their reactions when the vaccine is administered subcutaneously:

Patient:	A	B	C	D	E	F	G	H
Skin test, x:	10	19	17	9	5	4	8	16
Reaction, y:	22	26	22	18	20	17	15	30

The following sample statistics are computed:

$$S_{xx} = 224 \quad S_{xy} = 148 \quad \text{and} \quad S_{yy} = 170$$

and these are then used to compute the sample correlation coefficient

$$r = \frac{S_{xy}}{\sqrt{S_{xx}S_{yy}}} = \frac{148}{\sqrt{(224)(170)}} = 0.76$$

Because a positive association would be anticipated, the hypotheses would be H_0: $\rho = 0$ and H_a: $\rho > 0$. The critical value for an $\alpha = 0.05$ test is $t_{0.05,6} = 1.946$ and the test of significance is

$$t = \frac{0.76 - 0}{\sqrt{\dfrac{1 - (0.76)^2}{8 - 2}}} = 2.864$$

As may have been anticipated from the sizes of r and r^2, there is a significant linear association between skin test and vaccine reaction, and the relatively large value of $r^2 = 0.5776$ indicates that a fairly useful prediction of vaccine reaction can be made if based on the least squares equation in which the x variable is the result of the skin test.

Frequently in research papers we find that the correlation coefficient or the coefficient of determination will be computed and tested for significance even in situations where the x and y variables do not have a bivariate normal distribution. The t test, or its F test counterpart, will be valid with the usual assumptions (independent random sampling, normality, and equal variances) only for the y variable at each level of x in the experiment. The interpretation is different than for a bivariate normal population. If there is a bivariate normal population and an investigator wants to learn more about the relationship between the two variables (perhaps height and weight) in that population, he draws a random sample of members of the population and computes r as an estimate of ρ. In contrast to this, an agronomist may select 6 increasing levels of fertilizer x and then compute the correlation with yield of corn y. He is using the correlation coefficient as the square root of the coefficient of determination, or as an index of how well a linear relationship fits the experimental data. He can use the t test to determine whether the levels of fertilizer explain a significant portion of the variability in corn yield, but the value of r is not an estimate of correlation between yield and levels of fertilizer.

The experimenter who wishes to use correlation procedures needs to be aware of an unusual feature about ρ. This t test is valid only to decide whether x and y are independent or whether there is a useful linear relationship between x and y, that is, the specific null hypothesis $\rho = 0$. It cannot be used to test a hypothesis such as $\rho = 0.5$. Furthermore, the analogy between the t test and confidence interval, which we have observed in other situations, does not hold true with regard to the correlation coefficient.

This situation arises because the correlation coefficient is bounded between -1 and $+1$, and therefore the distribution of the sample estimates, the r's, is symmetrical only when $\rho = 0$. If the value of ρ is very close to $+1$, then the range of overestimates is small but the range of underestimates is relatively large. The opposite is true if ρ is closer to -1. Thus, when ρ is not zero, the sampling distribution will be skewed to the right or left depending upon whether ρ is negative or positive, respectively. Furthermore, the sample correlation coefficient r is a biased estimate of the parameter ρ when the latter is nonzero. Thus it is obvious that the sampling distribution of r is not a normal distribution when $\rho \neq 0$, and therefore a t test cannot be used because, as we have seen, such a test requires that the sampling distribution be normal.

A solution to the difficulty was first presented by R. A. Fisher (1890 to 1962), whose early theoretical research in statistics involved the sampling distribution of the correlation coefficient. Three of Fisher's findings are of particular use to us:

1. Although we assume a bivariate normal distribution of the x, y data points when we estimate the population correlation parameter ρ, when this parameter has a value of 0, the distribution of r does not depend on the distribution of x but only on that of y. This is important here because it means that, since y has a normal distribution, the two tests for a useful linear relationship are equivalent:

$$t = \frac{r}{\sqrt{\dfrac{1 - r^2}{n - 2}}} \quad \text{and} \quad t = \frac{b}{\sqrt{\dfrac{s_{y \cdot x}^2}{S_{xx}}}}$$

Thus we may use whichever is more convenient when testing $\rho = 0$.

2. No matter what the value of ρ, there is a transformation

$$\mathbf{z}_r = \log_e \sqrt{(1 + r)/(1 - r)}$$

that provides a near-normal sampling distribution and permits the use of procedures involving the normal distribution.

3. The variance of the transformed value $\mathbf{z}_r^{\dagger}$ is practically independent of ρ and r and can be considered a known parameter $\sigma^2 = 1/(n - 3)$. Because the variance is known, we use the normal distribution rather than the t distribution when dealing with the $\mathbf{z}_r$ transformation.

As a consequence of points 2 and 3, we can make the following kinds of statistical inference about the correlation coefficient.

Example 9.3.   Confidence Interval for ρ

In a study of obesity, the sample correlation coefficient for weights of 28 mature obese brother–sister pairs is computed to be $r = 0.64$. A nutritionist wishes to place a 95% confidence interval on the population correlation coefficient ρ.

---

†We use the symbol $\mathbf{z}_r$ to avoid confusion with the standard normal deviate.

A confidence interval is first found on the transformed parameter $\mathbf{z}_p$, using $\mathbf{z}_r$, and then the confidence limits are transformed back to r values:

$$CI_{1-\alpha}: \mathbf{z}_r - z_{\alpha/2}(1/\sqrt{n-3}) \leq \mathbf{z}_\rho \leq \mathbf{z}_r + z_{\alpha/2}(1/\sqrt{n-3})$$

Since $r = 0.64$ is transformed to $\mathbf{z}_r = \log_e \sqrt{(1+0.64)/(1-0.64)} = 0.758$ (see Table A.13a in the Appendix),

$$CI_{0.95}: 0.758 - 1.96(1/5) \leq \mathbf{z}_\rho \leq 0.758 + 1.96(1/5)$$
$$0.366 \leq \mathbf{z}_\rho \leq 1.150$$

Using Table A.13b, the corresponding r values are

$$\mathbf{z}_r = 0.366 \rightarrow r = 0.350$$
$$\mathbf{z}_r = 1.150 \rightarrow r = 0.818$$

Thus

$$CI_{0.95}: 0.350 \leq \rho \leq 0.818$$

A similar approach is used to test whether the population correlation coefficient is some nonzero value.

Example 9.4.   Test of $H_0: \rho = \rho_0$ with $\rho_0 \neq 0$

The nutritionist in the previous example wants to test $H_0: \rho = 0.5$ against $H_a: \rho \neq 0.5$ because of some prior theory or available evidence. The test is a z test with statistic

$$z = \frac{\mathbf{z}_r - \mathbf{z}_{\rho_0}}{1/\sqrt{n-3}}$$

Since $r = 0.64$, it follows that $\mathbf{z}_r = 0.758$, and $\rho_0 = 0.5$ is transformed to $\mathbf{z}_{\rho_0} = 0.549$ (Table A.13a). Thus

$$z = \frac{0.758 - 0.549}{1/5} = 1.048$$

The null hypothesis is rejected at $\alpha = 0.05$ if $|z| > 1.96$, so the nutritionist concludes that ρ may be 0.5.

Fisher's transformation can also be used to compare two correlation coefficients.

Example 9.5.   Testing $\rho_1 = \rho_2$

Suppose that the nutritionist has data on 23 brother–sister pairs of conventional mature weight in addition to the data above for obese pairs where $r_1 = 0.64$. For the conventional sample, $r_2 = 0.38$. To test whether the correlation is the same for both populations at $\alpha = 0.05$, the following test is used:

$$H_0: \rho_1 = \rho_2 \quad \text{against} \quad H_a: \rho_1 \neq \rho_2$$

is tested with

$$z = \frac{\mathbf{z}_{r_1} - \mathbf{z}_{r_2}}{\sqrt{\dfrac{1}{n_1 - 3} + \dfrac{1}{n_2 - 3}}}$$

Thus

$$z = \frac{0.758 - 0.400}{\sqrt{\dfrac{1}{25} + \dfrac{1}{20}}} = 1.193$$

Since $z_{\alpha/2} = 1.96$, there is no significant difference between the two correlation coefficients. The correlation between weights of brother–sister pairs may be the same for obese siblings as for those of conventional weight.

The various types of inference about correlation coefficients are summarized below.

Procedure. Inferences about Correlation Coefficients

Assumption: bivariate normal distribution

Tests of Hypotheses

Significance level: α

1. $H_0: \rho = 0$
   $H_a: \rho \neq 0$ or $\rho > 0$ or $\rho < 0$
   Test statistic:

$$t = \frac{r}{\sqrt{\dfrac{1 - r^2}{n - 2}}}$$

   Reject H_0 if $|t| \geq t_{\alpha/2, n-2}$ or $t \geq t_{\alpha, n-2}$ or $t \leq -t_{\alpha, n-2}$, respectively.

2. $H_0: \rho = \rho_0$ with $\rho_0 \neq 0$
   $H_a: \rho \neq \rho_0$ or $\rho > \rho_0$ or $\rho < \rho_0$
   Test statistic:

$$z = \frac{\mathbf{z}_r - \mathbf{z}_{\rho_0}}{1/\sqrt{n - 3}} \quad \text{using Table A.13a for } \mathbf{z}_r \text{ and } \mathbf{z}_{\rho_0}$$

   Reject H_0 if $|z| \geq z_{\alpha/2}$ or $z \geq z_\alpha$ or $z \leq -z_\alpha$, respectively.
3. $H_0: \rho_1 = \rho_2$
   $H_a: \rho_1 \neq \rho_2$ or $\rho_1 > \rho_2$ or $\rho_1 < \rho_2$

Test statistic:

$$z = \frac{\mathbf{z}_{r_1} - \mathbf{z}_{r_2}}{\sqrt{\dfrac{1}{n_1 - 3} + \dfrac{1}{n_2 - 3}}}$$

Reject H_0 if $|z| \geq z_{\alpha/2}$ or $z \geq z_\alpha$ or $z \leq -z_\alpha$, respectively.

Confidence Interval on ρ

Compute $CI_{1-\alpha}$: $\mathbf{z}_r \pm z_{\alpha/2}(1/\sqrt{n-3})$, then use Table A.13b to transform the lower and upper limits back to r values.

There are many other statistical tests of association or "correlation." Some of them employ data on the ordinal scale of perception, and to distinguish them from the method studied here, they are sometimes called *rank correlation* procedures (see Section 9.5). Conversely, the procedure to be used for bivariate normal data is sometimes called the *Pearson product moment correlation*, in recognition of Karl Pearson's original contributions. By convention, however, when the unmodified term "correlation" is seen, it is assumed that Pearson's procedure is the one under discussion.

EXERCISES

9.4.1. Given the scatter diagrams for x, y pairs in Figure 9.18, select the best answer for each diagram.

Statistic	Diagram 1	Diagram 2
a. Slope of trend line	$-2, -1, 0, +1, +2$	$-2, -1, 0, +1, +2$
b. Intercept of y axis	0, 2, 4, 8, 10	0, 1, 2, 3, 4
c. Correlation coefficient	$-0.9, -0.4, 0,$	$-0.9, -0.4, 0,$
	$+0.4, +0.9$	$+0.4, +0.9$
d. t test for $\rho = 0$	Significant,	Significant,
	nonsignificant	nonsignificant

FIGURE 9.18. Scatter diagrams for Exercise 9.4.1.

9.4.2. Test H_0 concerning the population correlation coefficient:

    a. H_0: $\rho = 0$, H_a: $\rho \neq 0$, $n = 20$, $r = 0.550$, $\alpha = 0.01$
        Would the H_0 be accepted or rejected? What does this mean?

    b. H_0: $\rho = 0$, H_a: $\rho > 0$, $n = 18$, $r = 0.43$, $\alpha = 0.05$
        Would the H_0 be accepted or rejected? What does this mean?

    c. H_0: $\rho = 0.4$, H_a: $\rho \neq 0.4$, $n = 28$, $r = 0.62$, $\alpha = 0.05$
        Would the H_0 be accepted or rejected? What does this mean?

9.4.3. Twenty-six newborn baby boys are weighed and measured for length. The standard deviation of weight is 2 lb, but usual linear regression techniques reveal that 40% of the variability in weight can be explained by the relationship between weight and length. Make a test to determine whether the relationship explains a significant ($\alpha = 0.05$) portion of the variability in weight.

9.4.4. In a study involving 25 dairy cattle, the correlation between milk yield from first and second lactations was found to be 0.42.

    a. Test the significance of the relationship ($\alpha = 0.05$).

    b. How useful do you think the relationship would be in predicting milk yield for second lactation?

9.4.5. Given the scatter diagrams in Figure 9.19:

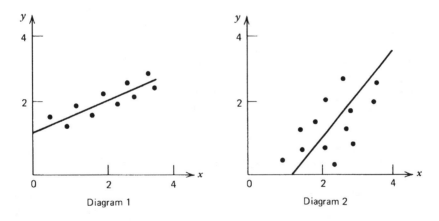

FIGURE 9.19. Scatter diagrams for Exercise 9.4.5.

    a. Which diagram has the greater b value?

    b. Which diagram has the greater r value?

    c. For diagram 1, does $\bar{y} = 1$, 2, 3, or 4?

    d. For diagram 2, does $\bar{y} = 1$, 2, 3, or 4?

An oncologist wants to evaluate the usefulness of the CAT scan for uterine tumor diagnosis. For 12 women with fibroid tumors, certain measurements are taken by CAT scan techniques prior to surgery and then compared with other measurements taken on the tumors in the pathology laboratory after they had been surgically removed. Suppose the paired measurements on tumor mass are

Patient	A	B	C	D	E	F	G	H	I	J	K	L
CAT scan, x	18	17	28	20	11	24	16	15	19	24	23	13
Pathology, y	20	4	25	16	19	21	22	10	23	27	18	11

and the statistics computed are

$$\sum(x - \bar{x})^2 = 278 \qquad \sum(y - \bar{y})^2 = 498$$

$$\frac{\sum(x - \bar{x})(y - \bar{y})}{\sum(x - \bar{x})^2} = 0.723 \qquad \frac{\left[\sum(x - \bar{x})(y - \bar{y})\right]^2}{\sum(x - \bar{x})^2} = 108.58$$

a. Find the sample correlation coefficient.

b. State the most logical hypotheses about the correlation between the CAT scan measurement of tumor mass and that obtained at pathology.

c. Give the critical value for an $\alpha = 0.05$ test of your null hypothesis.

d. Perform the test of significance.

e. Do you think the relationship would be useful in being able to use the CAT scan information to predict fibroid tumor mass prior to surgery?

Using the data in Exercise 9.1.7, place a 90% confidence interval on the correlation coefficient for the relationship between $x =$ patient load and $y =$ time available for records and reports.

9.5. NONPARAMETRIC STATISTICS: RANK CORRELATION

When we record data at the ordinal scale of measurement or reduce numerical data to the ordinal scale by transforming them to ranks, we can perform the computational procedures of correlation on the ranks. The resulting coefficient, which is given the symbol r_s and called *Spearman's rank correlation* in recognition of the psychologist C. E. Spearman, who popularized the procedure, has much the same meaning as the correlation coefficient we have already studied. It provides a measure of linear association between the ranks of the x variable and those of the y variable. The bounds on the coefficient are the same: $-1.0 \leq r_s \leq +1.0$. If r_s is fairly large and positive, then there is close positive agreement between the ranks of the two variables. If r_s is close to -1.0, then, when one variable has a high rank, its companion tends to have a low rank, and vice versa. Also, when r_s is near zero, the ranks of the x and y variables are nearly independent.

To demonstrate the computational procedures, we will designate r_x as the rank of an x variable and r_y as the rank of its companion y variable; then

$$r_s = \frac{\sum(r_x - \bar{r}_x)(r_y - \bar{r}_y)}{\sqrt{\sum(r_x - \bar{r}_x)^2 \sum(r_y - \bar{r}_y)^2}}$$

However, with respect to both r_x and r_y, we are dealing with the ranks from 1 to N, so

$$\bar{r}_x = \bar{r}_y = (N + 1)/2, \quad \text{and} \quad \sum(r_x - \bar{r}_x)^2 = \sum(r_y - \bar{r}_y)^2 = \frac{N(N^2 - 1)}{12}.$$

Therefore we can employ some moderately mundane mathematical manipulation and arrive at the following equation which simplifies the computations:

$$r_s = 1 - \frac{6 \sum d^2}{N(N^2 - 1)},$$

where $d = r_x - r_y$ is the difference in ranks assigned to an x, y pair.

Under the null hypothesis that r_x and r_y are independent,

$$E(r_s) = 0 \quad \text{and} \quad V(r_s) = \frac{1}{N-1}$$

and it is generally agreed that if there are 10 or more x, y pairs the distribution of r_s can be well approximated by a normal distribution. Therefore, we can test the null hypothesis H_0: $E(r_s) = 0$ with a z test:

$$z = \frac{r_s - 0}{\sqrt{\dfrac{1}{N-1}}} = r_s \sqrt{N - 1}$$

For samples smaller than 10, tables for the exact distribution of r_s or $\sum d^2$ can be found in most textbooks on nonparametric statistics.

Example 9.6.   Spearman's Correlation

Color indicators are frequently used to detect the level of certain chemical compounds in water or other liquids, and then further action is based on how dark the color becomes when the indicator is added to a sample of the liquid. Suppose that there are two chemists who regularly make decisions about the treatment of a city's water and they want to be sure that they are in close agreement about their evaluations of the darkness of a color indicator, which, depending on the level of the impurity in the water, will range from a light pink to a cherry red. So the two chemists prepare 10 bottles of water each containing different quantities of the impurity. Then they have a third person randomly assign identifying letters to the bottles so that they can independently sample the bottles, apply the color indicator, and rank their samples from lightest to darkest:

Bottle of water:	A	B	C	D	E	F	G	H	I	J
Rank by chemist 1, x:	5	2	1	7	3	6	9	8	10	4
Rank by chemist 2, y:	4	3	2	8	1	7	10	6	9	5
$d = r_x - r_y$:	1	−1	−1	−1	2	−1	−1	2	1	−1
d^2:	1	1	1	1	4	1	1	4	1	1

$$r_s = 1 - \frac{6 \sum d^2}{N(N^2 - 1)} = 1 - \frac{6(16)}{10(100 - 1)} = 1 - \frac{96}{990} = 0.903$$

and they can test the null hypothesis H_0: $E(r_s) = 0$, choosing H_a: $E(r_s) > 0$ as the alternative because they expect there to be agreement between the rankings of the two chemists. The test is

$$z = \frac{r_s - 0}{\sqrt{\dfrac{1}{N-1}}} = r_s \sqrt{N - 1} = 0.903(3) = 2.709$$

If the null hypothesis were true, the probability of a value of z as large or larger than 2.709 would be $P = 0.003$. Because the P value is much smaller than the conventional $\alpha = 0.05$, they would reject the null hypothesis and claim that there is a positive association between the ranks which they give to the water samples. They seem to agree quite well on the lightness or darkness of the color indicator in a water sample.

When data are on the ordinal scale, as in the previous example, we expect no ties to occur. However, when we use the rank transformation on numerical data and find that certain recorded numerical values are identical, we follow a procedure similar to that which we used before for ties. We need to remember that we are concerned only about ties which occur among the numerical values of the x variable and among those of the y variable. Thus, if two numerical values of the x variable are tied for the second and third rank, we use the average of the ranks to be assigned to the ties, and $r_x = (2 + 3)/2 = 2.5$ is assigned to each of the members of the tie. We also follow the same procedure in obtaining r_y when there are ties among the numerical values of the y variable.

For reasons other than just its computational simplicity, Spearman's rank correlation is a very useful nonparametric procedure. Even if paired x, y data have a bivariate normal distribution, and thus are suitable for conventional correlation procedures, r_s and r will be similar in numerical value, and the test of hypothesis for r_s will be almost as powerful as that for r. When data do not have a bivariate normal distribution, r_s is frequently superior to r in detecting association between the x and y variables.

Procedure. Spearman's Rank Correlation

H_0: $E(r_s) = 0$ (The ranking of the x variable is independent of that of the y variable.)
H_a: $E(r_s) \neq 0$ or $E(r_s) > 0$ or $E(r_s) < 0$
Significance level: α

Computation of the rank correlation coefficient:
The measurements on the x variable are ranked from 1 to N and designated as r_x.
The measurements on the y variable are ranked from 1 to N and designated as r_y.

$$r_s = 1 - \frac{6 \sum d^2}{N(N^2 - 1)}$$

with $d = r_x - r_y$, the difference in ranks which are assigned to an x, y pair.
Test statistic:

$$z = \frac{r_s - 0}{\sqrt{\dfrac{1}{N - 1}}} = r_s\sqrt{N - 1}$$

Region of rejection: $|z| \geq z_{\alpha/2}$ or $z > z_\alpha$ or $z < -z_\alpha$, respectively.

EXERCISES

9.5.1. An anthropologist has a choice of two different methods of determining the age of pottery fragments of ancient civilizations, and she wants to know if both procedures will yield the same results. Using each method, she determines the age (recorded in thousands of years) for 10 pottery fragments of different ages and then compares the results:

Fragment:	A	B	C	D	E	F	G	H	I	J
Method x:	10.5	15.3	12.4	12.9	14.4	11.6	12.9	13.6	10.8	14.6
Method y:	10.7	15.6	12.2	12.7	14.5	11.3	13.0	14.0	10.6	14.5

a. Compute Spearman's rank correlation.
b. If Spearman's rank correlation is to be tested for significance:
   i. What are the most logical null and alternative hypotheses?
   ii. What is the critical value for $\alpha = 0.05$?
c. Make the test of significance and draw inference.
d. Compute Pearson's correlation and compare its value to r_s.

A physician examines the blood constituents of 12 patients who have become sick from a toxic amount of heavy metal in their drinking water. Among several variables of interest are the following measurements of albumen and magnesium in their blood:

Patient:	A	B	C	D	E	F	G	H	I	J	K	L
Albumen:	4.5	5.0	5.2	4.8	4.9	4.6	4.9	3.5	5.1	3.7	4.7	4.3
Magnesium:	1.7	1.2	1.3	1.5	1.6	0.8	1.0	1.6	1.2	1.4	1.1	1.9

a. Show that $\sum d^2 = 405$.
b. What null and alternative hypotheses would you suggest for this study? Why?
c. Compute the rank correlation coefficient and perform the test of significance at $\alpha = 0.05$.

Use the data in Exercise 9.4.6 to perform Spearman's rank correlation.

a. How does the rank correlation coefficient compare to that obtained using conventional procedures?
b. Using $\alpha = 0.05$, is the decision about the respective null hypothesis the same for both test procedures?

9.6. COMPUTER USAGE

Scatter Plots

In Example 9.1 an efficient expert is investigating a possible linear relationship between the number of hours of instruction employees receive and the number of units they produce per hour. He enters the data into a JMP data table and names it "training":

To produce a scatter diagram the investigator uses the "Fit Y by X" item in the Analyze menu. He selects Units as the Y, Response variable and Hours as the X, Factor in the dialog box.

If there are enough points in the scatter diagram, they may indicate the general shape of the curve or line that can possibly be used as a model for the variables. A generalized random scatter may indicate that there is no relationship between the variables. Here the scatter plot indicates a linear relationship.

Regression

To find the regression line, test the slope, and produce a graph that contains the regression line, he uses the "Fit Line" item in a pop-up menu labeled "Bivariate Fit of Units by Hours." The output window is shown on the next page.

The values of interest are the F Ratio, Prob > F, and RSquare. The F Ratio is the statistic described in Section 9.5 and is used to test whether there is a significant linear relationship between hours and units. The Prob > F is the P value of the F statistic. In this case there is a significant linear relationship at the 0.05 level of significance because Prob > F is 0.0352. Rsquare is the coefficient of determination, that is, the square of the correlation coefficient.

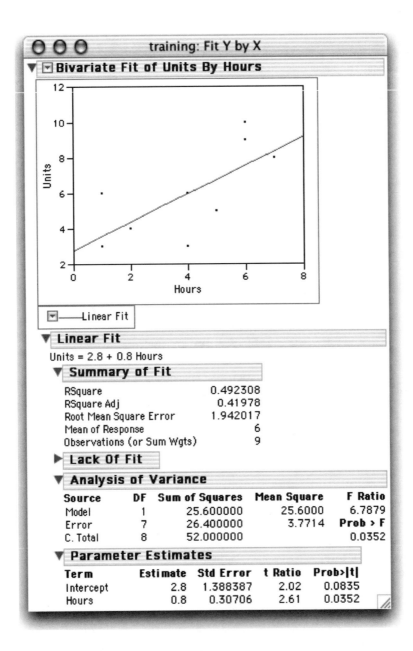

The estimates of the regression coefficients are found in the table of Parameter Estimates. The parameter estimate listed for Intercept is a, the estimate of the intercept, and the parameter estimate listed for Hours is b, the estimate of the slope. The t Ratio column gives the value of the test statistics for the t tests for $\alpha = 0$ and $\beta = 0$. Notice the t value of 2.61 is the square root of the F ratio in the Analysis of Variance table.

Correlation

A correlation analysis is done by choosing the "Density Ellipse" option in the "Bivariate Fit" pop-up menu. The output contains a graph and a correlation report.

The bivariate density ellipse plot views the relationship between hours and units as a bivariate normal probability distribution. The plot is an ellipse that encloses 0.95 of the probability. The **Correlation** text report gives the estimates of the five parameters of the bivariate normal distribution. The sample correlation coefficient is 0.701646 and the P value for the test of whether $\rho = 0$ is 0.0352. Notice that this number is also the P value for the F and t statistics.

9.7. ESTIMATING ONLY ONE LINEAR TREND PARAMETER

When we try to fit a trend line to data, especially for estimation, we generally use least-squares regression to obtain an estimate of b the slope and of a the intercept of the line. Then with these two estimates, we can predict the value of y for a specified value of x with the prediction equation

$$\hat{y} = a + bx$$

However, there are times when we can assume that either the intercept or the slope is known, and need not be estimated. For each of these situations, there are special statistical procedures that are used instead of the least-squares methods examined in earlier sections of this chapter.

The first of the special methods is familiar and commonly used even by those unfamiliar with least-squares estimation. It is *ratio estimation* and simply assumes that y increases proportionally with x. Suppose a recipe for a fruit punch calls for 2 quarts of fruit juice to prepare enough punch for 10 people, but 20 are expected to be at the picnic. Then we estimate that it will require 4 quarts of juice to have enough punch for 20 people. That is all there is to predicting y (quarts of juice) for a specified x (number of people). The slope of the line is the only parameter estimated when ratio procedures are used, for a ratio carries the automatic assumption that the intercept is zero. To say that the intercept is zero is to say that when $x = 0$, $y = 0$, and this seems reasonable in the case of quarts per person, for if no people attend the picnic no juice is needed.

The second procedure, called *difference estimation*, is also familiar and in common use. It is used when y is predicted simply by adding a constant to x. Everyone who watches television news has had to suffer through one or another commercial for a diet medication that promises, "You will lose seven pounds the first week!" According to that prediction, one's weight next week (y) will be this week's weight (x) less 7 lb. To test the advertiser's claim, only the intercept of the line has to be estimated, for difference estimation assumes that the slope of the linear relationship is equal to 1.0.

There are special advantages to ratio estimation and difference estimation besides their familiarity and ease of use. In practice, one of the most difficult conditions data must meet for the legitimate use of least-squares procedures is the assumption that the variance of y is the same no matter what x it is associated with. It was noted in the discussion of least squares in Section 9.2, that it is necessary to assume that variability of y from the trend line is the same for all values of x. However, in many areas of study, y is often more variable for large values of x than it is for smaller values. For example, the variability in weight (y) among people whose height is $x \leq 5$ ft will usually be less than that among those for $x \geq 6$ ft, and the variability in length of forearm will be greater for tall people than for short ones. This assumption is not required for statistical inference in ratio estimation, and in difference estimation it is part of the basic assumption about a common difference between x and y. However, all conditions except the third stated in Section 9.2 for least-squares line analyses (the equal-variance condition) must be met for inference based on either ratio estimation or difference estimation. For inference based on either ratio estimation or difference estimation, the fourth condition of linearity must be specified as a positive linear relationship.

For all methods of trend analysis, the variance of interest is that of the deviations of y values from the trend line, that is the variance of the e, where $e = y - \hat{y}$. The sample variance among these deviations is computed as

$$s_e^2 = \frac{\sum (y - \hat{y})^2}{n - 1}$$

The degrees of freedom are $n - 1$ rather than the $n - 2$ used for the sample variance in least-squares procedures. This is because only one parameter, either the slope or the intercept, of the trend line is being estimated, whereas both parameters are estimated for a least-squares trend line. To avoid confusion over the nature of the line or the degrees of freedom, we use different subscripts to designate the variance from the trend line when only one parameter is estimated.

As with least-squares methods, inference requires computation not only of the variance but also of the standard error of the estimates involved. Computational procedures will be shown in the examples explaining the use of each of these estimation procedures.

Example 9.7. Ratio Estimation

The threat of attacks by terrorists using anthrax spores is a concern to U.S. health officials. Because there are also health risks associated with the use of protective vaccines, health officials want to avoid mass vaccination of all citizens unless necessary. Instead, they keep the anthrax vaccine available at well-located health care facilities around the country, ready for use if needed.

At each facility, an inventory is kept on the number of vials of vaccine in storage. However, some are used for people who may have been exposed to naturally occurring anthrax, other vials are accidentally broken, and others are discarded when the vaccine in the vial becomes cloudy or otherwise appears to have spoiled. In all such cases the inventory should be changed to reflect the loss, but this can be forgotten when the demands of health care are more important than record keeping. So a public health worker conducts a study to learn how to use the number of vials shown in the inventory to estimate the actual number of vials available at a health care facility.

She takes a random sample of 20 facilities where anthrax vaccine is being kept. Then she visits each facility in the sample in order to record how many vials of vaccine are shown on the inventory (x) and to count the number of vials actually available (y) in the storage refrigerator. Her data and partial work are as follows:

Facility	Inventory (x)	In Storage (y)	$\hat{y} = 0 + 0.875x$	$e = y - \hat{y}$
a	36	33	31.500	1.500
b	78	67	68.250	−1.250
c	101	91	88.375	2.625
d	65	57	56.875	0.125
e	21	17	18.375	−1.375
f	84	73	73.500	−0.500
g	10	7	8.750	−1.750
h	13	9	11.375	−2.375
i	31	29	27.125	1.875
j	26	23	22.750	0.250
k	25	21	21.875	−0.875
l	11	11	9.625	1.375
m	82	72	71.750	0.250
n	22	22	19.250	2.750
o	96	84	84.000	0.000
p	88	78	77.000	1.000
q	52	45	45.500	−0.500
r	75	66	65.625	0.375
s	8	5	7.000	−2.000
t	36	30	31.500	−1.500
Sum	960	840	840.000[†]	0.000

So that it will not be mistaken as the least-squares slope, the public health worker may choose to symbolize the estimated slope for ratio estimation by b_r, and compute it as

$$b_r = \frac{\bar{y}}{\bar{x}} = \frac{42}{48} = 0.875, \text{ or equivalently } b_r = \frac{\sum y}{\sum x} = \frac{840}{960} = 0.875$$

[†]$b_r = \sum y / \sum x$, hence $\sum (b_r x)$ will always equal $\sum y$; this provides a check of arithmetic.

She sees that, on the average, 0.875 is the proportion of vials shown in inventory that are actually in storage, and she can estimate the number of vials in storage at any facility by using the equation for a straight line,

$$\hat{y} = a + b_r x$$
$$= 0 + 0.875x$$

To compute the variance from the ratio trend line, she first subtracts the expected number of vials ($\hat{y}$) at each facility from the observed number (y) to obtain the deviations (e) given in the last column of her work sheet. The desired variance is that among the 20 deviations,

$$s_e^2 = \frac{\sum (y - b_r x)^2}{n - 1} = \frac{43.0625}{20 - 1} = 2.2664$$

This method of computing is fairly easy here because there are only three decimal places associated with b_r and only 20 pairs of values, but instead she could have used algebra to obtain an equation some find more useful for calculators,

$$s_e^2 = \frac{\sum (y - b_r x)^2}{n - 1} = \frac{\sum y^2 + b_r^2 \sum x^2 - 2b_r \sum xy}{n - 1}$$

$$= \frac{50902 + (0.875)^2 65812 - 2(0.875)(57855)}{20 - 1}$$

$$= \frac{43.0625}{19} = 2.2664$$

Once s_e^2 is obtained, for statistical inference, she still must compute the standard error of the ratio, and this requires the equation

$$\text{s.e.}(b_r) = \sqrt{\frac{s_e^2}{n\bar{x}^2}} = \sqrt{\frac{2.2664}{20(48)^2}} = 0.007$$

A confidence interval is the statistical inference the public health worker likely wants to make, so she uses the estimate b_r and its standard error to compute a $\text{CI}_{0.95}$ in the usual fashion:

$$\text{CI}_{1-\alpha}: b_r \pm t_{\alpha/2, n-1} \frac{s_e}{\sqrt{n\bar{x}^2}}$$

$$0.875 \pm 2.093(0.007)$$
$$0.875 \pm 0.015$$

To express proportions as percentages, she would multiply values in the $\text{CI}_{0.95}$ by 100. Then based on her random sample, she can conclude that only 87.5% of the mean number of vials of anthrax vaccine shown on health center inventories are actually in storage. To include the width of the confidence band, she would give the *margin of sampling error* as $\pm 1.5\%$.

If she wanted to predict the number of vials available at a particular facility where the number on inventory is x^*, remembering the intercept is assumed to be zero, she would make the prediction

$$\hat{y} = b_r x^*$$

To compute a prediction interval for a single facility, she would use

$$PI_{1-\alpha}: \hat{y} \pm t_{\alpha/2,n-1}s_e$$

The mathematical procedure for difference estimation has already been studied in Example 8.3, where the matched-pair t-test was discussed. So we need only to look at how the same procedures can be used in linear estimation. The example pertained to a random sample of 12 students who each used two different types of calculators, and the study was to determine if the mean difference in speed of calculation on the two machines was significantly different from zero.

To reexamine that study as one in linear estimation, we remember that the equation for using a straight line for estimation is

$$\hat{y} = a + bx$$

Then, because in difference estimation we assume that the slope of the linear relationship is $\beta = 1.0$, only the intercept a needs to be estimated. The computation of a is the same as $\bar{y}_d$ in Example 8.3, and the sample variance around the trend line is the same as s_d^2 in that example. The same data are used again in Example 9.8 to demonstrate the difference estimation procedure.

Example 9.8. Difference Estimation

We want to see if we can use a student's speed of calculation on Calculator A (x) to predict his speed using Calculator B (y). The data are

Student	Machine A (x)	Machine B (y)	$d = (y - x)^\dagger$	$d^2 = (y - x)^2$
1	23	19	−4	16
2	18	18	0	0
3	29	24	−5	25
4	22	23	1	1
5	33	31	−2	4
6	20	22	2	4
7	17	16	−1	1
8	25	23	−2	4
9	27	24	−3	9
10	30	26	−4	16
11	25	24	−1	1
12	27	28	1	1
			$\sum d = -18$	$\sum d^2 = 82$

---

†The signs of the differences are reversed from Example 8.3 because the subtraction here is $B - A$.

If we wish to use a different symbol for the intercept to distinguish it from the least-squares intercept, we can give the equation to compute it as

$$a_d = \frac{\sum (y - x)}{n} = \frac{\sum d_i}{n} = \frac{-18}{12} = -1.5$$

The variance from the trend is computed as before in the matched-pair t test

$$s_d^2 = \frac{\sum d^2 - \left(\sum d\right)^2 / n}{n - 1} = \frac{82 - (-18)^2 / 12}{11} = 5$$

and the standard error of the estimate of the intercept is

$$\frac{s_d}{\sqrt{n}} = \sqrt{\frac{5}{12}} = 0.645$$

As we have seen before, once we have an estimate of a parameter and the standard error of the estimate, we have the two numerical values necessary for statistical inference, a test of hypothesis, confidence interval, or prediction interval.

Procedure. Linear Trend Estimation

Assumption: $y = \alpha + \beta x + \varepsilon$ with the ε's independently distributed as $N(0, \sigma^2)$

Estimation: A value of y can be estimated for a specific x^* with the linear equation

$$\hat{y} = a + bx^*$$

For each method of trend fitting, the intercept and slope must be estimated or assumed to be a specified value.
The variance of the ε's is estimated by $\hat{\sigma}^2 = \Sigma(y - \hat{y})^2 / v$, where v is the degrees of freedom

Method	Intercept	Slope	Estimated Variance
Least squares	$a = y - b\bar{x}$	$b = S_{xy}/S_{xx}$	$s_{y \cdot x}^2 = (S_{yy} - bS_{xy})/(n - 2)$
Ratio	$a = 0$	$b_r = \dfrac{\sum y}{\sum x}$	$s_e^2 = \sum(y - b_r x)^2 / (n - 1)$
Difference	$a_d = \sum(\bar{y} - \bar{x})$	$b = 1$	$s_d^2 = \dfrac{\sum d^2 - \left(\sum d\right)^2 / n}{n - 1}$

Standard errors of the estimates are as follows:

Method	Standard Error for Intercept	Standard Error for Slope
Least squares	$s_{y \cdot x}\sqrt{\dfrac{1}{n} + \dfrac{\bar{x}^2}{S_{xx}}}$	$\dfrac{s_{y \cdot x}}{\sqrt{S_{xx}}}$
Ratio	No estimate involved	$\dfrac{s_e}{\sqrt{n\bar{x}^2}}$
Difference	$s_d/\sqrt{n}$	No estimate involved

EXERCISES

9.6.1. Use algebra to verify that when $b_r = \sum y / \sum x$, for any set of bivariate data, $\sum b_r x$ will be equal to $\sum y$.

9.6.2. Using the same data set as in Examples 8.3 and 9.8:

   a. Compute the least-squares trend line and ratio trend line.

   b. Compare the values of $\sum (y - \hat{y})^2$ for each trend line; why should it be smallest for the least-squares trend line?

Using the data in Example 9.7:

   a. Compute the least squares trend line and difference trend line.

   b. Compare the numerical values of intercepts and slopes for each method.

   c. Which method would you use to estimate the number of vials of vaccine? Explain why.

U.S. attack helicopters are difficult to maintain in good flying condition in arid, sandy terrain. When based in such areas, there will usually be some that are not ready to fly until repaired. A general in command of 15 squadrons of helicopters at various bases in an arid, sandy region knows that on most days each squadron will have a few craft that are being repaired and not ready to fly. He wants to estimate the mean number per squadron that will not be flight-ready. On a randomly chosen day, the following data were obtained from these squadrons:

Squadron	1	2	3	4	5	6	7	8
Copters	20	26	24	22	28	27	25	25
Ready	13	21	18	15	21	25	25	24
Squadron	9	10	11	12	13	14	15	Sum
Copters	17	18	29	25	30	18	29	363
Ready	11	17	27	18	30	11	24	300

   a. What must be assumed about the data in order to make valid statistical inference about the mean number of helicopters that will not be flight-ready on a given day?

   b. Difference estimation is attractive because it is easy to use for estimating the mean number of unready helicopters per squadron. Estimate the mean number of helicopters that will not be ready to fly. Then estimate those that will be ready.

   c. Set a confidence interval for the mean number not ready to fly.

   d. The general feels that to wage a successful campaign at least 276 of the 363 helicopters under his command must be ready to fly on the day they are needed. At the 0.05 level, is there statistically significant evidence that he will have that minimum number ready to fly? *Hint:* What is the average number of flight-ready craft per squadron necessary for a total of 276?

REVIEW EXERCISES

Decide whether each of the following statements is true or false. If a statement is false, explain why.

9.1.   The sample regression line is called the least-squares trend line because for it $\sum (y - \hat{y})^2$ is smaller than for any other straight line fitted to the sample points.

9.2.   The trend line always passes through the origin (0,0).

9.3.   If the slope of the regression line relating cake volume to amount of baking powder is $3.22 \text{ cm}^3/\text{g}$, this means that for each additional gram of baking powder the mean increase in the volume of the cake will be a 3.22 cm^3.

9.4.   It is possible to fit a line other than the least-squares trend line so that $\sum (y - \hat{y}) = 0$.

9.5.   The experimenter would test $H_0: \beta > 0$ if he thought that the slope of the trend line was positive.

9.6.   Since $\sum (y - \hat{y})^2 \leq \sum (y - \bar{y})^2$, it follows that $s_{y \cdot x}^2 \leq s_y^2$.

9.7.   The better the line fits the sample points, the smaller $\sum (x - \bar{x})^2$ will be.

9.8.   Units of measurement can affect both the magnitude of the slope and the significance of the slope of the least-squares trend line.

9.9.   There can be a strong dependent relationship between y and x that will not be detected by linear regression analysis.

9.10. $s_{y \cdot x}^2 / \sum (x - \bar{x})^2$ is to b as s^2/n is to $\bar{x}$.

9.11. The phrase "regression of y on x" indicates a negative relationship between the y and x variables.

9.12. The confidence interval for $E(y$ if $x = x^*)$ will be greater at $x^* = \bar{x}$ than for any $x^* \neq \bar{x}$.

9.13. Confidence intervals can be set for the true slope of the regression line, the true intercept on the y axis, and the true mean of y for any given value of x.

9.14. When computing a correlation coefficient, the experimenter assumes that there is a cause-and-effect relationship between x and y.

9.15. If $\sum (y - \hat{y})^2$ is large relative to $\sum (y - \bar{y})^2$, this indicates that a large portion of the variability in y is attributed to the linear relationship between y and x.

9.16. The greater the magnitude of r, the stronger the relationship between x and y.

9.17. One of the assumptions made in regression analysis is that the dependent variable follows a normal distribution.

9.18. In testing b for significance, it is assumed that y has the same variance for each fixed value of x.

9.19. For the same data set, because it has $n - 1$ degrees of freedom, the variance around the ratio trend line can be smaller than that around the least-squares trend line.

9.20. As the strength of the relationship between two variables increases, the regression line becomes a better fit for the points.

SELECTED READINGS

Anscombe, F. J., and J. W. Tukey (1963). The examination of residuals. *Technometrics*, **5**, 141–160.

Bartlett, M. S. (1949). Fitting a straight line when both variables are subject to error. *Biometrics*, **5**, 207–212.

Behnken, D. W., and N. R. Draper (1972). Residuals and their variance patterns. *Technometrics*, **14**, 101–111.

Box, G. E. P. (1966). Use and abuse of regression. *Technometrics*, **8**, 625–629.

Châtillon, G. (1984). The balloon rule for a rough estimate of the correlation coefficient. *American Statistician*, **38**, 58–60.

Daniel, C., and F. S. Wood (1980). *Fitting Equations to Data*, 2nd. ed. Wiley, New York.

Denby, L., and D. Pregibon (1987). An example of the use of graphics in regression. *American Statistician*, **41**, 33–38.

Draper, N., and H. Smith (1998). *Applied Regression Analysis*, 3rd ed. Wiley, New York.

Gillingham, R., and D. Heien (1971). Regression through the origin. *American Statistician*, **25**, 54–55.

Joiner, B. L. (1981). Lurking variables: Some examples. *American Statistician*, **35**, 227–233.

Jurečková, J. (1971). Nonparametric estimate of regression coefficients. *Annals of Mathematical Statistics*, **42**, 1328–1338.

Kendall, M. G. (1970). *Rank Correlation Methods*, 4th ed. Griffin, London.

Kruskal, W. H. (1958). Ordinal measures of association *Journal of the American Statistical Association*, **53**, 814–861.

Madansky, A. (1959). The fitting of straight lines when both variables are subject to error. *Journal of the American Statistical Association*, **54**, 173–205.

Olkin, I., and J. W. Pratt (1958). Unbiased estimation of certain correlation coefficients. *Annals of Mathematical Statistics*, **29**, 201–211.

Prescott, P. (1975). An approximate test for outliers in linear regression. *Technometrics*, **17**, 129–132.

Rodgers, J. L., and W. A. Nicewander (1988). Thirteen watlys to look at the correlation coefficient. *American Statistician*, **42**, 59–66.

Sampson, A. R. (1974). A tale of two regressions. *Journal of the American Statistical Association*, **69**, 682–689.

Schilling, M. F. (1984). Some remarks on quick estimation of the correlation coefficient. *American Statistician*, **38**, 330.

Thigpen, C. C. (1987). A sample-size problem in simple linear regression. *American Statistician*, **41**, 214–215.

10 Techniques for One-Way Analysis of Variance

In Chapter 8 we discussed a group comparison test for *two* independent samples that came from normal populations with possibly different means but with the same variance. The hypothesis H_0: $\mu_1 = \mu_2$ was tested. In this chapter we test similar hypotheses for *three, four,* or *more* independent samples taken from normal populations with possibly different means but a common variance.

10.1. THE ADDITIVE MODEL

A psychologist studying factors that influence the amount of time mice require to solve a new maze might be observing 4 groups of 3 mice each. Each group has had a different amount of previous experience at maze solving, and the psychologist is looking for evidence of learning. The mice in the first group have had 1 previous experience in maze solving; those in the second group have solved 2 mazes; the third group has solved 3; and the fourth group has solved 4. Each mouse is now placed in a new maze, and the amount of time (in minutes) required to solve the maze is recorded.

The data (simplified for this example) might be as follows:

		Group	
1	2	3	4
11	7	6	5
9	9	5	3
10	8	7	4

Before a formal analysis of these data, we plot the values as in Figure 10.1 and add the sample averages ($\bar{y}_1, \bar{y}_2, \bar{y}_3, \bar{y}_4$) to the graph.

Learning would be indicated by a decrease in the time required to solve the maze. The graph does seem to indicate a decrease in time for increased experience. However, the apparent differences in the graph could be due to sampling variability rather than learning. We need a method for deciding whether the differences in the sample averages are significant. If there is no learning, the four populations from which the samples were taken will all have the same means, $\mu_1 = \mu_2 = \mu_3 = \mu_4$. The analysis of variance is a formal method for testing this hypothesis.

To be able to speak more precisely about these data, in this text the symbol y_{ij} is used for the jth observation from the ith group. The first subscript i is reserved for the treatment group

Statistics for Research, *Third Edition*, Edited by Shirley Dowdy, Stanley Weardon, and Daniel Chilko.
ISBN 0-471-26735-X © 2004 John Wiley & Sons, Inc.

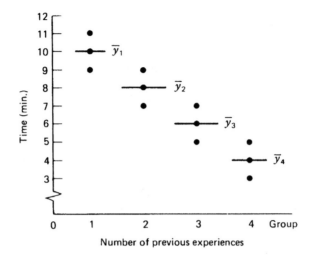

FIGURE 10.1. Data on time required to solve the maze.

number irrespective of whether groups are displayed in columns or in rows. Experimenters differ in how they display and label their data, so four groups of three observations each may be displayed as in Table 10.1 or as in Table 10.2. When reading books and articles, be careful to check how the subscripts are being used since the notation is not consistent.

In the example under consideration, the number of groups is $a = 4$ and the number of observations within each group is $n = 3$. We assume in all of the examples (until stated otherwise) that each group contains the same number of observations, n observations.

The psychologist in the present example wants to know if the amount of previous experience changes the time required to solve a maze. He wants to test H_0: $\mu_1 = \mu_2 = \mu_3 = \mu_4$ (that is, each of the samples comes from a population with the same mean) against H_a: *At least one inequality* (that is, $\mu_1 \neq \mu_2$ or $\mu_1 \neq \mu_3$ or $\mu_1 \neq \mu_4$ or $\mu_2 \neq \mu_3$ or $\mu_2 \neq \mu_4$ or $\mu_3 \neq \mu_4$). He is assuming that the four populations have a common variance σ^2.

It would be possible to test the equality of each pair of means by a t test; however, $\binom{4}{2} = 6$ separate t tests would be required for the null hypothesis under consideration. Besides being tedious, 6 separate t tests on the same data would have an α level much higher than the α used in each t test. A possible alternative procedure involves comparing the sample variance among

TABLE 10.1. Treatment Groups Displayed in Columns

	Group			
	y_{1j}	y_{2j}	y_{3j}	y_{4j}
	$y_{11} = 11$	$y_{21} = 7$	$y_{31} = 6$	$y_{41} = 5$
	$y_{12} = 9$	$y_{22} = 9$	$y_{32} = 5$	$y_{42} = 3$
	$y_{13} = \underline{10}$	$y_{23} = \underline{8}$	$y_{33} = \underline{7}$	$y_{43} = \underline{4}$
Total:	30	24	18	12	$\sum_i \sum_j y_{ij} = 84$

TABLE 10.2. Treatment Groups Displayed in Rows

Group				Total
y_{1j}	$y_{11} = 11$	$y_{12} = 9$	$y_{13} = 10$	30
y_{2j}	$y_{21} = 7$	$y_{22} = 9$	$y_{23} = 8$	24
y_{3j}	$y_{31} = 6$	$y_{32} = 5$	$y_{33} = 7$	18
y_{4j}	$y_{41} = 5$	$y_{42} = 3$	$y_{43} = 4$	12
			$\sum_i \sum_j y_{ij} = 84$	

groups with the sample variance within groups. This test is possible because if the null hypothesis is true, both of these statistics are estimates of σ^2.

To understand why the test is based on variance, it will be helpful if we consider the different types of averages associated with these data.

The grand average:
$$\bar{y} = \sum_i \sum_j y_{ij}/an = 84/12 = 7$$

The group averages:
$$\bar{y}_1 = \sum_j y_{1j}/n = 30/3 = 10$$
$$\bar{y}_2 = \sum_j y_{2j}/n = 24/3 = 8$$
$$\bar{y}_3 = \sum_j y_{3j}/n = 18/3 = 6$$
$$\bar{y}_4 = \sum_j y_{4j}/n = 12/3 = 4$$

The average of the group averages = The grand average = $\bar{y} = 7$

If we consider the population parameters related to these sample averages, each observation can be thought of in terms of an additive model consisting of three terms,

$$y_{ij} = \mu + \alpha_i + \varepsilon_{ij}$$

in which μ (estimated by $\bar{y}$) is the mean time for all mice, α_i (estimated by $\bar{y}_i - \bar{y}$) is the mean treatment effect, or adjustment, for all mice in the ith group, and ε_{ij} is a random effect due to the individual mouse. The data could then be written as

Group 1	Group 2
$11 = 7 + (10 - 7) + 1$	$7 = 7 + (8 - 7) + (-1)$
$9 = 7 + (10 - 7) + (-1)$	$9 = 7 + (8 - 7) + 1$
$10 = 7 + (10 - 7) + 0$	$8 = 7 + (8 - 7) + 0$

Group 3	Group 4
$6 = 7 + (6 - 7) + 0$	$5 = 7 + (4 - 7) + 1$
$5 = 7 + (6 - 7) + (-1)$	$3 = 7 + (4 - 7) + (-1)$
$7 = 7 + (6 - 7) + 1$	$4 = 7 + (4 - 7) + 0$

In terms of the additive model, the null hypothesis can be written in a different manner now:

$$H_0: \alpha_1 = \alpha_2 = \alpha_3 = \alpha_4 = 0, \text{ or } H_0: \alpha_i = 0 \text{ for all } i$$

with

$$H_a: \text{At least one inequalty, or } H_a: \alpha_i \neq 0 \text{ for some } i$$

The development of the F test that follows, comparing the variance among groups with the variance within groups to test the above hypothesis, assumes this additive model. It also assumes that all treatments of interest to the experimenter are being used, that each treatment group is normally distributed, that all groups have the same variance, and that the experimental units are randomly assigned to the treatment group. For example, in this experiment the 12 mice should be chosen at random from those available and randomly assigned to groups 1, 2, 3, and 4. This type of analysis of variance is called a *one-way completely randomized ANOVA* (analysis of variance). In symbols, the assumptions are written

$$y_{ij} = \mu + \alpha_i + \varepsilon_{ij}$$

with

$$\sum_i \alpha_i = 0$$

and

$$\varepsilon_{ij} \text{IND}(0, \sigma^2)$$

that is, the ε_{ij} are *independently normally distributed* with a mean of zero and a common variance of σ^2.

Returning now to the three types of sample averages, there are three types of sample variances that can be obtained by considering deviations from these sample averages. A sample variance is an average squared deviation from a sample average in which the averaging is achieved by dividing by the corresponding degrees of freedom. Thus the three types of sample variances are as given in Table 10.3.

The within-group variance is a pooled variance as in Chapter 8. The multiplication by n in the among-group variance is necessary if this variance is to be compared with the within-group variance. The among-group variance estimates the dispersion in the sampling distribution of averages of all samples of size n (that is, σ^2/n), so the among-group variance must be multiplied by n to estimate the dispersion of the original distribution.

The three types of deviations considered above are illustrated in Figure 10.2. The straight lines at right angles indicate the deviations of the observations from the grand average; these will be used for the total variance. The braces indicate the deviations of the observations from their respective group average; these will be used for the within-group variance. The dashed lines indicate the deviations of the group averages from the grand average, and these will be used for the among-group variance. If the null hypothesis is true, $\bar{y}_1$, $\bar{y}_2$, $\bar{y}_3$, and $\bar{y}_4$ are not significantly different from $\bar{y}$, and the within-group variance will be approximately the same as the among-group variance. However, if the null hypothesis is false, then the among-group variance will be larger because of the significant deviations of the group averages from the grand average.

TABLE 10.3. Three Types of Variance

Type of Variance	Formula	Meaning
Total variance	$$\dfrac{\sum_i \sum_j (y_{ij} - \bar{y})^2}{na - 1}$$	The average squared deviation of the observations from the grand average
Within-group variance	$$\dfrac{\sum_i \sum_j (y_{ij} - \bar{y})^2}{a(n - 1)}$$	The average squared deviation of the observations from their respective group average (the pooled variance)
Among-group variance	$$n\left[\dfrac{\sum_i (\bar{y}_i - \bar{y})^2}{a - 1}\right]$$	The average squared deviation of the group averages from the grand average multiplied by the number of observations in each group

In the maze example, the sum of squares (SS) or numerator of the variance in each case is as follows:

Total SS
$$\sum_i \sum_j (y_{ij} - \bar{y})^2 = 4^2 + 2^2 + 3^2 + \cdots + (-4)^2 + (-3)^2$$
$$= 68$$

Within SS
$$\sum_i \sum_j (y_{ij} - \bar{y}_i)^2 = [1^2 + (-1)^2 + 0^2] + \cdots + [1^2 + (-1)^2 + 0^2]$$
$$= 8$$

Among SS
$$n\sum_i (\bar{y}_i - \bar{y})^2 = 3[3^2 + 1^2 + (-1)^2 + (-3)^2]$$
$$= 60$$

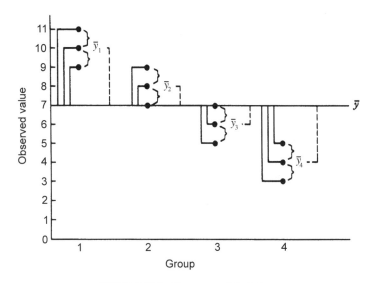

FIGURE 10.2. Three types of deviations.

This example illustrates that the total sum of squares can be partitioned into two parts, the among-group sum of squares and the within-group sum of squares.

Total SS	=	Among SS	+	Within SS
68	=	60	+	8

This relationship among the total, among-group, and within-group sum of squares leads to a shorter computational method, to be developed later. For now, the computation of the sum of squares just given will be used for the test. To change the sums of squares into variances (mean squares, or MS), they must be divided by their degrees of freedom.

The degrees of freedom are also partitioned as the sums of squares:

Total df	=	Among df	+	Within df
$na - 1$	=	$a - 1$	+	$a(n - 1)$
11	=	3	+	8

A conventional form used is a work table, as follows:

Source	df	SS	MS
Among groups	$a - 1 = 3$	60	$60/3 = 20$
Within groups	$a(n - 1) = 8$	8	$8/8 = 1$
Total	$an - 1 = 11$	68	

If the null hypothesis H_0: $\mu_1 = \mu_2 = \mu_3 = \mu_4$ is true, the among MS and the within MS are both estimates of σ^2. This is because we are sampling from the same population (Figure 10.3). The variance among the averages estimates σ^2/n so n times the variance among the averages, or the among-group variance, estimates σ^2.

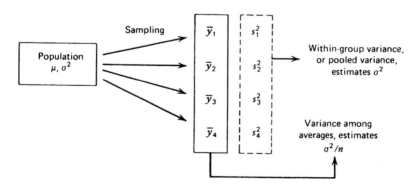

FIGURE 10.3. Within-group and among-group variances.

The test of hypothesis about the equality of means is therefore an F test for the equality of two variances:

$$F = \frac{\text{among MS}}{\text{within MS}} = \frac{20}{1} = 20$$

is computed. This F statistic is compared with the critical value $F_{0.05,3,8}$ and leads to rejection if $F \geq 4.066$. This is a one-sided F test since if the null hypothesis is false, the among MS is greater than the within MS. In this example, $F \geq 4.066$, so the null hypothesis is rejected and it is concluded that the sample came from 4 populations among which there is at least one inequality; that is, prior experience does affect the time required for the mice to solve a new maze.

EXERCISES

10.1.1. Compute the total sum of squares, among sum of squares, and within sum of squares for the following data:

Group		
1	2	3
1	2	3
1	1	2
0	1	2
0	0	3
0	1	1

Show that the total SS = among SS + within SS.

10.1.2. Four groups, each comprising 4 randomly selected persons, are asked to perform a simple mechanical task. Prior to the task, group A is given a strong depressant, group B a mild depressant, group C a mild stimulant, and group D a strong stimulant. The times (in seconds) required to complete the task are as follows:

Group				
A	4	2	3	2
B	2	3	3	2
C	2	2	3	1
D	1	2	1	1

a. Graph these data and add the group averages to the graph.

b. Do the drugs seem to affect the time required to complete the task?

c. Test the hypothesis $H_0: \mu_A = \mu_B = \mu_C = \mu_D$ using an F test.

10.1.3. Four pea plants of a certain variety are grown without fertilizer, and 4 plants of the same variety are grown with fertilizer. The mature heights (in feet) are recorded below:

Without:	0.9	1.0	0.8	1.2
With:	1.5	1.2	1.6	1.3

a. Test H_0: $\mu_1 = \mu_2$ by the ANOVA technique described in this section.

b. Test H_0: $\mu_1 = \mu_2$ by a two-sample t test.

c. What is the relationship between the F statistic and the t statistic?

10.1.4. In the maze example developed in this section, show that the average of the group averages is equal to the grand average. Why is this always true?

10.2. ONE-WAY ANALYSIS-OF-VARIANCE PROCEDURE

The procedure explained in Section 10.1 is a one-way ANOVA. In this section, we develop a shorter computational method for this procedure.

This short method depends on the fact already noted:

$$\text{Total SS} = \text{Within SS} + \text{Among SS}$$

This fact is used with an approach similar to the computational formula for the sample variance (Section 6.2):

$$s^2 = \frac{\sum y^2 - \left(\sum y\right)^2 / n}{n - 1}$$

In the computational formula, the sum of squares (the numerator) is found by considering the sum of the squared deviations from the origin, $\sum y^2$, and subtracting the correction factor, $\left(\sum y\right)^2 / n$, to get the sum of the squared deviations from the sample average. This method is used because it is simpler to compute with the deviations from the origin (the actual values) than with deviations from the average.

In ANOVA, a similar computational approach is used. We illustrate this using the mouse study of Section 10.1:

	y_{1j}	y_{2j}	y_{3j}	y_{4j}	
	11	7	6	5	
	9	9	5	3	
	10	8	7	4	
Totals	30	24	18	12	Grand total 84

When analyzing these data, we can consider three types of totals:

$$1 \text{ total of 12 observations} \quad \sum_i \sum_j y_{ij} \text{: } 84$$

$$4 \text{ totals of 3 observations} \quad \sum_j y_{ij} \text{: } 30, 24, 18, 12$$

$$12 \text{ totals of 1 observation} \quad y_{ij} \text{: } 11, 9, \ldots, 3, 4$$

For the short computational method, these totals will be squared, divided by the number of observations per total, and summed. Table 10.4 summarizes this procedure.

The ANOVA can then be computed from these uncorrected sums of squares as follows:

Source	df	SS	MS
Among groups	$a - 1 = 3$	$SS_a = A - CF = 60$	$60/3 = 20$
Within groups	$a(n - 1) = 8$	$SS_e = T - A = 8$	$8/8 = 1$
Total	$an - 1 = 11$	$SS_t = T - CF = 68$	

To aid memory, it should be noted that the degrees of freedom and the number of squared values (totals) can be used to determine the sum of squares in the ANOVA table. For example, the among SS has $a - 1$ degrees of freedom, and among $SS = A - CF$, in which A contains a squared values and CF contains 1 squared value. The within SS has $a(n - 1) = an - a$ degrees of freedom, and within $SS = T - A$, with T containing an squared values and A containing a squared values. Similarly for the total SS.

In articles in professional journals, the sums of squares column is not usually given, nor is the row for the total. However, the sums of squares are often used to compute a statistic that gives information similar to that of coefficient of determination discussed in Section 9.4. If it is useful for the experimenter to know how much of the variability among the maze-solving times of the 12 mice can be attributed to being grouped by experience it can be expressed as

$$1 - \frac{\text{unexplained variability}}{\text{total variability}} = \frac{SS_a}{SS_t} = \frac{60}{68} = 0.882$$

TABLE 10.4. Uncorrected Sums of Squares for Equal-Sized Groups

Name	Symbol	Number of Totals	Observations/ Total	Formula	Numerical Value
Uncorrected total SS	T	$an = 12$	1	$\sum_i \sum_j y_{ij}^2$	$11^2 + 9^2 + \cdots + 4^2 = 656$
Uncorrected group SS	A	$a = 4$	$n = 3$	$\sum_i \left(\sum_j y_{ij} \right)^2 \Big/ n$	$30^2/3 + 24^2/3 + 18^2/3 + 12^2/3 = 648$
Correction factor	CF	1	$an = 12$	$\left(\sum_i \sum_j y_{ij} \right)^2 \Big/ an$	$84^2/12 = 588$

Because this statistic serves a purpose similar to the coefficient of determination, it is identified as *Rsquare*.

Another, more realistic, example of a one-way ANOVA follows.

Example 10.1. One-Way Completely Randomized ANOVA with Equal Sample Sizes

In a study of the physiological stress resulting from operating hand-held chain saws, experimenters measured the kickback that occurs when a saw is used to cut a 3-in.-thick synthetic fiber board. The variable of interest was the angle (in degrees) to which the saw is deflected when it begins to cut the board. Below are the angles of deflection recorded for 5 random saws from each of 4 different manufacturers' models. A graph of the data and group averages appears in Figure 10.4.

<table>
<tr><th></th><th colspan="4">Chain Saw Model</th><th></th></tr>
<tr><th></th><th>A</th><th>B</th><th>C</th><th>D</th><th>Totals</th></tr>
<tr><td></td><td>42</td><td>28</td><td>57</td><td>29</td><td></td></tr>
<tr><td></td><td>17</td><td>50</td><td>45</td><td>40</td><td></td></tr>
<tr><td></td><td>24</td><td>44</td><td>48</td><td>22</td><td></td></tr>
<tr><td></td><td>39</td><td>32</td><td>41</td><td>34</td><td></td></tr>
<tr><td></td><td>43</td><td>61</td><td>54</td><td>30</td><td></td></tr>
<tr><td>$\sum_j y_{ij}$</td><td>165</td><td>215</td><td>245</td><td>155</td><td>780</td></tr>
<tr><td>$\sum_j y_{ij}^2$</td><td>5,999</td><td>9,965</td><td>12,175</td><td>4,981</td><td>33,120</td></tr>
<tr><td>$\left(\sum_j y_{ij}\right)^2$</td><td>27,225</td><td>46,225</td><td>60,025</td><td>24,025</td><td>157,500</td></tr>
</table>

FIGURE 10.4. Angles of deflection for four types of chain saws.

The hypothesis to be tested is

$$H_0: \alpha_A = \alpha_B = \alpha_C = \alpha_D$$

against

$$H_a: \text{At least one inequality}$$

$$T = 33.120$$

$$A = \frac{157,500}{5} = 31,500$$

$$CF = \frac{780^2}{20} = 30,420$$

Source	df	SS	MS
Among groups	$a - 1 = 3$	$SS_a = A - CF = 1080$	$MS_a = SS_a/(a-1)$ $= 360$
Within groups (error)	$a(n-1) = 16$	$SS_e = T - A = 1620$	$MS_e = SS_e/a(n-1)$ $= 101.25$

The test statistic is $F = 360/101.25 = 3.56$ and $F_{0.05,3,16} = 3.239$. The null hypothesis is rejected. There is a significant difference among the average kickbacks of the four types of saws. The proportion of variability in kickback that can be attributed to the different models of saws is

$$\text{Rsquare} = 1 - \frac{SS_e}{SS_t} = 1 - \frac{1620}{1620 + 1080} = 0.60$$

A significant portion of the variability among the data has been explained by the differences among the group means.

To determine which of the models are different with respect to kickback, a follow-up procedure will be needed. This procedure is developed in the next section.

We can summarize the one-way ANOVA procedure for equal group sizes as follows. The symbol SS_e is used for the within-group sum of squares because this quantity represents the variability due to random sampling, that is, the *sampling error*.

Procedure. One-Way Completely Randomized ANOVA with Equal Sample Sizes

$H_0: \alpha_1 = \alpha_2 = \cdots = \alpha_a = 0$, or $H_0: \alpha_i = 0$ *for all i*
$H_a:$ *At least one inequality*, or $H_a: \alpha_i \neq 0$ *for some i*
$y_{ij} = j$th observation in the ith treatment group

$i = 1, \ldots, a \quad j = 1, \ldots, n$
Compute:

$$T = \sum_i \sum_j y_{ij}^2$$

$$A = \sum_i \left(\sum_j y_{ij} \right)^2 \Big/ n$$

$$CF = \left(\sum_i \sum_j y_{ij} \right)^2 \Big/ an$$

Source	df	SS	MS	F
Among groups	$a - 1$	$SS_a = A - CF$	$MS_a = SS_a/(a-1)$	MS_a/MS_e
Within groups (error)	$a(n-1)$	$SS_e = T - A$	$MS_e = SS_e/a(n-1)$	
Total	$an - 1$	$SS_t = T - CF$		

Reject H_0 if $F \geq F_{\alpha, a-1, a(n-1)}$

Many times the experimenter has no control over sample size, and an unbalanced design is necessary. This can happen in a genetics experiment in which the experimenter has no control over the number of offspring, in wildlife experiments that depend on the number of animals trapped, in a botany experiment in which some plants die (from causes extraneous to the experiment), or in situations where cost restricts equalizing the sample sizes. The one-way ANOVA can also be used if the sample sizes are unequal, although there may be some loss of power. The sums of squares needed for the computations are as in Table 10.5.

TABLE 10.5. Uncorrected Sums of Squares for Unequal-Sized Groups

Source	Symbol	Number of Squared Values	Observations/Square	Formula
Uncorrected total SS	T	N	1	$\sum_i \sum_j y_{ij}^2$
Uncorrected group SS	A	a	n_i	$\sum_i \left(\sum_j y_{ij} \right)^2 \Big/ n_i$
Correction factor	CF	1	N	$\left(\sum_i \sum_j y_{ij} \right)^2 \Big/ N$

Note: n_i is the number of observations in the ith group and $N = \sum_i n_i$.

Example 10.2.   One-Way Completely Randomized ANOVA with Unequal Groups

A psychologist is studying several types of behavioral disorders in children and has reached a stage where she can classify children as belonging to one of 7 types, depending on certain behavioral characteristics. She has a feeling that the mean level of intelligence may differ in some of these groups, so she begins to examine the IQ scores of children in these 7 categories. In her files she finds cases of all 7 types. There is some question in her mind about the randomness of these data and also whether they meet the other assumptions of an ANOVA. However, as a preliminary investigation, she would like to test H_0: $\alpha_1 = \alpha_2 = \cdots = \alpha_7$; that is, that there is no difference among the mean IQ of children in the different categories. Since the psychologist has no control over the number of cases in her file, the groups have unequal sizes.

				Disorder			
	1	2	3	4	5	6	7
	105	115	103	124	115	85	79
	98	109	96	127	112	106	87
	110	121	105	118		98	
		130	107			111	
			112				
$\sum_j y_{ij}$	313	475	523	369	227	400	166
n_i	3	4	5	3	2	4	2
$\dfrac{\left(\sum_j y_{ij}\right)^2}{n_i}$	32,656.3	56,406.2	54,705.8	45,387.0	25,764.5	40,000.0	13,778.0
$\sum_j y_{ij}^2$	32,729	56,647	54,843	45,429	25,769	40,386	13,810

$\sum_i n_i = 23 \quad \sum_i \sum_j y_{ij} = 2473 \quad \sum_i \sum_j y_{ij}^2 = 269,613$

$T = \sum_i \sum_j y_{ij}^2 = 269,613.00$

$A = \sum_i \left(\sum_j y_{ij}\right)^2 \Big/ n_i = 268,697.80$

$CF = \left(\sum_i \sum_j y_{ij}\right)^2 \Big/ N = 265,901.26$

Source	df	SS	MS	F	Critical Value $\alpha = 0.05$
Among groups	$a - 1 = 6$	2,796.54	466.09	8.14	2.741
Within groups	$N - a = 16$	915.20	57.20		

   The null hypothesis is rejected, and the psychologist concludes that there seems to be a difference among the mean IQ of the children in the different categories.

The procedure for unequal groups can be summarized as follows.

Procedure. One-Way Completely Randomized ANOVA with Unequal Sample Sizes

H_0: $\alpha_1 = \alpha_2 = \cdots = \alpha_a = 0$ or H_0: $\alpha_i = 0$ *for all i*
H_a: *At least one inequality* or H_a: $\alpha_i \neq 0$ *for some i*

$y_{ij} = j$th observation in the ith treatment group

$i = 1, \ldots, a, j = 1, \ldots, n_i, \sum_i n_i = N$
Compute:

$$T = \sum_i \sum_j y_{ij}^2$$

$$A = \sum_i \left(\sum_j y_{ij} \right)^2 \Big/ n_i$$

$$CF = \left(\sum_i \sum_j y_{ij} \right)^2 \Big/ N$$

Source	df	SS	MS	F
Among groups	$a - 1$	$SS_a = A - CF$	$MS_a = SS_a/(a - 1)$	MS_a/MS_e
Within groups (error)	$N - a$	$SS_e = T - A$	$MS_e = SS_e/(N - a)$	
Total	$N - 1$	$SS_t = T - CF$		

Reject H_0 if $F \geq F_{\alpha, a-1, N-a}$.

EXERCISES

10.2.1. Five groups of 4 men each are randomly assigned diets. At the end of a week, the following changes in weight (in pounds) are observed.

		Diet		
1	2	3	4	5
+3	+2	+4	+3	+1
−2	0	0	0	−1
0	+2	+1	−1	−2
−2	+1	+2	+1	−1

Perform an ANOVA to see if there is any difference among the effects of these diets.

10.2.2. Five brands of lawnmowers are compared on the basis of hours of trouble-free operation. Eight randomly chosen mowers of each type are used in the study. Complete the following ANOVA table:

Source	df	SS	MS
Among brands	—	140	—
Within brands	—	—	11

Give the null and alternative hypotheses to be tested by these data. Draw conclusions concerning the hypotheses.

10.2.3. Given the information below about the life (in months) of 3 types of light bulbs, graph the data and complete the ANOVA table.

	Brand		
	A	B	C
	7.0	13.4	9.5
	11.8	15.0	13.6
	10.5	14.6	10.6
	12.6	17.3	13.5
$\sum_j y_{ij}$	41.9	60.3	47.2
$\bar{y}_i$	10.5	15.1	11.8
$\sum_j (y_{ij} - \bar{y}_i)^2$	18.35	7.99	12.86

$$41.9 + 60.3 + 47.2 = 149.4$$

$$(41.9)^2 + (60.3)^2 + (47.2)^2 = 7{,}619.54$$

$$(149.4)^2 = 22{,}320.36$$

What is the hypothesis about the means of the brands? Would the hypothesis be accepted? What conclusion do you draw about the light bulbs?

10.2.4. Tomato plants are treated with 5 different fertilizers, and the sum of the weight (in pounds) of the ripe fruit is recorded for each plant that matures:

Fertilizer:	A	B	C	D	E
Number of mature plants:	4	7	6	5	6
$\sum_j y_{ij}$:	81	111	138	96	101
$\sum_j y_{ij}^2$:	1649	1775	3184	1850	1715

Perform an ANOVA to test for equality of means. What assumptions are necessary for this analysis to be valid?

10.2.5. Three difference methods of processing orange juice are compared. The amount of vitamin C per 8-oz serving is the variable of interest (in milligrams). Five servings are chosen at random from each process.

Processing Method

	A	B	C	Totals
	96	123	76	
	87	115	78	
	85	122	79	
	92	118	77	
	90	122	80	
$\sum_j y_{ij}$	450	600	390	1,440
$\left(\sum_j y_{ij}\right)^2$	202,500	360,000	152,100	714,600
$\sum_j y_{ij}^2$	40,574	72,046	30,430	143,050

What null hypothesis can be tested? Graph the data. Does the null hypothesis appear to be true? If $\alpha = 0.05$, what is the critical value of the test statistic? Show that the correction factor for these data is 138,240. Complete the ANOVA table. Should the null hypothesis be rejected? What conclusion do you draw?

10.2.6. Given the following information, complete the analysis of variance to test for equality of group means:

Source	Number of Squared Values	Observations per Squared Value	Numerical Value
$\sum_i \sum_j y_{ij}^2$	30	1	1565
$\sum_i \left(\sum_j y_{ij}\right)^2 / n$	6	5	1325
$\left(\sum_i \sum_j y_{ij}\right)^2 / an$	1	30	1200

10.2.7. Live traps are set to capture samples of rabbits at 5 different locations in a large wooded area. The weights (in ounces) are as follows:

Area

1	2	3	4	5
37	29	49	40	50
40	33	47	38	46
46	34		42	49
	31		39	
			41	

Use box plots to graph the data and the group averages. Do the box plots and the size of Rsquare suggest that the mean weights of the rabbits differ at some of the locations? Test a hypothesis about locations at the 1% level of significance. What assumptions are necessary about the rabbits?

10.2.8. A dean at a small college believes there may be a difference in the mean age of his faculty in different departments. He obtains the following information about faculty ages:

Mathematics	28	35	31	32	
English	45	37	42	38	36
Foreign languages	27	32	29		
History	43	39			

a. Are there significant differences in the average ages for these 4 departments?

b. What assumptions must be made in order for ANOVA techniques to be valid for this study?

10.2.9. A forest entomologist has isolated 7 insecticides that are reasonably safe to the rest of the environment when used to control gypsy moths. She wants to determine whether any one of them produces significantly greater mortality than the others when applied topically to adult gypsy moths. Using standard bioassay techniques, she applies a given insecticide to the abdomen of each of 100 moths. This procedure is repeated 5 times for each insecticide, with new solutions being prepared each time. Per cent mortality is recorded after 24 hours for each insecticide trial. Assume that the data, although distributed in a binomial fashion, will approximate the normal distribution adequately for ANOVA procedures.

a. Although 3500 moths are used, why are there only 34 degrees of freedom associated with the experiment?

b. In using the y_{ij} notation, does the j subscript refer to the insecticide or the trial?

c. What are the assumptions for an ANOVA?

d. Use this information to complete the accompanying ANOVA table:

$$\frac{\left(\sum_i \sum_j y_{ij}\right)^2}{an} = 143{,}360 \qquad \sum_i \sum_j y_{ij}^2 = 144{,}334$$

Source	df	SS	MS
Insecticides	—	—	55
Trials within insecticides	—	—	—

e. Give the null and alternative hypotheses.

f. Give the critical value ($\alpha = 0.05$) for a test of the above hypothesis and draw conclusions about the experiment.

10.2.10. The following linear model is used in a study involving 5 artists and 4 paintings per artist:

$$u_{ij} = \lambda + v_i + \delta_{ij}$$

in which $i = 1, \ldots, a = 5$ and $j = 1, \ldots, n = 4$. The data below give the number of smudges per picture:

	Artist				
	A	B	C	D	E
	7	2	4	11	2
	6	4	6	7	0
	8	4	6	8	3
	7	4	2	4	5
Total	28	14	18	30	10

a. To perform an ANOVA on these data, what must be assumed about v_i and δ_{ij}?

b. What is the numerical value of u_{23} and $\sum_j u_{3j}$?

c. Given that $7^2 + 6^2 + 8^2 + \cdots + 5^2 = 630$ and $28^2 + 14^2 + 18^2 + 30^2 + 10^2 = 2304$, complete a table for the uncorrected sums of squares giving the number of squared values, the number of observations per squared value, and the numerical value of the uncorrected sum of squares.

10.2.11. Suppose that a building contractor wants to test 3 types of wooden beams for weight-bearing capacity. Five beams of each type are broken by stacking lead weights on them, and the weight required to break each beam is recorded.

a. Given the mathematical model

$$z_{hi} = \psi + \delta\_ + \xi\_$$

in which $z_{hi} =$ the breaking strength of beam i within type h

$$\delta\_ = \text{the symbol of the type effect}$$
$$\xi\_ = \text{the symbol of the beam within type effect,}$$

fill in the blanks with the appropriate subscripts.

b. What assumptions must be made about $\delta\_$ and $\xi\_$?

c. What is the largest numerical value that can be taken by each subscript in the model?

d. If the computations made on the experimental data are

$$\sum_h \sum_i z_{hi}^2 = 3,620,000$$

$$\sum_h \left(\sum_i z_{hi} \right)^2 = 18,040,000$$

$$\left(\sum_h \sum_i z_{hi} \right)^2 = 54,000,000$$

complete the ANOVA.

10.3. MULTIPLE-COMPARISON PROCEDURES

In Sections 10.1 and 10.2 of this chapter, the ANOVA procedure is developed to test H_0: $\mu_1 = \mu_2 = \cdots = \mu_a$ (or H_0: $\alpha_1 = \alpha_2 = \cdots = \alpha_a$). If the null hypothesis is rejected, we conclude that there is at least one inequality among the means of the treatment groups (or among the treatment effects).

If the treatment groups under consideration exhaust the cases that are of interest to the experimenter (as we have been assuming in this chapter) and the F test is significant, the experimenter may want to draw some further conclusions. She may want to decide which pairs of treatments are different or she may want to contrast one treatment effect with the average of some other treatment effects or she may want to estimate some of the parameters in the experiment.

In this section we discuss several procedures for deciding which pairs of means are different. In general, these techniques are called *multiple-comparison procedures*. Contrasts, estimation, and Bonferroni procedures, which have gained widespread use, are discussed in Sections 10.4 to 10.6.

Several multiple-comparison procedures are available to researchers. We discuss 5 different approaches and their relative merits for various experimental situations. In all cases we assume equal sample sizes for the treatment groups.

Some Multiple-Comparison Procedures

1. Fisher's least significant difference
2. Duncan's new multiple-range test
3. Student–Newman–Keuls' procedure
4. Tukey's honestly significant difference
5. Scheffé's method

1. Fisher's Least Significant Difference. R. A. Fisher's multiple-comparison procedure is known as the least significant difference. It is based on a t test. If the treatment groups are all of equal size n, then two sample averages, $\bar{y}_1$ and $\bar{y}_2$ for example, can be tested for a significant difference by the statistic

$$t = \frac{\bar{y}_1 - \bar{y}_2}{\sqrt{(s_p^2/n) + (s_p^2/n)}} = \frac{\bar{y}_1 - \bar{y}_2}{\sqrt{2s_p^2/n}}$$

in which s_p^2 is the pooled sample variance as in Chapter 8. Thus Fisher said the difference $\bar{y}_i - \bar{y}_j$ is significant if

$$|\bar{y}_i - \bar{y}_j| \geq t_{\alpha/2, a(n-1)} \sqrt{\frac{2\text{MS}_e}{n}}$$

since MS_e in the ANOVA is a pooled estimate of the common variance of the treatment groups and MS_e has $a(n-1)$ degrees of freedom.

In order to protect the overall Type I error rate for the experiment, Fisher's procedure requires a prior significant F test in the ANOVA. With this condition, the overall error rate has been shown by simulation to be approximately the α level of the F test.

Example 10.3.   Fisher's Least Significant Difference

In the chain saw study, Example 10.1 of Section 10.2, the sample averages are

$$\bar{y}_A = \frac{165}{5} = 33 \quad \bar{y}_C = \frac{245}{5} = 49$$

$$\bar{y}_B = \frac{215}{5} = 43 \quad \bar{y}_D = \frac{155}{5} = 31$$

The experimenter wants to test $\binom{a}{2} = \binom{4}{2} = 6$ hypotheses

$$H_0: \mu_A = \mu_B \qquad H_0: \mu_B = \mu_C$$
$$H_0: \mu_A = \mu_C \qquad H_0: \mu_B = \mu_D$$
$$H_0: \mu_A = \mu_D \qquad H_0: \mu_C = \mu_D$$

to locate the specific difference or differences he believes exist because of the prior significant F test.

If Fisher's test is used, the differences between all pairs of sample averages must be compared with

$$t_{\alpha/2,a(n-1)}\sqrt{\frac{2MS_e}{n}} = t_{0.025,16}\sqrt{\frac{2(101.25)}{5}}$$
$$= 2.120(6.36)$$
$$= 13.5$$

at $\alpha = 0.05$.

To keep track of all possible differences between sample averages, he arranges them in order according to size, from the smallest to the largest,

$$31 \quad 33 \quad 43 \quad 49$$

and forms a table listing the ordered averages on the left omitting the largest and across the top omitting the smallest:

		A	B	C
		33	43	49
D	31			
A	33			
B	43			

If the top average is larger than one on the left, he subtracts the average on the left from the average on the top and enters the difference in the table:

		A	B	C
		33	43	49
D	31	2	12	18
A	33		10	16
B	43			6

These differences are then compared with the least significant difference 13.5, which was computed earlier. He begins at the right of the top row of differences. There he finds 18, which is greater than 13.5, so he marks 18 with an asterisk and concludes that $\mu_D \neq \mu_C$. The next entry in the top row is 12, which is less than 13.5, so he goes no further in that row. He then treats the second and third rows in the same manner. The final table has the following form:

		A 33	B 43	C 49
D	31	2	12	18*
A	33		10	16*
B	43			6

The only pairs of means that are different are $\mu_D \neq \mu_C$ and $\mu_A \neq \mu_C$.

In a journal, in order to save space, he would report that at the $\alpha = 0.05$ level by Fisher's least significant difference any two averages not underlined by the same line segment are significantly different.

D	A	B	C
31	33	43	49

Since the middle line is already indicated by the first line, it can be omitted:

31	33	43	49

Fisher's test has a drawback; it requires that the null hypothesis be rejected in the ANOVA procedure. It is possible that the F test will fail to detect a single significant difference among several treatment groups. In a case like this, Fisher's least significant difference cannot be used. The other multiple-comparison procedures to be discussed do not require a significant F test; they protect the Type I error rate by different approaches.

2. Duncan's New Multiple Range Test[†]. We will not go into the details of Duncan's method for protecting the error rate. Briefly, he considers the error rate for each pairwise comparison (rather than an overall rage) and allows a higher rate for pairs of sample averages that are further apart when ordered by size. Thus, if

$$\bar{y}_1 \quad \bar{y}_2 \quad \bar{y}_3$$

are three sample averages arranged from smallest to largest, a test of $\mu_1 = \mu_3$ would have a higher error rate than the test of $\mu_1 = \mu_2$. Because of this, Duncan's procedure will involve several different critical differences, in contrast to Fisher's single least significant difference.

[†]Duncan (1955) is the most common reference to his test, and while hardly a recent publication, "New" is still retained in its title to avoid confusion with other tables in the literature.

To reject H_0: $\mu_i = \mu_j$ when $\bar{y}_i$, $\bar{y}_j$ span r ranked sample averages, it is necessary that

$$|\bar{y}_i - \bar{y}_j| \geq d_{\alpha,r,a(n-1)}\sqrt{\frac{MS_e}{n}}$$

in which $d_{\alpha,r,a(n-1)}$ is found in Tables A.14a and A.14b in the Appendix. The α is the significance level set by the experimenter; Duncan makes the necessary adjustments in his table. Note also that the radical does not contain the factor of 2 found in the t test; it has been absorbed into the d value. If we are dealing with adjacent sample averages,

$$d_{\alpha,2,v} = t_{\alpha/2,v}\sqrt{2}$$

Example 10.4. Duncan's New Multiple-Range Test

Using the table of differences of sample averages for the power saw data, we see that the lowest diagonal consists of differences of adjacent ranked averages,

		A 33	B 43	C 49	
D	31	2	12	18	
					↖ spans 4 ranked averages
A	33		10	16	
					↖ spans 3 ranked averages
B	43			6	
					↖ spans 2 ranked averages

that is, a span of two ranked averages. The second diagonal consists of differences of averages separated by one average, that is, the difference spans three ranked averages. The remaining difference spans four ranked averages. Using Table A.14a in the Appendix, the experimenter finds

$$d_{0.05,2,16}\sqrt{\frac{MS_e}{n}} = 2.998(4.50) = 13.5$$

$$d_{0.05,3,16}\sqrt{\frac{MS_e}{n}} = 3.144(4.50) = 14.1$$

$$d_{0.05,4,16}\sqrt{\frac{MS_e}{n}} = 3.235(4.50) = 14.6$$

Comparing the differences with these critical values, he finds two significant differences:

		A 33	B 43	C 49	
D	31	2	12	18*	Compare with
					↖ 14.6
A	33		10	16*	
					↖ 14.1
B	43			6	
					↖ 13.5

His conclusion would be identical with the one reached with Fisher's procedure.

Duncan's test is slightly more conservative than Fisher's; that is, it will sometimes find fewer significant differences. However, there is about 95% agreement between the two procedures. It may be tempting to use the $d_{\alpha,r,a(n-1)}$ values in the table for similarly conservative confidence intervals for differences between pairs of means, $\mu_i - \mu_j$, but it is inappropriate to do so. This is because, as noted before, α is allowed to increase as we compare averages farther apart in ranked order; hence there would not be a constant $1 - \alpha$ value for all confidence intervals. Proper procedures for *simultaneous confidence intervals*, intervals for all $\mu_i - \mu_j$ pairs, are discussed in Section 10.5.

3. Student–Newman–Keuls' Procedure. Student–Newman–Keuls' procedure is still more conservative than Duncan's. Like Duncan's test, different critical values are used depending on the span of the two ranked averages being compared. However, this test protects the Type I error rate using a constant level for each diagonal.

Two sample averages which span r ranked averages are significantly different if

$$|\bar{y}_i - \bar{y}_j| \geq q_{\alpha,r,a(n-1)}\sqrt{\frac{\mathrm{MS}_e}{n}}$$

in which the q values are found in Tables A.15a and A.15b in the Appendix, the Studentized range.

Example 10.5.   Student–Newman–Keuls' Procedure

Using the chain saw data of Example 10.3 and Table A.15a in the Appendix, the investigator finds:

$$q_{0.05,2,16}\sqrt{\frac{\mathrm{MS}_e}{n}} = 2.998(4.50) = 13.5$$

$$q_{0.05,3,16}\sqrt{\frac{\mathrm{MS}_e}{n}} = 3.649(4.50) = 16.4$$

$$q_{0.05,4,16}\sqrt{\frac{\mathrm{MS}_e}{n}} = 4.046(4.50) = 18.2$$

The table of differences is

		A	B	C	
		33	43	49	
D	31	2	12	18	Compare with
					⟍ 18.2
A	33		10	16	
					⟍ 16.4
B	43			6	
					⟍ 13.5

Thus, none of the differences are significant using this procedure.

This procedure is so conservative that it located no differences, whereas the F test in the ANOVA indicated that a difference exists. As mentioned in the discussion of Duncan's new multiple-range test, tabular values for a multiple-range test cannot validly be used to replace $\sqrt{2}t_{\alpha/2,a(n-1)}$ for conservative simultaneous confidence intervals. Using a q value in place of a t value, there is an appropriate confidence interval only for the difference between the largest and smallest averages, as is seen in Section 10.5.

4. Tukey's Honestly Significant Difference. Tukey's procedure is still more conservative. It uses a single critical difference:

$$q_{\alpha,a,a(n-1)}\sqrt{\frac{MS_e}{n}}$$

that is, the largest critical difference in Student–Newman–Keuls's procedure. The error rate is for the entire experiment.

Example 10.6.   Tukey's Honestly Significant Difference

For the chain saw data (see Example 10.3), two averages $\bar{y}_i, \bar{y}_j$ are significantly different if

$$|\bar{y}_i - \bar{y}_j| \geq q_{0.05,4,16}\sqrt{\frac{101.25}{5}}$$
$$= 4.046(4.50)$$
$$= 18.2$$

Thus, none of the pairs of averages is significantly different according to this procedure.

Multiple-range procedures discussed are designed as simultaneous tests of H_0: $\mu_i = \mu_j$ for pairwise comparison of all averages in the experiment. We have noted that for a single t test of the difference between two sample averages there is a correspondence between

$$t = \frac{\bar{y}_1 - \bar{y}_2}{\sqrt{2s_p^2/n}} \quad \text{and} \quad \text{CI}_{1-\alpha}: (\bar{y}_1 - \bar{y}_2) \pm t_{\alpha/2.2(n-1)}\sqrt{\frac{2s_p^2}{n}}$$

However, we have noted that, while this is true for the t test and confidence interval involving just two means, it may not hold for simultaneous tests and confidence intervals involving the $a > 2$ means in an ANOVA. Fisher's and Tukey's procedures are not really multiple-range tests because the same q value is used to test all averages, irrespective of relative rank, and with the same q value in all confidence intervals there is no question concerning the actual size of $1 - \alpha$. As we might suspect, and as we see in Section 10.5, confidence intervals using $q_{\alpha,a,a(n-1)}$ rather than $t_{\alpha/2.a(n-1)}\sqrt{2s_p^2/n}$ are very wide, hence very conservative.

5. Scheffé's Method. Scheffé's method can be used to compare means and also to make other types of contrasts. For example, we might want to test

$$H_0: \mu_1 = \frac{\mu_2 + \mu_3}{2}$$

that is, that treatment 1 is the same as the average of treatments 2 and 3. The error rate α in Scheffé's procedure applies to all possible contrasts.

To compare two means using this method, $\bar{y}_i$ and $\bar{y}_j$ are significantly different if

$$|\bar{y}_i - \bar{y}_j| \geq \sqrt{(a-1)F_{\alpha,a-1,a(n-1)}}\sqrt{\frac{2MS_e}{n}}$$

Example 10.7. Scheffé's Method for Comparing Means

In the chain saw study, the critical difference is

$$\sqrt{3F_{0.05,3,16}}\sqrt{\frac{2(101.25)}{5}} = \sqrt{3(3.239)}\sqrt{40.5} = 19.8$$

Again this yields no significant difference.

Scheffé's is the most conservative of the methods we have discussed. It is very likely to miss detecting a real difference that exists. Scheffé's approach is used more often for the other contrasts; in these cases an adjustment is needed in the standard error. For example, to test

$$H_0: \mu_1 = \frac{\mu_2 + \mu_3}{2} \quad \text{or the equivalent,} \quad H_0: \mu_1 - \frac{\mu_2}{2} - \frac{\mu_3}{2} = 0$$

the standard error is $\sqrt{3MS_e/2n}$. The coefficient $3/2$ is the sum of $1^2 + (-1/2)^2 + (-1/2)^2$, that is, the sum of the squares of the coefficients in the linear combinations of the μ's in the null hypothesis. Thus, in the chain saw example, if we wanted to test whether the kickback of model A was significantly different from the average of models B and C, we would compute the critical difference

$$\sqrt{3F_{0.05,3,16}}\sqrt{\frac{3MS_e}{2n}} = \sqrt{3(3.239)}\sqrt{\frac{3(101.25)}{2(5)}} = 17.180$$

Since

$$\left|\bar{y}_A - \frac{\bar{y}_B}{2} - \frac{\bar{y}_C}{2}\right| = \left|33 - \frac{43}{2} - \frac{49}{2}\right| = |-13| = 13$$

we would conclude that the difference is not significant.

The 5 procedures we have just outlined are only some of the multiple comparisons available to the researcher. Which procedure should be used depends upon which type of error is more serious. In the chain saw example, assume the prices are approximately the same. Then a Type I error is not serious; it would imply that we decide one model has less kickback than another when in fact the two models have the same amount of kickback. A Type II error would imply that a difference in kickback actually exists but we fail to detect it, a more serious error. Thus, in this experiment we want maximum power and we would probably use Fisher's least significant difference. The experimenter should decide before the experimentation which method will be used to compare the means.

Table 10.6 lists the five tests indicating decreasing power and increasing error rate. The five procedures can be summarized as follows.

TABLE 10.6. Comparison of Multiple-Comparison Procedures

Multiple-Comparison Procedure	Power		Type I Error Rate
Fisher's Duncan's Student–Newman– Keuls'	Highest ↓	Highest ↑ More conservative, less likely to detect real differences	Highest More likely to indicate false differences
Tukey's Scheffé's	Lowest	Lowest	

Procedure. Multiple-Comparison Procedures

H_0: $\mu_1 = \mu_2$, H_0: $\mu_1 = \mu_3$, and so on, for all pairs of group means, or in general terms, these hypotheses can be written as H_0: $\mu_i = \mu_j$ *for all* $i \neq j$.
H_a: $\mu_1 \neq \mu_2$, H_a: $\mu_1 \neq \mu_3$, ... or in general notation, H_a: $\mu_i \neq \mu_j$ *for some* $i \neq j$.
Compute $\bar{y}_1, \bar{y}_2, \ldots, \bar{y}_a$, the a sample averages, and arrange them in order from the smallest to the largest:

$$\bar{y}_{(1)}, \bar{y}_{(2)}, \ldots, \bar{y}_{(a)}$$

Form a table of differences:

	$\bar{y}_{(2)}$	$\bar{y}_{(3)}$	$\cdots$	$\bar{y}_{(a)}$
$\bar{y}_{(1)}$	$\bar{y}_{(2)} - \bar{y}_{(1)}$	$\bar{y}_{(3)} - \bar{y}_{(1)}$	$\cdots$	$\bar{y}_{(a)} - \bar{y}_{(1)}$
$\bar{y}_{(2)}$		$\bar{y}_{(3)} - \bar{y}_{(2)}$	$\cdots$	$\bar{y}_{(a)} - \bar{y}_{(2)}$
.			$\cdots$	
.			$\cdots$	
.			$\cdots$	
$\bar{y}_{(a-1)}$			$\cdots$	$\bar{y}_{(a)} - \bar{y}_{(a-1)}$

Determine the critical difference or differences:

Fisher's	$t_{\alpha/2,a(n-1)}\sqrt{\dfrac{2MS_e}{n}}$	Apply to all differences
Duncan's	$d_{\alpha,2,a(n-1)}\sqrt{\dfrac{MS_e}{n}}$	Apply to bottom diagonal
	$d_{\alpha,3,a(n-1)}\sqrt{\dfrac{MS_e}{n}}$	Apply to second lowest diagonal
	$\vdots$	
	$d_{\alpha,a,a(n-1)}\sqrt{\dfrac{MS_e}{n}}$	Apply to top diagonal
Student–Newman–Keuls'	$q_{\alpha,2,a(n-1)}\sqrt{\dfrac{MS_e}{n}}$	Apply to bottom diagonal
	$q_{\alpha,3,a(n-1)}\sqrt{\dfrac{MS_e}{n}}$	Apply to second lowest diagonal
	$\vdots$	
	$q_{\alpha,a,a(n-1)}\sqrt{\dfrac{MS_e}{n}}$	Apply to top diagonal
Tukey's	$q_{\alpha,a,a(n-1)}\sqrt{\dfrac{MS_e}{n}}$	Apply to all differences
Scheffé's	$\sqrt{(a-1)F_{\alpha,a,a(n-1)}}\sqrt{\dfrac{2MS_e}{n}}$	Apply to all differences

---

Only Fisher's procedure requires a prior significant F test for the ANOVA.
In each procedure, reject H_0 if $|\bar{y}_i - \bar{y}_j| \geq$ critical difference.

It is possible to modify Fisher's and Scheffé's procedures for unequal sample sizes. The standard error becomes

$$\sqrt{\frac{MS_e}{n_i} + \frac{MS_e}{n_j}}$$

For Duncan's, Student–Newman–Keuls', and Tukey's procedures an approximation approach is possible by letting n be

$$\tilde{n} = \frac{a}{1/n_1 + 1/n_2 + \cdots + 1/n_a}$$

This approximation is best when the n_i are similar in size.

EXERCISES

10.3.1. An ANOVA is conducted to compare the yields of several different varieties of blight-resistant corn.

Source	df	SS	MS
Among varieties	—	—	598
Within varieties	20	3600	—

Variety:	C	A	D	B	E
Average yield:	60	80	82	85	93

a. Complete the ANOVA table.

b. Show that the standard error of a sample average is 6.0.

c. Would it be appropriate to use Fisher's least significant difference to compare variety means in this experiment?

d. Perform Fisher's test at $\alpha = 0.05$.

10.3.2. Five kinds of insecticides are used in an effort to control insect damage to a certain crop. Damage is measured in terms of square centimeters of leaf area destroyed. The data are summarized as follows:

Insecticide:	1	2	3	4	5	Totals
Plants examined:	4	4	4	4	4	20
$\sum_j y_{ij}$:	24	19	29	67	34	173
$\sum_j y_{ij}^2$:	178	97	237	1313	342	2167
$\left(\sum_j y_{ij}\right)^2$:	576	361	841	4489	1156	7423
$\left(\sum_j y_{ij}\right)^2 / n$:	144.00	90.25	210.25	1122.25	289.00	1855.75

a. Show that the correction factor is 1496.45.

b. Perform an ANOVA and test H_0 at $\alpha = 0.05$.

c. Use Fisher's procedure to test for differences among the means.

10.3.3. A behavioral biologist subjected spiders to different stressful conditions and then measured the number of gaps in their webs.

Condition

1	2	3	4
11	13	21	10
4	9	18	4
6	14	15	19
21	36	54	33

$$\sum_i \sum_j y_{ij}^2 = 2086$$

$$\sum_i \left(\sum_j y_{ij}\right)^2 = 5742$$

$$\left(\sum_i \sum_j y_{ij}\right)^2 = 20{,}736$$

a. Complete the ANOVA at $\alpha = 0.01$.

b. Would it be valid to use Fisher's procedure to test for a difference between group means? Why or why not?

c. Use Scheffé's procedure to test for a difference between means.

10.3.4. Five male students are selected at random from each of 5 colleges in a study to determine whether there is an association between sentimentality and the selected field of study. They are shown a movie about a little crippled orphan, his blind dog, and a senile grandfather who is trying to care for them in his cabin, which is in the path of a strip-mine operation. Polygraph equipment is used to record emotional response to the picture. The F test for differences among colleges is

$$F = \frac{\text{among-college MS}}{\text{within-college MS}} = \frac{50.00}{11.25}$$

a. Show that the standard error of a college average is 1.5.

b. Use Duncan's procedure to test for differences in emotional response among the college means.

College:	Law	Business	Agriculture	Arts and Sciences	Engineering
Sample average:	3	7	14	15	21

10.3.5. To see whether 3 commonly used weed killers may have differential effects on the yield of rye, each is sprayed on 6 different plots of rye at the seedling stage. The within-spray MS is 96, and the average yields are

Weed killer:	I	II	III
Average:	10	20	30

    a. Use Student–Newman–Keuls' procedure to determine whether there are any differences in the mean yields.

    b. If the agronomist conducting the experiment wants to use Fisher's least significant difference, how large would the F value have to be in order for her to be justified in using the procedure? Does the computed F value exceed this critical value?

    c. How could the experimenter test whether the plot sprayed with weed killer III produces an average yield that is significantly different from the average of the other two?

10.3.6. Consider a significant result from ANOVA in which $a = 6$, $n = 5$, $MS_e = 33.78$, and the treatment averages are

Treatment average:	39.3	45.2	48.4	50.4	55.5	58.2
Treatment:	A	B	C	D	E	F

Use all five multiple-comparison procedures at $\alpha = 0.05$ on these data and form a table indicating the different conclusions reached by each test.

10.4. ONE-DEGREE-OF-FREEDOM COMPARISONS

The multiple-comparison procedures in Section 10.3 are known as *a posteriori* tests, that is, they are after the fact. After the experiment is completed, the investigator decides to look for possible pairwise differences.

    There is also an *a priori* approach, that is, contrasts that are planned before the experiment. The experimenter believes prior to the investigation that certain factors may be related to differences in treatment groups. For example, in the chain saw experiment (Example 10.1 of Section 10.2), suppose that models A and D are lightweight chain saws for home use and that B and C are heavy-duty industrial types. The investigator might want to know if the kickback from the home type is the same as the kickback from the industrial type. In addition, he might also be interested in any differences in kickback within types.

	Comparison	H_0 to Be Tested
1	Home vs. industrial	$\dfrac{\mu_A + \mu_D}{2} - \dfrac{\mu_B + \mu_C}{2} = 0$
2	Home model A vs. home model D	$\mu_A - \mu_D = 0$
3	Industrial model B vs. industrial model C	$\mu_B - \mu_C = 0$

Each of the null hypotheses is a linear combination of the treatment means:

	Linear Combination
1	$(1/2)\mu_A - (1/2)\mu_B - (1/2)\mu_C + (1/2)\mu_D$
2	$(1)\mu_A + (0)\mu_B + (0)\mu_C - (1)\mu_D$
3	$(0)\mu_A + (1)\mu_B - (1)\mu_C + (0)\mu_D$

A set of linear combinations of this type is called a set of *orthogonal contrasts* or orthogonal comparisons. A set of linear combinations must satisfy two mathematical properties in order to be orthogonal contrasts:

A. The sum of the coefficients in each linear combination must be zero; this makes the linear combination a contrast.

$$\text{In 1:} \quad 1/2 - 1/2 - 1/2 + 1/2 = 0$$
$$\text{In 2:} \quad 1 + 0 + 0 - 1 = 0$$
$$\text{In 3:} \quad 0 + 1 - 1 + 0 = 0$$

B. The sum of the products of the corresponding coefficients in any two contrasts must equal zero; this makes the contrasts orthogonal.

In contrasts 1 and 2: $\left(\dfrac{1}{2}\right)(1) + \left(\dfrac{-1}{2}\right)(0) + \left(\dfrac{-1}{2}\right)(0) + \left(\dfrac{1}{2}\right)(-1) = 0$

In contrasts 1 and 3: $\left(\dfrac{1}{2}\right)(0) + \left(\dfrac{-1}{2}\right)(1) + \left(\dfrac{-1}{2}\right)(-1) + \left(\dfrac{1}{2}\right)(0) = 0$

In contrasts 2 and 3: $(1)(0) + (0)(1) + (0)(-1) + (-1)(0) = 0$

In general, if

$$L = a_1\mu_1 + a_2\mu_2 + \cdots + a_a\mu_a$$

and

$$M = b_1\mu_1 + b_2\mu_2 + \cdots + b_a\mu_a$$

are two linear combinations, then L and M are orthogonal contrasts if

$$\sum_i a_i = 0, \quad \sum_i b_i = 0, \quad \text{and} \quad \sum_i a_i b_i = 0$$

A set of contrasts is mutually orthogonal if every pair of contrasts is orthogonal. An experiment involving a treatments can have several different sets of mutually orthogonal contrasts, but each set consists of at most $a - 1$ orthogonal contrasts.

If the experimenter is able to plan reasonable comparisons of this type prior to the experiment, then the tests can be done within the ANOVA procedure. If contrasts are not incorporated into the design of the experiment but are suggested during the data gathering or analysis, Scheffé's procedure can be used instead of the procedure discussed here. Also, Scheffé's procedure can be used when the contrasts of interest are not orthogonal. (In Section 10.6, there is discussion of Bonferroni techniques which serve purposes similar to Scheffé's procedure.) Generally, however, such tests will not be as powerful as those for planned orthogonal contrasts, and it seems reasonable that experiments which are well designed and which test specific hypotheses will have the greatest statistical power.

Example 10.8. One-Degree-of-Freedom Comparisons

Five toothpastes are being tested for their abrasiveness. The variable of interest is the time in minutes until mechanical brushing of a material similar to tooth enamel exhibits wear.

The 5 toothpastes are all the same except for the absence or presence of certain additives. The material is assigned randomly to the treatments.

Toothpaste	Additive
I	Whitener
II	None
III	Fluoride
IV	Fluoride with freshener
V	Whitener with freshener

Group totals and the basic ANOVA table are as follows for 4 observations per treatment group:

Toothpaste:	I	II	III	IV	V
$T_i = \sum_j y_{ij}$:	197.4	199.0	211.3	215.8	186.5

Source	df	SS	MS	F
Among toothpastes	4	136.8	34.20	39.8
Within toothpastes	15	13.0	0.86	

The investigator deliberately chose these 5 toothpastes so that the following $a - 1$ orthogonal contrasts could be made:

Comparison	H_0 to Be Tested
Additive vs. no additive	$\dfrac{\mu_1 + \mu_3 + \mu_4 + \mu_5}{4} - \mu_2 = 0$
Whitener vs. fluoride	$\dfrac{\mu_1 + \mu_5}{2} - \dfrac{\mu_3 + \mu_4}{2} = 0$
Whitener vs. whitener with freshener	$\mu_1 - \mu_5 = 0$
Fluoride vs. fluoride with freshener	$\mu_3 - \mu_4 = 0$

To test these comparisons within the ANOVA procedure, the among SS is partitioned into $a - 1$ components which are each sums of squares for a one-degree-of-freedom F test. The sum of squares for *additive vs. no additive* is found as follows. The null hypothesis is rewritten as

$$H_0: \mu_1 + \mu_3 + \mu_4 + \mu_5 - 4\mu_2 = 0$$

by multiplying by 4. The contrast is then in an equivalent form without fractions:

$$L_1 = \mu_1 + \mu_3 + \mu_4 + \mu_5 - 4\mu_2$$

The coefficients are

$$a_1 = a_3 = a_4 = a_5 = 1 \quad \text{and} \quad a_2 = -4$$

The sum of squares is

$$SS_{L_1} = \frac{\left[\sum_i a_i T_i \right]^2}{n \sum_i a_i^2} = \frac{[197.4 + 211.3 + 215.8 + 186.5 - 4(199)]^2}{4[1^2 + 1^2 + 1^2 + 1^2 + (-4)^2]} = 2.8$$

Similarly, the sum of squares can be found for the other three contrasts:
Whitener vs. fluoride:

$$H_0: L_2 = \mu_1 + \mu_5 - \mu_3 - \mu_4 = 0$$
$$SS_{L_2} = \frac{(197.4 + 186.5 - 211.3 - 215.8)^2}{4[1^2 + 1^2 + (-1)^2 + (-1)^2]} = 116.6$$

Whitener vs. whitener with freshener:

$$H_0: L_3 = \mu_1 - \mu_5 = 0$$
$$SS_{L_3} = \frac{(197.4 - 186.5)^2}{4[1^2 + (-1)^2]} = 14.9$$

Fluoride vs. fluoride with freshener:

$$H_0: L_4 = \mu_3 - \mu_4 = 0$$
$$SS_{L_4} = \frac{(211.3 - 215.8)^2}{4[1^2 + (-1)^2]} = 2.5$$

The ANOVA table is then enlarged as follows:

Source	df	SS		MS	F	$F_{0.05}$
Among toothpastes	4	136.8		34.20	39.8*	3.056
Additive vs. no additive	1		2.8	2.8	3.3	4.543
Whitener vs. fluoride	1		116.6	116.6	135.6*	4.543
Whitener vs. whitener and fluoride	1		14.9	14.9	17.4*	4.543
Fluoride vs. fluoride and freshener	1		2.5	2.5	2.9	4.543
Within toothpastes	15	13.0		0.86		

These comparisons show a significant difference between the abrasiveness of the whitener and the fluoride; the whitener is more abrasive. There is also a significant difference between the whitener alone and the whitener with freshener, the latter being still more abrasive.

It should be noted in the above example that the among SS has been partitioned, that is, divided into nonoverlapping parts, by the orthogonal contrasts. This has an advantage over the multiple-comparison procedures of the previous section in that the partition can be used to determine the percentage of variability that is due to the different factors. In this example, the difference between the whitener and the fluoride is responsible for $116.6/136.8 = 85\%$ of the sums of squares among toothpastes.

A significant F test is not a prerequisite for these one-degree-of-freedom tests. In fact, the ANOVA procedure need not be carried out. Also, if MS_e is used for s_p^2, five t tests can be used rather than the five F tests. It is essential, however, that the contrasts be planned before examining the data; otherwise the investigator may be biased by what he sees.

A priori tests of this type are not always possible because there may be insufficient information to set up reasonable contrasts. The experimenter needs a great deal of information to be able to choose treatment groups in such a way that a set of orthogonal contrasts relevant to the experiment will exist. When possible, these contrasts usually answer more relevant questions than multiple comparisons.

The one-degree-of-freedom comparisons can be summarized as follows.

Procedure. One-Degree-of-Freedom Comparisons

To test a set of $a - 1$ mutually orthogonal comparisons, write each contrast in the form $L = a_1\mu_1 \text{ plus;} a_2\mu_2 + \cdots + a_a\mu_a$ with integer coefficients. Then the sum of squares for each contrast is found by the formula

$$SS_L = \frac{\left[\sum_i a_i T_i\right]^2}{n \sum_i a_i^2}$$

in which T_i is the ith treatment group total and n is the number of observations in each group. This sum of squares has one degree of freedom. The contrast is tested with the statistic

$$F = \frac{MS_L}{MS_e}$$

and the comparison is significant if $F \geq F_{\alpha,1,a(n-1)}$.

The procedure described in this section applies only to groups of equal sample sizes.

If desired, the sums of squares for the one-degree-of-freedom tests can be computed from the group averages instead of the group totals. In that case the formula becomes

$$SS_L = \frac{\left[\sum_i a_i \bar{y}_i\right]^2 n}{\sum_i a_i^2}$$

EXERCISES

10.4.1. In the chain saw experiment, test the 3 comparisons proposed at the beginning of this section by means of one-degree-of-freedom F tests.

10.4.2. Certain people convicted of crimes return to prison over and over again while others seem to be rehabilitated. To determine whether this may be related to the nature of the first offense, a sociologist sampled prison records of former inmates of the same age. She recorded the nature of the first offense and the total number of times they were imprisoned:

Nature of crime:	Assault	Rape	Fraud	Embezzlement
Average number of imprisonments:	7.5	5.5	4.5	2.5

a. Make the following orthogonal comparisons if $n = 10$ and $MS_e = 15$:

Assault vs. rape

Fraud vs. embezzlement

Violent vs. nonviolent

b. What conclusions can be drawn from this analysis?

10.4.3. A study is done on the effectiveness of various types of analgesics. There are 6 treatment groups, one of which is a control group and receives a placebo. Five persons who have pain are chosen at random for each treatment. All patients take the medication in capsule form and do not know which of the 6 groups they are in. The capsules that contain aspirin (with or without something else) all contain the same amount of aspirin. The variable of interest is the amount of time (in hours) until relief from pain is felt.

Group	Treatment	$\sum_j y_{ij}$	$\left[\sum_j y_{ij}\right]^2$	$\sum_j y_{ij}^2$	$\bar{y}_i$
1	Placebo	20	400	105	4.0
2	Aspirin, brand 1	5	25	6	1.0
3	Aspirin with caffeine	10	100	19	2.0
4	Aspirin, brand 2	6	36	7	1.2
5	Aspirin with buffer	8	64	10	1.6
6	Aspirin with buffer and caffeine	11	121	22	2.2
	Totals	60	746	169	

a. State the null and alternative hypotheses.

b. Perform the ANOVA at $\alpha = 0.01$.

c. Make the following orthogonal comparisons:

Placebo vs. analgesic

Pure aspirin vs. aspirin with additives

Aspirin 1 vs. aspirin 2

Aspirin with caffeine (alone) vs. aspirin with buffer (with or without caffeine)

Aspirin with buffer vs. aspirin with buffer and caffeine

d. Show that the set of comparisons in part c are mutually orthogonal.

e. What part of the sum of squares among groups is caused by the difference between pure aspirin and aspirin with additives?

f. What should the experimenter conclude from the above analyses?

10.5. ESTIMATION

Often an investigator wants to obtain one or more estimates of parameters after an ANOVA. He may want to estimate μ (the overall mean), $\mu + \alpha_i$ (the ith treatment mean), or α_i (the ith treatment effect). He might also be interested in the difference of two parameters as $\alpha_1 - \alpha_2$ or some other linear combination of parameters as $\mu_1 - (\mu_2 + \mu_3)/2$. Usually he wants the estimate in the form of a confidence interval.

The following table summarizes the point estimators and the estimators of the standard errors needed to form these confidence intervals.

$$\text{CI}_{1-\alpha}: \text{Point Estimator} \pm t_{\alpha/2, N-a} \text{ (Standard Error)}$$

Parameter	Symbol	Point Estimator	Standard Error
Mean	μ	$\bar{y}$	$\sqrt{\text{MS}_e/N}$
Treatment mean	$\mu_i = \mu + \alpha_i$	$\bar{y}_i$	$\sqrt{\text{MS}_e/n_i}$
Treatment effect	α_i	$\bar{y}_i - \bar{y}$	$\sqrt{\text{MS}_e\left(\dfrac{N - n_i}{n_i N}\right)}$
Difference between treatment means	$\mu_i - \mu_{i'}$ or $\alpha_i - \alpha_{i'}$	$\bar{y}_i - \bar{y}_{i'}$	$\sqrt{\dfrac{\text{MS}_e}{n_i} + \dfrac{\text{MS}_e}{n_{i'}}}$
A linear combination of means	$\sum_i a_i \mu_i$ with $\sum_i a_i = 0$	$\sum_i a_i \bar{y}_i$	$\sqrt{\text{MS}_e\left[\sum_i \left(\dfrac{a_i^2}{n_i}\right)\right]}$

All of the standard errors except the one for the treatment effect can be seen to follow from the properties of the variance of a linear combination of random variables. The standard error for the treatment effect is different because $\bar{y}_i$ and $\bar{y}$ are dependent.

Example 10.9.   Confidence Intervals Related to ANOVA

In the chain saw study, Example 10.1 of Section 10.2, the averages are

$\bar{y}_A$	$\bar{y}_B$	$\bar{y}_C$	$\bar{y}_D$	$\bar{y}$
33	43	49	31	39

$n = 5$ and $\text{MS}_e = 101.25$. Some of the possible point estimates are given in Figure 10.5.

The experimenter wants to find 95% confidence intervals for the overall mean, for the mean of model B, for the model B effect, for the difference between models A and D, and for the difference between the oldest model, model A, and the average of the three newer models.

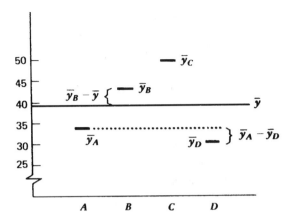

FIGURE 10.5. Point estimators of parameters in ANOVA.

Overall Mean, μ

$$\text{CI}_{0.95}: \bar{y} \pm t_{0.025,a(n-1)}\sqrt{\frac{\text{MS}_e}{N}}$$

$$39 \pm 2.120\sqrt{\frac{101.25}{20}}$$

$$39 \pm 4.77$$

Mean of Model B, μ_B

$$\text{CI}_{0.95}: \bar{y}_B \pm t_{0.025,a(n-1)}\sqrt{\frac{\text{MS}_e}{n}}$$

$$43 \pm 2.120\sqrt{\frac{101.25}{5}}$$

$$43 \pm 9.5$$

Model B Effect, α_B

$$\text{CI}_{0.95}: \bar{y}_B - \bar{y} \pm t_{0.025,a(n-1)}\sqrt{\text{MS}_e\frac{(N-n_B)}{n_B N}}$$

$$(43-39) \pm 2.120\sqrt{101.25\frac{(20-5)}{5(20)}}$$

$$4 \pm 8.27$$

Since this interval contains zero, model *B* does not differ significantly from the overall mean of all four models.

The Difference between the Means of Models A and D, $\mu_A - \mu_D$

$$CI_{0.95}: \quad \bar{y}_A - \bar{y}_D \pm t_{0.025,a(n-1)}\sqrt{\frac{MS_e}{n} + \frac{MS_e}{n}}$$

$$(33 - 31) \pm 2.120\sqrt{\frac{2(101.25)}{5}}$$

$$2 \pm 13.49$$

Since this interval contains zero, models A and D do not differ significantly with respect to kickback.

The Difference between the Mean of Model A and the Average of the Means of the Other Three Models, $\mu_A - (\mu_B + \mu_C + \mu_D)/3$

$$a_A = 1, \quad a_B = a_C = a_D = -\frac{1}{3} \quad \sum a_i = 0$$

$$CI_{0.95}: \quad \bar{y}_A - \frac{\bar{y}_B + \bar{y}_C + \bar{y}_D}{3} \pm t_{0.025,a(n-1)}\sqrt{MS_e \sum_i \frac{a_i^2}{n}}$$

$$33 - \frac{43 + 49 + 31}{3} \pm 2.120\sqrt{101.25\left[\frac{1^2 + 3(-1/3)^2}{5}\right]}$$

$$-8 \pm 11.0$$

Thus the older one (model A) does not seem to be significantly different from the average of the three newer ones.

The investigator should remember that repeated estimates within the same experiment will not preserve the original α level. By chance alone, one or more of the intervals may fail to cover the parameter. There are several ways to guard against this:

1. If an experiment-wide confidence no greater than $1 - \alpha$ is needed, Scheffé's procedure can be used rather than the conventional confidence interval based on $t_{\alpha/2,a(n-1)}$.

2. If confidence intervals for pairwise differences between all group averages are wanted, it is possible to use Tukey's honestly significant difference procedure wherein the confidence interval $t_{\alpha/2,a(n-1)}$ is replaced with $q_{\alpha,ma(n-1)}$, where m is the number of confidence intervals to be constructed. The formula for this procedure thus will be

$$(\bar{y}_i - \bar{y}_j) \pm q_{\alpha,ma(n-1)}\sqrt{\frac{MS_e}{n}}$$

but note that that it is appropriate only when the sample size n is the same for all samples.

3. If m confidence intervals are involved, then $t_{\alpha/2m,N-a}$ is used for each individual confidence interval. The set of intervals is then called *multiple-t confidence intervals*. A t table that lists very small values of α is necessary to find most multiple-t confidence intervals. This is one of the Bonferroni procedures discussed in greater detail in Section 10.6.

EXERCISES

10.5.1. In the insecticide study of Exercise 10.3.2:

    a. Place a 95% confidence interval on the overall experimental mean.

    b. Place a 99% confidence interval on the effect of the third insecticide.

    c. Place a 90% confidence interval on the difference between the second and fourth insecticides.

    d. Place a 95% confidence interval on the fifth treatment mean.

10.5.2. In the spider study of Exercise 10.3.3:

    a. Place a 95% confidence interval on the mean of the second treatment.

    b. Place a 95% confidence interval on the difference between the mean of the first and the third treatments.

    c. Place a 95% confidence interval on the difference between the first and second treatment effects.

10.5.3. Four normal populations with homogeneous variances give rise to the following data from random samples:

	Group		
1	2	3	4
52	40	38	48
41	28	33	36
52		27	38
39		33	38
		39	48
			49
			36
			38
			47

    a. Perform an ANOVA.

    b. Estimate $\mu_1 - \mu_3$ with a 90% confidence interval.

    c. Estimate μ with a 90% confidence interval.

    d. Estimate α_3 with a 90% confidence interval.

    e. Estimate $(\mu_1 + \mu_4)/2 - (\mu_2 + \mu_3)/2$ with a 90% confidence interval.

10.5.4. Use Tukey's procedure to place a set of simultaneous 95% confidence intervals on the differences between all pairwise kickback averages in Example 10.1. (This will require that $q_{0.05,16}$ be used rather than a t value.) How do the conclusions drawn from the confidence intervals compare to those for the pairwise tests of averages?

10.6. BONFERRONI PROCEDURES

The procedures discussed in this section are said to date from the middle of the last century when they were suggested by the Italian mathematician Carlo E. Bonferroni (1892 to 1960).

However, they gained little attention until 1961 when Olive J. Dunn published a table of t values with small α levels suitable for the procedures. This brief history is given to explain why we discuss a procedure attributed to someone but do not give a reference to his work. Readily available high-speed computing and sophisticated statistical computer packages have made the procedures readily accessible, so they are frequently mentioned in research papers and need to be part of the statistical arsenal with which researchers are armed. Luckily the added armor is light and not too difficult to use if the proper statistical software is available.

In Sections 10.3 to 10.5 we expressed the need for concern about the *global* α *level* (α_G), the overall α level for all hypotheses tested in an experiment. Although the consequences are not as drastic, the likelihood of mischance can be compared to playing Russian roulette. Whether justified or not, that adventure is attributed to young noblemen in Czarist Russia who tested their courage by placing a single cartridge in one of the six chambers in the cylinder of a revolver, spinning the cylinder, placing the handgun to their heads and pulling the trigger. Assuming the spinning process is random, the probability of an imminent funeral is $1/6$, and if the experiment is repeated after a new spin of the cylinder, it remains $1/6$ because the trials are independent. However, if one's courage needs to be tested m times in one evening, $P(\text{funeral}) = 1 - (5/6)^m$. When $m = 1$, the probability is 0.167, but if m is increased to 6, the probability increases to 0.665, and something unpleasant is most likely to occur. Similarly, if we have a research experiment with m independent t tests each with $\alpha = 0.05$, the probability that at least one will show significance by chance alone is $1 - (0.95)^m$. When $m = 1$, $P(\text{Type I error}) = 0.05$, but if m increases to 6, the probability of at least one chance difference is 0.265, so again something unpleasant is quite likely to occur.

The analogy used to explain the dire consequences of repeated testing, whether it be of courage or null hypotheses, is not perfect. We have no cylinder to spin between tests of hypotheses among the same set of averages, so the tests are not independent. In fact, in the chain saw experiment that is becoming tattered from overuse, $\bar{y}_A$, and every other group average, is used in three of the six pairwise tests of difference between averages. Yet even without complete independence it is intuitive that with repeated tests of hypotheses probability will increase for at least one difference being significant by chance alone. Thus the experiment-wide α level will be greater than the 0.05 customarily claimed by the experimenter. When it is important to maintain the global α_G level for all simultaneous tests or confidence intervals at a set level is when Bonferroni procedures are most useful.

The statistical procedures are the same as we are accustomed to using for t tests and confidence intervals; the only difference is that we change the value of $t_{\alpha,\nu}$ that will be used for statistical inference. If we revisit the chain saw experiment using Bonferroni procedures to perform $m = \binom{4}{2} = 6$ simultaneous t tests or construct $m = 6$ simultaneous confidence intervals, each test or confidence interval will have its own α_i level, but they must be chosen so that

$$\alpha_1 + \alpha_2 + \cdots + \alpha_m \leq \alpha_G$$

This requirement poses the greatest difficulty in using the procedure because it means that we will often need t tables for α levels that seem bizarre. In the case where $m = 6$ and $\alpha_G = 0.05$ is divided equally among the 6 t tests, the critical t value for each two-sided test will be one with $a(n - 1) = 16$ degrees of freedom and an $\alpha_i = \alpha_G/2m = 0.05/2(6) = 0.0042$. Tables of the t distribution for such a value likely do not exist, and that is why computers with sophisticated statistical programs are usually needed for Bonferroni procedures. How this can be done with such a statistical package (JMP) will be demonstrated in Example 10.10.

Example 10.10.   Simultaneous Bonferroni *t* tests

When the 4 models of chain saws are compared with Bonferroni *t* tests, 6 separate *t* tests are performed in the usual fashion:

$$t = \frac{\bar{y}_i - \bar{y}_j}{\sqrt{2MS_e/n}}$$

The critical value used is the only thing that is different. Rather than the $t_{0.025,a(n-1)}$ that would be used for Fisher's least significant difference at the 0.05 level of significance, to maintain a global α_G level of 0.05 for the 6 tests, we need a critical *t* value for $\alpha_i = 0.05/2{\cdot}6 = 0.0042$ for each test. Even if tables for such *t* values exist, they will likely be difficult to find. There are statistical computer packages that would allow us to compute $t_{0.0042,16}$, but since most statistical computer routines give the *P* values for tests, we can use the *P* value for each of the 6 *t* tests and see if it is equal to or less than $\alpha_i = 0.0042$. The averages for the models are ordered again and arranged in the same sort of table used for multiple comparisons, and within the table are the six *t* tests and their respective *P* values:

$\bar{y}_A = 33$	$\bar{y}_B = 43$	$\bar{y}_C = 49$
$\bar{y}_D = 31$	$t = 0.3134$	$t = 1.8856$	$t = 2.8284$
	$P = 0.7574$	$P = 0.0776$	$P = 0.0121$
$\bar{y}_A = 33$		$t = 1.5713$	$t = 2.5142$
		$P = 0.1357$	$P = 0.0230$
$\bar{y}_B = 43$			$t = 0.9428$
			$P = 0.3598$

None of the *P* values is equal to or smaller than $\alpha_i = 0.0042$, so none of the differences between model averages can be considered statistically significant.

The Bonferroni *t* tests just considered are the usual *a posteriori* multiple-comparison tests for differences among all averages. This set of tests is required for multiple-comparison procedures such as Duncan's or Student–Newman–Keuls', but not for Bonferroni *t* tests. The experimenter is free to use whatever set of *m* tests he chooses; the *t* tests need not be the $\binom{a}{2}$ set used for multiple comparisons; they need not be an orthogonal set; they do not require equal sample sizes; they can be single sample *t* tests of a hypothesized μ; and after computing the appropriate standard error, they can be for comparing averages of several groups with those of others. However, the set of tests should be chosen in advance of the experiment. The researcher will be violating the intent of maintaining a global α level if he looks at the data and then decides what tests might lead to significance. To demonstrate some of the versatility, the *t* tests and *P* values that have already been attained can be used for a different set of *m* simultaneous *t* tests.

Suppose even before any data were gathered the experimenter knew that model *C* had such strong kickback it might become a safety risk if used by frail or elderly people. Thus he chose the other three models as possibly safer alternatives. The set of *m* tests of interest to him would be the comparison of each of the averages of the other models to that for model *C* to see if one may have significantly less average kickback. He would need only three *t* tests to test the three hypotheses

$$H_0: \mu_C = \mu_A \quad \text{with} \quad H_a: \mu_C > \mu_A$$
$$H_0: \mu_C = \mu_B \quad \text{with} \quad H_a: \mu_C > \mu_B$$
$$H_0: \mu_C = \mu_D \quad \text{with} \quad H_a: \mu_C > \mu_D$$

Thus, if he wishes to maintain a global $\alpha_G = 0.05$, each Bonferroni t test would have an $\alpha_i = \alpha_{G/m} = 0.05/3 = 0.0167$. He would not use $\alpha_i = \alpha_G/2m$ because the alternative hypotheses are one sided; he wants to find a model with significantly less kickback. The tabulation of t tests and P values is

	$\bar{y}_A = 33$	$\bar{y}_B = 43$	$\bar{y}_D = 31$
$\bar{y}_C = 49$	$t = 2.5142$	$t = 0.9428$	$t = 2.8284$
	$P = 0.0230$	$P = 0.3598$	$P = 0.0121$

The P value for the t test of the difference between models C and D averages is less than $\alpha_i = 0.0167$, so those two models differ significantly with respect to kickback and he can recommend model D for people who need a saw with significantly less kickback.

Example 10.10 demonstrated that Bonferroni t tests are computed in the same fashion as we have computed other t tests. The only difference is in the critical value of t that is used for inference. There may be no table with the t values we need, but if we have a computer program that gives the P values for t tests, we can use them to make tests of significance. Another idea to be gained from the example is the extreme versatility of Bonferroni t tests; they can be used for any set of m tests with their respective α_i, values which may even be of different sizes so long as the global α is maintained by

$$\alpha_1 + \alpha_2 + \cdots + \alpha_m \le \alpha_G$$

When multiple-comparison procedures were discussed in Section 10.3, it was noted that the d values for Duncan's tests could not be substituted for the t value to construct the confidence interval for the difference between group averages. It was similarly noted that the q value could be used in place of a t value for a confidence interval only for the difference between means largest and smallest in rank. This is because the comparison of largest and smallest is the same whether one uses Student–Newman–Keuls' or Tukey's procedure. As mentioned in Section 10.5, Tukey's procedure uses only $q_{\alpha/2,a,a(n-1)}$ for all statistical inferences involving differences between group averages, hypothesis testing, or interval estimation and thereby provides a known global α. Bonferroni simultaneous confidence intervals, like their t-test counterparts, offer greater versatility as well as familiarity. We can choose a set of m confidence intervals among those given in Section 10.5 or any other sensible intervals and also choose the α_i level we want to use for each interval, with the only condition that $\alpha_1 + \alpha_2 + \cdots + \alpha_m \le \alpha_G$. Then, again, if we refer to Section 10.5 and compute the appropriate standard errors (s.e.), each confidence interval will be

$$\pm t_{\alpha_i,\nu}(\text{s.e.})$$

So the only difference between a Bonferroni interval and those demonstrated in Section 10.5 is the t value that is used to compute the interval. Finding the t value for an unusual α_i is no longer a problem with those who have access to sophisticated statistical computer packages. Example 10.11 will demonstrate the use of Bonferroni simultaneous confidence intervals for the ubiquitous chain saw data.

Example 10.11. Simultaneous Bonferroni Confidence Intervals

Suppose that in the experiment described in Example 10.1 the experimenter wants to maintain a global α of 0.05 while constructing simultaneous confidence intervals for the mean kickback of each of the four models. If he wants all the $m = 4$ intervals to have the same $1 - \alpha_i$ confidence, α_i would be $\alpha_G/2m = 0.05/2\cdot4 = 0.00625$. Because this is not one of the probability levels found in conventional t tables, the experimenter would have to interpolate between the t values given in Table A.11 for $\alpha = 0.01$ and $\alpha = 0.005$ or else use a statistical computer program. Using JMP, the necessary t value is found to be ± 2.813 for two-sided confidence intervals, and each of the simultaneous confidence intervals is

$$\pm t_{\alpha_i, \nu}(\text{s.e.}) = \pm t_{0.00625, 16}\sqrt{\frac{\text{MS}_e}{n}} = \pm 2.813\sqrt{\frac{101.25}{5}} = \pm 12.66$$

The common interval is quite wide, so reporting that the estimated mean kickback for model C is 49 ± 12.66, or between 36.34 and 61.66 degrees, is not especially useful, but the experimenter must remember that he has a relatively small experiment and only 5 observations on model C. Interval estimates with narrow bounds almost always require large sample sizes.

Because of his concern about the safety of model C saws, suppose he wants a narrower bound for his interval estimate of the mean kickback for that model. However, to accomplish that using the same data, he would have an α_i for model C that is different from that used for other saws. Thus he sets $\alpha_C = 0.02$ and $\alpha_A = \alpha_B = \alpha_D = 0.01$ in order for the four α_i levels to sum to the desired global α of 0.05. Because the α_i are not the same for all simultaneous intervals, he must compute two confidence intervals, one for model C using $t_{0.02/2, 16} = t_{0.01, 16}$ and $t_{0.01/2, 16} = t_{0.005, 16}$ for the other three intervals. Fortunately both of the desired t values can be found in Table A.11 and do not have to be computed.

The confidence interval for mean kickback of model C saws is computed as

$$\bar{y}_C \pm t_{0.01, 16}\sqrt{\frac{\text{MS}_e}{n}} = 49 \pm 2.583(4.50) = 49 \pm 11.62$$

That for each of the other saws is

$$\pm t_{0.005, 16}\sqrt{\frac{\text{MS}_e}{n}} = \pm 2.921(4.50) = \pm 13.14$$

The confidence intervals in Example 10.11 may seem disappointingly wide, but we need to remember that asking, "Is there a significant difference between the means of two groups?" is quite different from asking, "How great is the difference between the two means?" The first question is answered by hypothesis testing and the second by interval estimation, and large sample sizes are usually needed for a narrow confidence interval.

We have seen that Bonferroni simultaneous t tests and confidence intervals are not new computational procedures to be learned. They employ the same computations as the t tests and confidence intervals we have encountered before. The difference lies in the t values needed for statistical inference, and these usually must be computed rather than obtained from a table. We learned in Chapter 8 that when degrees of freedom increase, the t distribution converges to the standard normal z distribution. Thus we might believe that with moderate degrees of freedom we could use a z value from Table A.10 to approximate the t value we would otherwise have to compute. Unfortunately, this is another instance where we can cite the old proverb about the danger of a little learning.

The t distributions are said to have "fat tails," meaning that there is a greater area under the extreme tails of a t distribution than under the standard normal distribution. Because it is in these tails that we find Bonferroni t values, they do not agree well with z values for the same α. For example, in the standard normal distribution $P(z > 2.5758) = 0.005$, so 2.5758 should be the approximate numerical value of $t_{0.005,\nu}$ when ν is large enough to substitute $z_{0.005}$ for $t_{0.005,\nu}$. So we can examine the last column of Table A.11 to see how large the degrees of freedom must be before $t_{0.005,\nu}$ is near 2.5758. We find that it is only when an experiment has $\nu = 120$ degrees of freedom that $t_{0.005,120}$ and $z_{0.005}$ both can be rounded to 2.6. If we wish to maintain a global α for an experiment and choose to do so with Bonferroni procedures, it seems that we cannot avoid the need of a computer program to compute the t values that are required.

EXERCISES

10.6.1. Given that $t_{0.0042,16} = 3.0045$, use the data in Example 10.1 to compute the minimum significant difference between kickback averages when the Bonferroni procedure is used. (Remember that the t value is multiplied by the standard error of the difference between two means.) Compare the computed value with that for Tukey's test and tell which procedure is more conservative.

10.6.2. An experiment is performed to compare the economy of operation of three types of "hybrid" automobiles that operate by both a gasoline engine and electricity. Six autos of each type are driven for 500 miles in the same city, and the variable of analysis is total costs of gasoline, electricity, and maintenance. The data and some of the analysis are given below:

Hybrid car:	D	E	F
	20.3	24.5	21.0
	19.8	20.8	17.8
	21.1	22.0	18.1
	18.7	23.1	19.4
	20.0	23.5	17.5
	20.1	24.1	20.2
Sum	120.0	138.0	114.0

a. If the uncorrected sums of squares are $T = 7762.7$, $A = 7740.0$, and $CF = 7688$, show that $MS_e = 1.51$.

b. If average costs of operation of hybrid car types are to be compared by Bonferroni t tests with a global α of $\alpha_G = 0.06$, what will be α_i for each Bonferroni t test?

c. Perform the tests and decide which types are significantly different from each other.

d. Construct simultaneous confidence intervals for each of the three types.

e. Suppose we knew in advance that type E cars had a more powerful gasoline engine than cars of the other two types. So, using MS_e, we want to perform two Bonferroni t tests: (1) the average of type E compared to the combined average of the other two and (2) the average of type D compared to that of type F. The global

α can be maintained at $\alpha_G = 0.06$ if we choose $\alpha_1 = 0.05$ and $\alpha_2 = 0.01$. Why might we want a greater α_1 for testing the average of type E compared to the combined average of the other two?

f. Perform the two Bonferroni t tests and draw conclusions.

10.7. NONPARAMETRIC STATISTICS: KRUSKAL–WALLIS ANOVA FOR RANKS

In Section 10.1, we noted that the sample variance among group averages is an estimate of σ^2/n under the null hypothesis. Because the within MS also estimates σ^2, we obtained the two independent estimates of variance which are necessary for an F test from the ratio

$$F = \frac{n[\text{Variance among sample averages}]}{\text{Pooled variance within groups}} = \frac{n\left[\sum_i (\bar{y}_i - \bar{y})^2/(a-1)\right]}{\sum_i \sum_j (y_{ij} - \bar{y}_i)^2/a(n-1)}$$

W. H. Kruskal and W. A. Wallis have shown that a very similar analysis can be performed on rank data. Thus, once again, after examining a procedure designed for normally distributed data, we are able to discuss a similar nonparametric procedure for ordinal data or numerical data which have been transformed to the ordinal scale. However, this procedure is not simply a matter of replacing original observations with ranks and then performing the ANOVA and an F test. Because ordinal data consist of the integer values from 1 to N, under the null hypothesis, the E(within MS) $= N(N + 1)/12$. It may be recalled from Chapters 7 and 8 that an F statistic is the ratio of two independent estimates of the same variance, whereas chi square is the ratio of a sample sum of squares divided by a known variance. Thus, because the within MS for ranks is known, we employ the chi-square distribution in the Kruskal–Wallis test. The test statistic, usually symbolized as H, is the among-group SS computed from the rank data divided by $N(N + 1)/12$:

$$H = \frac{n[\text{Sum of squares among sample rank averages}]}{N(N + 1)/12} = \frac{n\left[\sum_i (\bar{r}_i - (N + 1)/2)^2\right]}{N(N + 1)/12}$$

and H is compared to $\chi^2_{\alpha, a-1}$ for the test of significance.

The chain saw data in Example 10.1 may have become somewhat tiresome, but they lend themselves very well for a demonstration of the Kruskal–Wallis test. First, the original data must be transformed into ranks, as is done in the table shown below.

Model: Measurement:	A		B		C		D	
	Degrees	Rank	Degrees	Rank	Degrees	Rank	Degrees	Rank
	42	12	28	4	57	19	29	5
	17	1	50	17	45	15	40	10
	24	3	44	14	48	16	22	2
	39	9	32	7	41	11	34	8
	43	13	61	20	54	18	30	6
Sum of ranks (R_i):		38		62		79		31
Average rank ($\bar{r}_i$):		7.6		12.4		15.8		6.2

For the rank data, the hypotheses are

$$H_0: E(\bar{r}_i) = \frac{N+1}{2} \text{ for all } i$$

and

$$H_a: E(\bar{r}_i) \neq \frac{N+1}{2} \text{ for some } i$$

Because the mean of the 20 ranks is $(20 + 1)/2 = 10.5$, the sum of squares among groups for the rank data is

$$5[(7.6 + 10.5)^2 + (12.4 - 10.5)^2 + (15.8 - 10.5)^2 + (6.2 - 10.5)^2] = 293.0$$

This sum of squares can also be obtained by using the rank data to perform the computational procedures introduced in Section 10.2. For the ranks:

$$A = \frac{\sum R_i^2}{n} = \frac{38^2 + 62^2 + 79^2 + 31^2}{5} = 2498$$

and

$$CF = \frac{[N(N+1)/2]^2}{N} = 2205$$

We note, again, that because the ranked data consist of the integers from 1 to N, under the null hypothesis the within MS for the ranked data estimates $N(N+1)/12 = 20(21)/12 = 35.0$, which is the denominator in the computation of the test statistic:

$$H = \frac{n\left[\sum_i(\bar{r}_i - (N+1)/2)^2\right]}{N(N+1)/12} = \frac{293.0}{35.0} = 8.371$$

When $H = 8.371$ is compared to $\chi^2_{0.05,3} = 7.815$, we reject the null hypothesis and conclude that at least one model of chain saw tends to outrank another with respect to degree of kickback.

If we wish to determine which models of chain saws are different from others, it is suggested that we utilize mean separation techniques similar to those discussed in Sections 10.3 and 10.4. These procedures differ only in that $E(\text{within MS}) = N(N+1)/12$ under the null hypothesis, and since we are dealing with a known variance, we employ the normal and chi-square distributions rather than the t and F distributions, which are used when σ^2 is estimated rather than known.

For an *a posteriori* procedure similar to Fisher's least significant difference, we can use

$$z_{0.025}\sqrt{\frac{2[N(N+1)]}{12n}} = z_{0.025}\sqrt{\frac{N(N+1)}{6n}} = 1.96\sqrt{\frac{20(21)}{6(5)}} = 7.33$$

Thus we may conclude that there is a significant difference between any two models of the chain saws if the difference between their average ranks is 7.33 or greater. This test may be somewhat conservative because when the null hypothesis is false, the ranks within groups will be of similar magnitude hence the within MS for the rank will be less than $N(N + 1)/12$. Still, for the chain saw data we find the same significant differences among models that were obtained when Fisher's and Duncan's procedures were used on the original data.

Orthogonal contrasts can be used in an *a priori* procedure for finding significant differences among the models of chain saws. When there is sufficient information in advance of the experiment, one can construct $a - 1$ sets of orthogonal contrasts which can be used to partition the test statistic H and its $a - 1$ degrees of freedom into $a - 1$ orthogonal H statistics each with one degree of freedom. Each of the orthogonal H values is computed by

$$H = \frac{\left[\sum a_i R_i\right]^2 / n \sum a_i^2}{N(N + 1)/12}$$

Thus, if we knew prior to the experiment that models A and D were chain saws designed for home use and that models B and C were intended for industrial use, we could compare the average rank of the two "home" models to that of the two "industrial" models with the original contrast:

$$H = \frac{[(-1)38 + (+1)62 + (+1)79 + (-1)31]^2 / 5[(-1)^2 + (+1)^2 + (+1)^2 + (-1)^2]}{20(21)/12}$$

$$= \frac{5184/20}{35} = 7.406$$

When test statistic H is compared to $\chi^2_{0.05,1} = 3.841$, we see that there is a significant average difference in rank between home and industrial saws. This result agrees closely to the results obtained when the same orthogonal contrasts are used in the analysis of the original numerical data. Such will frequently be the case, because even when a rank transformation is performed on data which are normally distributed, these rank test procedures will usually lead to the same conclusions that one would obtain from an analysis of the original data with the ANOVA procedures discussed earlier in this chapter. Furthermore, rank procedures should be superior when data are not normally distributed; however, the other assumptions of ANOVA must still hold, namely (1) random, independent samples, (2) a linear model, and (3) equal variances within groups.

Procedure. Kruskal–Wallis One-Way ANOVA for Ranked Data

$H_0: E(\bar{r}_i) = \dfrac{N + 1}{2}$ *for all i*

$H_a: E(\bar{r}_i) \neq \dfrac{N + 1}{2}$ *for some i*

Rank the data from 1 (the smallest observation) to N (the largest), irrespective of the group in which they are found. If two or more observations are tied for the same numerical value, assign to each the average rank for which they are tied.

Let $\bar{r}_i =$ the average rank of group i; $i = 1, \ldots, N$.

Compute:

$$H = \frac{n\left[\sum_i \left(\bar{r}_i - \frac{N+1}{2}\right)^2\right]}{N(N+1)/12}$$

Reject H_0 if $H \geq \chi^2_{\alpha,a-1}$

EXERCISES

10.7.1. A clockmaker is designing a decorative clock which will require only a small battery-powered motor and a flexible strip of metal for its operation. There are three types of alloys (labeled A, B, and C here) which seem to fit all requirements for the strip of metal, so the one to be used in the design will be the alloy which can be flexed for the longest period of time without breaking. A random sample of four strips of each type of alloy is obtained and all 12 strips are placed on a device which will continue to flex them until all break. They are observed periodically, and a record is kept, by alloy, of the order in which the strips break:

(First) *A B A A B A B C B C C C* (Last)

a. State the null and alternative hypotheses?

b. What is the critical value of the test statistic for an $\alpha = 0.05$ test?

c. Compute the test statistic H and make the test of significance.

d. Use the procedure similar to Fisher's least significant difference to determine whether any alloy tends to outrank another with respect to length of time it can be flexed before breaking.

10.7.2. Business school students often have difficulty in their first course in accounting. The instructor thinks this is because of differences in the students' mathematics achievements in high school. To test whether this is the case, the instructor takes a random sample of four students from among those receiving each of the letter grades in the accounting course and then compares them on the basis of their high-school grade point averages in mathematics courses. The data are given below:

Grade in Accounting	High-School GPA in Math			
A	3.5	3.0	3.6	4.0
B	3.2	2.8	3.8	3.1
C	2.8	3.0	3.4	3.3
D	2.2	2.8	2.9	3.1
F	2.5	2.6	2.7	2.9

a. Transform the 20 math grade point averages to ranks.

b. Use the rank data to compute the among SS using ANOVA procedures and as the numerator of H and show that it is 367.25 for both procedures.

c. Make the test of significance.

10.7.3. Use the Kruskal–Wallis procedure to analyze the data in Exercise 10.2.3.

    a. When would a nonparametric procedure be preferred for data such as these?

    b. Suppose it is known in advance of the experiment that bulbs of brands A and C both contain the same kind of filament but brand B bulbs have a different kind of filament. Use orthogonal contrasts to complete the analysis of the rank data.

10.7.4. In addition to his interests in science, Francis Galton was a social reformer, but surprisingly he did not consider the castelike social classes of his time to be unjust. Instead, he said they were "ordained by evolution." He believed the number and quality of "abilities" a man had determined the class in which he belonged. His descendants would remain in that class because they would inherit his skills. Galton believed a man could rise above the class in which he was born, but only by the improbable luck of inheriting nearly all of the abilities of both his father and mother. (On the other hand, a woman was of the class into which she married, and Sir Francis expressed concern because so many politicians married the daughters of wealthy merchants. He feared the consequence on the next generation would be deterioration in Britain's commerce rather than an improvement in its politics.) To see if class status is genetically determined, suppose Galton's scale for measuring abilities given in Exercises 1.1.3 and 8.5.3 is used to compare eight children from each of three classes and the results are

Class:	x	g	f	e	d	c	b	a	A	B	C	D	E	F	G	X
Nobility:	0	1	0	1	0	0	1	0	1	1	0	1	0	1	1	0
Merchants:	1	0	1	0	1	0	0	1	1	0	1	0	1	0	0	1
Laborers:	0	0	1	1	1	1	0	0	0	1	0	0	1	1	1	0

The scale is ordered with lowercase x the smallest possible score and uppercase X the greatest possible score.

    a. Why would it be appropriate to analyze using the Kruskal–Wallis test rather than an ANOVA?

    b. Give the null and alternative hypotheses.

    c. Is there a statistically significant difference among the classes?

    d. In Galton's time the nobility and merchants were probably more similar than either was with laborers, so what is the most sensible set of orthogonal contrasts.

    e. Perform the contrasts and draw conclusions about Galton's experimental hypothesis.

REVIEW EXERCISES

Decide whether each of the following statements is true or false. If a statement is false, explain why.

10.1. In an ANOVA, there is a degree of freedom associated with each squared total in the uncorrected sums of squares.

10.2. The standard deviation among sample averages is called the standard error and is computed from an ANOVA procedure by (within MS)/n.

10.3. Either a t test or an ANOVA may be used if only two treatment groups are being compared.

10.4. In ANOVA the uncorrected total sum of squares will be equal to or greater than any other corrected or uncorrected sum of squares.

10.5. An ANOVA uses both sides of the F distribution for critical values because the alternative hypothesis contains $\neq$.

10.6. An ANOVA cannot be done if the treatment groups are unequal in size.

10.7. An ANOVA requires that all treatment groups have the same variance, and this variance is estimated by MS_e.

10.8. If the null hypothesis is rejected in an ANOVA, we can conclude that the group with the smallest sample average has a mean that is different from all of the other group means.

10.9. In an ANOVA, the data from a control group are handled in a manner different from the treatment groups.

10.10. Fisher's least significant difference requires equal treatment group sizes.

10.11. When sample sizes are unequal Fisher's procedure is the only multiple-comparison procedure available to the researcher.

10.12. A confidence interval on the difference between two treatment means is the same as a confidence interval on the difference between two treatment effects.

10.13. The method of one-degree-of-freedom comparisons is an example of a multiple-comparison procedure.

10.14. The correction factor is the average variability from the overall average.

10.15. Multiple-comparison procedures and orthogonal contrasts are both methods for drawing conclusions from experiments in which H_0 is not true.

10.16. It is common to imbed a set of multiple comparisons into the design of an experiment for which ANOVA will be used.

10.17. A set of mutually orthogonal contrasts can be used to make all pairwise contrasts among a set of group means.

10.18. Although the F test involves variances, when it is used in ANOVA, it is to test hypotheses about means.

10.19. An F test is used to decide whether Duncan's test should be used to find significant differences among group means.

10.20. Orthogonal comparisons can be used to divide the treatment mean square into independent parts the sum of which equals the treatment mean square.

SELECTED READINGS

Anderson, R. L. (1965). Negative variance estimates. *Technometrics*, **7**, 75–76.

Andrews, H. P., and R. D. Snee (1980). Graphical display of means. *American Statistician*, **34**, 195–199.

Bernhardson, C. S. (1975). Type I error rates when multiple comparison procedures follow a significant F test of ANOVA. *Biometrics*, **31**, 229–232.

Carmer, S. G., and M. R. Swanson (1971). Detection of differences between means: A Monte Carlo study of five pairwise multiple comparison procedures. *Agronomy Journal*, **63**, 940–945.

Carmer, S. G., and M. R. Swanson (1973). Evaluation of ten pairwise multiple comparison procedures by Monte Carlo methods. *Journal of the American Statistical Association*, **68**, 66–74.

Chew, V. (1977). *Comparisons among Treatment Means in an Analysis of Variance*. Agricultural Research Service, U.S. Department of Agriculture, Washington, D.C.

Cobb, G. W. (1984). An algorithmic approach to elementary ANOVA. *American Statistician*, **38**, 120–123.

Duncan, D. B. (1955). Multiple range and multiple *F* tests. *Biometrics*, **11**, 1–42.

Dunn, O. J. (1959). Confidence intervals for the means of dependent, normally distributed variables. *Journal of the American Statistical Association*, **54**, 613–621.

Dunn, O. J. (1961). Multiple comparisons among means. *Journal of the American Statistical Association*, **56**, 52–64.

Dunnett, C. W. (1955). A multiple comparison procedure for comparing several treatments with a control. *Journal of the American Statistical Association*, **50**, 1096–1121.

Keuls, M. (1952). The use of the "Studentized range" in connection with an analysis of variance. *Euphytica*, **1**, 112–122.

Kramer, C. Y. (1956). Extension of multiple range tests to group means with unequal numbers of replications. *Biometrics*, **12**, 307–310.

Light, R. J., and B. H. Margolin (1971). An analysis of variance for categorical data. *Journal of the American Statistical Association*, **66**, 534–544.

Miller, R. G., Jr. (1996). *Simultaneous Statistical Inference*. McGraw-Hill, New York.

Saville, D. J. (1990). Multiple comparison procedures: The practical solution. *American Statistician*, **44**(May), 174–180.

Scheffé, H. (1953). A method for judging all contrasts in the analysis of variance. *Biometrika*, **40**, 87–104.

Scheffé, H. (1959). *The Analysis of Variance*. Wiley, New York.

Shaffer, J. P. (1977). Multiple comparisons emphasizing selected contrasts: An extension and generalization of Dunnett's procedure. *Biometrics*, **33**, 293–303.

Sirotnik, K. (1971). On the meaning of the mean in ANOVA (or, The case of the missing degree of freedom. *American Statistician*, **25**(Oct.), 36–37.

Tukey, J. W. (1949). Comparing individual means in the analysis of variance. *Biometrics*, **5**, 99–114.

11 The Analysis-of-Variance Model

Now that we are familiar with the basic ANOVA procedure, we need to look more closely at the underlying model and its assumptions.

11.1. RANDOM EFFECTS AND FIXED EFFECTS

The one-way ANOVA discussed in Chapter 10 can be applied to many different experiments. For example, it could be used to pick the least corrosive chemical from among 6 chemicals that are all effective for melting ice. Or it could be used to test whether there is significant variability among the achievements of introductory economics classes when they use the same method and materials but are taught by different teachers.

In Chapter 10 we assumed experimental situations similar to the ice-melting chemical example. That is, we assumed that all treatments of interest, the 6 chemicals, were included in the experiment. This type of ANOVA is based on a model called the *fixed-effects model (FEM)*. In this model the experimenter—usually in the latter stages of experimentation—narrows down the possible treatments to several in which he has a special interest. In the case of the chemicals, for example, tests would already have been completed to determine that these 6 were all available, suitable for melting ice, and economically feasible. Now a final choice is to be made on the basis of corrosiveness. In the FEM we are usually trying to pick the best of several possibilities. The inference made is restricted to the treatments used in the experiment.

The fixed effects model is sometimes called Model I. It is referred to as fixed because if the investigator decided to repeat the experiment he would use the same treatments in the repetition.

The achievement of economics classes taught by different teachers is an example of Model II, or the *random-effects model (REM)*; it is also called the components of-variance model. The random effects model assumes that the treatments are a random sample of all of the treatments of interest. It does not look for differences among the group means of the treatments being tested, but rather asks whether there is significant variability among all possible treatment groups. For example, if 5 teachers were used in the study, these 5 teachers would be the treatments and the grades of their students on some standardized test might be the variable of interest. The investigator would be interested in the variability among *all* economics teachers using this method and these materials. The 5 teachers in the experiment are a random sample from all of the treatments of interest. If the experiment were to be repeated, 5 different teachers chosen at random would be used.

Statistics for Research, Third Edition, Edited by Shirley Dowdy, Stanley Weardon, and Daniel Chilko. ISBN 0-471-26735-X © 2004 John Wiley & Sons, Inc.

When the REM is used, the investigator is interested in σ_A^2, the variance among all possible treatment groups. The ANOVA procedure can be used to test $H_0: \sigma_A^2 = 0$. If this null hypothesis is rejected, there is evidence of variability among groups. In the teacher example, if the null hypothesis is rejected, teachers do have an effect on the achievements of introductory economics classes. The inference is to all economics teachers, not just the five involved in the study.

In Chapter 10 we did not consider examples that follow the REM model. The assumptions for the underlying mathematical additive model $y_{ij} = \mu + \alpha_i + \varepsilon_{ij}$ differ for fixed effects and random effects. However, the numerical procedure for the one-way ANOVA is identical for both models.

The following table summarizes the two models.

$$y_{ij} = \mu + \alpha_i + \varepsilon_{ij}$$
$$i = 1, 2, \ldots, a$$
$$j = 1, 2, \ldots, n$$

Fixed-Effects Model (FEM)	**Random-Effects Model** (REM)
H_0: $\alpha_1 = \alpha_2 = \cdots = \alpha_a$	H_0: $\sigma_A^2 = 0$
H_a: At least one inequality	H_a: $\sigma_A^2 > 0$
μ: A constant, the mean of all possible experiments using the a designated treatments	μ: A constant, the population mean for all experiments involving all possible treatments of the type being considered
α_i: A constant for the ith treatment group, the deviation from the mean due to the ith treatment: $\sum_i \alpha_i = 0$	α_i: A constant for the ith treatment group, a random deviation from the population mean. The α_i's are normal, with $E(\alpha_i) = 0$ and $V(\alpha_i) = \sigma_A^2$
ε_{ij}: A random effect containing all uncontrolled sources of variability. The ε_{ij}'s are IND $(0, \sigma^2)$, that is, they are normally distributed with a mean of zero and a variance σ^2 and they are independent of each other and of the α_i's.	ε_{ij}: Same as for FEM
MS_a: Estimates $\sigma^2 + n \sum_i \alpha_i^2/(a-1)$	MS_a: Estimates $\sigma^2 + n\sigma_A^2$
MS_e: Estimates σ^2	MS_e: Estimates σ^2

In both models we assume that the experimental units are chosen at random from the population and assigned at random to the treatments. Frequently these assumptions are not completely met. Sometimes it is almost impossible to obtain a random sample from the entire population of interest. For example, the investigator may want to make inference about all white mice but must use a random sample of the white mice received from distributors. Or a researcher may be studying the effect of exercise on blood pressure in human males and may want to make inference to all males but may have to use volunteers with no opportunity to choose subjects at random. In both of these examples, however, it is possible to assign the subjects at random to the treatments. In some other investigations, even this second stage of randomization is not possible. For example, in a study of the effect of different teaching methods on the learning of college students, the investigator may have to utilize for the treatment groups the classes in which the students have enrolled. In this example there is no opportunity for a random choice of students or for a random assignment of the students to the treatments.

The ANOVA procedure is reliable if the assumptions are met. The more the experiment deviates from the assumptions, the less reliable are the conclusions. An investigator should mention any shortcomings of this type in the report of the study.

The follow-up procedures after ANOVA will differ depending upon whether the FEM or REM is being used. For the FEM we use multiple comparisons, orthogonal contrasts, or estimation of parameters (or linear combinations of parameters). For REM we are interested in the *intraclass correlation*, sometimes identified by the acronym ICC, as an estimate of the percentage of the total variance that is due to the differences among the treatments.

The ICC serves a function similar to that of the coefficient of determination which we examined in our study of linear trend or to the Rsquare statistic given in Chapter 10. The ICC gives the proportion of the variance that is explained by the groups or treatments in the model. If the effects α_i are on the numerical scale, we could compute the coefficient of determination r^2, but it would never be greater than the intraclass correlation r_I. That is because r^2 gives the percentage of variance explained by a linear relationship, whereas r_I provides the percentage explained by any relationship. Another advantage of r_I is that the groupings or α_i effects can be on the nominal scale and we can still obtain a statement of relationship of the treatments to the y variable and have an estimate of the variance explained by the method of groupings employed in the experiment.

Example 11.1.   One-way ANOVA for the Random Effects Model

As with the rest of the U.S. population, obesity is a major health problem in Appalachia. In a preliminary investigation, a nutritionist is looking for familial differences in body shape and plans to use body mass index (BMI) as the variable of interest. She selects 30 three-child families at random and then weighs and measures the height of each child in each family in order to obtain the 90 measures needed for her ANOVA. Gender differences among the children in her study are a lesser concern because BMI measures body density allowing for examination of weight while accounting for height. Still it would be better if she could undertake two studies, one of families with 3 girls and the other of families with 3 boys.

The 30 families are her treatment groups. Each group has a sample size of 3. These 30 families are a random sample of all families in Appalachia, so this is the REM.

The ANOVA is carried out as in Chapter 10 except that the null hypothesis is H_0: $\sigma_A^2 = 0$.

Source	df	SS	MS	F
Among families	$a - 1 = 29$	4,779.2	164.8	7.01
Within families	$a(n - 1) = 60$	1,410.0	23.5	

Since $F_{0.05,29,60} = 1.656$, H_0 is rejected and there is significant variability among the BMIs of the families; that is, there is some evidence of familial differences in body density.

Since MS_a estimates $\sigma^2 + n\sigma_A^2$ and MS_e estimates σ^2, the investigator computes the intraclass correlation r_I as follows:

$$\hat{\sigma}^2 = \mathrm{MS}_e = 23.5$$

$$\hat{\sigma}_A^2 = \frac{\mathrm{MS}_a - \mathrm{MS}_e}{n} = \frac{164.8 - 23.5}{3} = 47.1$$

$$r_I = \frac{\hat{\sigma}_A^2}{\hat{\sigma}_A^2 + \hat{\sigma}^2} = \frac{47.1}{47.1 + 23.5} = 0.667$$

that is, 66.7% of the total variance in BMIs is due to the differences among the families. The causes may be heredity, environment, or both, but a significant percentage of the variance of BMIs can be attributed to family differences.

The nutritionist could not have used bivariate correlation r and the coefficient of determination r^2 instead of the ICC because she had 3 rather than 2 members. Furthermore, even if the number per family had been $n = 2$, only the variability due to linear association between family members would have been obtained.

She would not have used Rsquare because its computation is

$$\mathrm{Rsquare} = 1 - \frac{\mathrm{SS}_e}{\mathrm{SS}_t}$$

which gives the percentage of the explainable sums of squares among the $na = 90$ individuals in the study. To show the difference between Rsquare and the ICC, for a one-way ANOVA the ICC can be expressed as

$$r_I = 1 - \frac{\hat{\sigma}}{\hat{\sigma}_t}, \text{ where } \hat{\sigma}_t = \hat{\sigma}_A + \hat{\sigma}$$

Thus it estimates the percentage of the BMI variance explained by families in the *target population*, the population of Appalachian families.

There are many experimental situations in which the random effects model is used and the intraclass correlation is calculated. For example, in an environmental study on the amount of lung damage in wild animals in a heavily industrial region, the region is divided into sections, random sections chosen, and traps set to capture a sample of animals. The random sections are the treatments and the intraclass correlation indicates the amount of variability in lung damage due to the different sections.

For another example, the REM would be used in a preliminary study to see if bees are attracted to color. Alfalfa blossoms range in color from dark purple to yellow to white. A random sample of alfalfa plants with different colored blossoms is chosen. The number of visits of bees to the different plants is the variable of interest. If the null hypothesis is rejected, plans can be made to conduct experiments that would reveal the specific color or colors that attract bees.

When the ICC is computed, the investigator is interested in the percentage of the total variance due to the treatments. The specific percentage that is meaningful depends on the experiment. If the investigator is looking for evidence of repeatability, as in a lab test to measure blood sugar where the treatment groups are different samples of blood, he will want a high ICC, perhaps 95%. In many other situations a lower value is meaningful; an example might be a study to see if a high level of low density cholesterol (LDC) runs in families. In an ANOVA where families are treatment groups, it is possible that the procedure leads to a significant F value, but at the same time r_I is small. However, because there is strong clinical evidence that LDC is associated with the blood clotting that leads to coronary artery blockage, even a small but significant r_I could be of value. It could suggest lines of further study on LDC, or at least alert physicians to the need for frequent blood tests for LDC among those with a relative who suffered coronary blockage.

As we examine more complex models, it will become very important to know what is estimated by each mean square in an ANOVA. We must know what a mean square estimates in order to determine what is a valid F test for the hypothesis we wish to test, and as we have seen, we need this information in order to obtain the ICC. The value or linear combination of values estimated by a mean square is called the *expected value of the mean square* and is symbolized as $E(MS)$ with a subscript identifying the MS under consideration.

We have seen that in the FEM, we want to test the hypothesis

$$H_0: \ \alpha_1 = \alpha_2 = \cdots = \alpha_a = 0, \quad \text{or} \quad H_a: \ \alpha_i = 0 \ \textit{for all } i$$

If the null hypothesis is true, all the $\alpha_i = 0$, meaning that group averages are not significantly different from the overall mean, then also $\sum \alpha_i^2 = 0$, and $E(MS_a) = \sigma^2 + n \sum \alpha_i^2/(a-1) = \sigma^2$. Similarly for the REM, we want to test the hypothesis that $\sigma_A^2 = 0$; thus under the null hypothesis $E(MS_a) = \sigma^2 + n\sigma_A^2 = \sigma^2$ because $\sigma_A^2 = 0$. We can see, then, for either model, when the null hypothesis is true, both MS_a and MS_e are independent estimates of the same variance and thus can be validly tested using the F distribution.

The mean square which estimates random variability will always be given as MS_e, and $E(MS_e) = \sigma^2$. Other $E(MS)$ will contain σ^2 plus terms representing other sources of variability, and the final term will be one about which we want to make a test of hypothesis. Thus, depending on the model, $E(MS_a)$ is written as $\sigma^2 + n \sum \alpha_i^2/(a-1)$ or as $\sigma^2 + n\sigma_A^2$, and when the null hypothesis is true, the last term in $E(MS_a)$ becomes zero. We can see that when we want to test the hypothesis that there is only random variability among group averages, we need an F test which is the ratio of two mean squares whose expectations are the same except for the term which becomes zero when the null hypothesis is true:

Expectations of Mean Squares

Source	Fixed Model	Random Model	If Null Hypothesis Is True
Among groups	$\sigma^2 + n \sum \alpha_i^2/(a-1)$	$\sigma^2 + n\sigma_A^2$	$n \sum \alpha_i^2/(a-1)$ and $n\sigma_A^2$ are 0
Within groups	σ^2	σ^2	

For either model, $F = MS_a/MS_e$ is the appropriate F test.

Because the notation $\sum \alpha_i^2/(a-1)$ is awkward to write, we will use $\theta_A^2 = \sum \alpha_i^2/(a-1)$ instead. With this symbolism, the expectations of mean squares will look more nearly alike, but it must be remembered that σ_A^2 represents the variance among a large population of groups which has been randomly sampled, whereas θ_A^2 represents the sum of a set of constants.

The ICC procedure can be summarized as follows.

Procedure. Intraclass Correlation

Perform the ANOVA as in Chapter 10.
  Estimate σ_A^2 and σ^2 as follows:

$$\hat{\sigma}^2 = \text{MS}_e$$

$$\hat{\sigma}_A^2 = \frac{\text{MS}_a - \text{MS}_e}{n}$$

Then r_I, the ICC, is

$$r_I = \frac{\hat{\sigma}_A^2}{\hat{\sigma}_A^2 + \hat{\sigma}^2}$$

$$0 \leq r_I \leq 1$$

The ICC can be interpreted as the proportion of the total variability due to the differences in all possible treatments of this type.

EXERCISES

11.1.1. Decide whether each of the following is using the FEM or the REM.

  a. A professor is trying to select a textbook for a sociology course from 4 different ones which are available. He divides his students at random into 4 groups and assigns the textbooks to the groups at random. After using the different books for the course, all students still enrolled take the same examination. ANOVA is used to analyze the results.

  b. A manufacturer builds a piece of equipment to turn out machined parts. To study the performance of her machines, she selects 8 machines at random and then selects 10 parts at random from the production of each of these machines. She measures the lengths of the 80 pieces and performs an ANOVA.

  c. An educator wishes to study the competence in algebra of all New York City students who have just completed the ninth grade. Five junior high schools are selected at random, and within each school a random sample of ninth-grade students are given examinations. Using these scores, the hypothesis that there is variability among the schools is tested.

  d. Worms are classified into three groups by a structural characteristic: small, medium, or large ventral flap. Three random samples of 11 worms are taken from each group and the weight of each worm is recorded. The hypothesis is tested that the mean weight of each group is the same.

  e. A psychologist devises an examination in such a way that the final score depends almost entirely upon the ability of the subject to follow instructions. The test is given to 40 students who have been divided into 4 equal groups at random. The instructions are given in the following 4 ways:

Group I	written and brief
Group II	oral and brief
Group III	written and detailed
Group IV	oral and detailed

An ANOVA is performed.

11.1.2. An ANOVA is used to study the effect of seam differences on variability in the sulfur content of coal. Seams and samples from seams are taken at random.

Source	df	SS	MS
Among coal seams	24	2400	100
Within coal seams	125	5000	40

a. Do differences among seams contribute significantly to the variability in the sulfur content of coal?

b. What percentage of the variability in the sulfur content of coal is attributable to seam differences?

c. Would you advise coal producers in search of low-sulfur coal to seek low-sulfur seams or to seek other factors that might affect variability? Justify your answer on the basis of the above analyses.

11.1.3. The following data are from a (fictional) study of obesity on 10 families each of which have 3 brothers:

Family	Brothers Pounds Overweight			Total
A	50	58	72	180
B	80	96	100	276
C	60	72	84	216
D	89	80	77	246
E	82	95	90	267
F	96	75	78	249
G	102	88	86	276
H	79	100	85	264
I	85	72	89	246
J	98	79	84	261

$$\sum\sum y_{ij}^2 = 209{,}769 \qquad \sum\left(\sum y_{ij}\right)^2/n = 207{,}849 \qquad \sum\sum y_{ij} = 2481$$

a. Complete the ANOVA.

b. Compute the ICC.

c. What is the target population, the population about which inference is to be made?

d. What conclusions do you draw about obesity being a characteristic of some families?

11.1.4. Given the following ANOVA, compute the ICC.

Source	df	SS	MS
Among treatments	10	4368	436.8
Within treatments	33	4320	130.9

11.1.5. Suppose a physiologist is working on a new method to measure blood sugar. Blood samples are taken from 10 people, and two assays are done on each sample.

Source	df	SS	MS
Among persons	9	1710	190
Within persons	10	100	10

    a. Which model is being used?
    b. What is the null hypothesis?
    c. Should the null hypothesis be rejected?
    d. Compute the ICC.
    e. Does this new method seem to be reliable?

11.1.6. Fifteen varieties of corn are chosen at random from all available varieties, and plots are planted of each variety. At maturity, five random plants are chosen from each plot and yield is measured, leading to the following analysis:

Source	Df	SS	MS
Among corn varieties	14	4368	—
Within corn varieties	—	—	72

    a. Complete the ANOVA.
    b. Compute the ICC.
    c. Interpret the ICC.

11.2. TESTING THE ASSUMPTIONS FOR ANOVA

In both the fixed effects and random effects models we assume the observations fit the additive model

$$y_{ij} = \mu + \alpha_i + \varepsilon_{ij}$$

in which the ε_{ij}'s are IND(0, σ^2). In practice, this means:

1. The treatment groups are normally distributed (this is required so that the ε_{ij}'s will be normally distributed).
2. The treatment groups all have the same variance (this is required so that the ε_{ij}'s will have the same variance for each i).
3. The experimental units are picked at random and assigned at random to the treatment groups (this is required so that the ε_{ij}'s are independent of each other and the α_i's).

We discuss each of these conditions in turn.

Normality   The normality of the treatment groups can be roughly checked by constructing histograms of the sample from each treatment group. Histograms reveal skewness and bimodality. Another approach is to plot the cumulative percentages on normal probability paper; a normal distribution leads to a straight line. Unfortunately, a large number of observations are needed for both of these procedures. The ANOVA, however, leads to valid conclusions in some cases where there are departures from normality. For small sample sizes the treatment groups should be symmetric and unimodal. For large samples, more radical departures are acceptable since the central limit theorem comes into play. Thus if there is doubt about normality, one solution is to use a large number of observations.

Some traditionally small experiments lead to nonnormal distributions:

1. Data composed of small counts, even into the hundreds, such as the number of parasites on wildlife
2. Data composed of very large counts, such as bacterial counts
3. Proportions, or percentage data
4. Arbitrary scales, such as a 10-point taste test
5. Weights of very small things

In the first three cases, not only is the assumption of normality invalid but the variances of the treatment groups may be unequal and there may be a lack of independence between the random effect and the treatment effect. One approach in these cases is to transform the data and perform the ANOVA on the transformed values; this is discussed in Section 11.3.

In experiments involving arbitrary scales, as the taste test, normality can be approximately achieved by using several tasters (5 or more) and recording their average ratings.

Weights of very small things are often not normally distributed because of the limits of the accuracy of the weighing process. Weighing objects in groups can sometimes overcome this difficulty.

Equality of Variances.   An ANOVA assumes homogeneity of variances (*homoscedasticity*); that is, all of the treatment groups have the same variance. The F tests are *robust* with respect to departures from homogeneity; that is, moderate departures from equality of variances do not greatly affect the F statistic. If the experimenter fears a large departure from homogeneity, several procedures are available to test

$$H_0: \quad \sigma_1^2 = \sigma_2^2 = \cdots = \sigma_\alpha^2$$

Unfortunately, most of these tests rely on the assumption of normality.

We discuss here only one test for homogeneity of variances, the F_{max} test developed by Hartley (1950). Hartley's test is one of the simplest; it may be used when all treatment groups are the same size and involves comparing the largest sample variance with the smallest sample variance.

Example 11.2. F_{max} Test for Homogeneity of Variances

In the chain saw study (Example 10.1), the investigator wants to test

$$H_0: \quad \sigma_A^2 = \sigma_B^2 = \sigma_C^2 = \sigma_D^2$$

He first computes the sample variance for each treatment group:

	Group			
	D	A	B	C
$\sum\limits_{j} y_{ij}$	155	165	215	245
$\sum\limits_{j} y_{ij}^2$	4981	5999	9965	12,175
$\left(\sum\limits_{j} y_{ij}\right)^2 \big/ n$	4805	5445	9245	12,005
s_i^2	44.0	138.5	180.0	42.5

The F_{max} statistic is

$$F_{max} = \frac{\text{largest treatment variance}}{\text{smallest treatment variance}}$$

$$= \frac{180}{42.5} = 4.24$$

Here, F_{max} is significant if it exceeds the value given in the table computed by Hartley, Table A.16 in the Appendix of Useful Tables. This table is entered by a, the number of treatment groups, and $v = n - 1$, in which n is the number of observations per treatment group. In this example

$$F_{max_{0.05,a,v}} = F_{max_{0.05,4,4}} = 20.6$$

Thus the null hypothesis of homogeneity of variances is accepted.

Hartley's procedure can be summarized as follows.

Procedure. Hartley's Test for Homogeneity of Variances

To test:

$$H_0: \ \sigma_1^2 = \sigma_2^2 = \cdots = \sigma_a^2 \quad \text{against} \quad H_a: \ \text{At least one inequality}$$

when each of the a populations is normal and there is a random sample of size n from each population, compute

$$s_1^2, s_2^2, \ldots, s_a^2$$

and calculate

$$F_{max} = \frac{\text{largest } s_i^2}{\text{smallest } s_i^2}$$

Here, F_{max} is significant if it equals or exceeds the value $F_{max}\alpha,a,v$ in Hartley's table, Table A.16 in the Appendix, with a the number of populations and $v = n - 1$.

Because of the sensitivity of this test to departures from normality, if $F_{\max}$ is significant, it indicates either unequal variances or a lack of normality.

Two other commonly used tests of homogeneity of variances are those of Cochran (1947) and Bartlett (1937). In most situations, Cochran's test is equivalent to Hartley's. Bartlett's test has a more complicated test statistic but has two advantages over the other two: It can be applied to groups of unequal sample sizes, and it is more powerful. Scheffé has a test that is less sensitive to departures from normality. For a discussion of these tests see Winer (1971, pp. 205–220).

If the experimenter finds that only one or two of the treatment groups have a different variance, he might discard these samples and work only with the remaining ones. However, if discarding these treatment groups makes it impossible to answer the experimental questions, another approach may be needed. One possibility is to transform the data as described in Section 11.3; another would be a nonparametric technique in place of ANOVA. This does not imply that there are no assumptions to be met for nonparametirc anaļ ses. For instance, in addition to random and independent samples, rank order tests such as ʻ,ᵢe Kruskal—Wallis test require that all the sampled distributions have the same shape. When that assumption is met, they are more powerful than ANOVA for a number of nonnormal distributions.

Independence. The random effects (ε_{ij}'s) in the additive model must be

1. independent of each other and
2. independent of the treatment effects (α_i's).

If these conditions are missing, it will be difficult to detect real differences that may exist.

The first condition is usually satisfied if the experimental units are randomly chosen and randomly assigned to the treatments. If the treatment groups already exist, such as members of a certain profession, the experimenter does not have the opportunity to assign the subjects at random to the treatments. In such cases he uses random samples from each treatment group.

It is not usually acceptable to use ANOVA on repeated observations on the same subject unless precautions are taken to avoid a systematic effect caused by the repetition of the experiment, for example, learning by the subject who repeats the same task. Sometimes lack of independence occurs because of instrument wear or drift. This type of dependence within groups can be detected by plotting the data in the order in which they were collected.

The second condition, that the random effect is independent from the treatment effect, can be checked by plotting the sample means against the sample variances (Figure 11.1). Independence will lead to an unpatterned scatter around a horizontal line, while dependence

ε_{ij} independent of α_i ε_{ij} dependent on α_i ε_{ij} dependent on α_i

FIGURE 11.1. Visual test for the independence of the error term and the treatment effect.

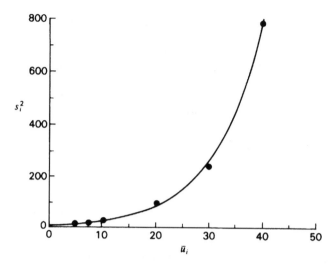

FIGURE 11.2. Data that may be improved by a log transformation.

usually takes the form of some curve. A transformation can sometimes be used to remove this type of dependence.

EXERCISES

11.2.1. Given below are the calculations from an experiment involving the breaking strengths of 6 different fabrics:

	Nylon	Rayon	Linen	Dacron	Cotton	Silk
$\sum_j y_{ij}$	144	96	119	168	98	140
n	10	10	10	10	10	10
$\sum_j y_{ij}^2$	2080.8	1063.8	1449.4	2904.4	1018.0	1979.8
$\left(\sum_j y_{ij}\right)^2 / n$	2073.6	921.6	1416.1	2822.4	960.4	1960.0

  a. Test to decide whether the different fabrics have a common variance for breaking strength.
  b. Which variances are significantly different from each other? (*Hint:* Test all pairs of variances by using a two-way table similar to the table for multiple comparisons; however, use the ratios of the variances and $F_{\max}$ tests along each diagonal.)

11.2.2. In the light bulb experiment, Exercise 10.2.3, test whether the variances of the 3 brands are equal.

11.2.3.   In the orange-juice experiment, Exercise 10.2.5, show that there is no evidence that the variances of vitamin C are different among the 3 methods of processing orange juice.

11.3.   TRANSFORMATIONS

If we find that the variances are not homogeneous, or if we find a lack of normality, or if there is a dependence between the treatment effects and the random effects, it is sometimes possible to use a transformation to get the data into a form for which the ANOVA is valid. A transformation replaces each observed value u_{ij} by another value y_{ij} according to a certain rule, for example, $y_{ij} = \log u_{ij}$. It is essential that any transformation preserve the order of the data values; thus, if u_1 and u_2 are transformed to y_1 and y_2, respectively, and $u_1 < u_2$, then $y_1 < y_2$. Since the order of the observations is not changed by the transformations we use, any conclusion about differences in the transformed data are true for the original data. This technique, however, has the disadvantage that we must report results in unusual units of measure, as the log of a length or the square root of the number of fish.

Various transformations are available, and sometimes the nature of the data, together with a plot of sample averages against sample variances, will provide clues to help the experimenter decide which transformation to use. If the data span several log scales, that is, if they contain both relatively small and relatively large observed values, one usually looks at the graph for an exponential relationship between sample means and variances (Figure 11.2). This relationship frequently occurs when the data arise from large counts (such as blood cells or bacterial counts). Each observation u_{ij} is transformed to $y_{ij} = \log(u_{ij})$ or to $y_{ij} = \log(u_{ij} + c)$ with $c > 0$ if zero or negative numbers are in the data. Logs with either base 10 or base e may be used. Table A.17 in the Appendix is a table of logs base 10. A log transformation will preserve the order of the data and the order of the averages, but the log transformation can make the variances more nearly alike and thereby break up the strong relationship between sample averages and sample variances. The ANOVA is carried out as usual, except that the transformed values y_{ij} replace the corresponding original observations u_{ij}. Before performing the ANOVA, however, it is wise to check the transformed data for the properties of normality, homogeneity of variance, and independence.

Example 11.3.   The Log Transformation

As an alternative to dangerous insecticides, a chemist is working on a synthetic pheromone (a type of hormone involved in mating behavior) to be used as a bait to attract destructive insect into traps. In a field experiment, 6 different levels of the synthetic hormone are used, with 10 traps per level. The 60 traps are placed at random in a peach orchard, and the observed values below (u_{ij}) represent the number of Mediterranean fruit flies trapped during the same 4-hour period.

Level:	1	2	3	4	5	6
u_{ij}	2	12	22	28	24	17
	4	9	12	17	25	54
	10	5	11	9	36	24
	15	10	7	39	17	33
	3	3	4	15	38	27
	2	7	7	33	19	41

Level:	1	2	3	4	5	6
	4	5	8	11	65	76
	2	16	17	12	18	109
	5	6	6	15	42	36
	3	2	11	21	16	33
Average $(\bar{u}_i)$:	5.0	7.5	10.5	20.0	30.0	45.0
Variance (s_i^2):	18.00	18.50	30.6	102.22	240.00	785.78

The plot of sample averages against sample variances for these data is given in Figure 11.2. As can be seen, the data closely fit a curvilinear relationship suggesting a log transformation:

Level:	1	2	3	4	5	6
$y_{jj} = \log(u_{izzzzz} + 1)$	0.4771	1.1139	1.3617	1.4624	1.3979	1.2553
	0.6990	1.0000	1.1139	1.2553	1.4150	1.7404
	1.0414	0.7782	1.0792	1.0000	1.5682	1.3979
	1.2041	1.0414	0.9031	1.6021	1.2553	1.5315
	0.6021	0.6021	0.6990	1.2041	1.5911	1.4472
	0.4771	0.9031	0.9031	1.5315	1.3010	1.6232
	0.6990	0.7782	0.9542	1.0792	1.8195	1.8865
	0.4771	1.2304	1.2553	1.1139	1.2788	2.0414
	0.7782	0.8451	0.8451	1.2041	1.6335	1.5682
	0.6021	0.4771	1.0792	1.3424	1.2304	1.5315
Average:	0.7057	0.8769	1.0194	1.2795	1.4491	1.6023
Variance:	0.0605	0.0533	0.0393	0.0403	0.0384	0.0545

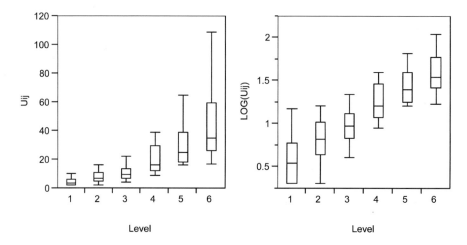

FIGURE 11.3. Box plots of data before and after log transformation.

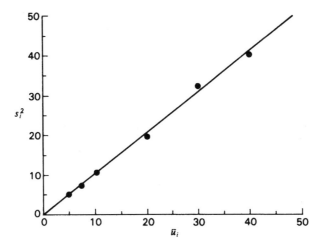

FIGURE 11.4. Data that may be improved by a square-root transformation.

After the log transformation has been performed on the data, the sample averages of the y_{ij} have the same order as averages of the u_{ij} but the variances are very similar from one group to another, and not even in the same order as the averages. Thus averages and variances for the y_{ij} appear to be independent. Figure 11.3 shows box plots of the data for each level before and after transformation. The box plots of the y_{ij} provide evidence that necessary conditions are satisfied, or nearly, so an ANOVA of the transformed data should provide an approximate, but reasonably good, test of a hypothesis about the effect of different concentrations of the synthetic pheromone in attracting insects.

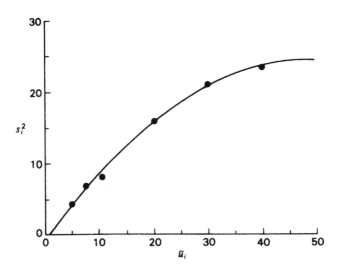

FIGURE 11.5. Data that may be improved by an angular transformation.

A graph that frequently appears when sample averages are plotted against sample variances for small counts is a straight line with a $45°$ angle (Figure 11.4). The graph often indicates a Poisson distribution in which $\mu_i = \sigma_i^2 = \lambda_i$. The transformation that often helps is to replace u_{ij} with $y_{ij} = \sqrt{u_{ij}}$ or $y_{ij} = \sqrt{u_{ij} + c}$. The data from which Figure 11.4 was plotted can be found in Exercise 11.3.3. Those with further interest in this transformation may want to examine these data to verify the straight-line relationship between sample average and variance for the original data, and to observe how this relationship is affected by the square-root transformation.

If the data are from a population with a binomial distribution (percentage or proportion data), the mean and the variance are not independent,

$$\mu_i = n\pi_i \quad \text{and} \quad \sigma_i^2 = n\pi_i(1 - \pi_i)$$

The diagram in this case has the form found in Figure 11.5. A transformation often used in this case, especially if $\pi < 0.2$ or $\pi > 0.8$, is arc-sin $\sqrt{u_{ij}}$ in which u_{ij} is expressed as a proportion. Tables are available for this transformation. Table A.18 in the Appendix is one such table. In Table A.18, u_{ij} is entered as a percentage and the transformed value is in degrees.

Since ANOVA was designed for *continuous* variables and proportions arise from *discrete* variables, the investigator should remember that ANOVA may not be the best way to analyze data of this type. In fact, an F test with or without a transformation may be less powerful than the appropriate procedure. Sometimes the investigator may decide to use ANOVA because of its convenience or for reporting results in a uniform way when ANOVA is being used on other variables in the study. This approach, however, is at most second best.

Many transformations are available in addition to the ones discussed in this section. Some computer packages offer several to the investigator. It is invalid to transform the data by each available transformation and perform ANOVA in order to pick out the transformation that leads to significant results. However, several transformations can be used on the data, and the one that best equalizes the ranges of the samples can be used for ANOVA since the ranges are closely related to the variances. If the ranges are not very different, then the variances may be homogeneous.

In the discussions of nonparametric procedures found earlier in the text, data which were measured on the numerical scale were transformed either to the nominal or to the ordinal scale. It can be noted that the rank transformations used in some of these nonparametric tests often have the same benefits sought here. The rank transformation will not change the order of two observations; the group means of the ranks will usually have the same order as those of the original observations; also the variances of the ranks are usually of similar magnitude, and the plot of sample averages and variances does not tend to show a strong relationship between the two. Consequently, the rank transformation may also be considered as one which will make data suitable for ANOVA as well as nonparametric procedures. As before, the observations are ranked from smallest to largest, and observations having the same numerical value are assigned the average of the ranks for which they tie. After the observations in each group have been replaced by ranks, rather than a nonparametric test, an ANOVA procedure is performed on the ranks. Also, the null hypothesis is tested by F rather than chi-square. This is because in many complex designs (see Chapter 12) it is difficult or impossible to know the value of the variance needed for chi-square. Although ranks do not have a normal distribution, the procedure is considered to be robust, meaning that the true level of significance is reasonably close to that obtained from the F table after the ANOVA.

To use this procedure, however, all assumptions for the ANOVA (except for normal distribution) must still be met by the rank data. For further discussion of this technique see Conover and Iman (1976).

EXERCISES

11.3.1. Using the data from Example 11.3:

    a. Show how to obtain the transformed value $y_{14} = 1.2041$.

    b. Compare the F_{max} test performed on the original observations (u_{ij}) with that performed on their transformed values (y_{ij}).

    c. Using the transformed values, plot the sample averages against the sample variances and compare your plot to that in Figure 11.2 to see if there is still an obvious relationship between averages and variances.

11.3.2. In a certain experiment in graph reading, subjects take the following amounts of time (in seconds divided by 10) to answer a set of questions:

	Group		
	A	B	C
	28	16	31
	17	13	22
	18	16	16
	21	12	21
	13	13	13
	29	12	16
$\sum_j u_{ij}$	126	82	119
$\sum_j u_{ij}^2$	2848	1138	2567

    a. Show that the variances of the groups are unequal using Hartley's F_{max} test.

    b. Use a square-root transformation on the times.

    c. Does the transformation correct the lack of homogeneity of variances?

    d. Perform ANOVA on the transformed data.

    e. How would the results of the ANOVA be reported?

11.3.3. A dermatologist wants to study the effectiveness of sunscreens in providing protection for the skin of inveterate sunbathers. Six different formulations of sunscreens are to be compared, and sufficient random sampling is done among volunteers in order to have 10 sunbathers for each formulation. The volunteers are examined every two weeks and at the end of the summer, and for each the dermatologist has the total number per subject (u_{ij}) of skin lesions attributable to exposure to the sun. These are given below, along with the transformed values (y_{ij}) to be used in the ANOVA:

	Formulation					
	A	B	C	D	E	F
			u_{ij}			
	4	12	12	18	30	49
	4	9	11	17	33	34
	8	5	10	19	29	45
	10	9	7	29	32	31
	3	4	14	24	36	36
	5	7	7	23	27	41
	4	5	9	15	32	46
	3	11	17	22	18	39
	5	6	7	15	38	46
	4	7	11	18	25	33
Average ($\bar{u}_i$)	5.0	7.5	10.5	20.0	30.0	40.0
Variance (s_i^2)	5.11	7.17	10.72	19.78	32.89	40.22

	Formulation					
	A	B	C	D	E	F
			y_{ij}			
	2.236	3.606	3.606	4.359	5.568	7.071
	2.236	3.162	3.464	4.243	5.831	5.916
	3.000	2.449	3.317	4.472	5.477	6.782
	3.317	3.162	2.828	5.477	5.745	5.657
	2.000	2.236	3.873	5.000	6.083	6.083
	2.449	2.828	2.828	4.899	5.292	6.481
	2.236	2.449	3.162	4.000	5.745	6.856
	2.000	3.464	4.243	4.796	4.359	6.325
	2.449	2.646	2.828	4.000	6.245	6.856
	2.236	2.828	3.464	4.359	5.099	5.831
Average ($\bar{y}_i$)	2.416	2.883	3.361	4.560	5.544	6.386
Variance (s_i^2)	0.181	0.208	0.224	0.225	0.291	0.248

a. Identify the transformation which was used, and tell why you think it was chosen.

b. What evidence is there that the transformation has changed the strong relationship between sample average and sample variance which can be seen in Figure 11.4?

c. If for the original data $\sum\sum u_{ij} = 1130$, why is it that with this transformation $\sum\sum y_{ij}^2 = 1130 + 60 = 1190$?

11.3.4. Four groups of subjects were given a certain task to perform. The number of mistakes out of 18 trials is recorded.

Group	Errors Out of 18 Trials							
1	0	0	0	0	1	3	0	0
2	5	3	2	11	3	0	0	0
3	1	1	0	0	2	3	0	0
4	3	0	1	0	4	1	1	2

a. Convert the number of errors to percentage of errors.

b. Show that the groups have unequal variances when the variable is percentage of errors.

c. Use arcsin $\sqrt{\% \times 0.01}$ to transform the data.

d. Check the transformed data for homogeneity of variance.

11.3.5. Holly is a broadleaf evergreen that is very attractive in landscaping, but many nurseries do not attempt to raise it because of the difficulty in getting its seed to germinate. In an effort to improve germination, a horticulturist uses 6 different seed treatments. For each treatment she prepares 10 seed beds with a hundred seeds in each bed. The data below represent the number of seeds which germinate in each of the seed beds.

	Seed Treatment					
	I	II	III	IV	V	VI
				u_{ij}			
	6	6	12	20	27	37
	5	5	8	14	24	38
	3	9	11	27	32	36
	4	4	6	17	29	42
	5	8	11	18	34	50
	3	13	16	19	30	35
	2	6	8	24	33	38
	6	7	10	22	39	45
	9	10	13	16	27	36
	7	7	10	23	25	43
Average $(\bar{u}_i)$	5.0	7.5	10.5	20.0	30.0	40.0
Variance (s_i^2)						
(Figure 11.5)	4.44	6.94	8.06	16.00	21.11	23.56

	Seed Treatment					
	I	II	III	IV	V	VI
				y_{ij}			
$y_{ij} = $ arcsin	14.18	14.18	20.27	26.57	31.31	37.46
$\sqrt{u_{ij} \times 0.01}$	12.92	12.92	16.43	21.97	29.33	38.06
	9.97	17.46	19.37	31.31	34.45	36.87
	11.54	11.54	14.18	24.35	32.58	40.40
	12.92	16.43	19.37	25.10	35.67	45.00
	9.97	21.13	23.58	25.84	33.21	36.27
	8.13	14.18	16.43	29.33	35.06	38.06
	14.18	15.34	18.43	27.97	38.65	42.13
	17.46	18.43	21.13	23.58	31.31	36.87
	15.34	15.34	18.43	28.66	30.00	40.98
Average $(\bar{y}_i)$	12.66	15.70	18.76	26.47	33.16	39.21
Variance (s_i^2)	7.907	7.841	7.103	8.221	8.166	7.986

a. For the transformed data, plot the sample averages against the sample variances to see if there is any evidence of a relationship between the two.

b. The SAS output for an ANOVA and Fisher's least significant difference on the transformed data is as follows.

What conclusions can you draw from this output?

                          The SAS System

                        The GLM Procedure

                    Class Level Information

          Class   Levels   Values

          Treat   6   1 2 3 4 5 6

              Number of observations = 60

                        The GLM Procedure

Dependent Variable: Y

Source	DF	Sum of Squares	Mean Square	F Value	Pr > F
Model	5	5455.745683	1091.149137	138.63	<.0001
Error	54	425.021775	7.870774		
Corrected Total	59	5880.767458			

R-Square	Coeff Var	Root MSE	Y Mean
0.927727	11.53305	2.805490	24.32566

Source	DF	Type I SS	Mean Square	F Value	Pr > F
Treat	5	5455.745683	1091.149137	138.63	<.0001

Source	DF	Type III SS	Mean Square	F Value	Pr > F
Treat	5	5455.745683	1091.149137	138.63	<.0001

                        The GLM Procedure

                      t Tests (LSD) for Y

NOTE: This test controls the Type I comparisonwise error rate, not the experimentwise error rate.

```
Alpha                                0.05
Error Degrees of Freedom               54
Error Mean Square                7.870774
Critical Value of t              2.00488
Least Significant Difference      2.5154
```

Means with the same letter are not significantly different.

t Grouping	Mean	N	Treat
A	39.209	10	6
B	33.157	10	5
C	26.468	10	4
D	18.763	10	3
E	15.696	10	2
F	12.661	10	1

REVIEW EXERCISES

Decide whether each of the following statements is true or false. If a statement is false, explain why.

11.1. The REM could be called the component-of-variance model because the experimenter is more interested in causes of variation than in comparing means.

11.2. Because of a general lack of knowledge about the nature of effects, the REM is probably more common than the FEM.

11.3. The experimenter does not test for homogeneity of variance unless he has reason to doubt this customary assumption for the ANOVA.

11.4. If Hartley's test is significant when performed on the original data, a suitable transformation will result in nonsignificance when the test is performed on the transformed data.

11.5. The proper transformation should provide a more powerful F test than one based on the original data that do not meet the conditions for an ANOVA.

11.6. If in a scientific journal an ANOVA is based on the additive model $y_{ij} = \pi + \theta_i + \delta_{ij}$, the reader has enough information to distinguish whether or not it was a FEM.

11.7. When the model is $y_{ij} = \mu + \alpha_i + \varepsilon_{ij}$, the same F test will be performed whether the α_i's are fixed or random.

11.8. Multiple-comparison procedures such as Tukey's honestly significant differences are used to determine differences among fixed effects, but for random effects the investigator is more interested in whether there is variability among the effects than in making comparisons among them.

11.9. If the sample sizes are large, the experimenter should always check for normality prior to an ANOVA.

11.10. Transformations can correct nonnormality, unequal variances, and lack of independence between the ε_{ij}'s and the α_i's.

11.11. In an ANOVA, if the overall average of the experiment is zero, the numerical value of the correction factor will be zero.

11.12. Heterogeneity of variance is more likely in a REM, in which groups are randomly drawn from a large population, than in a FEM, in which groups are carefully selected.

11.13. When means are correlated with variances in an experiment, a suitable transformation can result in homogeneity of variance but still permit heterogeneity of means.

11.14. Transformations are used as second-best procedures when certain conditions such as homogeneous variances, independent effects, and random sampling do not occur in the experiment.

11.15. A significant negative ICC means that there are marked dissimilarities among individuals in the same group.

11.16. If the null hypothesis is true, $E(\mathrm{MS}_a) = E(\mathrm{MS}_e)$.

11.17. In the FEM, if the null hypothesis is true, $E(\mathrm{MS}_a) = \sigma^2$ because $\sigma_A^2 = 0$.

11.18. When using the log transformation, it must be remembered that the log of a negative number is obtained by subtraction.

11.19. After a transformation is used, the group averages and variances for the transformed data should be plotted to see if the problem of dependence has been solved.

11.20. One does not need to be concerned about the assumption of equal variances when the data are transformed to ranks and a nonparametric procedure is used.

SELECTED READINGS

Bartlett, M. S. (1936). The square root transformation in the analysis of variance. *Journal of the Royal Statistical Society*, Supplement, Series B, **3**, 68–78.

Bartlett, M. S. (1937). Some examples of statistical methods of research in agriculture and applied biology. *Journal of the Royal Statistical Society*, **4**, 137–169.

Bartlett, M. S. (1947). The use of transformations. *Biometrics*, **3**, 39–52.

Box, G. E. P. (1953). Non-normality and tests on variances. *Biometrika*, **40**, 318–335.

Box, G. E. P. (1954). Some theorems on quadratic forms applied in the study of analysis of variance problems: I. Effect of inequality of variance in the one-way classification. *Annals of Mathematical Statistics*, **25**, 290–302.

Box, G. E. P., and D. R. Cox (1964). An analysis of transformations. *Journal of the Royal Statistical Society*, Series B, **26**, 211–243.

Chernoff, H., and G. L. Lieberman (1954). Use of normal probability paper. *Journal of the American Statistical Association*, **49**, 778–785.

Cochran, W. G. (1947). Some consequences when the assumptions for the analysis of variance are not satisfied. *Biometrics*, **3**, 22–38.

Conover, W. J., and R. L. Iman (1976). On some alternative procedures using ranks for the analysis of experimental design, *Communications in Statistics—Theory and Methods*, **A5**, 1349–1368.

Crump, S. L. (1946). The estimation of variance components in analysis of variance. *Biometrics*, **2**, 7–11.

Crump, S. L. (1951). The present status of variance component analysis. *Biometrics*, **7**, 1–15.

Dolby, J. L. (1963). A quick method for choosing a transformation. *Technometrics*, **5**, 317–325.

Draper, N. R., and D. R. Cox (1969). On distributions and their transformation to normality. *Journal of the Royal Statistical Society*, Series B, **31**, 472–476.

Eisenhart, C. (1947). The assumptions underlying the analysis of variance. *Biometrics*, **3**, 1–21.

Freeman, M. F., and J. W. Tukey (1950). Transformations related to the angular and the square root. *Annals of Mathematical Statistics*, **21**, 607–611.

Hartley, H. O. (1950). The maximum *F*-ratio as a short-cut test for heterogeneity of variance. *Biometrika*, **37**, 308–312.

Lord, F. M. (1969). Statistical adjustments when comparing preexisting groups. *Psychological Bulletin*, **72**, 336–337.

Mage, D. T. (1982). An objective graphical method for testing normal distributional assumptions using probability plots. *American Statistician*, **36**, 116–120.

Mosteller, F., and C. Youtz (1961). Tables of the Freeman–Tukey transformations for the binomial and Poisson distributions. *Biometrika*, **48**, 433–440.

Scheffé, H. (1959). *Analysis of Variance*. Wiley, New York.

Searle, S. R. (1971). Topics in variance component estimation. *Biometrics*, **27**, 1–76.

Shapiro, S. S., M. B. Wilk, and H. J. Chen (1988). A comparative study of variance tests for normality. *Journal of the American Statistical Association*, **63**, 1343–1372.

Welch, B. L. (1947). The generalization of "Student's" problem when several different population variances are involved. *Biometrika*, **34**, 28–35.

Wilk, M. B., and O. Kempthorne (1955). Fixed, mixed, and random models. *Journal of the American Statistical Association*, **50**, 1144–1167.

Winer, B. J. (1971). *Statistical Principles of Experimental Design*, 2nd ed. McGraw-Hill, New York.

12 Other Analysis-of-Variance Designs

The one-way analysis of variance described in Chapters 10 and 11 is only one of many designs for an experiment. Many experiments have a more complex design than the one-way completely randomized design. The investigator may be using replications or subsamples. There may be a need to control extraneous factors or there may be interest in more than one set of treatments. In this chapter, we illustrate several different designs. In each case we discuss when they should be used and how the analysis is carried out.

12.1. NESTED DESIGN

A *nested design* (or hierarchal design) is used for experiments in which there is interest in one set of treatments and the experimental units are measured more than once or are subsampled. For example, if 3 diets are being tested for their effect on blood cholesterol level and 4 volunteers are assigned at random to each diet (a total of 12 volunteers), the investigator might want to obtain 2 lab determinations of cholesterol level for each volunteer (24 determinations) because of variability in the measurement of this variable (Figure 12.1). In this example, there are repeated observations of the subjects.

If 4 dyes are being tested for colorfastness on cotton, each dye might be used on 2 bolts of material (a total of 8 bolts) and then 6 swatches of material from each bolt selected at random (48 swatches) for the test. In this example the experimental units (bolts) are subsampled.

Other examples of nested designs:

1. Three drugs are each used at 2 different clinics (a total of 6 clinics) and are given to 5 patients at each clinic.
2. Ten roosters are each mated to 5 different hens, and a random sample of 6 chicks from each hen is examined for a certain genetic characteristic.
3. Four fungicides are used on a certain type of tree. Each fungicide is applied to 3 trees, and 10 leaves are examined from each tree.
4. Each of 3 methods of teaching geometry is used by 2 teachers (6 teachers are in the experiment), and a random sample of 10 students of each teacher is tested.

The additive model for these nested designs is

$$y_{ijk} = \mu + \alpha_i + \beta_{ij} + \varepsilon_{ijk}$$

Statistics for Research, Third Edition, Edited by Shirley Dowdy, Stanley Weardon, and Daniel Chilk.
ISBN 0-471-26735-X © 2004 John Wiley & Sons, Inc.

FIGURE 12.1. A nested design.

with

$$i = 1, \ldots, a$$
$$j = 1, \ldots, b$$
$$k = 1, \ldots, n$$

The terms in this model have the following meanings:

μ: A constant, the mean for all experiments of this type.

α_i: A constant for the ith treatment group, the effect of the ith treatment. If the treatments are fixed effects, $\sum_i \alpha_i = 0$; if the treatments are random effects, α_i is IND(0, σ_A^2).

β_{ij}:   A random effect due to the ijth experimental unit; β_{ij} is IND(0, σ_B^2) for each i.

ε_{ijk}:   A random effect due to the ijkth observation. It contains all uncontrolled variability; ε_{ijk} is IND(0, σ^2).

In the examples given above, all of the treatments are fixed effects except the roosters in example 2. The ANOVA is computationally the same whether the treatments are fixed or random. We consider only cases in which the experimental units are random effects (if they are fixed, the F test is different).

The ANOVA for the nested design is an extension of the one-way design. The main hypothesis to be tested is H_0: $\alpha_1 = \alpha_2 = \cdots = \alpha_a = 0$ for the FEM and H_0: $\sigma_A^2 = 0$ for the REM. A secondary hypothesis can be tested to determine if there is variability among the experimental units, H_0: $\sigma_B^2 = 0$.

Subscripts ijk are used in the following manner. The first subscript i refers to the treatment group. The second subscript j refers to the jth experimental unit within a treatment group. The third subscript k refers to the kth subsample or replicate within an experimental unit.

In the diet example at the beginning of this section, the diets are the treatments, so $i = 1, 2, 3$. The volunteers are the experimental units, so $j = 1, 2, 3, 4$. The lab determinations are replications, so $k = 1, 2$. Thus y_{241} is the cholesterol level from the first determination for the fourth person on diet 2.

	Diet		
Volunteer	1	2	3
1	y_{111}	y_{211}	y_{311}
	y_{112}	y_{212}	y_{312}
	$T_{11.}$	$T_{21.}$	$T_{31.}$

	Diet		
Volunteer	1	2	3
2	y_{121} y_{122} $T_{12.}$	y_{221} y_{222} $T_{22.}$	y_{321} y_{322} $T_{32.}$
3	y_{131} y_{132} $T_{13.}$	y_{231} y_{232} $T_{23.}$	y_{331} y_{332} $T_{33.}$
4	y_{141} y_{142} $T_{14.}$	y_{241} y_{242} $T_{24.}$	y_{341} y_{342} $T_{34.}$
	$T_{1..}$	$T_{2..}$	$T_{3..}$ $T_{...} = \sum_i T_{i..}$

There are four types of totals:

y_{ijk} = the individual observations, a total of one observation
$T_{ij.}$ = the subsample of replicate totals
$T_{i..}$ = the treatment group totals
$T_{...}$ = the grand total

These four types of totals lead to four uncorrected sums of squares, as shown:

Uncorrected Sums of Squares

Sum of Squares	Formula	Symbol	Number of Totals	Observations/Total
Uncorrected total	$\sum_i \sum_j \sum_k y_{ijk}^2$	T	abn	1
Uncorrected treatment	$\sum_i (T_{i..}^2/bn)$	A	a	bn
Uncorrected experimental unit	$\sum_i \sum_j (T_{ij.}^2/n)$	B	ab	n
Correction factor	$T_{...}^2/abn$	CF	1	abn

The corrected sum of squares, as for the one-way ANOVA, are found by computational formulas in which the number of totals in the uncorrected sums of squares correspond to the degrees of freedom.

Corrected Sums of Squares

Sum of Squares	Symbol	df	Definition	Computional Formula
Total	SS_t	$abn - 1$	$\sum_i \sum_j \sum_k (y_{ijk} - \bar{y})^2$	$T - CF$
Among treatments	SS_a	$a - 1$	$bn \sum_i (\bar{y}_i - \bar{y})^2$	$A - CF$

Corrected Sums of Squares

Sum of Squares	Symbol	df	Definition	Computional Formula
Among units within treatments	SS_b	$a(b-1)$	$n\sum_i\sum_j(\bar{y}_{ij}-\bar{y}_i)^2$	$B-A$
Among samples (or replicates) within units	SS_e	$ab(n-1)$	$\sum_i\sum_j\sum_k(y_{ijk}-\bar{y}_{ij})^2$	$T-B$

In the definitions,

$$\bar{y} = T_{...}/abn \text{ is the overall experimental average}$$
$$\bar{y}_i = T_{i..}/bn \text{ is the } i\text{th treatment average}$$
$$\bar{y}_{ij} = T_{ij.}/n \text{ is the } ij\text{th experimental unit average}$$

Example 12.1.   Nested ANOVA

A taxicab company is going to choose among 5 types of cars for its fleet. The company has already determined that these 5 are comparable in initial cost and maintenance, and it wants to make a decision based on gas mileage in heavy city traffic. Ten cars are available for the experiment, 2 of each type. Each car is to be tested 3 times. Thus $a = 5$, $b = 2$, and $n = 3$.

	Car	A	B	C	D	E	
		15.8	18.5	12.3	19.5	16.0	
	1	15.6	18.0	13.0	17.5	15.7	
		16.0	18.4	12.7	19.1	16.1	
$T_{i1.}$		47.4	54.9	38.0	56.1	47.8	
		13.9	17.9	14.0	18.7	15.8	
	2	14.2	18.1	13.1	19.0	15.6	
		13.5	17.4	13.5	18.8	16.3	
$T_{i2.}$		41.6	53.4	40.6	56.5	47.7	Total
$T_{i..}$		89.0	108.3	78.6	112.6	95.5	484.0
$\sum_j\sum_k y_{ijk}^2$		1326.10	1955.59	1031.44	2115.44	1520.39	7948.96

Type of Car

Uncorrected SS

$$T = \sum_i\sum_j\sum_k y_{ijk}^2 = 7948.96 \quad B = \sum_i\sum_j T_{ij.}^2/n = 7944.95$$

$$A = \sum_i T_{i..}^2/bn = 7937.81 \quad CF = T_{...}^2/abn = 7808.53$$

Source	df	SS	MS
Among types	$a - 1 - = 4$	$SS_a = A - CF = 129.28$	$MS_a = SS_a/(a - 1) = 32.32$
Among cars within types	$a(b - 1) = 5$	$SS_b = B - A = 7.14$	$MS_b = SS_b/a(b - 1) = 1.43$
Among trials within cars	$ab(n - 1) = 20$	$SS_e = T - B = 4.01$	$MS_e = SS_e/ab(n - 1) = 0.20$
Total	$abn - 1 = 29$	$SS_t = T - CF$	

In this design,

$$MS_a \text{ estimates } \sigma^2 + n\sigma_B^2 + bn \sum_i \alpha_i^2/(a - 1)$$

$$MS_b \text{ estimates } \sigma^2 + n\sigma_B^2$$

$$MS_e \text{ estimates } \sigma^2$$

so the F tests take the following form:

Source	F	$F_{0.05}$	H_0
Among types	$MS_a/MS_b = 22.60$	5.192	$\alpha_1 = \alpha_2 = \cdots = \alpha_5 = 0$
Among cars within types	$MS_b/MS_e = 7.15$	2.711	$\sigma_B^2 = 0$

Thus, there is at least one significant difference among the average mileages for the types. A secondary conclusion is that there is significant variability among the different cars within types.

The term *expected mean square*, $E(MS)$, is used to indicate the parameter being estimated by the mean square. These expected values will differ for treatments that are fixed or random (they are fixed in the car example). However, in both cases MS_b estimates everything in $E(MS_a)$ except for the term that is being tested in the null hypothesis, so the main F test has the form MS_a/MS_b.

Expected Mean Squares		
Source	FEM (Treatments)	REM (Treatments)
Among treatments	$\sigma^2 + n\sigma_B^2 + bn \sum_i \alpha_i^2/(a - 1)$	$\sigma^2 + n\sigma_B^2 + bn\sigma_A^2$
Among units within treatments	$\sigma^2 + n\sigma_B^2$	$\sigma^2 + n\sigma_B^2$
Among trials within units	σ^2	σ^2

If desired, multiple comparisons can be done following ANOVA to find specific differences among the treatment means. Only one modification is necessary: The standard error of the difference of two means is $\sqrt{2MS_b/bn}$ instead of $\sqrt{2MS_e/n}$. Estimation of parameters or linear combinations of parameters can also be carried out, again substituting MS_b for MS_e. The degrees of freedom are $a(b - 1)$.

The procedure for ANOVA for a nested design is summarized as follows.

Procedure. Nested ANOVA for Equal Sample Sizes

Main Hypothesis
H_0: $\alpha_1 = \alpha_2 = \cdots = \alpha_a = 0$ or H_0: $\sigma_A^2 = 0$ against
H_a: At least one inequality or H_a: $\sigma_A^2 > 0$

Secondary Hypothesis
H_0: $\sigma_B^2 = 0$ against H_a: $\sigma_B^2 > 0$
Model:

$$y_{ijk} = \mu + \alpha_i + \beta_{ij} + \varepsilon_{ijk}$$
$$i = 1, \ldots, a$$
$$j = 1, \ldots, b$$
$$k = 1, \ldots, n$$

Compute:

$$T = \sum_i \sum_j \sum_k y_{ijk}^2$$

$$A = \sum_i T_{i..}^2 / bn$$

$$B = \sum_i \sum_j T_{ij.}^2 / n$$

$$CF = T_{...}^2 / abn$$

Source	df	SS	MS	F
Among treatments	$a - 1$	$SS_a = A - CF$	$MS_a = SS_a/(a-1)$	MS_a/MS_b
Among units within treatments	$a(b-1)$	$SS_b = B - A$	$MS_b = SS_b/a(b-1)$	MS_b/MS_e
Among trials with units	$ab(n-1)$	$SS_e = T - B$	$MS_e = SS_e/ab(n-1)$	

Reject the main H_0 if $F = MS_a/MS_b \geq F_{\alpha,\ a-1,a(b-1)}$. Reject the secondary hypothesis if $F = MS_b/MS_e \geq F_{\alpha,\ a(b-1),ab(n-1)}$.

It is possible to analyze a nested design with unequal sample sizes. Modifications are necessary in the uncorrected sums of squares and the degrees of freedom.

Many statistical packages will contain procedures for various types of ANOVA. In the SAS System, PROC ANOVA can be used to analyze data collected using a nested design. The following program will perform the analysis of the mileage data in Example 12.1:

```
DATA TAXIS;
    DO TYPE = 1 TO 5;
        DO CAR = 1 TO 2;
            DO REP = 1 TO 3; INPUT MILES @@; OUTPUT
                END;
            END;
        END;

CARDS;
15.8    15.6    16.0    13.9    14.2    13.5
18.5    18.0    18.4    17.9    18.1    17.4
12.3    13.0    12.7    14.0    13.1    13.5
19.5    17.5    19.1    18.7    19.0    18.8
16.0    15.7    16.1    15.8    15.6    16.3
;
PROC ANOVA;
    CLASS TYPE CAR REP;
    MODEL MILES=TYPE CAR (TYPE);
    TEST H = TYPE E=CAR (TYPE);
```

The data set created by this program contains four variables, TYPE, CAR, REP, and MILES with 30 observations. The variable TYPE has five values, 1, 2, 3, 4, and 5, corresponding respectively to types A, B, C, D, and E in the experiment. CAR has values 1 and 2 for the two cars of each type which were used. REP has values 1, 2, and 3 for the three repetitions on each car.

The SAS program uses the PROC ANOVA procedure to perform the analysis of variance. The CLASS statement identifies the variables which correspond to the treatments, the experimental units, and the repetitions—TYPE, CAR, and REP, respectively, in this example. The MODEL statement indicates that the variable of interest is MILES, that the variable TYPE will identify the treatment groups, and that CAR is nested within TYPE [indicated by the notation CAR (TYPE)].

The SAS System

The ANOVA Procedure

Class Level Information

Class	Levels	Values
TYPE	5	1 2 3 4 5
CAR	2	1 2
REP	3	1 2 3

Number of observations 30

The ANOVA Procedure

Dependent Variable: MILES

Source	DF	Sum of Squares	Mean Square	F Value	Pr > F
Model	9	136.4133333	15.1570370	75.53	<.0001
Error	20	4.0133333	0.2006667		
Corrected Total	29	140.4266667			

R-Square	Coeff Var	Root MSE	MILES Mean
0.971420	2.776601	0.447958	16.13333

Source	DF	Anova SS	Mean Square	F Value	Pr > F
TYPE	4	129.2766667	32.3191667	161.06	<.0001
CAR(TYPE)	5	7.1366667	1.4273333	7.11	0.0006

Tests of Hypotheses Using the Anova MS for CAR(TYPE) as an
Error Term

Source	DF	Anova SS	Mean Square	F Value	Pr > F
TYPE	4	129.2766667	32.3191667	22.64	0.0021

EXERCISES

12.1.1. Ring-necked pheasants establish breeding colonies, each consisting of one male (cock), several hens per cock, and several chicks per hen. If adult males and females can be identified by wing band, a wildlife biologist can locate the nests of female pheasants in a hunting reserve, and he can collect eggs through random sampling in such a manner that they will represent the breeding colonies of 5 cocks, 3 hens per cock, and 2 eggs per hen. The eggs will be marked and incubated, and chicks are weighed at 28 days of age.

a. Given that the linear model for this study is

$$y_{ijk} = \mu + \alpha_i + \beta_{ij} + \varepsilon_{ijk}$$

  i. What does α_i represent? Is it a fixed or a random effect?

  ii. What does β_{ij} represent? Is it a fixed or a random effect?

b. Given the computations

$$\sum_i T_{i..}^2/6 = 918.0 \qquad \sum_i \sum_j T_{ij.}^2/2 = 1833.0$$

$$\sum_i \sum_j \sum_k y_{ijk}^2 - \sum_i \sum_j T_{ij.}^2/n = 7.5 \qquad T_{...}^2/30 = 900$$

complete the ANOVA and test for significance of variability due to males.

12.1.2. Soda crackers lose their crispness in damp climates unless they are packaged in containers that protect them from humidity. A bakery firm wishes to compare 5 methods of packaging (including a cardboard box control). Four boxes are selected at random from each method of packaging, assigned numbers, and placed in a chamber in which the humidity is maintained at 80% for 24 hours. The boxes are opened and 3 crackers are selected from each box at random to be measured for moisture content. The measurements on the 60 crackers are below given in milligrams:

	Control Box				Wax Paper Box		
11	12	13	14	21	22	23	24
73	81	70	67	60	64	63	53
75	77	62	69	61	67	59	50
77	73	63	62	65	61	55	56
225	231	195	198	186	192	177	159

	Metal Foil Box				Plastic Box		
31	32	33	34	41	42	43	44
46	49	54	59	60	49	39	52
49	54	60	53	66	43	40	55
46	56	57	53	60	52	44	49
141	159	171	165	186	144	123	156

	Metal Foil and Plastic Box		
51	52	53	54
38	45	60	50
36	46	55	47
40	50	53	44
114	141	168	141

a. Give the linear model and the assumptions.

b. State the null hypothesis of greatest concern.

c. Given that $\sum_i \sum_j \sum_k y_{ijk}^2 = 195{,}988$, perform the ANOVA.

d. Are there significant differences among the methods of packaging?

e. Which method of packaging do you recommend?

f. Is there significant variability among boxes receiving the same method of packaging?

12.1.3. In the taxicab study in this section, Example 12.1, use Fisher's least significant difference to locate the pairs of means that are different. Which type or types would you recommend?

12.1.4. In the taxicab study of this section, Example 12.1, estimate μ, $\alpha_4 - \alpha_5$, and $\mu_4 - (\mu_1 + \mu_2 + \mu_3)/3$ with 95% confidence intervals.

12.1.5. Prior to reforestation projects, provenance studies are performed in an effort to find the best source of seeds to be used in reforestation. In such a study, a forester selects forests at a different locations as possible sources of seeds. In each forest, b seed-bearing trees are selected at random, and enough seeds are selected at random from each tree to produce n seedlings for planting. The seeds are germinated in a greenhouse and the resulting seedlings planted in a completely random design at the reforestation site. Suppose in such a study the SAS analysis below is that of the first year's growth of the seedlings:

Source	DF	Sum of Squares	Mean Square	F Value	Pr > F
Model	41	107.748	2.628	4.171	<.0001
Error	168	105.840	0.630		
Corrected Total	209	213.588			

R-Square	Coeff Var	Root MSE	GROWTH Mean
0.504467	0.46	0.7937	1.715

Source	DF	Anova SS
FORESTS	5	32.333
TREES(FORESTS)	36	75.415

Use the computer output to answer the following questions:

a. Assuming this is a balanced experiment, give the numerical values for the number of forests (a) sampled, the number of trees (b) per forest, and the number of seedlings (n) used from each tree.

b. What percentage of the sums of squares among the 210 seedlings can be attributed to differences among forests or differences among trees within forests?

c. Show how to compute the value $F = 3.087$, which tests for differences among forests.

d. Give the numerical value for $\sum_i \sum_j \sum_k (y_{ijk} - \bar{y})^2$.

12.2. RANDOMIZED COMPLETE BLOCK DESIGN

An experimenter uses a *randomized complete block* design if he is interested in one set of treatments and wants to control an extraneous source of variability. For example, a physiologist studying the effect of 4 different drugs A, B, C, and D on mice might feel that the responses will be influenced by the particular litter from which the mice came. He would not want this litter effect to interfere with the analysis of the drug effect. To remove this nuisance variability, he can use litters as *blocks*, an extension of matched pairs. He chooses 4 mice at random from each litter, and each drug is assigned at random to 1 mouse from each litter (Figure 12.2). The design is called complete because each treatment appears in each block exactly once.

FIGURE 12.2. Four treatments assigned at random within three blocks.

Other examples of a randomized complete block design:

1. Four varieties of corn are each planted on sections of 5 different farms (the farms are chosen at random and the sections assigned at random), and yields are measured. The farms are the blocks. This design makes it possible to remove any differences in yield due to differences in fertilities.
2. Five dyes are each applied to portions of 8 random strips of cloth from a bolt (the strips are chosen at random and the portions assigned at random to the dyes), and the dyes are tested for permanence. The strips are the blocks. This design makes it possible to remove any differences due to variability of the cloth.
3. Three social studies textbooks are used in 3 classes at each of 4 different schools (the assignment of textbook to a class is random), and average class performance is measured. Schools are the blocks.
4. Four formulas for sun protection are tested on the skin of 5 subjects. Each formula is applied to different randomly chosen portions of skin of each subject. The subjects are the blocks.
5. Six different bacteria to be treated with a drug are cultured in a medium which is prepared in 4 batches. Each type of bacterium is cultured once in a portion of each batch of medium. The batches are the blocks.

In all of these examples the investigator is primarily interested in the treatment effects (varieties of corn, dyes, textbooks, formulas, bacteria), and the blocking is done to avoid extraneous variability (from different fertilities on the farms, from differences in the cloth in different parts of the bolt, from differences in schools, from differences in skin types, from differences in batches of medium). If this extraneous variability is not removed, it will show up in the MS_e, making it difficult to detect treatment differences.

The additive model for a randomized complete block design is

$$y_{ij} = \mu + \alpha_i + \beta_j + \varepsilon_{ij}$$
$$i = 1, \ldots, a$$
$$j = 1, \ldots, b$$

in which the terms have the following meanings:

μ: A constant, the overall mean of experiments of this type.

α_i: A constant for the ith treatment group, the deviation from the mean due to the ith treatment; $\sum_i \alpha_i = 0$ if the treatments are fixed effects or α_i IND$(0, \sigma_A^2)$ if the treatments are random.

β_j: A constant for the jth block, the deviation from the mean caused by the jth block; $\sum_j \beta_j = 0$ if the blocks are fixed effects or β_j IND$(0, \sigma_B^2)$ if they are random.

ε_{ij}:   A random deviation associated with the ijth observation, containing all uncontrolled sources of variability; ε_{ij} IND$(0, \sigma^2)$.

Data for a randomized complete block design are arranged as follows, in which i designates the treatment and j the blocks:

Treatment (*i*)

		1	2	3	4	Totals
	1	y_{11}	y_{21}	y_{31}	y_{41}	$T_{.1}$
Block (*j*)	2	y_{12}	y_{22}	y_{32}	y_{42}	$T_{.2}$
	3	y_{13}	y_{23}	y_{33}	y_{43}	$T_{.3}$
Totals		$T_{1.}$	$T_{2.}$	$T_{3.}$	$T_{4.}$	$T_{..} = $ Grand Total

Sometimes rows and columns are interchanged for convenience of presentation, but we will continue to use *i* for the treatment and *j* for the blocks even in that case. Treatment group totals are represented by $T_{i.}$, indicating that the summation was over *j*. Block totals are $T_{.j}$ and the grand total is $T_{..}$. The corresponding averages are $\bar{y}_{i.}$, $\bar{y}_{.j}$, and $\bar{y}_{..}$.

    The uncorrected sums of squares, corrected sums of squares, and ANOVA procedure are as follows. In a block design the error sum of squares is sometimes called the *residual sum of squares*.

Uncorrected Sums of Squares

Sum of Squares	Formula	Symbol	Number of Totals	Observations/Total
Uncorrected total	$\sum_i \sum_j y_{ij}^2$	T	ab	1
Uncorrected treatment	$\sum_i T_{i.}^2/b$	A	a	b
Uncorrected block	$\sum_j T_{.j}^2/a$	B	b	a
Residual	$T_{..}^2/ab$	CF	1	ab

Corrected Sums of Squares

Sum of Squares	df	Symbol	Definition	Computational Formula
Total	$ab - 1$	SS_t	$\sum_i \sum_j (y_{ij} - \bar{y}_{..})^2$	$T - CF$
Treatment	$a - 1$	SS_a	$b\sum_i (\bar{y}_{i.} - \bar{y}_{..})^2$	$A - CF$
Block	$b - 1$	SS_b	$a\sum_j (\bar{y}_{.j} - \bar{y}_{..})^2$	$B - CF$
Residual	$(a - 1)(b - 1)$	SS_e	$\sum_i \sum_j (y_{ij} - \bar{y}_{i.} - \bar{y}_{.j} + \bar{y}_{..})^2$	$T - A - B + CF$

As in the one-way design, the short computational formulas correspond to the degrees of freedom. For example, the residual degrees of freedom are $(a - 1)(b - 1) = ab - a - b + 1$, and the terms $T - A - B + CF$ contain ab, a, b, and 1 total, respectively.

Procedure. Randomized Complete Block ANOVA

Main Hypothesis

$H_0: \alpha_1 = \alpha_2 = \cdots = \alpha_a = 0$ or $H_0: \sigma_A^2 = 0$ against

$H_a:$ *At least one inequality*   or $H_a: \sigma_A^2 > 0$

Model:

$$y_{ij} = \mu + \alpha_i + \beta_j + \varepsilon_{ij}$$
$$i = 1, \ldots, a$$
$$j = 1, \ldots, b$$

Compute

$$T = \sum_i \sum_j y_{ij}^2 \qquad B = \sum_j T_{.j}^2 / a$$

$$A = \sum_i T_{i.}^2 / b \qquad CF = T_{..}^2 / ab$$

Source	df	SS	MS	F
Among treatments	$a - 1$	$SS_a = A - CF$	$MS_a = SS_a/(a - 1)$	MS_a/MS_e
Among blocks	$b - 1$	$SS_b = B - CF$	$MS_b = SS_b/(b - 1)$	MS_b/MS_e
Residual	$(a - 1)(b - 1)$	$SS_e = T - A - B + CF$	$MS_e = SS_e/(a - 1)(b - 1)$	
Total	$ab - 1$	$SS_t = T - CF$		

It is also possible to test for a block difference, $H_0: \beta_1 = \beta_2 = \cdots = \beta_b = 0$ or $H_0: \sigma_B^2 = 0$. These hypotheses are tested by $F = MS_b/MS_e$ with the corresponding degrees of freedom. The form of the F test can be determined in each case by the expected mean squares. The denominator of the F test must estimate everything except the term being tested.

Expected Mean Squares for Randomized
Complete Block Design

MS	$E(MS)$	
	Fixed	Random
MS_a	$\sigma^2 + b \sum_i \alpha_i^2/(a - 1)$	$\sigma^2 + b\sigma_A^2$
MS_b	$\sigma^2 + a \sum_j \beta_j^2/(b - 1)$	$\sigma^2 + a\sigma_B^2$
MS_e	σ^2	σ^2

Example 12.2.   Randomized Complete Block ANOVA

A psychology experiment involving 3 treatments is planned with a randomized complete block design, the random subjects being the blocks. The 3 treatments are administered on 3 difference days and the order in which each subject receives the treatment is random. There are 4 subjects and the random variable is the length of time required to complete a certain task.

		Treatment			
		1	2	3	$T_{.j}$
	1	4.7	9.4	6.3	20.4
Subject	2	3.5	7.6	5.1	16.2
	3	0.1	5.3	1.8	7.2
	4	1.6	6.2	3.6	11.4
	$T_{i.}$	9.9	28.5	16.8	$55.2 = T_{..}$

$$a = 3$$
$$b = 4$$
$$\sum_i \sum_j y_{ij}^2 = 331.46$$

$$T = 331.460 \qquad B = 286.800$$
$$A = 298.125 \qquad CF = 253.920$$

Source	df	SS	MS	F	$F_{0.05}$
Among treatments	$a - 1 = 2$	$A - CF = 44.205$	22.102	290.8	5.143
Among blocks	$b - 1 = 3$	$B - CF = 32.880$	10.960	144.2	4.757
Residual	$(a - 1)(b - 1) = 6$	$T - A - B + CF = 0.455$	0.076		

Since the F statistic for treatment is significant, there is evidence of differences among the treatments.

Although the psychologist in the example above is not interested in block differences for their own sake, the fact that the F for blocks is significant shows that this design is appropriate for the experiment. The decision to use a block design must come before the experiment. The experimenter knows from previous experience that an extraneous source of variability is present and designs the experiment so that this effect can be removed and the statistical procedure can be more powerful.

It is not always advantageous to use a block design instead of a completely random design. When a block design is appropriate, along with the reduction of the error sum of squares there is also a reduction in the associated degrees of freedom, but the F value is still larger. However, if blocking is used when there is really no block effect, the reduction in the error sum of squares will not be sufficient to offset the reduction in power due to the loss in degrees of freedom in the denominator.

The power of the randomized complete block design will also be reduced if the treatment effect and block effect are not simply additive, as implied in the model:

$$y_{ij} = \mu + \alpha_i + \beta_j + \varepsilon_{ij}$$

Additivity is not present if there is an interaction between treatments and blocks. An interaction is an additional boost or reduction due to the particular combination of a block and treatment. For example, in the psychologist's experiment, subject 1 may be much faster than the average person under treatment 1 but much slower than average under treatment 2, whereas subject 2 may be just the opposite. An absence of interactions means that although there are different reaction times for individuals, the general pattern is the same. If an interaction effect is present, there is no specific term for it in the block design. Since the variability due to the interaction will be in the total sum of squares and will not be removed by the treatment sum of squares or the block sum of squares, it will be left in the error sum of squares:

$$SS_t - SS_a - SS_b = SS_e$$

Thus the error sum of squares may contain not only variability due to sampling but also variability due to the interaction effect. (This is the reason for calling SS_e the *residual* sum of squares.) If an interaction is present, the power of the test is reduced because of the inflated SS_e, which contributes to the denominator of the F statistic. If interactions are suspected, the randomized complete block design should not be used. The two-factor model described in Section 12.4 makes specific provision for an interaction effect.

A randomized complete block design with fixed treatment effects can be followed by multiple comparisons, one-degree-of-freedom F tests, or estimation of the fixed effects. The MS_e is used in the standard error, and n must be replaced by a or b, whichever is appropriate in the formulas given in Chapter 10. Intraclass correlations can be computed for the random effects. The total variance is $\sigma^2 + \sigma_A^2 + \sigma_B^2$. Example 12.3 shows how this is done.

Example 12.3. Intraclass Correlation in a Two-Way ANOVA

Following a shoulder injury, even after corrective surgery, patients must undergo physical therapy to regain use of the injured member. One sign of success of the therapy is how well patients can elevate the arm that was injured, so this may be one of the first measurements a physical therapist makes when a patient returns for treatment. There are gauges to measure how many degrees above horizontal the patient can elevate his or her arm, but there is still a certain amount of subjectivity in how the therapist reads a gauge. Thus it is possible that one therapist will make measurements that consistently tend to be high and another consistently low. This could create a problem in evaluating patients' progress if a patient does not have the same therapist at every visit for therapy.

The chief therapist at a medical center wants to see if there is significant variability among the center's many therapists in the way they read the gauge. This would reduce the reliability of measures taken by different therapists. She takes a random sample of $a = 5$ of the therapists, explains the problem to the patients, and asks if they will volunteer to participate in an experiment to provide data. Nearly all do, so she takes a second sample of $b = 6$ patients, and each therapist measures the arm elevation of each of the patients in random order. The $ab = 30$ measurements are

		Therapist				
Patient	A	B	C	D	E	$T_{.j}$
1	69	58	60	61	52	300
2	85	77	78	84	72	396
3	81	71	74	81	73	380
4	48	42	43	51	41	225
5	59	46	51	52	44	252
6	60	51	54	61	54	280
$T_{i.}$	402	345	360	390	336	$1833 = T_{..}$

Degrees of freedom and sums of squares are computed in the same way as for all two-way designs, and the ANOVA, F tests, and expectations of mean squares are

Source	df	SS	MS	F	P value	E(MS)
Therapists	4	541.2	135.30	27.95	<0.0001	$\sigma^2 + 6\sigma_A^2$
Patients	5	4752.7	950.54	196.39	<0.0001	$\sigma^2 + 5\sigma_B^2$
Residual	20	96.8	4.84			σ^2

She wants to know if the variance among therapists is significant, so the hypothesis of is H_0: $\sigma_A^2 = 0$, and that hypothesis is rejected with a P value <0.0001. In addition to this test of hypothesis, however, she is also interested in the size of r_I, the intraclass correlation (ICC), to know the reliability of different measures on the same patient when the measures are taken by different therapists. To compute the ICC, she must first estimate the three variances associated with measurements:

$$\hat{\sigma}^2 = \text{Residual MS} = 4.84$$

$$\hat{\sigma}_A^2 = \frac{\text{MS}_A - \text{Residual MS}}{b} = \frac{27.95 - 4.84}{6} = 21.74$$

$$\hat{\sigma}_B^2 = \frac{\text{MS}_B - \text{Residual MS}}{a} = \frac{196.39 - 4.84}{5} = 189.14$$

With these estimates she can compute the intraclass correlation for her experiment:

$$r_I = \frac{\hat{\sigma}_B^2}{\hat{\sigma}_B^2 + \hat{\sigma}_A^2 + \hat{\sigma}} = \frac{189.14}{189.14 + 21.74 + 4.84} = 0.877$$

Because 87.7% of the total variance is among patients and the remaining 12.3% is attributable to differences among therapists or unexplained causes, the reliability of a measurement is reasonably good irrespective of other factors, including the therapist who made the measurement. However, she must decide whether that is sufficient reliability. With training and experience the therapists can learn to make measures that are independent but still more similar than those in this experiment. That would lessen the size of the estimate of σ_A^2 and thereby increase the size of the intraclass correlation.

Sometimes, in carrying out a blocked experiment, an observation is missing for reasons extraneous to the experiment. For example, a plant dies because of an accident in the greenhouse, a subject leaves town or is ill and cannot complete the experiment (assuming the illness is not related to the treatment), or the data are lost or erased. One way to handle this situation is to remove the entire block that contains the missing value. The analysis is then carried out with $b - 1$ blocks.

Another approach is to estimate the missing value y_{ij} by

$$\hat{y}_{ij} = \frac{aT_{i.} + bT_{.j} - T_{..}}{(a-1)(b-1)}$$

and to decrease the residual degrees of freedom by 1.

For example, in the psychology example in this section (Example 12.2), if y_{23} were missing, it could be estimated as follows:

Treatment

		1	2	3	$T_{.j}$
	1	4.7	9.4	6.3	20.4
Subject	2	3.5	7.6	5.1	16.2
	3	0.1	—	1.8	1.9
	4	1.6	6.2	3.6	11.4
	$T_{i.}$	9.9	23.2	16.8	49.9 = $T_{..}$

$$\hat{y}_{23} = \frac{3(23.2) + 4(1.9) - 49.9}{(3-1)(4-1)} = 4.55$$

The residual degrees of freedom would be 5.

If there are several missing values, an iterative procedure may be used. For example, if there are three missing values a, b, and c, we guess values for b and c and then approximate a as above. Using the approximation of a and the original guess of c, b is approximated as above. Finally, c is approximated using the approximated values of a and b. The cycle is then repeated to obtain second approximations of each of the three values. Repetition of the cycle continues until there are no noticeable changes in the approximations. The total degrees of freedom and residual degrees of freedom are reduced by 1 for each missing value. For further details, see Cochran and Cox (1957).

EXERCISES

12.2.1. Four varieties of hybrid corn have been developed for resistance to the fungal infection known as smut. However, nothing is known about their potential for grain yield. Each hybrid is planted at each of 5 locations within the state, and the following yields are obtained.

			Location		
Hybrid	NW	NE	C	SE	SW
FR-11	62.3	64.0	64.3	65.0	66.4
BCM	63.3	62.7	66.2	66.8	64.5
DBC	60.8	64.3	65.2	62.2	65.1
RC-3	55.4	56.0	59.8	58.0	58.8

  a. Give the linear model and the assumptions.
  b. Perform the appropriate ANOVA.
  c. Are there differences in yield among the means of the hybrids?
  d. Are there differences that can be attributed to location?
  e. If a smut-resistant hybrid is used, which do you recommend?

12.2.2. In a study of reaction time under the influence of alcohol, age is thought to be another variable that could affect the time. A randomized complete block design is used, and reaction time is measured in seconds.

		Amount of Alcohol			
		None	1 oz	2 oz	$T_{.j}$
	20–39	0.42	0.47	0.65	1.54
Age	40–59	0.51	0.62	0.66	1.79
	60 or over	0.57	0.73	0.79	2.09
	$T_{i.}$	1.50	1.82	2.10	$5.42 = T_{..}$

$$\sum_i \sum_j y_{ij}^2 = 3.3818$$

  a. Complete the ANOVA table.
  b. Is there any difference in reaction time among the alcohol groups?
  c. Use the Student–Newman–Keuls' procedure to compare the alcohol means.
  d. Is there a significant difference in reaction time due to age?

12.2.3. A large company is going to buy cars to be used by employees on business trips. Five models of cars are tested for mileage per gallon in 5 different randomly chosen cities. Five cars of each model are used and assigned to the cities in random order:

City

Model	1	2	3	4	5	Totals
A	15.83	17.56	21.11	20.48	26.04	101.02
B	14.80	16.22	21.30	20.84	19.27	92.43
C	17.43	19.54	17.67	22.58	19.86	97.08
D	16.60	16.34	17.01	15.82	16.57	82.34
E	21.24	21.29	20.34	19.43	25.05	107.35
Totals	85.90	90.95	97.43	99.15	106.79	480.22

a. What is the ANOVA model for this investigation?

b. Is the model effect random or fixed?

c. Is the city effect random or fixed?

d. What is the hypothesis of main interest to the investigator?

e. Complete the ANOVA.

f. Are there any differences in mileage among the models?

g. Which mean separation procedure seems appropriate for this investigation? Why?

h. Use Fisher's least significant difference to find the best model or models.

i. Is there significant variability due to cities?

j. What percentage of the total variability is due to the cities?

12.2.4. An experiment was conducted involving 6 schools and 3 teaching methods per school.

a. Identify the sources of variability represented by the sums of squares.

Source	Number of Squared Values	Observations/Squared Value	Numerical Value
_____	1	18	125
_____	3	6	151
_____	18	1	236
_____	6	3	180

b. Complete the uncorrected sum of squares table and the ANOVA table.

c. Could Fisher's least significant difference be used to test for differences among teaching methods? Justify your answer.

12.2.5. Given the following ANOVA:

Source	df	SS	MS
Treatment	3	150.0	50.0
Block	4	56.0	14.0
Residual	12	86.4	7.2

a. What are the values of a and b?

b. What is the numerical value of the standard error of a treatment average?

c. Use Duncan's procedure to compare the treatment means.

Treatment:	1	2	3	4
$\bar{y}_{i.}$:	6	9	12	13

12.2.6. a. Estimate the missing value in the block design.

b. Complete the ANOVA.

Blocks

Treatments			
	3	4	5
	1	2	2
	3	—	7
	5	8	2
	4	6	5

12.3. LATIN SQUARE DESIGN

Sometimes the investigator is aware of two causes of nuisance variability, and a blocked design is not adequate for the experiment. For example, in addition to a litter effect in a drug experiment on mice, there may also be a size-of-mouse effect. If there are no interactions present, and the experimenter is working with 4 drugs (A, B, C, D), 4 litters, and 4 sizes of mice, then a Latin square design may be used (Figure 12.3).

In a Latin square, each treatment appears exactly once in each row and column. This is a very economical design because it avoids the necessity of working with every combination possible. For example, in the mouse experiment, if all combinations of drug, litter, and size of

FIGURE 12.3. A Latin square design.

mouse were used, 64 mice would be needed. In addition, litters of the proper number and with the needed assortment of sizes probably would not exist.

The smallest Latin square that can be analyzed is 3×3. Squares larger than 9×9 are rarely used because of the difficulty of finding equal numbers of categories for the rows, columns, and treatments.

Standard Latin squares can be found in Fisher and Yates (1963). If more than one is available, the standard square should be selected by a random process, and the rows and columns should be randomized. For example, if

A	B	C
B	C	A
C	A	B

is the standard Latin square, two random sequences of the digits 1, 2, 3 are chosen, say (2, 1, 3) and (3, 1, 2). Then the columns are rearranged by the first sequence and the rows by the second (Figure 12.4).

Latin squares were originally used for agricultural experiments. Treatments were applied to a field in a Latin square design in order to randomize for any differences in fertility in different sections of the field. However, the design is very useful in other disciplines, and it is not necessary that the treatments be applied physically in a Latin square design. The mouse experiment which controls for litter and size is a typical nonagricultural application.

Other examples of a Latin square design are the following:

1. Yield is measured for 4 varieties of wheat that were planted on 4 different farms and in 4 different corners of the farms, NE, NW, SE, and SW.
2. Miles per gallon are measured on 6 models of cars using 6 brands of gasoline, each model used in 6 different cities.
3. The strength of coated paper is measured for 4 different coatings applied at 4 positions down the roll and 4 positions across the roll to control for variability in the strength of the uncoated paper.
4. A psychological experiment consists of 6 treatments given to 6 subjects in 6 different orders to control for learning.

FIGURE 12.4. Randomizing columns and rows.

5. Drug response is measured for 3 drugs given in 3 dosages and analyzed by 3 different lab technicians.
6. Time of assembly is measured for 4 products, 4 assemblers, and 4 positions in the assembly line.

The additive model for the Latin square design is

$$y_{ijk} = \mu + \alpha_i + \beta_j + \gamma_k + \varepsilon_{ijk}$$
$$i = 1, \ldots, a$$
$$j = 1, \ldots, a$$
$$k = 1, \ldots, a$$

in which the terms have the following meanings:

μ: A constant, the overall mean for all experiments of this type.

α_i: A constant for the ith treatment; $\sum_i \alpha_i = 0$ if this effect is fixed or α_i IND$(0, \sigma_A^2)$ if it is random.

β_j: A constant for the jth first extraneous effect; $\sum_j \beta_j = 0$ if this effect is fixed or β_j IND$(0, \sigma_B^2)$ if it is random.

γ_k: A constant for the kth second extraneous effect; $\sum_k \gamma_k = 0$ if this effect is fixed or γ_k IND$(0, \sigma_C^2)$ if it is random.

ε_{ijk}:   A random effect due to sampling; ε_{ijk} IND$(0, \sigma^2)$.

To use this model, we must be able to assume that there are no interactions between the α_i's and β_j's, α_i's and γ_k's, and β_j's and γ_k's.

Data for a Latin square design are arranged as in Figure 12.5, with the indicated notation. Treatments are indicated in parentheses within the cells. It does not matter which effect is placed in the rows, in the columns, or across the face of the table or which symbol, α_i, β_j, or γ_k, is assigned to a particular effect. The arrangement in Figure 12.5 is traditional because of the agricultural origins of this design, but other arrangements are common.

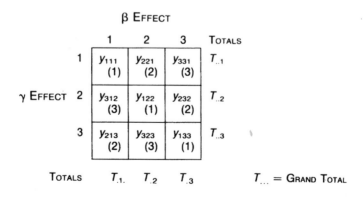

FIGURE 12.5. Notation for the Latin square design.

Averages are indicated by a notation corresponding to the totals, for example, $\bar{y}_{.2.} = T_{.2.}/a$ and $\bar{y}_{...} = T_{...}/a^2$.

Uncorrected Sums of Squares

Sum of Squares	Symbol	Formula	Number of Totals	Observations/ Total
Uncorrected total	T	$\sum_j \sum_k y_{ijk}^2$	a^2	1
Uncorrected treatment	A	$\sum_i T_{i..}^2/a$	a	a
Uncorrected β effect	B	$\sum_j T_{.j.}^2/a$	a	a
Uncorrected γ effect	C	$\sum_k T_{..k}^2/a$	a	a
Correction factor	CF	$T_{...}^2/a^2$	1	a^2

Corrected Sums of Squares

Source	df	Symbol	Definition	Computational Formula
Total	$a^2 - 1$	SS_t	$\sum_i \sum_j (y_{ijk} - \bar{y}_{...})^2$	$T - CF$
Treatment	$a - 1$	SS_a	$a \sum_i (\bar{y}_{i..} - \bar{y}_{...})^2$	$A - CF$
β effect	$a - 1$	SS_b	$a \sum_j (\bar{y}_{.j.} - \bar{y}_{...})^2$	$B - CF$
γ effect	$a - 1$	SS_c	$a \sum_k (\bar{y}_{..k} - \bar{y}_{...})^2$	$C - CF$
Residual	$(a-1)(a-2)$	SS_e	$\sum_i \sum_j \sum_k (y_{ijk} - \bar{y}_{i..}$ $-\bar{y}_{.j.} - \bar{y}_{..k} + 2\bar{y}_{...})^2$	$T - A - B - C + 2CF$

Note that in the definition of SS_e not all the combinations of ijk exist. The missing terms can be thought of as having zero value.

Procedure. Latin Square ANOVA

Main Hypothesis
H_0: $\alpha_1 = \cdots = \alpha_a = 0$ or H_0: $\sigma_A^2 = 0$

Secondary Hypotheses

H_0: $\beta_i = \cdots = \beta_a = 0$ or H_0: $\sigma_B^2 = 0$
H_0: $\gamma_1 = \cdots = \gamma_a = 0$ or H_0: $\sigma_C^2 = 0$

Model:

$$y_{ijk} = \mu + \alpha_i + \beta_j + \gamma_k + \varepsilon_{ijk}$$
$$i = 1, \ldots, a$$
$$j = 1, \ldots, a$$
$$k = 1, \ldots, a$$

Compute:

$$T = \sum_j \sum_k y_{ijk}^2 \qquad C = \sum_k T_{k..}^2 / a$$

$$A = \sum_i T_{i..}^2 / a \qquad CF = T_{...}^2 / a^2$$

$$B = \sum_j T_{.j.}^2 / a$$

Source	df	SS	MS	F
Among treatments	$a - 1$	$SS_a = A - CF$	$MS_a = SS_a/(a - 1)$	MS_a/MS_e
Among β effects	$a - 1$	$SS_b = B - CF$	$MS_b = SS_b/(a - 1)$	MS_b/MS_e
Among γ effects	$a - 1$	$SS_c = C - CF$	$MS_c = SS_c/(a - 1)$	MS_c/MS_e
Residual	$(a - 1)(a - 2)$	$SS_e = T - A - B - C + 2CF$	$MS_e = SS_e/(a - 1)(a - 2)$	
Total	$a^2 - 1$	$SS_t = T - CF$		

The F tests take the form given above because of the expectations of the mean squares:

Expected Mean Squares

MS	$E(MS)$ Fixed	Random
MS_a	$\sigma^2 + a \sum_i \alpha_i^2/(a - 1)$	$\sigma^2 + a\sigma_A^2$
MS_b	$\sigma^2 + a \sum_j \beta_j^2/(a - 1)$	$\sigma^2 + a\sigma_B^2$
MS_c	$\sigma^2 + a \sum_k \gamma_k^2/(a - 1)$	$\sigma^2 + a\sigma_C^2$
MS_e	σ^2	σ^2

Example 12.4. Latin Square ANOVA

An audiologist is studying 3 difference devices that help hearing in a certain type of deficiency. Three subjects with this type of hearing loss take hearing tests using each of the 3 devices. To control for learning, a Latin square design is used. Scores on the test are recorded. Devices are given in parentheses.

Order of Test (γ_k)

		First	Second	Third	
	1	74 (1)	57 (2)	50 (3)	$T_{.1.} = 181$
Subject (β_j)	2	6 (3)	94 (1)	78 (2)	$T_{.2.} = 178$
	3	40 (2)	29 (3)	112 (1)	$T_{.3.} = 181$
		$T_{..1} = 120$	$T_{..2} = 180$	$T_{..3} = 240$	$T_{...} = 540$

Device totals: $T_{1..} = 280$ $T_{2..} = 175$ $T_{3..} = 85$
The uncorrected sums of squares are:

$$T = 41{,}166 \quad B = 32{,}402 \quad CF = 32{,}400$$
$$A = 38{,}750 \quad C = 34{,}800$$

Source	df	SS	MS	F	H_0
Among devices	2	6350	3175	453.6	$\alpha_1 = \alpha_2 = \alpha_3 = 0$
Among subjects	2	2	1	0.1	$\sigma_B^2 = 0$
Among orders	2	2400	1200	171.4	$\gamma_1 = \gamma_2 = \gamma_3 = 0$
Residual	2	14	7		

Since $F_{0.01,2,2} = 99.000$, the audiologist concludes that there is a significant difference among the devices and there is a significant learning effect at the 0.01 level.

EXERCISES

12.3.1. A marketing expert for a publishing house wants to measure reader preference for 5 different covers of the same paperback novel. Five newsstands are selected at random and the novel is displayed at each newsstand for 5 weeks, one for each cover. One week is sufficient to determine sales potential because a new cover makes its impact immediately, followed by a pattern of diminishing returns. The number of sales are listed below with the cover given in parentheses:

Newsstand	Week				
	1	2	3	4	5
I	(D) 200	(C) 290	(A) 280	(E) 230	(B) 265
II	(C) 260	(B) 280	(E) 245	(A) 285	(D) 245
III	(A) 250	(D) 245	(C) 280	(B) 250	(E) 180
IV	(B) 260	(E) 190	(D) 230	(C) 205	(A) 200
V	(E) 340	(A) 335	(B) 265	(D) 270	(C) 230

a. Give the most logical null hypothesis with respect to covers.

b. Perform the ANOVA.

c. What should be concluded about covers?

d. Comment on the usefulness of the design employed.

e. Make simultaneous interval estimates of the means of the 5 covers using Bonferroni procedures when $\alpha_G = 0.05$.

12.3.2. A test is done on the miles per gallon for 5 models of cars using 5 brands of gasoline and tested in 5 different cities.

		Brand of Gasoline					
		A	B	C	D	E	Totals
Model	I	30.8 (1)	30.9 (4)	32.9 (2)	32.3 (5)	28.3 (3)	155.2
	II	33.1 (2)	32.5 (5)	33.5 (1)	33.5 (3)	31.3 (4)	163.9
	III	33.5 (3)	33.2 (2)	32.9 (5)	32.1 (4)	34.2 (1)	165.9
	IV	28.9 (4)	27.8 (1)	31.1 (3)	31.9 (2)	31.7 (5)	151.4
	V	26.1 (5)	27.6 (3)	26.5 (4)	32.7 (1)	29.8 (2)	142.7
	Totals:	152.4	152.0	156.9	162.5	155.3	
	City Total:	(1) 159.0	(2) 160.9	(3) 154.0	(4) 149.7	(5) 155.5	

$$\sum_i \sum_j y_{ijk}^2 = 24,413.35$$

a. What is the treatment of interest?

b. Why might the cities cause nuisance variability?

c. Carry out the ANOVA and compute Rsquare.

d. Test for differences among the models of cars.

e. Use Fisher's least significant difference to find the best car or cars.>

12.3.3. The National Occupational Safety and Health Act was a comprehensive effort to improve industrial health and safety in this country. Part of this act requires detailed reporting of industrial accidents. The data gained thereby can lead to the identification and elimination of unsafe practices in industry. With such a goal in mind, a safety engineer in a large chemical plant finds that the plant carries out 5 basic operations. Because he has to monitor each operation personally to record the number of unsafe incidents within a 5-day work week, he decides to take a random sample of 5 weeks in order to have a Latin square design.

a. Give the additive model for the experiment, using subscripts i for weeks, j for days, and k for operations.

b. List the assumptions of this design and tell whether you feel it is appropriate in this case.

c. Given the following computations, complete the ANOVA:

$$\sum_i \sum_j y_{ijk}^2 = 10{,}990 \qquad \sum_i T_{i..}^2/5 = 2{,}750$$

$$T_{...}^2/25 = 2{,}250 \qquad \sum_j T_{.j.}^2/5 - T_{...}^2/25 = 710$$

$$\left(\sum_k T_{..k}^2/5 - T_{...}^2/25\right)/4 = 195$$

d. What hypothesis can be tested about the operations?

e. Are weeks random or fixed? Days? Operations?

f. What conclusions can the safety engineer draw from this analysis?

12.3.4. An apiarist conducts an experiment to determine the best method of insulating hives for winter survival of bee colonies. She has 16 hives and decides to expose 4 to each direction of the compass. She has colonies of 4 different origins and she compares 4 different insulating materials. She uses a design in which each combination of direction, colony, and material is assigned once and only once to the 16 hives.

a. What design is the apiarist using?

b. What special assumption is necessary for this ANOVA design?

c. What is the null hypothesis for material effects?

d. What is the expected mean square for colonies?

e. What is the critical value at $\alpha = 0.05$ for a test of direction?

f. Complete the ANOVA table.

Source	df	SS	MS	F
Directions	3	105	35	___
Colonies	___	90	___	___
Materials	3	75	25	___

Source	df	SS	MS	F
Residual	___	___	___	___
Total	15	330		

12.3.5. Why is it impossible to analyze a 2×2 Latin square?

12.4. $a \times b$ FACTORIAL DESIGN

Often an investigator is interested in the combined effect of two types of treatments. For example, a study might be about weight loss for various diets combined with various levels of jogging per day (Figure 12.6).

This design differs from blocking in that neither of the treatments (diet or jogging) is considered extraneous to the experimental question. Subjects are assigned at random to each of the 12 combinations, and interest is in the combined effect as well as diet considered separately and jogging considered separately. This is an economical design since it accomplishes several things at once.

The sets of treatments are called *factors* or *main effects*, and the different treatments within the sets are called *levels*. If diet is factor A, it has $a = 4$ levels, and if jogging is factor B, it has $b = 3$ levels. (The levels need not be quantitative; the diets in this case have the same calories but different food group proportions.) A design of this type is called a two-factor design or, more precisely, an $a \times b$ *factorial design.*[†] In this example, the design is 4×3 factorial. (In this text the first number, 4, refers to the number of levels of factor A. It could refer to either the number of rows or the number of columns in the diagram, depending upon how the diagram is specified.)

In a factorial design, the factors may be treatments in the strict sense or they may be certain classifications of existing populations. The following examples illustrate some of the many different types of study that follow this design:

1. In the jogging–diet example, both factors are treatments; the factor *diet* is qualitative and the factor *jogging* is quantitative.

FIGURE 12.6. A two-factor design.

[†]Some statisticians prefer to call this a factorial experiment because combinations of treatments can be assigned in any kind of design.

2. If change in blood sugar level is measured for various dosages of vitamin C combined with various dosages of aspirin, both factors—*vitamin C* and *aspirin*—are quantitative treatments.
3. If sales of a certain product are recorded in several Standard Metropolitan Statistical Areas and at several different types of chain stores, the factors—*area* and *chain*—are classifications and they are both qualitative.
4. If the lifetimes of tires made by different companies are measured on several different road surfaces, the factor *manufacturer* is a qualitative classification and the factor *roads* is a qualitative treatment.

In all cases, randomization is necessary. In the jogging–diet and vitamin C–aspirin examples, subjects must be assigned at random to each combination of levels. In the sales example, stores must be chosen at random from the chain stores in the areas. In the tire example, tires from the companies are assigned at random to the type of road.

The tire example is not clearly distinct from a randomized complete block design; in fact, it can be thought of as a block design. However, if the investigator is interested in differences caused by various surfaces as well as differences in brand, and especially if he is interested in any interactions between road surface and brand, then it is a factorial design.

An interaction is an additional effect due to the particular combination of the two levels. For example, certain combinations of level of diet and level of jogging may produce a weight loss in excess of the sum of the effects of the two levels involved. Or a particular combination may produce less weight loss than expected. To be able to analyze the data for possible interactions, the investigator must observe more than one subject at each combination of levels.

Geometrically, the absence of interactions yields parallel lines when the means of the response variable are graphed for the various combinations of levels of the factors. Interactions are indicated by deviations from parallelism; Figure 12.7 illustrates the effect of interactions in the blood sugar experiment.

In the jogging–diet study, $n = 2$ subjects are assigned to each combination of levels, and the data are represented by the scheme and notation in Figure 12.8.

FIGURE 12.7. Effect of interaction on subclass means.

		FACTOR A (*Diet*)			
	N	HP	HF	HC	TOTALS
0 mi.	y_{111} y_{112} $T_{11.}$	y_{211} y_{212} $T_{21.}$	y_{311} y_{312} $T_{31.}$	y_{411} y_{412} $T_{41.}$	$T_{.1.}$
FACTOR B (*Jogging*) **1 mi.**	y_{121} y_{122} $T_{12.}$	y_{221} y_{222} $T_{22.}$	y_{321} y_{322} $T_{32.}$	y_{421} y_{422} $T_{42.}$	$T_{.2.}$
2 mi.	y_{131} y_{132} $T_{13.}$	y_{231} y_{232} $T_{23.}$	y_{331} y_{332} $T_{33.}$	y_{431} y_{432} $T_{43.}$	$T_{.3.}$
TOTALS	$T_{1..}$	$T_{2..}$	$T_{3..}$	$T_{4..}$	$T_{...}$ = GRAND TOTAL

FIGURE 12.8. Notation for a two-factor design.

The model for an $a \times b$ factorial design is

$$y_{ijk} = \mu + \alpha_i + \beta_j + \alpha\beta_{ij} + \varepsilon_{ijk}$$
$$i = 1, \ldots, a$$
$$j = 1, \ldots, b$$
$$k = 1, \ldots, n$$

μ: The overall mean for all experiments of this type.

α_i: The effect of the ith level of factor A; the levels may be fixed or random.

β_j: The effect of the jth level of factor B; the levels may be fixed or random.

$\alpha\beta_{ij}$: The interaction effect between the ith level of factor A and the jth level of factor B. ($\alpha\beta$ is a single symbol and is not a product.)

ε_{ijk}: A random effect due to sampling; ε_{ijk} IND($0, \sigma^2$).

Uncorrected Sums of Squares

Sum of Squares	Symbol	Formula	Number of Totals	Observations/ Total
Uncorrected total	T	$\sum_i \sum_j \sum_k y_{ijk}^2$	abn	1
Uncorrected A factor	A	$\sum_i T_{i..}^2/bn$	a	bn
Uncorrected B factor	B	$\sum_j T_{.j.}^2/an$	b	an
Uncorrected subclass	S	$\sum_i \sum_j T_{ij.}^2/n$	ab	n
Correction factor	CF	$T_{...}^2/abn$	1	abn

Corrected Sums of Squares

Source	df	Symbol	Definition	Computational Formula
Total	$abn - 1$	SS_t	$\sum_i \sum_j \sum_k (y_{ijk} - \bar{y}_{...})^2$	$T - CF$
Factor A	$a - 1$	SS_a	$bn \sum_i (\bar{y}_{i..} - \bar{y}_{...})^2$	$A - CF$
Factor B	$b - 1$	SS_b	$an \sum_j (\bar{y}_{.j.} - \bar{y}_{...})^2$	$B - CF$
$A \times B$	$(a - 1)(b - 1)$	SS_{ab}	$n \sum_i \sum_j (\bar{y}_{ij.} - \bar{y}_{i..} - \bar{y}_{.j.} + \bar{y}_{...})^2$	$S - A - B + CF$
Error	$ab(n - 1)$	SS_e	$\sum_i \sum_j \sum_k (y_{ijk} - \bar{y}_{ij.})^2$	$T - S$

Procedure. $a \times b$ Factorial ANOVA

Hypotheses

$$H_0: \alpha_1 = \cdots = \alpha_a = 0 \quad \text{or} \quad H_0: \sigma_A^2 = 0$$
$$H_0: \beta_1 = \cdots = \beta_b = 0 \quad \text{or} \quad H_0: \sigma_B^2 = 0$$
$$H_0: \alpha\beta_{11} = \alpha\beta_{12} = \cdots = \alpha\beta_{ab} = 0 \quad \text{or} \quad H_0: \sigma_{AB}^2 = 0$$

Compute

$$T = \sum_i \sum_j \sum_k y_{ijk}^2 \qquad S = \sum_i \sum_j T_{ij.}^2 / n$$

$$A = \sum_i T_{i..}^2 / bn \qquad CF = T_{...}^2 / abn$$

$$B = \sum_j T_{.j.}^2 / an$$

Source	df	SS	MS
Factor A	$a - 1$	$SS_a = A - CF$	$MS_a = SS_a/(a - 1)$
Factor B	$b - 1$	$SS_b = B - CF$	$MS_b = SS_b/(b - 1)$
$A \times B$	$(a - 1)(b - 1)$	$SS_{ab} = S - A - B + CF$	$MS_{ab} = SS_{ab}/(a - 1)(b - 1)$
Error	$ab(n - 1)$	$SS_e = T - S$	$MS_e = SS_e/ab(n - 1)$
Total	$abn - 1$	$SS_t = T - CF$	

The appropriate F test depends upon whether factors A and B are fixed or random. The F test can be determined by the expected mean squares. The denominator of the F test must estimate everything that the numerator estimates except for the term being tested.

Expected Mean Squares

MS	A and B Fixed	F
A	$\sigma^2 + nb \sum_i \alpha_i^2/(a-1)$	MS_a/MS_e
B	$\sigma^2 + na \sum_j \beta_j^2/(b-1)$	MS_b/MS_e
$A \times B$	$\sigma^2 + n \sum_i \sum_j \alpha\beta_{ij}^2/(a-1)(b-1)$	MS_{ab}/MS_e

MS	A and B Random	F
A	$\sigma^2 + n\sigma_{AB}^2 + nb\sigma_A^2$	MS_a/MS_{ab}
B	$\sigma^2 + n\sigma_{AB}^2 + na\sigma_B^2$	MS_b/MS_{ab}
$A \times B$	$\sigma^2 + n\sigma_{AB}^2$	MS_{ab}/MS_e

MS	A Fixed, B Random	F
A	$\sigma^2 + n\sigma_{AB}^2 + nb \sum_i \alpha_i^2/(a-1)$	MS_a/MS_{ab}
B	$\sigma^2 + na\sigma_B^2$	MS_b/MS_e
$A \times B$	$\sigma^2 + n\sigma_{AB}^2$	MS_{ab}/MS_e

MS	A Random, B Fixed	F
A	$\sigma^2 + nb\sigma_A^2$	MS_a/MS_e
B	$\sigma^2 + n\sigma_{AB}^2 + na \sum_j \beta_j^2/(b-1)$	MS_b/MS_{ab}
$A \times B$	$\sigma^2 + n\sigma_{AB}^2$	MS_{ab}/MS_e

Example 12.5. $a \times b$ Factorial ANOVA

In times of energy shortages, oil companies consider secondary and even tertiary recovery methods for obtaining more petroleum from exhausted oil wells. These methods attempt to free the oil from porous rock so that it can be pumped from the ground. To compare 3 such methods, an oil company takes a random sample of 4 exhausted oil fields and tries each method on 2 different wells randomly selected from each field. The results (in barrels of oil per day) are given below.

		Oil Field (Factor B)				
		1	2	3	4	Totals
Method (Factor A)	Mechanical fracture	2 1 — 3	4 2 — 6	3 1 — 4	1 1 — 2	15
	Carbon dioxide	4 5 — 9	3 3 — 6	6 7 — 13	6 5 — 11	39
	Pressurized steam	6 4 — 10	8 8 — 16	7 8 — 15	5 6 — 11	52
	Totals	22	28	32	24	106

$T = 596$ $B = 478$ $CF = 468.17$
$A = 556.25$ $S = 587$

Methods are fixed because there are only three methods of interest. Fields are random, and whatever inference can be made from this experiment is to be extended to the entire population of exhausted oil fields from which this random sample was drawn.

Source	df	SS	MS	F	$F_{0.05}$
Method	2	88.08	44.04	$MS_a/MS_{ab} = 12.62$	5.143
Field	3	9.83	3.28	$MS_b/MS_e = 4.37$	3.490
$M \times F$	6	20.92	3.49	$MS_{ab}/MS_e = 4.65$	2.996
Error	12	9.00	0.75		

All three *F* values are significant. At least one method is superior to another in all the fields, but because of the significant interaction, the degree of superiority varies from field to field.

By modifying the CLASS, MODEL, and TEST statements, many different types of analysis of variance can be carried out by the SAS System. The following program and output is for the *a* × *b* factorial design in Example 12.4:

```
DATA OIL;
    DO METHOD = 1 TO 3;
        DO FIELD = 1 TO 4;
            DO REPS = 1 TO 2; INPUT BARRELS @@; OUTPUT;
            END;
        END;
    END;
CARDS;
```

```
2  1  4  2  3  1  1  1
4  5  3  3  6  7  6  5
6  4  8  8  7  8  5  6
;
PROC ANOVA;
CLASS METHOD FIELD;
MODEL BARRELS = METHOD FIELD METHOD*FIELD;
TEST H = METHOD E = METHOD*FIELD;
```

In the MODEL statement for a factorial design the main effects are listed, METHOD and FIELD in this example, as well as the interaction METHOD*FIELD. If both effects are fixed, then the TEST statement would be unnecessary; however, if one or both effects are random, then a separate TEST statement is needed for each of the F tests which require a special denominator.

The SAS System
The ANOVA Procedure
Class Level Information

Class	Levels	Values
METHOD	3	1 2 3
FIELD	4	1 2 3 4

Number of observations 24

The ANOVA Procedure

Dependent Variable: BARRELS

Source	DF	Sum of Squares	Mean Square	F Value	Pr > F
Model	11	118.8333333	10.8030303	14.40	<.0001
Error	12	9.0000000	0.7500000		
Corrected Total	23	127.8333333			

R-Square	Coeff Var	Root MSE	BARRELS Mean
0.929596	19.60812	0.866025	4.41667

Source	DF	Anova SS	Mean Square	F Value	Pr > F
METHOD	2	88.08333333	44.04166667	58.72	<.0001
FIELD	3	9.83333333	3.27777778	4.37	0.0268
METHOD*FIELD	6	20.91666667	3.48611111	4.65	0.0115

Tests of Hypotheses Using the Anova MS for METHOD*FIELD as an Error Term

Source	DF	Anova SS	Mean Square	F Value	Pr > F
METHOD	2	88.08333333	44.04166667	12.63	0.0071

EXERCISES

12.4.1. Twenty-four men, each approximately 40 lb overweight, are assigned to the 24 treatments that arise from 4 diets and 3 levels of jogging. Each man consumes the same number of calories per day, but the diets differ in their proportions of protein, fat, and carbohydrate.

		Normal	HP	HF	HC	Totals
			Diet			
	0 mi.	8.5	15.5	8.5	15.5	
		11.5	16.5	7.5	13.5	
		20.0	32.0	16.0	29.0	97.0
	1 mi.	14.0	20.0	13.0	21.0	
Jogging		16.0	23.0	11.0	18.0	
		30.0	43.0	24.0	39.0	136.0
	2 mi.	24.5	27.0	22.0	24.5	
		19.5	24.0	27.0	27.5	
		44.0	51.0	49.0	52.0	196.0
	Totals	94.0	126.0	89.0	120.0	429.0

a. Are the diets random or fixed?

b. Are the jogging levels random or fixed?

c. Carry out the ANOVA.

d. What hypotheses can be tested?

e. Are there significant differences related to the diets?

f. Are there significant differences related to jogging?

g. Are interactions present?

h. Which regimen should be recommended for maximum weight loss?

12.4.2. The Council of Graduate Schools is an organization representing more than 700 U.S. institutions with graduate programs. Its member schools are used in a study of the difference in verbal Graduate Record Examination scores between males and females in mathematics graduate programs in the United States. Twelve institutions and 6 students of each gender are sampled in the study.

a. Are the effects due to the gender of student random or fixed?

b. Are the effects due to institution of student random or fixed?

c. Complete the ANOVA table.

Source	df	MS	E(MS)	F
Institution	___	132,250	_____	___
Gender	___	52,900	_____	___
$I \times S$	___	26,450	_____	___
Error	___	13,225	_____	___

d. Are any of the effects significant?

e. What is the final conclusion?

12.4.3. The State Road Commission decides to make a study of the soil erosion on hillsides that have been cut into in order to prepare roadbeds. A random sample is taken of native species of plants that can serve as ground cover. A random sample is selected among the affected hillsides around the state, and each species is planted on each hillside. After the plants are established, 5 observations on erosion are made on each plant and hillside combination.

a. Complete the following table:

Source	df	MS	E(MS)	F
Plant species	5	410	___	___
Hillside	___	416	___	___
$P \times H$	20	80	___	___
Error	___	12	___	___

b. Test the interaction variance for significance.

c. Compute Rsquare.

d. Which contributes more to the total variability, plant species or hillside? Give numerical values to support your answer.

12.5. $a \times b \times c$ FACTORIAL DESIGN

The $a \times b$ factorial design can be generalized to three or more factors. In this section, we discuss the case of the $a \times b \times c$ factorial design, that is, the three-factor design.

The weight loss problem of Exercise 12.4.1 becomes a three-factor design if we add an exercise program to the diet and jogging factors (Figure 12.9). Diet is factor A, and there are $a = 4$ levels. Amount of jogging is factor B, with $b = 3$ levels. Exercise is factor C, with $c = 2$ levels. Thus, this is a $4 \times 3 \times 2$ factorial design.

Some other examples of designs with 3 factors:

1. The amount of sales of a certain product at several different times of the year, both before and after an advertising campaign, using several different advertising media.

FACTOR C (*Exercise*)

FACTOR B (*Jogging*)	No			Yes		
	0 mi.	1 mi.	2 mi.	0 mi.	1 mi.	2 mi.
FACTOR A (*Diet*) N						
HP						
HF						
C						

FIGURE 12.9. A three-factor design.

2. The achievement of foreign language classes taught by 4 different instructors using 2 different methods and involving 3 different workbooks.
3. The yield of a certain crop with various amounts of fertilizer, various amounts of water, and using various amounts of spacing between plants.
4. The quality of a certain product when inspected by 3 different inspectors using 2 different methods and at 3 different times of the day.

There is some resemblance between this diagram and a Latin square design. However, in the $a \times b \times c$ factorial design, it is not necessary that $a = b = c$; multiple observations are made at each combination of the three factors, and it is possible to test for interactions.

FIGURE 12.10. Notation for a three-factor design.

Each of the 3 factors may be fixed or random. The model may be entirely fixed, entirely random, or mixed with one or two random factors. In mixed models, it is not always possible to use the usual F test to test for the effect of each factor; in some cases, an exact F test does not exist.

We consider here only $a \times b \times c$ factorial designs in which the same number of subjects n are assigned at random to each combination of levels of the three factors. In the weight loss problem, if $n = 2$, then the data are represented as in Figure 12.10.

The model for an $a \times b \times c$ factorial design is as follows:

$$y_{ijkl} = \mu + \alpha_i + \beta_j + \gamma_k + \alpha\beta_{ij} + \alpha\gamma_{ik} + \beta\gamma_{jk} + \alpha\beta\gamma_{ijk} + \varepsilon_{ijkl}$$

$$i = 1, \ldots, a$$
$$j = 1, \ldots, b$$
$$k = 1, \ldots, c$$
$$l = 1, \ldots, n$$

μ:	The overall mean for all experiments of this type.
α_i:	The effect of the ith level of factor A; the levels may be fixed or random.
β_j:	The effect of the jth level of factor B; the levels may be fixed or random.
γ_k:	The effect of the kth level of factor C; the levels may be fixed or random.
$\alpha\beta_{ij}$:	The interaction effect between the ith level of factor A and the jth level of factor B.
$\alpha\gamma_{ik}$:	The interaction effect between the ith level of factor A and the kth level of factor C.
$\beta\gamma_{jk}$:	The interaction effect between the jth level of factor B and the kth level of factor C.
$\alpha\beta\gamma_{ijk}$:	The interaction effect among the ith level of factor A, the jth level of factor B, and the kth level of factor C.
ε_{ijkl}:	A random effect due to sampling, ε_{ijkl} IND$(0, \sigma^2)$

Uncorrected Sum of Squares

Sum of Squares	Symbol	Formula	Number of Totals	Observations/Total
Uncorrected total	T	$\sum_i \sum_j \sum_k \sum_l y_{ijkl}^2$	$abcn$	1
Uncorrected subclass	S	$\sum_i \sum_j \sum_k T_{ijk.}^2 / n$	abc	n
Uncorrected $B \times C$	BC	$\sum_j \sum_k T_{jk.}^2 / an$	bc	an
Uncorrected $A \times C$	AC	$\sum_i \sum_k T_{i.k.}^2 / bn$	ac	bn
Uncorrected $A \times B$	AB	$\sum_i \sum_j T_{ij..}^2 / cn$	ab	cn
Uncorrected C	C	$\sum_k T_{..k.}^2 / abn$	c	abn
Uncorrected B	B	$\sum_j T_{.j.}^2 / acn$	b	acn
Uncorrected A	A	$\sum_i T_{i...}^2 / bcn$	a	bcn
Correction factor	CF	$T_{....}^2 / abcn$	1	$abcn$

Corrected Sums of Squares

Source	df	Symbol	Definition	Computational Formula
Total	$abcn - 1$	SS_t	$\sum_i \sum_j \sum_k \sum_l (y_{ijkl} - \bar{y}_{....})^2$	$T - CF$
A	$a - 1$	SS_a	$bcn \sum_i (y_{i...} - \bar{y}_{....})^2$	$A - CF$
B	$b - 1$	SS_b	$acn \sum_j (y_{.j..} - \bar{y}_{....})^2$	$B - CF$
C	$c - 1$	SS_c	$abn \sum_k (y_{..k.} - \bar{y}_{....})^2$	$C - CF$
$A \times B$	$(a-1)(b-1)$	SS_{ab}	$cn \sum_i \sum_j (\bar{y}_{ij..} - \bar{y}_{i...} - \bar{y}_{.j..} + \bar{y}_{....})^2$	$AB - A - B + CF$
$A \times C$	$(a-1)(c-1)$	SS_{ac}	$bn \sum_i \sum_k (\bar{y}_{i.k.} - \bar{y}_{i...} - \bar{y}_{..k.} + \bar{y}_{....})^2$	$AC - A - C + CF$
$B \times C$	$(b-1)(c-1)$	SS_{bc}	$an \sum_j \sum_k (\bar{y}_{.jk.} - \bar{y}_{.j..} - \bar{y}_{..k.} + \bar{y}_{....})^2$	$BC - B - C + CF$
$A \times B \times C$	$(a-1)(b-1)(c-1)$	SS_{abc}	$n \sum_i \sum_j \sum_k (\bar{y}_{ijk.} - \bar{y}_{ij..} - \bar{y}_{i.k.}$ $-\bar{y}_{.jk.} + \bar{y}_{i...} + \bar{y}_{.j..} + \bar{y}_{..k.} - \bar{y}_{....})^2$	$S - AB - AC - BC +$ $A + B + C - CF$
Error	$abc(n-1)$	SS_e	$\sum_i \sum_j \sum_k \sum_l (\bar{y}_{ijkl} - \bar{y}_{ijk.})^2$	$T - S$

Procedure. $a \times b \times c$ Factorial ANOVA

Hyotheses

$H_0: \alpha_1 = \cdots = \alpha_a$	or	$H_0: \sigma_A^2 = 0$
$H_0: \beta_1 = \cdots = \beta_b$	or	$H_0: \sigma_B^2 = 0$
$H_0: \gamma_1 = \cdots = \gamma_c$	or	$H_0: \sigma_C^2 = 0$
$H_0: \alpha\beta_{11} = \cdots = \alpha\beta_{ab}$	or	$H_0: \sigma_{AB}^2 = 0$
$H_0: \alpha\gamma_{11} = \cdots = \alpha\gamma_{ac}$	or	$H_0: \sigma_{AC}^2 = 0$
$H_0: \beta\gamma_{11} = \cdots = \beta\gamma_{bc}$	or	$H_0: \sigma_{BC}^2 = 0$
$H_0: \alpha\beta\gamma_{111} = \cdots = \alpha\beta\gamma_{abc}$	or	$H_0: \sigma_{ABC}^2 = 0$

Compute:

$$T = \sum_i \sum_j \sum_k \sum_l y_{ijkl}^2$$

$$A = \sum_i T_{i...}^2 / bcn$$

$$B = \sum_j T_{.j..}^2 / acn$$

$$C = \sum_k T_{..k.}^2 / abn$$

$$AB = \sum_i \sum_j T_{ij..}^2 / cn$$

$$AC = \sum_i \sum_k T_{i.k.}^2 / bn$$

$$BC = \sum_j \sum_k T_{.jk.}^2 / an$$

$$S = \sum_i \sum_j \sum_k T_{ijk.}^2 / n$$

$$CF = T_{....}^2 / abcn$$

Source	df	SS	MS
A	$a - 1$	$A - CF$	$SS_a/(a - 1)$
B	$b - 1$	$B - CF$	$SS_b/(b - 1)$
C	$c - 1$	$C - CF$	$SS_c/(c - 1)$
$A \times B$	$(a - 1)(b - 1)$	$AB - A - B + CF$	$SS_{ab}/(a - 1)(b - 1)$
$A \times C$	$(a - 1)(c - 1)$	$AC - A - C + CF$	$SS_{ac}/(a - 1)(c - 1)$
$B \times C$	$(b - 1)(c - 1)$	$BC - B - C + CF$	$SS_{bc}/(b - 1)(c - 1)$
$A \times B \times C$	$(a - 1)(b - 1)(c - 1)$	$S - AB - AC - BC$ $+ A + B + C - CF$	$SS_{abc}/(a - 1)(b - 1)(c - 1)$
Error	$abc(n - 1)$	$T - S$	$SS_e/abc(n - 1)$
Total	$abcn - 1$	$T - CF$	

Expected mean squares vary depending upon whether the factors are fixed or random. The expectations can be found by constructing a table such as Table 12.1. For convenience, the variability among fixed effects is symbolized by θ^2, but it must be remembered that if α, β, and γ are all fixed

$$\theta_A^2 \text{ represents } \frac{\sum_i \alpha_i^2}{(a - 1)} \quad \theta_B^2 \text{ represents } \frac{\sum_j \beta_j^2}{(b - 1)} \quad \theta_C^2 \text{ represents } \frac{\sum_k \gamma_k^2}{(c - 1)}$$

$$\theta_{AB}^2 \text{ represents } \frac{\sum_i \sum_j \alpha\beta_{ij}^2}{(a - 1)(b - 1)} \quad \theta_{AC}^2 \text{ represents } \frac{\sum_i \sum_k \alpha\gamma_{ik}^2}{(a - 1)(c - 1)}$$

$$\theta_{BC}^2 \text{ represents } \frac{\sum_j \sum_k \beta\gamma_{jk}^2}{(b - 1)(c - 1)} \quad \text{and} \quad \theta_{ABC}^2 \text{ represents } \frac{\sum_i \sum_j \sum_k \alpha\beta\gamma_{ijk}^2}{(a - 1)(b - 1)(c - 1)}$$

The rules followed in constructing a table such as 12.1 are as follows:

1. σ^2 is found in every $E(MS)$.
2. The coefficient for any σ^2 or θ^2 will contain n and a, b, or c if those letters are *not* also found in the subscript of the σ^2 or θ^2.
3. The coefficient for an interaction σ^2 or θ^2 will also contain $f(a)$, $f(b)$, or $f(c)$ if the letter is found in the subscript of the σ^2 or θ^2 but *not* in the subscript of the MS.

In the coefficients $f(a) = 0$ if A is fixed and $f(a) = 1$ if A is random; similarly for $f(b)$ and $f(c)$. An interaction term is written as the fixed form (θ^2) only if all factors in the interaction are fixed. One of the principal purposes for obtaining the expected mean squares is to determine the

TABLE 12.1. Expected Mean Squares for $a \times b \times c$ Factorial Design

Terms:	Fixed: θ^2 / Random: σ^2	θ^2_{ABC} / σ^2_{ABC}	θ^2_{BC} / σ^2_{BC}	θ^2_{AC} / σ^2_{AC}	θ^2_{AB} / σ^2_{AB}	θ^2_{C} / σ^2_{C}	θ^2_{B} / σ^2_{B}	θ^2_{A} / σ^2_{A}
Coefficients $\quad$ MS$_a$	1	$nf(b)f(c)$	—	$nbf(c)$	$ncf(b)$	—	—	bcn
MS$_b$	1	$nf(a)f(c)$	$naf(c)$	—	$ncf(a)$	—	acn	
MS$_c$	1	$nf(a)f(b)$	$naf(b)$	$nbf(a)$	—	abn		
MS$_{ab}$	1	$nf(c)$	—	—	nc			
MS$_{ac}$	1	$nf(b)$	—	nb				
MS$_{bc}$	1	$nf(a)$	na					
MS$_{abc}$	1	n						
MS$_e$	1							

appropriate F tests, if they exist. Remember, the MS in the denominator of an F test must estimate everything that the numerator MS estimates except for the term being tested.

Example 12.6. $a \times b \times c$ Factorial ANOVA

The freezing of bull semen became a commercial possibility in the 1950's when it was found, by accident, that a solution of egg parts, glycerin, and a buffer provided protection during the freezing and thawing process. The investigators wanted to try other "antifreezes" besides glycerin, and they wanted to know whether the same level of buffer should be used with each of them. Suppose they designed an $a \times b \times c$ factorial experiment which involved 3 levels of the buffer (a fixed effect), semen from 2 bulls (a random effect), 3 randomly chosen antifreezes (a random effect), and samples of size $n = 4$. The design would enable them to test for interactions along with main effects. (Small samples of random effects are used to keep computations manageable in this example but would not be appropriate in a real experiment.)

The model is

$$y_{ijkl} = \mu + \alpha_i + \beta_j + \gamma_k + \alpha\beta_{ij} + \alpha\gamma_{ik} + \beta\gamma_{jk} + \alpha\beta\gamma_{ijk} + \varepsilon_{ijkl}$$

in which y_{ijkl} is a measure of viability, α_i is the buffer effect, β_j the bull effect, γ_k the antifreeze effect, and the other terms are the interactions.

Buffer Factor A (fixed)

		A_1			A_2			A_3			
Antifreeze Factor C (random)		C_1	C_2	C_3	C_1	C_2	C_3	C_1	C_2	C_3	
		3	2	12	7	17	10	10	7	9	B
		1	6	6	3	7	9	4	10	11	Totals
	B_1	8	1	11	1	13	5	14	5	6	
		4	7	3	6	8	3	17	6	9	
Bull Factor B (random)		$\overline{16}$	$\overline{16}$	$\overline{32}$	$\overline{17}$	$\overline{45}$	$\overline{27}$	$\overline{45}$	$\overline{28}$	$\overline{35}$	261
		3	6	14	14	15	11	8	8	1	
		2	2	8	4	14	6	15	10	3	
	B_2	8	4	16	10	9	8	4	3	8	
		1	10	10	2	11	12	10	7	2	
		$\overline{14}$	$\overline{22}$	$\overline{48}$	$\overline{30}$	$\overline{49}$	$\overline{37}$	$\overline{37}$	$\overline{28}$	$\overline{14}$	279
A Totals			148			205			187			540 Grand Total

	AC Total		
	C_1	C_2	C_3
A_1	30	38	80
A_2	47	94	64
A_3	82	56	49
	159	188	193

C Totals

	AB Total	
	B_1	B_2
A_1	64	84
A_2	89	116
A_3	108	79

C Totals

	BC Total		
	C_1	C_2	C_3
B_1	78	89	94
B_2	81	99	99

$$a = 3, \quad b = 2, \quad c = 3, \quad n = 4$$

$$T = \sum_i \sum_j \sum_k \sum_l y_{ijkl}^2 = 5360.00$$

$$A = \sum_i T_{i...}^2/bcn = \frac{148^2 + 205^2 + 187^2}{24} = 4120.75$$

$$B = \sum_j T_{.j..}^2/acn = \frac{261^2 + 279^2}{36} = 4054.50$$

$$C = \sum_k T_{..k.}^2/abn = \frac{159^2 + 188^2 + 193^2}{24} = 4078.08$$

$$AB = \sum_i \sum_j T_{ij..}^2/cn = \frac{64^2 + \cdots + 79^2}{12} = 4202.83$$

$$AC = \sum_i \sum_k T_{i.k.}^2/bn = \frac{30^2 + \cdots + 49^2}{8} = 4518.25$$

$$BC = \sum_j \sum_k T_{.jk.}^2/an = \frac{78^2 + \cdots + 99^2}{12} = 4083.67$$

$$S = \sum_i \sum_j \sum_k T_{ijk.}^2/n = \frac{16^2 + \cdots + 14^2}{4} = 4654.00$$

$$CF = T_{...}^2/abcn = \frac{540^2}{72} = 4050.00$$

Source	df	SS	MS
A	$a - 1 = 2$	$A - CF = 70.75$	35.38
B	$b - 1 = 1$	$B - CF = 4.50$	4.50
C	$c - 1 = 2$	$C - CF = 28.08$	14.04
$A \times B$	$(a - 1)(b - 1) = 2$	$AB - A - B + CF = 77.58$	38.79
$A \times C$	$(a - 1)(c - 1) = 4$	$AC - A - C + CF = 369.42$	92.35
$B \times C$	$(b - 1)(c - 1) = 2$	$BC - B - C + CF = 1.09$	0.54
$A \times B \times C$	$(a - 1)(b - 1)(c - 1) = 4$	$S - AB - AC - BC + A$ $+ B + C - CF = 52.58$	13.15
Error	$abc(n - 1) = 54$	$T - S = 706.00$	13.07
Total	$abcn - 1 = 71$		

Mean Squares	E(MS)	F	Critical Value F
MS_a	$\sigma^2 + 4\sigma_{abc}^2 + 8\sigma_{ac}^2 + 12\sigma_{ab}^2 + 12 \sum \alpha_i^2$	No appropriate F test	—
MS_b	$\sigma^2 + 12\sigma_{bc}^2 + 36\sigma_b^2$	$MS_b/MS_{bc} = 8.333$	18.513
MS_c	$\sigma^2 + 12\sigma_{bc}^2 + 24\sigma_c^2$	$MS_c/MS_{bc} = 26.000$	19.000
MS_{ab}	$\sigma^2 + 4\sigma_{abc}^2 + 12\sigma_{ab}^2$	$MS_{ab}/MS_{abc} = 2.950$	6.944
MS_{ac}	$\sigma^2 + 4\sigma_{abc}^2 + 8\sigma_a^2$	$MS_{ac}/MS_{abc} = 7.023$	6.388
MS_{bc}	$\sigma^2 + 12\sigma_{bc}^2$	$MS_{bc}/MS_e = 0.041$	3.170
MS_{abc}	$\sigma^2 + 4\sigma_{abc}^2$	$MS_{abc}/MS_e = 1.006$	2.544
MS_e	σ^2		

The buffer effect cannot be tested with this design. In an $a \times b \times c$ factorial experiment in which there is more than one random effect, there will always be main effects which cannot be tested. However, an experimenter usually knows before the experiment whether or not there are significant differences among the levels of a main effect; hence the principal use of the factorial experiment is to study interactions. There are no significant differences between the bulls, but in a factorial experiment such as this, the goal is to learn whether there are interactions involving bulls and the other factors in the experiment. Since no interaction involving bulls is significant, there is evidence that the semen of all bulls can be treated the same. There are significant differences among antifreezes, portending further experimentation to find the best one, and there is a significant interaction involving buffers and antifreezes, indicating that the optimal level of buffer can differ from one antifreeze to another.

EXERCISES

12.5.1. When land is in continuous production, it needs to be treated with a complete fertilizer, that is, one combining nitrogen (chemical symbol N), phosphorus (P), and potassium (K, from the Latin *kalium*). So, shortly after a new variety or hybrid is developed, an NPK factorial experiment is conducted in order to learn something about its response to fertilizers. Suppose there is developed a fescue grass hybrid which is resistant to white grubs, and it will be sold for use on lawns and golf courses. Before marketing, however, an NPK experiment is conducted so that fertilizer recommendations can be made. Forty-eight plots containing mature stands of grass are assigned at random to each of 24 different combinations of fertilizer, 2 plots to each combination. The fertilizer is applied and given time to have an effect. Each plot is mowed, and the clippings are dried and weighed to provide the data below:

		Potassium			
		0 cwt/acre		3 cwt/acre	
		Plot			
Nitrogen	Phosphorus	1	2	1	2
0 cwt	0 cwt	91	54	80	85
	3 cwt	56	72	62	90
	6 cwt	103	154	158	175
3 cwt	0 cwt	254	266	262	258
	3 cwt	173	252	238	317
	6 cwt	383	392	340	465
6 cwt	0 cwt	243	303	239	345
	3 cwt	238	303	287	252
	6 cwt	389	394	384	403
9 cwt	0 cwt	252	175	114	229
	3 cwt	263	281	205	241
	6 cwt	295	244	271	380

a. Give the linear model.

b. Which effects are fixed and which are random?

c. Compute a one-way ANOVA with the following sources of variation and degrees of freedom:

Source	df
Fertilizer	23
Within	24

d. From the sum of squares for fertilizer, break out the effects of N, P, and K and all of their interactions.

e. Give the expectations of mean squares for the three-factor ANOVA above.

f. Make F tests that are valid and draw conclusions.

12.5.2. In an effort to learn more about the shrinkage of cotton knit undershirts when washed and dried at military base laundries, the U.S. Army Quartermaster Corps takes a random sample of 4 brands of shirts from several hundred available for purchase. They further randomly sample enough shirts to have 2 from each brand to be washed at each of 2 water temperatures and dried at each of 3 temperatures. The results, measured by shrinkage of length (in centimeters), are given below.

	Cold-Water Wash Drying Temperature			Hot-Water Wash Drying Temperature		
Brand	210°F	218°F	226°F	210°F	218°F	226°F
A	1.9, 2.1	3.3, 3.7	7.5, 7.9	3.4, 3.6	8.0, 7.6	7.5, 7.7
B	2.2, 2.4	4.8, 5.0	9.8, 9.2	4.6, 4.4	9.3, 9.5	10.1, 9.7
C	2.8, 3.2	6.5, 6.6	13.2, 13.0	5.7, 6.3	12.9, 13.3	13.1, 13.3
D	3.1, 3.7	4.5, 4.8	10.8, 11.2	5.6, 5.0	10.9, 10.7	11.4, 11.7

    a. Which effects are random and which are fixed?

    b. Give the expectations of mean square.

    c. Perform the ANOVA and make all valid F tests.

    d. Draw conclusions about the washing and drying procedures that minimize shrinkage.

12.5.3. Holly trees are attractive and desirable for landscaping, but their propagation presents many problems. Individual trees are either male or female, so there is no production of seed through self-fertilization. Furthermore, once seed are produced, they lie in the ground for about two years before the germination and emergence of the seedlings that begin the next generation of trees. In an effort to find ways to speed up the process, a horticulturist takes a random sample of 4 male trees and another of 4 female trees and makes all possible cross-pollinations. When seeds are produced, he divides the seeds from each of the 16 crosses into 2 groups at random. The seeds in one group are used as a control, and those in the other are scarified because it is claimed this process frequently promotes germination. Seeds are then planted in individual pots. Three years later, two healthy seedlings are selected at random from each cross and treatment and measured for height. The data (in inches) are recorded below.

Control

	F_1	F_2	F_3	F_4
M_1	4.6, 4.9	5.1, 6.1	4.4, 4.8	5.2, 6.3
M_2	8.6, 7.8	5.2, 5.4	3.4, 4.6	4.2, 3.8
M_3	8.7, 8.5	6.6, 7.4	2.0, 2.8	3.7, 4.3
M_4	7.6, 8.4	5.1, 5.4	5.3, 7.7	8.0, 7.5

Scarified

	F_1	F_2	F_3	F_4
M_1	5.3, 4.7	7.7, 8.5	5.3, 5.3	7.7, 6.5
M_2	7.3, 8.5	5.8, 5.4	7.7, 6.9	4.4, 4.6

		Scarified		
	F_1	F_2	F_3	F_4
M_3	6.6, 6.9	6.0, 7.0	8.0, 8.5	6.8, 7.2
M_4	6.9, 7.1	8.8, 8.2	8.9, 9.1	6.7, 7.3

a. Which effects are fixed and which are random?

b. Compute a two-way ANOVA with the following sources of variation and degrees of freedom:

Source	df
Cross	15
Treatment	1
Cross × treatment	15
Error	32

c. From the Cross SS, break out the effects of Male Tree, Female Tree, and the Male × Female interaction.

d. From the Cross × Treatment SS, break out the effects of the three interactions Treatment × Male, Treatment × Female, and Treatment × Male × Female.

e. Give the expectations of mean squares for the three-factor ANOVA, and make all valid F tests.

f. Estimate the percentage of total variability in height due to Male Tree, Female Tree, and Male × Female.

g. What conclusions should be drawn from this study?

12.5.4. A common factorial experiment is the 2^k *factorial* in which there are two different levels of each of k different main effects. To demonstrate this design, suppose that an orthopedic surgeon is uncertain about what, if any, rehabilitation therapy should be used after a certain kind of orthoscopic knee surgery. She can prescribe a rehabilitation regimen which includes (or does not include) walking on a treadmill, lifting weights with the injured leg, and hydrotherapy with swirling water. Thus there are $2^3 = 8$ different treatment combinations, with the control consisting of patients for whom none of these is prescribed (they receive complete rest). Because the surgeon believes none of these is harmful and the benefit to be derived is uncertain, she feels that no patient will be deliberately disadvantaged by whatever rehabilitation regimen is prescribed. The situation is discussed with the patients, and 40 give their consent to participate in an experiment and are assigned at random and in equal numbers to the treatment combinations. The measurement variable is the number of days until a certain level of mobility is attained, and below is a portion of the SAS analysis.

```
                 Dependent Variable: DAYS
Source           DF  Sum of Squares   Mean Square   F Value   Pr > F
Model             7      79.860000      11.408571     3.32     0.009
Error            32     109.940000       3.435625
Corrected Total  39     189.800000
```

	R-Square	Coeff Var	Root MSE	DAYS Mean
	0.420759	12.1544	1.853544	15.250000

Source	DF	Anova SS	F Value	Pr > F
WALK	1	14.550000	4.23	0.0397
LIFT	1	16.400000	4.77	0.0290
HYDRO	1	8.250000	2.40	0.1213
WALK*LIFT	1	7.800000	2.27	0.1319
WALK*HYDRO	1	15.400000	4.48	0.0343
LIFT*HYDRO	1	14.750000	4.29	0.0383
WALK*LIFT*HYDRO	1	2.710000	0.79	0.3741

a. Why should α_i, β_j, and γ_k all be considered to be fixed?

b. With respect to the hypothesis about the effect of hydrotherapy, H_0: $\gamma_1 = \gamma_2$:

   i. Show how the hypothesis is tested.

   ii. Tell how one knows whether or not the null hypothesis should be rejected.

c. What is the numerical value of MS_{AB}?

d. Compute the least significant difference which would be used to make comparisons among the eight treatment means.

12.6. SPLIT-PLOT DESIGN

In this section we discuss a split-plot design that involves randomized complete blocks and two fixed factors; this is probably the most commonly encountered split-plot design. Another one is discussed in Section 12.7 where the experimental units are nested within one fixed factor (as in a completely random design) but factorial to the second. Many other variations of the split-plot design exist, and the reader should consult a reference such as Steel and Torrie (1960) and Cochran and Cox (1957) if one of these other variations is needed.

An example of a split-plot design that involves randomized complete blocks is a marketing experiment in which the investigator wants to study the effectiveness of different incentives used in buy-by-mail advertising for different types of products.

Four large cities are randomly selected for the experiment. From the city directories, 100 households are selected to receive mailings for each of 3 products (a total of 300 households in each city). The 3 products are ladies' hosiery, men's underwear, and household linens. Half of each group receives a mailing that offers an extra discount on an order placed within a short time, and the other half is offered a free pen-and-pencil set with each order (Figure 12.11). Total sales are recorded for each category.

This design differs from the $a \times b \times c$ factorial design discussed in Section 12.5, although the diagrams appear to be similar. Cities in this experiment are randomized complete blocks rather than a factor. The investigator is not interested in cities as such but is using them to control for extraneous variability caused by different locations. Within cities, 3 samples of 100 are assigned at random to the products, which are the *main-unit treatment*, or whole-unit treatment. Then, within these samples, half are assigned at

CITY (*block*)

PRODUCT	A			B			C			D		
(*main unit*)	H	U	L	H	U	L	H	U	L	H	U	L

INCENTIVE   DISCOUNT
(*subunit*)   PREMIUM

FIGURE 12.11. A split-plot design.

random to each incentive group, the *subunit treatment.* The investigator's first interest is the incentive factor, but at the same time, he wishes to gather information about how the incentives work with different products.

Other examples of this split-plot design:

1. A study of vitamin C content of oranges grown in 6 different orchards (blocks) using 4 trees from each orchard which are each treated with a different spray (main-unit treatment) and 2 oranges picked from each tree and stored at different temperatures (subunit treatment).

2. A study of yield of soybeans using different types of seed with different fertilizer treatments. Farms are used for blocks, fertilizer is applied to large plots (whole units), and the different types of seed are planted on sections within the fertilizer plots (subunits).

3. A study of medications for reducing high blood pressure in males involving 4 different drugs (main-unit treatment), each assigned at random to 3 males from each of several ethnic groups (blocks), and within each medication group the drugs are administered once a day but at 3 different times of day (subunit treatment).

4. A study of the retention of historical facts in which students are blocked by schools, two techniques of teaching are used (main-unit treatment), and retention is measured on the same student after several different time periods (subunit treatment).

Here is a summary of the blocks, main-unit factor, and subunit factor for each of the examples above:

Example	Blocks	Treatment on Whole Units	Treatment on Subunits
Buy-by-mail	Cities	Products	Incentives
Vitamin C	Orchards	Sprays	Storage temperatures
Yield	Farms	Fertilizers	Seed types
Blood pressure	Ethnic groups	Drugs	Times of day
Retention	Schools	Techniques	Time periods

An example of the statistical analysis used for a split-plot design is helpful at this point.

Example 12.7. Split-Plot Design

A food scientist wishes to study the effects of tenderizer and length of cooking time on meat. Six beef carcasses are obtained at random from a meat packaging plant. The right rib-eye muscle is excised from each carcass; from the midportion of each muscle, 3 rolled roasts are prepared as nearly alike as possible. Each of the roasts is assigned at random to a tenderizing treatment: control, vinegar marinade, or papain marinade. After treatment, a coring device is used to make 4 cores of meat near the center of each roast. The cores, however, are left in place, and the 3 roasts from the same carcass are placed together in an oven preheated to 300°F and allowed to cook. After 30 minutes of roasting, 1 of the cores is taken at random from each roast, another randomly drawn set of 3 cores is taken after 36 minutes, a third set after 42 minutes, and the final set at 48 minutes. As each set is taken, the cores are allowed to cool to serving temperature and are then measured for tenderness using the Warner–Bratzler device, an instrument similar to a guillotine. The measurement is a number on the Warner–Bratzler scale. A large number indicates a tough piece of meat. The measurements from the 6 carcasses (blocks), 3 tenderizing treatments (on whole units), and 4 lengths of roasting time (on subunits) are the variables of analysis.

In this experiment, combinations of tenderizer and roasting time could not be assigned at random to the cores of meat; the nature of the experiment does not allow for that kind of assignment of treatment combinations. Instead, there were 3 distinct levels of randomization. Six carcasses were taken at random from a very large number of available carcasses. The right rib-eye muscle from each carcass (block) was divided into 3 roasts (whole units), to which 3 tenderizer treatments were assigned at random. Finally, 4 cores of

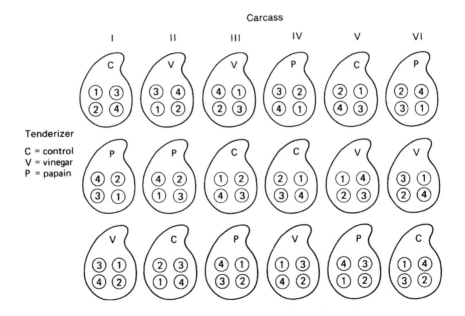

FIGURE 12.12. The split-plot design for a tenderizer study.

meat (subunits) were taken to measure the interior tenderness of each roast at a specified time of cooking, but at the specified time a core was drawn randomly from each roast. The experiment can be visualized as in Figure 12.12. The random variable is tenderness (average of 4 determinations) of meat prepared with one of three tenderizers and roasted for one of 4 lengths of time.

		Roasting Time (Factor B)	Carcass (Factor C)					
			I	II	III	IV	V	VI
	Control	30	8.25	8.00	7.75	8.25	7.50	7.75
		36	7.50	7.00	6.75	6.25	6.75	6.25
		42	4.25	3.25	3.75	4.00	3.25	3.00
		48	3.50	3.75	3.75	3.25	3.00	3.25
			23.50	22.00	22.00	21.75	20.50	20.25
Tenderizer (Factor A)	Vinegar	30	7.25	7.00	6.75	6.75	6.50	6.25
		36	6.25	6.00	6.00	5.50	5.25	5.00
		42	3.50	3.50	4.00	3.50	3.25	3.25
		48	3.50	3.25	3.25	3.50	3.50	3.00
			20.50	19.75	20.00	19.25	18.50	17.50
	Papain	30	6.50	6.00	6.25	5.75	5.25	5.25
		36	4.50	4.75	5.00	4.50	4.50	4.25
		42	3.50	4.00	3.50	3.50	3.25	3.25
		48	2.50	2.50	2.75	2.25	2.00	3.00
			17.00	17.25	17.50	16.00	15.00	15.75
	Totals		61.00	59.00	59.50	57.00	54.00	53.50

Roasting Time

Tenderizer	30	36	42	48	Totals
Control	47.50	40.50	21.50	20.50	130.00
Vinegar	40.50	34.00	21.00	20.00	115.50
Papain	35.00	27.50	21.00	15.00	98.50
Totals	123.00	102.00	63.50	55.50	344.00

Uncorrected Sum of Squares	Symbol	Number of Squared Values	Observations per Squared Value	Calculations	Numerical Value
Total	T	$abc = 72$	1	$(8.25)^2 + (7.50)^2$ $+ \cdots + (3.00)^2$	1852.25
Whole unit (roast)	W	$ac = 18$	$b = 4$	$[(23.50)^2 + \cdots$ $+ (15.75)^2]/4$	1668.72

Uncorrected Sum of Squares	Symbol	Number of Squared Values	Observations per Squared Value	Calculations	Numerical Value
Factor A (tenderizer)	A	$a = 3$	$bc = 24$	$[(13.0)^2 + (115.5)^2 + (98.5)^2]/24$	1664.27
Block (carcass)	C	$c = 6$	$ab = 12$	$[(61.0)^2 + \cdots + (53.5)^2]/12$	1647.46
Factor B (roasting time)	B	$b = 4$	$ac = 18$	$[(123.0)^2 + \cdots + (55.5)^2]/18$	1813.64
$A \times B$ (tenderizer by time)	AB	$ab = 12$	$c = 6$	$[(47.5)^2 + \cdots + (15.0)^2]/6$	1843.92
Correction factor	CF	1	$abc = 72$	$(344)^2/72$	1643.56

The analysis can initially be approached as though the experiment involved nothing more than 18 roasts and four roasting times. One could then conduct a two-way ANOVA, which we call the preliminary analysis.

Preliminary Analysis

Source	df	SS
Roast	$ac - 1 = 17$	$W - CF = 25.16$
Roasting time	$b - 1 = 3$	$B - CF = 170.08$
Residual	$(ac - 1)(b - 1) = 51$	$T - W - B + CF = 13.45$

But the roasts (whole units) are not independent; some are associated because they came from the same carcass and others because they received the same tenderizing treatment. Consequently, the variability due to these effects can be accounted for in the roast sum of squares in the following manner:

Source	df	SS
Roast (whole unit)	$ac - 1 = 17$	$W - CF = 25.16$
Tenderizer	$a - 1 = 2$	$A - CF = 20.71$
Carcass	$c - 1 = 5$	$C - CF = 3.90$
Whole-unit remainder	$(a - 1)(c - 1) = 10$	$W - A - C + CF = 0.55$

Other variability in the preliminary analysis can be accounted for, and that is the variability due to interaction between tenderizer and roasting time ($A \times B$). This variability is, perforce, part of the residual sum of squares, so it should be computed and removed, as shown below.

Source	df	SS
Residual in		
preliminary analysis	$(ac - 1)(b - 1) = 51$	$T - W - B + CF = 13.45$
$A \times B$	$(a - 1)(b - 1) = 6$	$AB - A - B + CF = 9.57$
Subunit remainder	$a(b - 1)(c - 1) = 45$	$T - W - AB + A = 3.88$

The complete ANOVA for the split-plot design can be obtained by putting together the sums of squares that have been broken out of the preliminary analysis. The final analysis is

Source	df	SS	MS	F
Whole units				
Tenderizer	$a - 1 = 2$	$A - CF = 20.71$	10.36	207.20*[†]
Carcass	$c - 1 = 5$	$C - CF = 3.90$	0.78	15.60*
Whole-unit remainder	$(a - 1)(c - 1) = 10$	$W - A - C + CF = 0.55$	0.05	
Subunits				
Roasting time	$b - 1 = 3$	$B - CF = 170.08$	56.69	629.89
Time × tenderizer	$(a - 1)(b - 1) = 6$	$AB - A - B + CF = 9.57$	1.59	17.67*
Subunit remainder	$a(b - 1)(c - 1) = 45$	$T - W - AB + A = 3.88$	0.09	

Not too surprisingly, the analysis results in claiming significance for all effects tested. This is largely due to the nature of the experiment. It has probably been known from the time of the cavemen that longer time of cooking can make meat more tender. Similarly, the benefits of marinating were discovered without benefit of statistical analysis. However, it is not uncommon in the split-plot design for the experimenter to know in advance of the experiment that the whole-unit treatments (tenderizers) and even subunit treatments (roasting times) are significant. The principal concern in the design is usually the interaction. Here, the food scientist wants to know about the best combinations of tenderizer and roasting time. Because the interaction term also proved to be significant, the food scientist will pay particular interest to a mean separation technique that allows for further examination of the interaction. This can be done with a two-way table of averages:

	Factor B (Roasting time, min)			
Factor A	30	36	42	48
Control	7.9167	6.7500	3.5833	3.4167
Vinegar	6.7500	5.6667	3.5000	3.3333
Papain	5.8333	4.5833	3.5000	2.5000

[†]An asterisk traditionally is used to indicate significance, in this case at $\alpha = 0.05$.

The Warner–Bratzler score employed here is a function of the force necessary to shear a piece of meat of a given size. Consequently, the greater the average score, the less tender is the meat. The interactions can be best understood by comparing average of roasting times for the same tenderizer or, conversely, the average of tenderizers at the same roasting time. Multiple comparisons within a split-plot design differ from those in other designs (see Steel and Torrie, 1960).

To compare means for roasting times (B means) with the same tenderizer (same A level), the least significant difference is

$$t_{\alpha/2,a(b-1)(c-1)}\sqrt{\frac{2\text{MS}_{\text{subunit remainder}}}{c}} = 2.014\sqrt{\frac{2(0.09)}{6}}$$

$$= 0.3488 \text{ at } \alpha = 0.05$$

To compare two tenderizer means (A means) at the same roasting time (same B level), an approximate test must be used because the two A means contain both A effects and AB interactions. The least significant difference is

$$t_{\alpha/2*}\sqrt{\frac{2[(b-1)\text{MS}_{\text{subunit remainder}} + \text{MS}_{\text{whole unit remainder}}]}{cb}}$$

in which

$$t_{\alpha/2*} = \frac{(b-1)\text{MS}_{sr}t_{\alpha/2,a(b-1)(c-1)} + \text{MS}_{wr}t_{\alpha/2,(a-1)(c-1)}}{(b-1)\text{MS}_{sr} + \text{MS}_{wr}}$$

Thus

$$t_{0.025*} = \frac{3(0.09)2.014 + (0.05)2.228}{3(0.09) + 0.05}$$

$$= 2.047$$

and the least significant difference is

$$2.047\sqrt{2[3(0.09) + 0.05]/24} = 0.334$$

In interpreting the significant interaction in this experiment, we can conclude that no matter what the precooking tenderizer treatment, the longer a roast is cooked, the more tender it will be. However, the degree of tenderness for any cooking time will depend on the kind of tenderizer used. This is an indication that the use of tenderizing marinade is especially important for those who prefer roasts rare or medium rare, because the differences between all tenderizer treatments are significant for roasts cooked 30 or 36 minutes. The reappearance of a significant difference between the papain marinade and the other two whole-plot treatments at 48 minutes may be an anomaly (that is, a Type I error). However, it could also represent a reproducible phenomenon which the food scientist might want to examine further. Even if it is a real difference, the gain in tenderness may not merit the added roasting time if there is offsetting loss in meat texture, juiciness, or other components of palatability.

In general, a split-plot design may be arranged similar to Figure 12.13, in which 3 blocks, 2 whole-unit treatments, and 4 subunit treatments are used.

			Factor C (Block)				
	Factor B (Subunit Treatment)		C_1	C_2	C_3		
		B_1	y_{111}	y_{112}	y_{113}		
		B_2	y_{121}	y_{122}	y_{123}		
	A_1	B_3	y_{131}	y_{132}	y_{133}		
		B_4	y_{141}	y_{142}	y_{143}		
			$T_{11.}$	$T_{1.2}$	$T_{1.3}$	$T_{1..}$	
Factor A (Whole-Unit Treatment)		B_1	y_{211}	y_{212}	y_{213}		
		B_2	y_{221}	y_{222}	y_{223}		
	A_2	B_3	y_{231}	y_{232}	y_{233}		
		B_4	y_{241}	y_{242}	y_{243}		
			$T_{21.}$	$T_{2.2}$	$T_{2.3}$	$T_{2..}$	
	C Totals		$T_{..1}$	$T_{..2}$	$T_{..3}$	$T_{...}$ Grand Total	

	B_1	B_2	B_3	B_4
A_1	$T_{11.}$	$T_{12.}$	$T_{13.}$	$T_{14.}$
A_2	$T_{21.}$	$T_{22.}$	$T_{23.}$	$T_{24.}$
B Totals	$T_{.1.}$	$T_{.2.}$	$T_{.3.}$	$T_{.4.}$

FIGURE 12.13. Notation for a split-plot design.

The model for this split-plot design, in which the whole-unit treatment is randomized within complete blocks, is

>

$$y_{ijk} = \mu + \alpha_i + \beta_j + \alpha\beta_{ij} + \gamma_k + \delta_{ik} + \varepsilon_{ijk}$$
$$i = 1, \ldots, a$$
$$j = 1, \ldots, b$$
$$k = 1, \ldots, c$$

The terms in this model have the following meanings:

μ: The overall mean for all experiments of this type.

α_i: The effect of the ith level of factor A, the whole unit treatment; a fixed effect, $\sum_i \alpha_i = 0$.

β_j: The effect of the jth level of factor B, the subunit treatment; a fixed effect, $\sum_j \beta_j = 0$.

$\alpha\beta_{ij}$:   The interaction effect between the ith level of factor A and the jth level of factor B.

γ_k: The kth block effect; blocks are random.

δ_{ik}:   The whole-unit random component, δ_{ik} IND$(0, \sigma_D^2)$.

ε_{ijk}:   The subunit random component, ε_{ijk} IND$(0, \sigma^2)$.

Uncorrected Sums of Squares

Sum of Squares	Symbol	Formula	Number of Totals	Observations/Total
Uncorrected total	T	$\sum_i \sum_j \sum_k y_{ijk}^2$	abc	1

Uncorrected whole unit	W	$\sum_i \sum_k T^2_{i.k}/b$	ac	b
Uncorrected A factor	A	$\sum_i T^2_{i..}/bc$	a	bc
Uncorrected B factor	B	$\sum_j T^2_{.j.}/ac$	b	ac
Uncorrected block	C	$\sum_k T^2_{..k}/ab$	c	ab
Uncorrected $A \times B$	AB	$\sum_i \sum_j T^2_{ij.}/c$	ab	c
Correction factor	CF	$T^2_{...}/abc$	1	abc

Procedure. Split-Plot ANOVA with Randomized Complete Block (Factors A and B Fixed Effects)

Hypotheses:
$H_0: \alpha_1 = \cdots = \alpha_a = 0$ (no difference among secondary treatments)
$H_0: \beta_1 = \cdots = \beta_b = 0$ (no difference among main treatments)
$H_0: \alpha\beta_{11} = \cdots = \alpha\beta_{ab} = 0$ (no interactions)
$H_0: \sigma^2_C = 0$ (no block effect)

Source	df	SS	E(MS)
Whole units			
Factor A	$a - 1$	$A - CF$	$\sigma^2 + b\sigma^2_D + cb\sum_i \alpha_i^2/(a-1)$
Block	$c - 1$	$C - CF$	$\sigma^2 + b\sigma^2_D + ab\sigma^2_C$
Whole-unit remainder	$(a-1)(c-1)$	$W - A - C + CF$	$\sigma^2 + b\sigma^2_D$
Subunits			
Factor B	$b - 1$	$B - CF$	$\sigma^2 + ca\sum_j \beta_j^2/(b-1)$
$A \times B$	$(a-1)(b-1)$	$AB - A - B + CF$	$\sigma^2 + c\sum_i \sum_j \alpha\beta_{ij}^2/(a-1)(b-1)$
Subunit remainder	$a(c-1)(b-1)$	$T - W - AB + A$	σ^2

Mean squares are found by dividing the sums of squares by the corresponding degrees of freedom. The appropriate F tests can be determined from the expected mean square. A split-plot experiment is actually two experiments conducted at the same time. The whole-plot experiment has an estimate of whole-plot experimental error based on the whole-plot residuals MS_{wr}. The subplot experiment has an estimate of subplot experimental error based on the subplot residuals MS_{sr}. Standard errors of estimates of whole-plot differences between A means are calculated from MS_{wr}, and standard errors of estimates of subplot differences between B means are calculated from MS_{sr}. Standard errors of estimates of differences between A means at the same or different B levels are caluclated from a weighted average of

MS_{wr} and MS_{sr}. The standard errors needed for estimates and for multiple comparison are given in the following table:

Difference Between	Standard Error	df for t
Two overall A means	$\sqrt{\dfrac{2MS_{wr}}{bc}}$	$(a-1)(c-1)$
Two overall B means	$\sqrt{\dfrac{2MS_{sr}}{ac}}$	$a(b-1)(c-1)$
Two B means at the same A level	$\sqrt{\dfrac{2MS_{sr}}{c}}$	$a(b-1)(c-1)$
Two A means at the same B level or different B levels	$\sqrt{\dfrac{2[(b-1)MS_{sr}+MS_{wr}]}{bc}}$	Use t_α^* below

$$t_\alpha^* = \frac{(b-1)MS_{sr}t_{\alpha,a(b-1)(c-1)} + MS_{wr}t_{\alpha,(a-1)(c-1)}}{(b-1)MS_{sr} + MS_{wr}}$$

It is appropriate to use a split-plot design if:

1. One of the treatments requires large quantities of material (such as the fertilizer in the yield example) and the whole units are used for this treatment.
2. An additional factor is to be incorporated into the experiment (such as the products in the buy-by-mail example). The main factor (incentives) is applied to the subunits and the additional factor to the whole units.
3. Larger differences are expected among the levels of one factor than among the levels of the other factor (as in the blood pressure example). The factor with the larger differences (drugs) is used for the whole units and the factor with small differences (time of day) for the subunits.
4. Greater precision is desired for comparisons among the levels of one factor than the other factor. The factor requiring the greater precision is used for the subunits.

Some split-plot designs could be laid out as an $a \times b \times c$ factorial design. For example, the achievements of foreign language classes taught by 4 different instructors using 2 different methods and 3 different workbooks is a $4 \times 2 \times 3$ factorial design if groups of students are assigned at random to each combination of teacher, method, and workbook. However, this could be planned as a split-plot design. If the students pick the teachers and each teacher is offering two classes, the teachers are the blocks. The classes are the whole units, and they are randomly assigned a method. Within classes, equal numbers of students (subunits) are randomly assigned to the three different workbooks.

The overall precision of the two experiments is probably the same. However, the split-plot design gives increased precision for subunit comparisons and a lower precision for whole-unit comparisons. Thus, if the experimenter wants to be able to detect differences among the workbooks, the split-plot design increases the probability of detecting these differences if they exist.

EXERCISES

12.6.1. Analyze the shrinkage data in Exercise 12.6.2 as if they arose from a split-plot design in which brands are the blocks, wash temperatures are applied to groups of shirts together (whole units), and the drying temperatures are randomly assigned within the whole units. Let the random variable be the average centimeters of shrinkage in length of the two shirts in each subgroup.

12.6.2. Crop rotation is recommended as a good farming practice, especially when a nitrogen-fixing legume is used as an alternative crop. To demonstrate the validity of this recommendation, an agricultural extension specialist set up an experiment in which alfalfa, clover, and a nonlegume grass were planted in five blocks according to a randomized complete block design. These plantings later served as the whole plots for the second year when 4 varieties of grain were planted in random subplots on each main plot. Thus each split plot can be identified by the crop which was planted on it in the first year and that which was planted on it in the second year. The extension specialist wants to be able to demonstrate whether the use of a piece of land during the previous year will affect the yield of the following year's crop. The yields of the varieties of the grain crop are given below:

First Crop (Whole Plot)	Second Crop (Split Plot)	Blocks				
		1	2	3	4	5
Alfalfa	A	21.7	20.8	18.2	25.2	17.8
	B	18.8	14.5	14.2	17.9	14.5
	C	25.0	18.1	18.7	20.9	15.9
	D	24.3	22.0	20.3	23.0	18.6
		89.8	75.4	71.4	87.0	66.8
Clover	A	26.3	23.1	20.0	20.3	17.3
	B	19.8	16.0	22.5	13.7	14.4
	C	21.6	20.0	21.2	18.0	19.8
	D	25.7	21.1	23.1	17.0	16.3
		93.4	80.2	86.8	69.0	67.8
Grass (Control)	A	17.5	18.5	21.2	18.6	13.0
	B	15.2	14.6	17.7	13.5	10.0
	C	15.5	17.2	19.9	15.5	15.0
	D	15.6	17.2	19.9	15.5	16.3
		63.8	67.5	78.7	63.1	54.3

Given that $\sum\sum\sum y_{ijk}^2 = 21,439.3$:
  a. Complete the ANOVA.
  b. Compute the least significant difference for comparing:
      i. Whole-plot means
      ii. Split-plot means

12.6.3. At a university's horticulture farm, an experimental orchard was originally established according to a randomized complete block design consisting of *a* varieties of apple trees and c replicates. Because of this original layout, the orchard is frequently used for experiments with a split-plot design in which the varieties are whole plots. Given below is a portion of the SAS analysis of an environmental experiment in which different levels of topically applied chemical pollutants are used as split-plot treatments. The measurement variable is pounds of the pollutant per ton of apples.

```
Dependent Variable: LBS
                            Sum of         Mean
Source             DF      Squares        Square     F Value    Pr > F
Model              20    216.173800    10.808690      4.23     <.0001
Error              84    214.284000     2.551000
Corrected Total   104    430.457800
```

	R-Square	Coeff Var	Root MSE	LBS Mean
	0.502195	35.9969	1.597185	4.437000

Source	DF	Anova SS	Mean Square	F Value	Pr > F
VAR	4	17.750000	4.437500	1.74	0.1468
REP	6	25.642400	4.273733	1.67	0.1384
VAR*REP	24	87.481200	3.645047	1.43	0.1185
LEVEL	2	20.356800	10.179000	3.99	0.0221
VAR*LEVEL	8	64.943400	8.117750	3.18	0.0036

Tests of Hypotheses Using the Anova MS for VAR*REP as an Error Term

Source	DF	Anova SS	Mean Square	F Value	Pr > F
VAR	4	17.750000	4.437500	1.22	0.3285
REP	6	25.642400	4.273733	1.17	0.3547

Use the SAS output to answer the following questions:

a. Give the numerical values for:

    i. The number of varieties of apples used in the experiment

    ii. The number of levels of the chemical pollutant

b. On the average, how many pounds of chemical pollutant are found per ton of apples taken from this orchard?

c. In the model for this experiment, which effects are most likely fixed? Tell why.

d. Why are there two F tests in which the VAR*REP MS is used as Error?

e. Which null hypotheses are rejected in this analysis?

12.7. SPLIT PLOT WITH REPEATED MEASURES

The split-plot design examined is Section 12.6 involved *complete blocks* that enable an experimenter to compare all main effects in the same block or replicate. Such was the case in Example 12.7; all 4 main effects (tenderizers) were used on the same carcass (block). However, sometimes this is difficult or even impossible. Suppose that in the nested design diagrammed in Figure 12.1, the investigator had wanted the two determinations taken at different times, one when the volunteers first began their diets and the second after they had been on their diets for a month (Figure 12.14). This would show whether there was a change in cholesterol level between times 1 and 2. This would add a second factor to an existing design, but it would not be the usual split-plot design because there are no complete blocks. For a volunteer to be a complete block, he would have to be on all 3 diets, and while he might have the appetite to enjoy them all, it would be impossible to measure their independent effects on his cholesterol level. There are simply some situations where a main effect can be replicated but not in a complete block. Thus we need to examine the analysis of experimental data that arise from such experiments.

It is convenient to think of a spit-plot design as the adding of a new factor, or split-plot effect, to an already existing design. We have seen how the conventional split-plot design can be obtained by adding a second factor, the split-plot effect, to a randomized block design. Such designs involve two randomized complete block designs. The same idea holds for the type of design now under discussion, except, as in the case of the cholesterol experiment, there it is a nested design to which a second factor is added. The analysis is sometimes called *repeated-measures analysis* because measures of cholesterol level are taken at two different times while volunteers are on their diets. However, the design is a split plot, just different in that it involves a nested design and a randomized complete block (RCB) design rather than two RCBs.

Other examples of this sort of split-plot design are as follows:

1. To see how effectively increased levels of corn in rations will fatten cattle, a feedlot experiment involving 3 different rations (main-effect treatments) is conducted. Cattle are a random effect nested within rations, and the split-plot effect will be the 4 times cattle are weighed while being fattened.
2. Because they can become very dirty during a game, football jerseys must be washed with detergents so strong that colors may fade. Thus a manufacturer of jerseys wants to test the colorfastness of 3 different dyes (main effects). Each dye is used to color 10 different jerseys (random experimental units), and all jerseys are washed 6 times (split-plot effect). Color fading is measured after each washing.
3. The effectiveness of hip replacement surgery is measured by how well over time bone tissue adheres to the prosthesis (artificial replacement of the head of the original bone). In such a study, 4 different prostheses (main-plot effects) are to be compared. Patients selected at random from a data base of hundreds of hip replacement surgeries are experimental units nested within main effects, and postoperation X-ray measurements taken at 5, 10, and 15 years provide the levels of the split-plot effect.

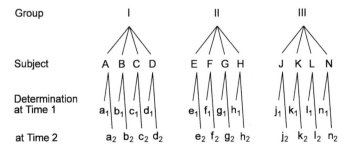

FIGURE 12.14. A nested design with repeated measures.

4. The angle of reentry into Earth's atmosphere greatly affects the temperature of NASA spacecrafts. Suppose there are appropriate data from previous flights of a certain kind of craft to compare the amount of heat generated for 2 different angles of reentry (main-plot effects). The experimental units are the 5 different flights nested in one or the other of the reentry angles. Three important distances from Earth during reentry are the levels of the split-plot effects. Measurements are the respective temperatures recorded for each flight at each distance of reentry.

A summary of the main effects, experimental units, and split-plot effects for each of the examples is

Example	Treatments on Whole plots	Experimental Units within Treatments	Measures on Each Experimental Unit
Cattle rations	Amount of Corn	Cattle	Length of time fed
Football jerseys	Dye	Jerseys	Times washed
Hip surgery	Prosthesis	Surgery patients	Years after surgery
Space flight	Reentry angles	Space craft flights	Distance from earth

Example 12.8 demonstrates the statistical analysis of this second kind of split-plot design.

Example 12.8.   Split Plot Involving Nested and Complete Block Designs

Certain strains of the bacterium *Escherichia coli* often found in undercooked foods become a serious health risk if they enter the blood stream. The organism is covered with a chemical compound called a lipopolysaccharide (LPS) that has a toxic effect on the hearts of infected animals. When LPS enters the circulatory system, heart function is affected and heart rate becomes highly elevated. A medical scientist wants to know if the residual effect on heart rate is different for LPS than for other compounds also known to increase heart rate.

An experiment, simplified for this example, is designed to see how heart rate decreases over time after it has been elevated either with LPS or another compound that will serve as a control. LPS is used on 3 rats and the control compound on another 3. A monitor records continuous measurements (one per second) of the rats' heart rates, but the measures to be used in the analysis are when each rat's heart rate reaches a maximum and every 20 minutes thereafter. The experimenter wants to compare the effect of the two compounds on heart rate during the hour after it has reached the maximum number of beats per minute. The $2 \times 3 \times 4 = 24$ measures for this experiment are in the table below:

Time	LPS Rat 11	Rat 12	Rat 13	Control Rat 21	Rat 22	Rat 23	Total
0	416	455	422	465	439	443	2640
20	404	448	411	395	366	373	2397
40	361	396	368	339	320	328	2112
60	307	348	317	290	266	278	1806
Total	1488	1647	1518	1489	1391	1422	8955

A tabulation of treatment sums at each time (factor C × factor A) is also needed:

Treatment	Times				Total
	0 min	20 min	40 min	60 min	
LPS	1293	1263	1125	972	4653
Control	1347	1134	987	834	4302
Total	2640	2397	2112	1806	8955

Then with these values, the uncorrected sums of squares can be computed:

Uncorrected Sum of Squares	Symbol	Number of Squared Values	Observations Per Squared Value	Calculations	Numerical Value
Total	T	$abc = 24$	1	$416^2 + 404^2$ $+ \cdots + 278^2$	3,420,759.000
Factor A (treatment)	A	$a = 2$	$bc = 12$	$[4653^2 + 4302^2]/12$	3,346,467.750
Experimental unit B (rat)	B	$ab = 6$	$c = 4$	$[1488^2 + 1647^2$ $+ \cdots + 1422^2]/4$	3,351,290.750
Factor C (time)	C	$c = 4$	$ab = 6$	$[2640^2 + 2397^2$ $+ \cdots + 1806^2]/6$	3,406,231.500
$A \times C$ (treatment by time)	AC	$ac = 8$	$b = 3$	$[1293^2 + 1347^2$ $+ \cdots + 834^2]/3$	3,415,839.000
Correction factor	CF	1	$abc = 24$	$8955^2/24$	3,341,334.375

As was done with the previous split-plot design, the experimenter can perform the analysis in two stages. The preliminary analysis is that for a nested design:

Source	df	SS
Treatment (whole unit)	$a - 1 = 1$	$A - CF = 5,133.375$
Among rats within units	$a(b - 1) = 4$	$B - A = 4,823.000$
Among measurements within rats (residual)	$ab(c - 1) = 18$	$T - B = 69,468.250$

The measurements within rats, however, are not independent because of the times at which they are taken. There is an association among those taken at 0, 20, 40, and 60 min, respectively. Furthermore, times are factorial to treatments rather than nested within them. So the variability due to time and the treatment by time interaction needs to be removed from the sums of squares for among measurements in the preliminary analysis:

Source	df	SS
Among measurements within rats	$ab(c - 1) = 18$	$T - B = 69{,}468.250$
Time	$c - 1 = 3$	$C - CF = 64{,}897.125$
$A \times C$	$(a - 1)(c - 1) = 3$	$AC - A - C + CF = 4{,}474.125$
Remainder	$a(b - 1)(c - 1) = 12$	$T - B - AC + A = 97.000$

Once again, the final ANOVA is obtained by replacing the residual sums of squares in the preliminary analysis with those broken out in the analysis just shown. The complete analysis then is

Source	df	SS	MS	F test
Whole units				
Treatment	$a - 1 = 1$	5133.375	5133.375	$MS_a/MS_b = 4.25$
Rat within treatment	$a(b - 1) = 4$	4823.000	1205.750	$MS_b/MS_e = 149.16^*$
Subunits				
Times	$c - 1 = 3$	64,897.125	21,632.375	$MS_c/MS_e = 2676.17^*$
Treatment × time	$(a - 1)(c - 1) = 3$	4,474.125	1,491.375	$MS_{ac}/MS_e = 184.50^*$
Subunit remainder	$a(b - 1)(c - 1) = 12$	97.000	8.083	

In the subunit analysis, both times and the interaction between treatments and times are significant, with $P < 0.0001$ for each F test. So the medical scientist sees that, irrespective of

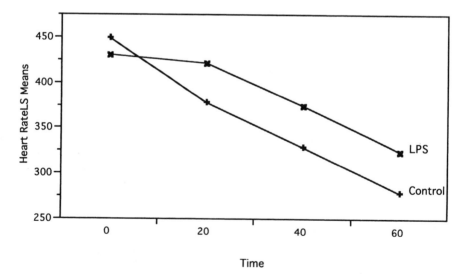

FIGURE 12.15. Rate of decrease in heart rates for LPS and control.

treatment group, heart rate decreases significantly during the hour after maximum rate has been reached. However, the significant interaction means that the rate of decrease is not the same for the LPS group as for the control. When JMP is used to compute and plot treatment averages at each time (Figure 12.15), the experimenter sees that the heart rate of LPS group returns toward normal significantly more slowly than for the control group. The bacterial toxin continues to elevate heart rate long after its initial effect on the heart, and this will need to be kept in mind by physicians treating patients infected by E. *coli* or similar bacteria with the LPS covering.

The model for the repeated-measures analysis is

$$y_{ijk} = \mu + \alpha_i + \beta_{ij} + \gamma_k + \alpha\gamma_{ik} + \varepsilon_{ijk}$$

$$i = 1, \ldots, a$$
$$j = 1, \ldots, b$$
$$k = 1, \ldots, c$$

where the symbols are defined as follows:

μ: The overall mean for experiments of this type.

α_i: The effect of the ith level of factor A, the whole unit treatment; a fixed effect, $\sum_i \alpha_i = 0$.

β_{ij}:   A random effect due to the (ij)th experimental unit; β_{ij} is IND$(0,\sigma_B^2)$ for each i.

γ_k: The effect of the kth level of factor C, the subunit treatment; a fixed effect, $\sum_k \gamma_k = 0$.

$\alpha\gamma_{ik}$:   The interaction effect between ith level of factor A and the kth level of factor C.

ε_{ijk}:   The subunit random component, ε_{ijk} IND$(0,\sigma^2)$.

Uncorrected Sums of Squares

Sum of Squares	Symbol	Formula	Number of Totals	Observations/ Total
Uncorrected total	T	$\sum_i \sum_j \sum_k y_{ijk}^2$	abc	1
Uncorrected A factor	A	$\sum_i T_{i..}^2/bc$	a	bc
Uncorrected experimental unit	B	$\sum_i \sum_j (T_{ij.}^2/c)$	c	
Uncorrected C factor	C	$\sum_k T_{..k}^2/ab$	c	ab
Uncorrected $A \times C$	AC	$\sum_i \sum_k T_{i.k}^2/b$	ac	b
Correction factor	CF	$T_{...}^2/abc$	1	abc

Procedure. Split-Plot ANOVA with Repeated Measures

Hypotheses:

H_0: $\alpha_1 = \cdots = \alpha_a = 0$ (no difference among main treatments)
H_0: $\sigma_B^2 = 0$ (no experimental unit effect)
H_0: $\gamma_1 = \cdots = \gamma_c = 0$ (no difference among secondary treatments)
H_0: $\alpha\gamma_{11} = \cdots = \alpha\gamma_{ac} = 0$ for all i and k (no interactions)

Source	df	SS	E(MS)
Whole units Factor A	$a - 1$	$A - CF$	$\sigma^2 + b\sigma_B^2 + cb \sum_i \alpha_i^2/(a-1)$
Experimental units Within Factor A	$a(b-1)$	$B - A$	$\sigma^2 + b\sigma_B^2$
Subunits Factor C	$c - 1$	$C - CF$	$\sigma^2 + ab \sum_k \gamma_k^2/(c-1)$
$A \times C$	$(a-1)(c-1)$	$AC - A - C + CF$	$\sigma^2 + b\dfrac{\sum_i \sum_k \alpha\gamma_{ik}^2}{(a-1)(c-1)}$
Subunit remainder	$a(c-1)(b-1)$	$T - B - AC + A$	σ^2

Mean squares are found by dividing the sums of squares by the corresponding degrees of freedom. The appropriate F tests can be determined from the expected mean square. The standard errors needed for estimates and for multiple comparison are given in the following table:

Difference Between	Standard Error	df for t
Two overall A means	$\sqrt{\dfrac{2MS_b}{bc}}$	$a(b-1)$
Two overall C means	$\sqrt{\dfrac{2MS_{sr}}{ab}}$	$a(b-1)(c-1)$
Two C means at the same A level	$\sqrt{\dfrac{2MS_{sr}}{b}}$	$a(b-1)(c-1)$

This split-plot design is appropriate under the same circumstances as those discussed in Section 12.6. The difference between the two designs lies in whether or not whole-plot effects can be replicated in a single experimental unit or a complete block. Here experimental units are nested within whole-plot effects rather than complete blocks that would be factorial to whole-plot treatments.

EXERCISES

12.7.1. When Francis Galton was determining how to measure boredom (as mentioned in Exercise 4.3.5), he felt it would be rude to use a watch while counting number of signs of boredom per minute. Instead, he trained himself to know, without a timepiece, when 60 seconds had passed. Primarily out of curiosity, a number of scientists have confirmed that people do have the ability to know when a constant period of time has passed, although that period may not be exactly 60 seconds. Suppose a graduate student decides to study whether the ability is affected by gender or periods of time other than 60 seconds. She decides on a repeated-measures design and solicits 4 male and 4 female colleagues to participate. She asks them to try to train themselves to know when 60, 120, and 180 seconds have passed, and after they feel they are ready, she tests them independently at her computer, where they make a keystroke when they feel that 1, 2, and 3 minutes have passed. The computer provides the exact number of seconds that have passed at each keystroke and the results are

	Males				Females				
Target Time	MI	M2	M3	M4	F1	F2	F3	F4	Total
60 sec	52	66	57	55	62	58	62	54	466
120 sec	112	127	126	122	121	123	133	118	982
180 sec	178	173	189	183	177	179	177	176	1432
Total	342	366	372	360	360	360	372	348	

a. Give the linear model and identify all of the symbols.

b. Given that $\sum\sum\sum y_{ijk}^2 = 404{,}616$:
   i. Complete the ANOVA.
   ii. Give the numerical values of Rsquare.

c. Are there significant gender differences in ability to tell when a certain period of time has passed? Explain.

d. For the actual times when males and females made the keystroke thinking 180 seconds had passed:
   i. Compute the average time when each gender thought 180 seconds had passed. Are these average times significantly different? Explain.
   ii. For males, how would you test H_0: $\mu = 180$. *Hint:* You will need the standard error of the average of 4 values.

12.7.2. An endodontist is interested in assessing the effects of 2 medications to provide pain relief for his patients following a root canal procedure. Two patients are randomly assigned to each medication. The procedure is performed and the patients given medication. The patients are asked to indicate their level of pain on a scale from 1 to 10 twice, 4 hours after the procedure and 8 hours after the procedure.

Medication	Patient	Time	Pain
1	1	4	0
		8	3
	2	4	6
		8	5
2	3	4	3
		8	8
	4	4	9
		8	10

a. Perform a repeated-measures ANOVA and interpret the results.

b. Critique the experiment in light of your analysis. Indicate two specific changes that would improve the experiment

12.7.3. In the Great Plains, controlled spring burning of pastures is a common practice. Burning destroys the weed seed, eliminates thatch, and may promote early emergence of the grass. However, it reduces critical soil moisture, so an agronomist wants to study the effect on deeper levels of soil moisture. He has three similar pastures and randomly selects early spring burning for one and late spring burning for another and leaves the other unburned (control). Then he drives a metal tube at 2 random locations in each pasture to obtain a 4-foot-long core of the soil at each location. From each core he takes the soil at 1-, 2-, and 3-feet depths below the surface and finds the moisture content at each depth. Simplified for ease of computation, the data are provided below. (Larger measures indicate greater moisture.)

	Early Spring		Late Spring		Control		
Depth (ft)	Core 11	Core 12	Core 21	Core 22	Core 31	Core 32	Total
1	2.1	1.4	0.9	0.8	2	2.5	9.7
2	2.3	1.6	1.2	1.4	2.9	2	11.4
3	2.5	2.7	2.4	2.6	3.2	2.7	16.1
Total	6.9	5.7	4.5	4.8	8.1	7.2	

a. Give the linear model for this experiment, identifying all symbols.

b. Give the null hypotheses that can be tested.

c. Perform the ANOVA and make all appropriate F tests.

d. The agronomist is especially interested in testing the average moisture for each burning treatment with that of the control at 1-, 2-, and 3-feet depths.

    i. Why would it be inappropriate for him to use Fisher's least significant difference?

    ii. JMP allows him to use MS_e to make the t tests of interest to him and provides the following P values for two-sided tests:

Comparison at Depth

	1 ft	2 ft	3 ft
Control vs. early	$P = 0.225$	$P = 0.225$	$P = 0.380$
Control vs. late	$P = 0.009$	$P = 0.021$	$P = 0.202$

Use Bonferroni procedures to determine what conclusions he should draw about the effect of burning on soil moisture at the depths he has chosen to compare

REVIEW EXERCISES

Decide whether each of the following is true or false. If a statement is false, explain why.

12.1. The Latin square design is appropriate for pilot experiments in new areas of research, because it provides an economical design for measuring three different kinds of variability.

12.2. The model $y_{ijk} = \mu + \alpha_i + \beta_{ij} + \varepsilon_{ijk}$ does not indicate whether the β is a block effect or a nested effect.

12.3. If an interaction exists in experimental data and no provision is made for it in the model and analysis, the interaction variability will be confounded with the estimate of random variation.

12.4. The chief advantage of the Latin square design is that it permits the analysis of main effects without any concern for interaction.

12.5. Because the residual mean squares from a blocked design will have fewer degrees of freedom than the within mean square of a one-way analysis of the same data, one could obtain a poorer F test of treatments in a blocked design if the block effects are nonsignificant.

12.6. When performing a randomized complete block ANOVA, the experimenter is usually as interested in finding differences among the blocks as among the treatments, so he uses some sort of multiple comparison technique on both sets of means.

12.7. Whether an effect is nested or factorial has no bearing on whether it is random or fixed.

12.8. In ANOVA, it may be possible to estimate a particular variance component, but still not be possible to have an exact test for significance.

12.9. In an experiment involving 3 effects in a factorial arrangement, if all 3 main effects are fixed, the interaction term drops out of the expectations of all mean squares.

12.10. The nested classification is a continued one-way classification of subgroups within the major groups.

12.11. Missing value techniques may be employed even when all observations in a row or column are missing.

12.12. To use a missing value technique does not cause the loss of one degree of freedom; the degree of freedom was lost when the observation was lost.

12.13. A repeated measures design consists for one randomized complete block design nested within another randomized complete block.

12.14. A linear model may contain both factorial and nested effects.

12.15. A linear model may contain both random and fixed effects.

12.16. In ANOVA, if the row mean square is nonsignificant and the column mean square is also nonsignificant, it is unlikely that the row × column mean square will be significant.

12.17. Because the Latin square design does not permit a treatment to be found twice in the same row or column, it is impossible to randomize treatments in that design.

12.18. There are four types of interactions in an $a \times b \times c$ factorial design.

12.19. Data collected for an $a \times b \times c$ factorial design may be analyzed as a split-plot design.

12.20. Approximate tests must be used for some follow-up procedures after a split-plot analysis.

SELECTED READINGS

Bliss, C. I. (1967). *Statistics in Biology*. McGraw-Hill, New York.

Box, G. E. P. (1954). Some theorems on quadratic forms applied in the study of analysis of variance problems: II. Effect of inequality of variance and of correlation between errors in the two-way classification. *Annals of Mathematical Statistics*, **25**, 484–498.

Brown, B. M. (1975). A short-cut test for outliers using residuals. *Biometrika*, **62**, 623–629.

Cochran, W. G., and G. M. Cox (1957). *Experimental Design*. Wiley, New York.

Daniel, C. (1960). Locating outliers in factorial experiments. *Technometrics*, **2**, 149–196.

Daniel, C. (1978). Patterns in residuals in the two-way layout. *Technometrics*, **20**, 385–395.

Davies, O. L., Ed. (1956). *The Design and Analysis of Industrial Experiments*, 2nd ed. Hafner, New York.

DeLury, D. B. (1946). The analysis of Latin squares when some observations are missing. *Journal of the American Statistical Association*, **41**, 370–389.

Fisher, R. A., and F. Yates (1963). *Statistical Tables for Biological, Agricultural, and Medical Research*. Hafner, New York.

Geisser, S. (1959). A method for testing treatment effects in the presence of learning. *Biometrics*, **15**, 389–395.

Glenn, W. A., and C. Y. Kramer (1958). Analysis of variance of a randomized block design with missing observations. *Applied Statistics*, **7**, 173–185.

Harter, H. L. (1970). Multiple comparison procedures for interactions. *American Statistician*, **24** (Dec.), 30–32.

Kramer, C. Y., and S. Glass (1960). Analysis of variance of a Latin square design with missing observations. *Applied Statistics*, **9**, 43–50.

Monlezun, C. J. (1979). Two-dimensional plots for interpreting interactions in the three-factor analysis of variance model *American Statistician*. **33**, 63–69.

Murray, L. W. (1986). Estimation of missing cells in randomized block and Latin square designs. *American Statistician*, **40**, 289–293.

Schultz, E. F., Jr. (1955). Rules of thumb for determining expectations of mean squares in analysis of variance. *Biometrics*, **11**, 123–135.

Snedecor, G. W., and W. G. Cochran (1973). *Statistical Methods*. 6th ed. Iowa State University Press.

Steel, R. G. D., and J. H. Torrie (1960). *Principles and Procedures of Statistics*. McGraw-Hill, New York.

Taylor, W. H., Jr., and H. G. Hilton (1981). A structure diagram symbolization for analysis of variance. *American Statistician*, **35**, 85–93.

Wilk, M. B., and O. Kempthorne (1957). Nonadditives in a Latin square design. *Journal of the American Statistical Association*, **52**, 218–236.

Winer, B. J. (1971). *Statistical Principles in Experimental Design*, 2nd ed. McGraw-Hill, New York.

13 Analysis of Covariance

The analysis of covariance is a combination of regression analysis with an ANOVA. Covariance is used when the response variable y, in addition to being affected by the treatments, is also linearly related to another variable x. In this chapter we discuss the analysis of covariance in which simple linear regression is combined with a one-way ANOVA. More complex designs exist but are beyond the scope of this book.

13.1. COMBINING REGRESSION WITH ANOVA

The analysis of covariance is useful in several types of research situations. For example, it can be used to

1. increase precision in an experiment,
2. control for an extraneous variable in a survey, and
3. compare regressions within several groups.

Specific examples of these three types of applications follow.

Increasing precision in an experiment is illustrated by the use of covariance analysis in a study of weight loss y under 3 different diets (the treatments). Ordinary ANOVA may fail to detect a significant difference among the treatment effects because the within-treatment-group variability is too large. Covariance sharpens the ANOVA on y by utilizing a related variable x, called a *covariate*, or *concomitant variable*. Pounds lost, y, is linearly related to x, pounds overweight at the beginning of the experiment. By combining the regression of y on x with the ANOVA on y, the within-treatment variability is reduced, making it more likely that treatment differences will be detected. Intuitively we can think of the analysis of covariance as removing that portion of the within-treatment variability which is accounted for by the regression. (Blocking by overweight classes could also be used to reduce within-group variability, but this cannot always be done since it requires equal numbers of subjects in each overweight class.)

Controlling for an extraneous variable in a survey is illustrated by a study of teachers' salaries y in 3 different school systems (treatment groups) in which the educational level in years attained by the teachers is an extraneous variable x. If y is linearly related to x, then the analysis of covariance can be used to adjust for differences in the educational attainment of the teachers. In this application, we can think of the analysis of covariance as transforming each of the data points (x_{ij}, y_{ij}) to $(\bar{x}_{..}, y'_{ij})$, a point on the vertical line at the overall average x value, by means of a translation parallel to the regression line (Figure 13.1).

Intuitively, this means that all the subjects are made average with respect to educational attainment, and then the corresponding adjusted y values are analyzed for significant

Statistics for Research, Third Edition, Edited by Shirley Dowdy, Stanley Weardon, and Daniel Chilko.
ISBN 0-471-26735-X © 2004 John Wiley & Sons, Inc.

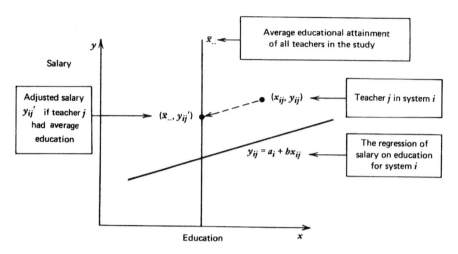

FIGURE 13.1. Adjusting observations by covariance analysis.

differences due to school systems. Group averages are also transformed in this process; sometimes the adjusted averages $(\bar{y}'_{i.})$ are further apart, sometimes they are closer together than the original averages (Figure 13.2). Because the regression lines are estimated from the data, the actual analysis is more complex than finding the lines, transforming all data points, and performing the ANOVA on the transformed points. However, the adjusted group averages can be found by this method.

In the third type of application of covariance, *comparing regressions within several groups*, the classifications (treatments) are not of primary concern, but rather the relationship of y to x within each classification is of main interest. In this case, the experimental hypothesis is that the treatments affect slopes differently. For example, it is known that high blood pressure is more common in some racial groups than in others. Data on the relationship of salt intake x and blood pressure y may be classified by racial groups and covariance used to determine whether the relationship between salt and blood pressure is the same for all the racial groups in the study.

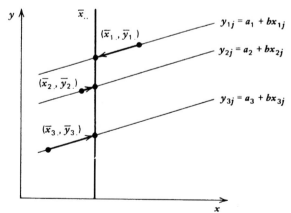

FIGURE 13.2. Adjusting group averages.

The additive model for the analysis of covariance is

$$y_{ij} = \mu + \alpha_i + \beta(x_{ij} - \bar{x}_{..}) + \varepsilon_{ij}$$
$$i = 1, \ldots, a$$
$$j = 1, \ldots, n_i$$
$$N = \sum_i n_i$$

The terms in this model have the following meanings:

μ: The true overall y mean for all studies of this type involving the specified treatments.

α_i: The deviation due to the ith treatment after allowance for the relationship of y to x; $\sum_i \alpha_i = 0$. (*Note:* α_i are the treatment effects and not the y intercepts.)

β: The true common slope of the a regression lines.

$\bar{x}_{..}$: The overall average of the covariate for the observations in the study.

ε_{ij}: A random effect for the jth element in the ith treatment group; ε_{ij} IND$(0, \sigma^2)$.

The model assumes that all of the regression lines have the same slope, that the variances about the regression lines are equal, and that the covariate x_{ij} is unaffected by the treatments, and it makes the usual assumptions for the ANOVA. Figure 13.3 may be helpful in understanding the terms in the model.

In the study of teachers' salaries, μ is the true mean salary for all teachers in the 3 school systems. The fixed effect α_2 is the true deviation from the mean salary in the second school system after making allowance for the educational attainment of the teachers in that system. The common slope β is the change in salary per additional year of teachers' education. The

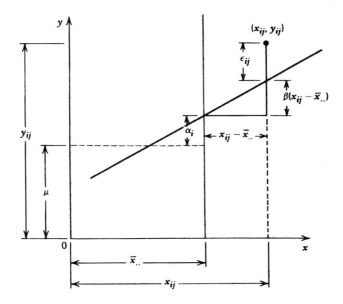

FIGURE 13.3. Terms in the covariance model.

average educational attainment for the teachers in the three samples is $\bar{x}_{..}$. The random effect ε_{32} is the deviation of the second teacher in the sample from the third school system from the regression line for the third system.

In an analysis of covariance, we are usually interested in testing for differences in the treatment effects:

$$H_0: \alpha_1 = \alpha_2 = \cdots = \alpha_a = 0 \quad \text{against} \quad H_a: \textit{At least one inequality}$$

If the inequality of slopes is of primary interest, it can also be tested within the covariance procedure. Since the equality of slopes is an assumption of the model, it is usually tested to verify that the proper model is being used.

EXERCISES

13.1.1. Samples of 3 varieties of wheat, A, B, and C, result in the following (artificial) data for yield y in bushels per acre and rainfall x in inches:

	A		B		C
x	y	x	y	x	y
1	2	2	3	3	2
2	6	3	7	4	6
4	10	5	11	6	10
5	10	6	11	7	10

a. Draw the scatter plot for each variety on a common graph, keeping the varieties separate by using different colors or symbols.

b. Find the unadjusted group means $(\bar{x}_{i.}, \bar{y}_{i.})$ and add them to the graph.

c. Draw the vertical line at $x = \bar{x}_{..}$.

d. Estimate the regression equation for each variety and add these lines to the graph. (Note that the estimates of the slopes are the same.)

e. Compute $\bar{y}'_{i.}$ for each variety from the regression equations. Locate the adjusted means on the graph.

f. Will the analysis of covariance increase or decrease the differences among the variety averages? Does it change the rank order of the group averages?

13.1.2. The diagrams in Figure 13.4 show the unadjusted treatment averages and the regression lines for the treatment groups in experiments in which covariance is being considered as a method of analysis. In which case or cases can covariance be justified?

13.1.3. Match the following statistical symbols with the indicated distances on the graph in Figure 13.5:

(1) $y_{ij} - \mu$ (5) $y_{ij} - \hat{y}_{ij}$
(2) y_{ij} (6) $\bar{y}'_{i.}$
(3) μ (7) $\bar{y}'_{i.} - \mu$
(4) $\bar{y}_{i.}$

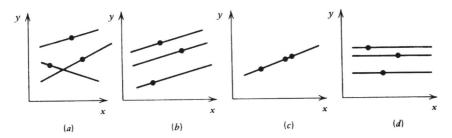

FIGURE 13.4. Regression lines and unadjusted treatment averages.

13.2. ONE-WAY ANALYSIS OF COVARIANCE

Let the data for a one-way analysis of covariance with $a = 3$ treatments and $n_1 = n_2 = n_3 = 4$ observations per treatment group be arranged as follows:

		Treatment			
I		II		III	
x	y	x	y	x	y
x_{11}	y_{11}	x_{21}	y_{21}	x_{31}	y_{31}
x_{12}	y_{12}	x_{22}	y_{22}	x_{32}	y_{32}
x_{13}	y_{13}	x_{23}	y_{23}	x_{33}	y_{33}
x_{14}	y_{14}	x_{24}	y_{24}	x_{34}	y_{34}
$T_{1.(x)}$	$T_{1.(y)}$	$T_{2.(x)}$	$T_{2.(y)}$	$T_{3.(x)}$	$T_{3.(y)}$

Totals: $T_{..(x)}\ T_{..(y)}$

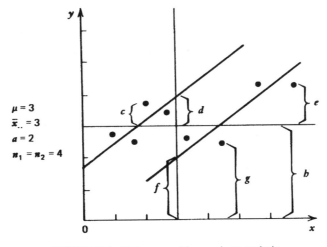

FIGURE 13.5. Distances used in covariance analysis.

Using a similar layout for a treatment groups and n_i observations per group, the general analysis-of-covariance procedure can be summarized as follows.

Procedure. Analysis of Covariance

Test of Hypothesis

$H_0: \alpha_1 = \alpha_2 = \cdots = \alpha_a$ against H_a: *At least one inequality*

Model:
$$y_{ij} = \mu + \alpha_i + \beta(x_{ij} - \bar{x}_{..}) + \varepsilon_{ij}$$
$$i = 1, 2, \ldots, a$$
$$j = 1, 2, \ldots, n_i$$
$$N = \sum_i n_i$$

Uncorrected Sums of Squares and Products

x	xy	y
$T_{(x)} = \sum_i \sum_j x_{ij}^2$	$T_{(xy)} = \sum_i \sum_j x_{ij}y_{ij}$	$T_{(y)} = \sum_i \sum_j y_{ij}^2$
$A_{(x)} = \sum_i T_{i.(x)}^2/n_i$	$A_{(xy)} = \sum_i T_{i.(x)}T_{i.(y)}/n_i$	$A_{(y)} = \sum_i T_{i.(y)}^2/n_i$
$CF_{(x)} = T_{..(x)}^2/N$	$CF_{(xy)} = T_{..(x)}T_{..(y)}/N$	$CF_{(y)} = T_{..(y)}^2/N$

Corrected Sums of Squares and Products

Source	df	$SS_{(x)}$	SP	$SS_{(y)}$
Treatment	$a - 1$	$SS_{a(x)} = A_{(x)} - CF_{(x)}$	$SP_a = A_{(xy)} - CF_{(xy)}$	$SS_{a(y)} = A_{(y)} - CF_{(y)}$
Error	$N - a$	$SS_{e(x)} = T_{(x)} - A_{(x)}$	$SP_e = T_{(xy)} - A_{(xy)}$	$SS_{e(y)} = T_{(y)} - A_{(y)}$
Total	$N - 1$	$SS_{t(x)} = T_{(x)} - CF_{(x)}$	$SP_t = T_{(xy)} - CF_{(xy)}$	$SS_{t(y)} = T_{(y)} - CF_{(y)}$

Adjusted Sums of Squares

Source	df'	$SS'_{(y)}$	$MS'_{(y)}$
Treatment	$a - 1$	$SS'_{a(y)} = SS'_{t(y)} - SS'_{e(y)}$	$MS'_{a(y)} = SS'_{a(y)}/(a-1)$
Error	$N - a - 1$	$SS'_{e(y)} = SS_{e(y)} - SP_e^2/SS_{e(x)}$	$MS'_{e(y)} = SS'_{e(y)}/(N-a-1)$
Total	$N - 2$	$SS'_{t(y)} = SS_{t(y)} - SP_t^2/SS_{t(x)}$	

Reject H_0 if $F = MS'_{a(y)}/MS'_{e(y)} > F_{\alpha, a-1, N-a-1}$ at the α level of significance.

The procedure is illustrated by the following example.

Example 13.1.   One-Way Analysis of Covariance

An experiment was conducted involving 3 different advertising media, each used for 5 fast food restaurants of a certain franchise. The 15 restaurants were located in different but comparable cities, and they were randomly assigned to the 3 advertising media: radio, newspaper, television. All advertising took place during the same time period. Profits y in thousands of dollars were recorded for the same time period. Although all restaurants were of the same size, they employed different numbers of workers. Since additional employees may affect profits, the number of employees was used as a concomitant variable x.

Medium

	I		II		III		
	x	y	x	y	x	y	
	10	30	21	24	34	17	$a = 3$
	14	18	26	20	39	11	$n_1 = n_2 = n_3 = 5$
	19	13	31	7	43	3	$N = 15$
	25	6	36	4	47	-6	
	27	3	41	-5	52	-10	
							Totals
$T_{i.(x)}$	95		155		215		$T_{..(x)} = 465$
$T_{i.(y)}$		70		50		15	$T_{..(y)} = 135$
$\sum x_{ij}^2$	2011		5055		9439		16,505
$\sum x_{ij}y_{ij}$		1030		1180		334	2,544
$\sum y_{ij}^2$		1438		1066		555	3,059

Uncorrected Sums of Squares and Products

	x	xy	y
T	16,505	2544	3059
A	$(95^2 + 155^2 + 215^2)/5$	$[95(70) + 155(50) + 215(15)]/5$	$(70^2 + 50^2 + 15^2)/5$
	$= 15,855$	$= 3525$	$= 1525$
CF	$465^2/15 = 14,415$	$465(135)/5 = 4185$	$135^2/15 = 1215$

Corrected Sums of Squares and Products

Source	df	$SS_{(x)}$	SP	$SS_{(y)}$
Treatment	$3 - 1$ $= 2$	$15,855 - 14,415$ $= 1440$	$3525 - 4185$ $= -660$	$1525 - 1215$ $= 310$
Error	$15 - 3$ $= 12$	$16,505 = 15,885$ $= 650$	$2544 - 3525$ $= -981$	$3059 - 1525$ $= 1534$
Total	14	2090	-1641	1844

The analysis of covariance uses both ANOVA and regression techniques. The corrected sums of squares of the y variable are obtained in the usual manner for ANOVA; the corrected total sums of squares is computed as $SS_{t(y)} = T_{(y)} - CF_{(y)}$ and the corrected among-treatment sums of squares as $SS_{a(y)} = A_{(y)} - CF_{(y)}$. Then the same mathematical procedure is performed on the x variable and the xy cross-products because they are needed for the aspect of the analysis of covariance that uses regression techniques.

Since $SS_{t(y)} = SS_{a(y)} + SS_{e(y)}$, the error sum of squares could be computed as

$$SS_{e(y)} = SS_{t(y)} - SS_{a(y)}$$

Note what is being done. We have the corrected total sums of squares for the experiment and we subtract from that the sums of squares due to differences among groups. What is left can be called *the variability after accounting for groups*, and this, of course, is the random variability making up the error sums of squares. Recalling this may help in understanding the different computations used for the adjusted sums of squares.

The adjusted sum of squares uses the corrected sums of squares and products that were computed in the previous table. The "adjusting" is along the trend line, sliding each y_{ij} along the parallel trend lines to $\bar{x}_{..}$, as shown in Figure 13.2. We do this mathematically with regression techniques. First we compute the sum of squares due to regression, SS_{xy}^2/SS_x, and then we adjust the corrected sums of squares by subtracting the sums of squares due to regression:

$$SS'_{(y)} = SS_{(y)} - SS_{(xy)}^2/SS_{(x)}$$

We perform this operation on the total sums of squares and the error sums of squares. However, we do not adjust the among-treatment sums of squares in this manner. To attempt to do so would result in trying to fit a straight line through the three points $(\bar{x}_i, \bar{y}_i)$. Instead, we compute the adjusted sum of squares among groups $SS'_{a(y)}$ in a different manner. Because the total sum of squares is the sum of both the among-treatment sum of squares and the error sum of squares, we obtain the adjusted-treatment sum of squares by subtraction

$$SS'_{a(y)} = SS'_{t(y)} - SS'_{e(y)}$$

which we can call the variability after accounting for regression. In $SS'_{a(y)}$ we have the variability in profit among the three different kinds of media after accounting for different numbers of employees per restaurant. The numerical operations can be seen in the table of adjusted sums of squares.

Adjusted Sums of Squares

Source	df'	$SS'_{(y)}$	$MS'_{(y)}$	F
Treatment	2	$555.54 - 53.44 = 502.10$	251.050	51.68
Error	11	$1534 - (-981)^2/650 = 53.44$	4.858	
Total		$1844 - (-1641)^2/2090 = 555.54$		

Since $F_{0.05,2,11} = 3.982$, the null hypothesis is rejected. There is a significant difference among the media effects on average profits after adjusting for number of employees.

The SAS output for this analysis would be

```
                       The GLM Procedure
                   Class Level Information
Class             Levels                    Values
MEDIUM               3                      I  II  III


               Number of observations            15
```

Dependent Variable: PROFIT

Source	DF	Sum of Squares	Mean Square	F Value	Pr > F
Model	3	1790.555385	596.851795	122.84	<.0001
Error	11	53.444615	4.858601		
Corrected Total	14	1844.000000			

R-Square	Coeff Var	Root MSE	PROFIT Mean
0.971017	24.49137	2.204224	9.000000

Source	DF	Type I SS	Mean Square	F Value	Pr > F
MEDIUM	2	310.000000	155.000000	31.90	<.0001
EMPLOY	1	1480.555385	1480.555385	304.73	<.0001

Source	DF	Type III SS	Mean Square	F Value	Pr > F
MEDIUM	2	502.095576	251.047788	51.67	<.0001
EMPLOY	1	1480.555385	1480.555385	304.73	<.0001

```
                   Least Squares Means
                          PROFIT      LSMEAN
        MEDIUM            LSMEAN      Number
        I              -4.1107692       1
        II             10.0000000       2
        III            21.1107692       3
        Least Squares Means for effect MEDIUM
        Pr > |t| for H0: LSMean(i) = LSMean(j)
```

Dependent Variable: PROFIT

i/j	1	2	3
1		<.0001	<.0001
2	<.0001		<.0001
3	<.0001	<.0001	

NOTE: To ensure overall protection level, only probabilities associated with pre-planned comparisons should be used.

PROC GLM produces two different sets of sums of squares, Type I and Type III. For the analysis of covariance the Type I sums of squares are the unadjusted $SS_{(y)}$ and the

Type III sums of squares are the adjusted $SS'_{(y)}$. The F value for medium in the Type III output is the value in the analysis of covariance. The other F values are not useful in this analysis.

EXERCISES

13.2.1. A certain airplane part must withstand extremes of temperature. The part can be made from a number of metal alloys; the one to be chosen must have the greatest strength y for a given density x. An experiment is designed involving 5 alloys and 5 parts per alloy. In hopes of obtaining a lighter part, the density of each alloy is deliberately varied within a safe range. The data are analyzed by covariance procedures to yield the following information:

Source	df	$SS_{(x)}$	SP	$SS_{(y)}$
Alloys	4	200	300	2500
Error	20	300	1200	7500

a. What is the linear model?

b. What assumptions must be made in order to perform the analysis of covariance?

c. Complete the analysis of covariance.

13.2.2. Complete the analysis of covariance for the data given in Exercise 13.1.1.

13.3. TESTING THE ASSUMPTIONS FOR ANALYSIS OF COVARIANCE

For an analysis of covariance to be valid, we may need to verify that:

1. All the treatment groups have the same variance about their regression lines, $\sigma_1^2 = \cdots = \sigma_a^2 = \sigma^2$.
2. All the regression lines have the same slope, $\beta_1 = \beta_2 = \cdots = \beta_a = \beta$.
3. The common slope β is not equal to 0; that is, the regression lines are not horizontal.

In this section, we illustrate these tests using the advertising media study, Example 13.1. We begin by estimating the individual regression lines for each treatment group:

Medium

	I		II		III	
	x	y	x	y	x	y
S_{xx}	206		250		194	
S_{xy}		-300		-370		-311
S_{yy}			458		566		510
$\bar{x}_{i.}$	19		31		43	
$\bar{y}_{i.}$		14		10		3
b_i		-1.46		-1.48		-1.60
a_i		41.74		55.88		71.80

In this table, the sums of squares and cross-products are computed as for simple linear regression. Thus, for medium I,

$$S_{xx} = \sum_j x_{1j}^2 - T_{1.(x)}^2/n_1 = 2011 - (95)^2/5 = 206$$

$$S_{xy} = \sum_j x_{1j}y_{1j} - T_{1.(x)}T_{1.(y)}/n_1 = 1030 - (95)(70)/5 = -300$$

$$S_{yy} = \sum_j y_{1j}^2 - T_{1.(y)}^2/n_1 = 1438 - (70)^2/5 = 458$$

and so on.

The slope and y intercept are also computed as in simple linear regression. For example, for medium I,

$$b_1 = \frac{S_{xy(1)}}{S_{xx(1)}} = \frac{-300}{206} = -1.46$$

and

$$a_1 = \bar{y}_1 - b_1\bar{x}_1 = 14 - (1.46)19 = 41.74$$

To test for the equality of variances about the trend lines, we may use the F_{max} test or Bartlett's test (see Section 11.2). The variability about each line is computed using

$$s_i^2 = \frac{\sum_j (y_{ij} - \hat{y}_{ij})^2}{n_i - 2} = \frac{S_{yy(i)} - S_{xy(i)}^2/S_{xx(i)}}{n_i - 2}$$

Using the sums of squares and cross-products above, we have

Medium	df	$S_{yy(i)} - S_{xy(i)}^2/S_{xx(i)}$	s_i^2
I	$n_1 - 2 = 3$	21.11	7.04
II	$n_2 - 2 = 3$	18.40	6.13
III	$n_3 - 2 = 3$	11.44	3.81
		50.95	

$$F_{max} = \frac{\text{largest } s_i^2}{\text{smallest } s_i^2} = \frac{7.04}{3.81} = 1.85$$

and $F_{max\ \alpha,a,n-2} = F_{max\ 0.05,3,3} = 27.8$ from Table A.16 in the Appendix with $a = 3$ groups, and $n_i - 2 = 3$ degrees of freedom for each estimated variance. Since F_{max} is not significant, we conclude that the variances are the same, and we proceed to test the other assumptions necessary for an analysis of covariance.

The equality of the slopes $\beta_1 = \beta_2 = \beta_3$ is tested by comparing the sum of squared deviations from the regression lines $\sum\sum (y_{ij} - \hat{y}_{ij})^2$ when the lines are found two different ways. First, using the individual estimates of the slopes

$$b_1 = -1.46 \quad b_2 = -1.48 \quad b_3 = -1.60$$

and, second, using a pooled estimate of the slope

$$b = \frac{SP_e}{SS_{e(x)}} = \frac{-981}{650} = -1.51$$

If the three separate estimates b_1, b_2, b_3 are all estimates of the same parameter, the difference between these two sums of squared deviations should not be significant.

The sum of squared deviations about the regression lines using b, the pooled estimate of the slope, is

$$G = SS'_{e(y)} = 53.44$$

and the sum of squared deviations using the individual estimates of the b_i's is

$$H = \sum_i \left[S_{yy(i)} - \frac{S^2_{xy(i)}}{S_{xx(i)}} \right]$$

$$= 50.95$$

The test can be summarized in the following table:

$$H_0: \quad \beta_1 = \beta_2 = \beta_3 \quad \text{against} \quad H_a: \quad \text{At least one inequality}$$

Source	df	SS	MS
About regression lines using one b	$N - a - 1 = 11$	$G = SS_{e(y)}' = 53.44$	
About regression lines using three b_i	$N - 2a = 9$	$H = 50.95$	5.661
Difference	$a - 1 = 2$	$G - H = 2.49$	1.245

$$F = \frac{(G - H)/(a - 1)}{H/(N - 2a)} = \frac{1.245}{5.661} = 0.220$$

Reject H_0 if $F > F_{0.05,\ 2,\ 9} = 4.256$, so there is no evidence of unequal slopes.

Sometimes an experimenter expects differences among regression lines, and his experimental hypothesis may even be that the treatments will affect the slopes. An example of this might be an experiment comparing aspirin substitutes for how quickly they reduce the fever of babies. The temperature of the baby is the y variable, and the time at which temperature is taken during fever is the x variable. Since aspirin is not recommended for babies, the experimenter wants to compare safe substitutes on the basis of the slopes of their lines for the regression of temperature on time. The more negative the slope, the quicker the drug reduces fever.

When an ANOVA detects significant differences, to determine which averages are significantly different from others, we use Fisher's least significant difference or one of the other mean separation techniques discussed in Chapter 10. Unfortunately, there are no similar procedures that are generally accepted for finding significant differences among the b_i. Rather than rely only on the relative sizes of the numerical values of the b_i, we could perform $\binom{a}{2}$

F tests comparing all slopes two at a time. In the example of the regression of profit on number of employees there are 3 media, I, II, and III, and we tested H_0: $\beta_I = \beta_{II} = \beta_{III}$. However, by leaving data from medium III out of the analysis, we could perform a test of H_0: $\beta_I = \beta_{II}$. Then in a second analysis we could omit data for medium II and test H_0: $\beta_I = \beta_{III}$ and finally ignore data for medium I in a third analysis to test H_0: $\beta_{II} = \beta_{III}$. However, the experiment involved $an = 15$ observations and it seems intuitively unattractive to leave a third of them out of each analysis. Furthermore, there is the problem of the global α when repeated tests of hypotheses are performed, so if this is a concern, we can consider them simultaneous tests and adjust the α_i for each F test according to Bonferroni procedures. This suggestion is not considered optimum but rather is based on the feeling that it is better to try to determine which trend lines are different from others rather than to claim there are significant differences and not identify them.

For the analysis of covariance to be a significant improvement over a simple one-way ANOVA, the common slope β must not be zero:

$$H_0: \beta = 0 \quad \text{against} \quad H_a: \beta \neq 0$$

is tested by

$$F = \frac{SP_e^2/SS_{e(x)}}{MS'_{e(y)}} = \frac{(-981)^2/650}{4.858} = 304.77$$

with 1 and $N - a - 1$ degrees of freedom. Since $F_{0.05,\ 1,\ 11} = 4.844$, we reject $\beta = 0$, and it is appropriate to use an analysis of covariance.

There are times when the hypothesis H_0: $\beta = 0$ is not rejected, but covariance still provides a more powerful test of H_0: $\alpha_1 = \alpha_2 = \cdots = \alpha_a$ than would a comparable ANOVA of the y variable. If the experimenter has reason to suspect that a sizable portion of the variability in y is attributable to a covariate x, the experiment should be designed and data collected with covariance analysis in mind. The worst that can happen is the loss of one degree of freedom attributable to a nonsignificant b. But even with that loss, $MS'_{e(y)}$ may still be sufficiently smaller than $MS_{e(y)}$ for covariance analysis to be more powerful than ANOVA.

EXERCISES

13.3.1. In Exercise 13.2.1

    a. What is the pooled estimate of the slope?

    b. Test that the slope is not zero.

13.3.2. Given that $n_1 = n_2 = n_3 = 10$, y is the yield of a certain crop, x is the amount of limestone added to the soil, and

Soil Type	S_{xx}	S_{xy}	S_{yy}
A	4500	4200	4300
B	5800	3600	2400
C	5100	5100	5300

    a. Estimate the individual slopes for each type of soil.

    b. Estimate the variances about the trend lines.

    c. Test for homogeneity of variances.

    d. Estimate the common slope.

    e. Test that the three slopes are equal.

13.3.3. See Exercise 13.2.2. Was the analysis of covariance on the data from Exercise 13.1.1 justified?

13.3.4. Darwin's theory of evolution postulates that there is a struggle for existence and only the fittest survive. Using these two principles, experimental geneticists can quantify the relative fitness of different species by comparing their survival under some stressful conditions. Suppose a researcher wishes to compare the relative survival of 3 species of *Drosophila* under increasing levels of organic phosphorous insecticide. Four batches of medium are prepared and all batches are identical except for the level of insecticide they contain. One hundred eggs from each species are deposited on each preparation. The variables recorded for each container are level of insecticide x in parts per million (ppm) and number of flies that survive to adulthood y. The researcher knows that the experiment may show either of two results: The mean number of survivors is not the same from species to species or the effect of increasing the level of insecticide is not the same for all species.

    a. Give the null and alternative hypotheses used to test each of these responses.

    b. Give the null and alternative hypotheses used to test each of these responses.

    c. Which null hypothesis should be tested first?

    d. Given the following data:

	Species		
Level of Insecticide (ppm)	*Drosophila melanogaster*	*Drosophila pseudoobscura*	*Drosophila serrata*
0.0	91	89	87
0.3	71	77	43
0.6	23	12	22
0.9	5	2	8

    i. Test the hypothesis that all species show that same response to increasing levels of insecticide in the medium.

    ii. Should the researcher compute adjusted average survival for an average level of insecticide? Why?

    iii. Draw a graph to show how each species responds to increased levels of insecticide.

    iv. If the 3 species were competing for existence in an environment in which insecticide is accumulating, which species seems to have the best advantage, that is, the greatest relative fitness?

13.4. MULTIPLE-COMPARISON PROCEDURES

If the analysis of covariance is justified and leads to a significant F test for differences among the adjusted averages, then we will want to follow this procedure with a test that compares the means, a *multiple-comparison procedure*.

The adjustments must be performed, of course, before we can test the adjusted averages. (These are symbolized as $\bar{y}'_{i.}$ earlier but as adj $\bar{y}_{i.}$ here to identify them clearly as adjusted rather than "raw" averages.) Intuitively, the original group averages $(\bar{x}_{i.}, \bar{y}_{i.})$ are transformed along the regression lines to the vertical line $x = \bar{x}_{..}$ (see Figure 13.6). Algebraically the transformed y averages can be found by the formula

$$\text{adj } \bar{y}_{i.} = \bar{y}_{i.} - b(\bar{x}_{i.} - \bar{x}_{..})$$

Thus, in the advertising media experiment (Example 13.1),

$$\text{adj } \bar{y}_{1.} = 14 - (-1.51)(19 - 31) = -4.1$$
$$\text{adj } \bar{y}_{2.} = 10 - (-1.51)(31 - 31) = 10.0$$
$$\text{adj } \bar{y}_{3.} = 3 - (-1.51)(43 - 31) = 21.1$$

If desired, confidence intervals can be found for the adjusted means:

$$\text{CI}_{1-\alpha}: \text{ adj } \bar{y}_{i.} \pm t_{\alpha/2, N-a-1}\sqrt{\text{MS}'_{e(y)}}\sqrt{\frac{1}{n_i} + \frac{(\bar{x}_{i.} - \bar{x}_{..})^2}{\text{SS}_{e(x)}}}$$

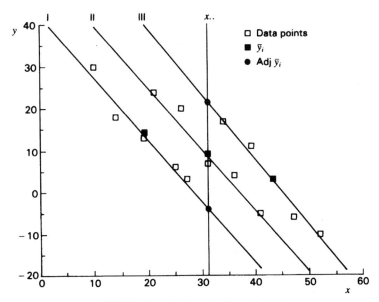

FIGURE 13.6. Media study, Example 13.1.

For example, for the adjusted mean of the third group, we have

$$
\text{CI}_{0.95}: \quad 21.1 \pm t_{0.025,11}\sqrt{4.858}\sqrt{\frac{1}{5} + \frac{(43-31)^2}{650}}
$$

$$
21.1 \pm 3.15
$$

If the treatment groups are the same size n, comparisons of two adjusted averages adj $\bar{y}_{i.} -$ adj $\bar{y}_{i'}$ can be made with the significant difference at the α level being given by

$$
t_{\alpha/2,N-a-1}\sqrt{\text{MS}'_{e(y)}}\sqrt{\frac{2}{n} + \frac{(\bar{x}_{i.} - \bar{x}_{i'})^2}{\text{SS}_{e(x)}}}
$$

In the advertising media example adj $\bar{y}_{2.} -$ adj $\bar{y}_{1.} = 10.0 - (-4.1)$, and at the 0.05 level of significance the critical difference is

$$
2.201\sqrt{4.858}\sqrt{\frac{2}{5} + \frac{(31-19)^2}{650}} = 3.82
$$

Thus $\alpha_1 \neq \alpha_2$ after adjusting for the number of employees. Similarly, adj $\bar{y}_{3.} -$ adj $\bar{y}_{2.}$ $= 21.1 - 10.0 = 11.1$, and the critical difference is

$$
2.201\sqrt{4.858}\sqrt{\frac{2}{5} + \frac{(43-31)^2}{650}} = 3.82
$$

Thus $\alpha_2 \neq \alpha_3$ after adjusting for the number of employees. Finally, adj $\bar{y}_{3.} -$ adj $\bar{y}_{1.}$ $= 21.1 - 4.1 = 17.0$, and the critical difference is

$$
2.201\sqrt{4.858}\sqrt{\frac{2}{5} + \frac{(43-19)^2}{650}} = 5.50
$$

Thus $\alpha_1 \neq \alpha_3$ after adjusting for the covariate. The SAS printout for these comparisons would be

Least Squares Means

MEDIUM	PROFIT LSMEAN	LSMEAN Number
I	-4.1107692	1
II	10.0000000	2
III	21.1107692	3

Least Squares Means for effect MEDIUM
Pr $>$ |t| for H0: LSMean(i) $=$ LSMean(j)

```
Dependent Variable: PROFIT
i/j      1       2       3
 1             <.0001  <.0001
 2     <.0001          <.0001
 3     <.0001  <.0001
```
Note: To ensure overall protection level, only probabilities associated with pre-planned comparisons should be used.

The adjusted means for the media are obtained by the LSMEAN statement. The name LSMEAN stands for *least-square mean*, or the adjusted average in this case because SAS output uses the term mean for both population and sample values.

The final conclusion of the media study is that each of the media used has a different effect on profits, and medium III has the greatest positive effect. We would not have come to this conclusion if we had not adjusted for the number of employees—before the adjustment, medium III had the lowest of the group averages.

EXERCISES

13.4.1. Given the following information from a one-way analysis of covariance involving 3 groups and 8 observations per group:

Source	$SS_{(x)}$	SP	$SS_{(y)}$	Group	$\bar{x}_{i.}$	$\bar{y}_{i.}$
Group	144	120	208	1	27	20
Within	175	140	132	2	30	18
				3	33	25

   a. Graph the unadjusted group averages.

   b. Find the estimate of the common slope, b.

   c. Graph the trend lines using the common slope.

   d. Find the adjusted y averages graphically.

   e. Find the adjusted y averages algebraically.

   f. Find the 95% confidence intervals on the adjusted means.

   g. Test the adjusted group means for significant differences.

13.4.2. It is possible to isolate genetic material in one species and transfer it to another species, and the genetic mechanism which permits North American fruit trees to resist cold weather can be transferred to tropical fruit trees, which could then be grown in more northern climes. Suppose that a horticulturist has had some limited, preliminary success in attempting this genetic transfer technique with several varieties of mango trees. The genetically altered trees are grown in an experimental orchard along the Gulf Coast, and the first year in which fruit is produced, there are significant differences in yield among the varieties, but the horticulturist wants to know whether the difference in yield is due to different numbers of fruit per tree or due to different weights of fruit. Therefore it is decided that the yield data should be analyzed by covariance. Suppose the data are as given below:

Variety

	V_1		V_2		V_3		V_4	
	x	y	x	y	x	y	x	y
	5	17	7	24	5	20	10	30
	7	21	7	26	4	13	9	28
	5	18	8	23	3	14	8	22
	4	11	6	23	7	22	7	20
	3	6	5	18	6	23	11	31
	6	23	9	30	5	16	9	25
$T_{i.(x)}$	30		42		30		54	
$T_{i.(y)}$		96		144		108		156

a. Which is the x variable? Is it the number of mangos per tree or the weight of mango fruit per tree?

b. Perform the analysis of covariance.

c. Compute the adjusted means and test to determine the significant differences among varieties.

13.4.3. Babies who are "undersize" at birth have a reduced chance of survival, and those who do survive tend to remain small for the rest of their lives. A Public Health Service physician is making a study of adolescents who were undersize at birth. Because such births are especially common among very young mothers, it is desirable to study the effect of mother's age. Thus two random samples are taken, one from among those born to mothers under 15 years old (group A) and those born to mothers who were older (group B). Data recorded for each group include birth weights (x) and adolescent weights (y) of the children in the study in kilograms. The results of the SAS analysis follow.

```
DATA BABIES;
    INPUT GROUP $ BIRTHWT ADOLESWT @@;
    CARDS;
B    1.1   29.4    A    1.7   66.1    B    1.9   61.1    A    2.1   56.3
B    2.2   58.6    A    0.7   36.7    B    1.2   59.0    A    1.9   57.2
B    1.5   25.8    A    1.5   60.2    B    2.3   55.9    A    2.1   59.4
B    2.4   61.8    A    1.6   33.8    B    2.2   76.9    A    0.7   29.2
B    1.9   47.9    A    1.2   29.9    B    2.4   73.0    A    1.1   28.2
B    1.1   31.8    A    1.7   56.6    B    2.4   55.3    A    1.5   64.0
B    1.4   34.7    A    1.7   54.0    B    1.3   37.3    A    1.7   35.3
B    1.1   35.3    A    1.3   47.7    B    1.3   38.5    A    2.0   42.5
B    1.3   31.4    A    1.8   56.1    B    1.7   60.9    A    1.5   54.4
B    1.8   52.1    A    0.8   28.9    B    1.6   47.4    A    0.9   28.7
```

```
B   1.8  41.3   A    2.0  60.1   B   1.6  52.3   A   1.1  43.6
B   2.0  53.7   A    1.9  43.2   B   2.1  76.7   A   1.8  55.2
B   2.6  85.1   A    1.4  66.7
    ;
    PROC GLM;
       CLASS GROUP;
       MODEL ADOLESWT = GROUP BIRTHWT;
       LSMEANS GROUP/PDIFF;
```

The SAS System

The GLM Procedure

Class Level Information

Class	Levels	Values
GROUP	2	A B

Number of observations   50

The GLM Procedure

Dependent Variable: ADOLESWT

Source	DF	Sum of Squares	Mean Square	F Value	Pr > F
Model	2	5655.11343	2827.55671	26.85	<.0001
Error	47	4950.44977	105.32872		
Corrected Total	49	10605.56320			

R-Square	Coeff Var	Root MSE	ADOLESWT Mean
0.533221	20.71488	10.26298	49.54400

Source	DF	Type I SS	Mean Square	F Value	Pr > F
GROUP	1	159.132800	159.132800	1.51	0.2251
BIRTHWT	1	5495.980626	5495.980626	52.18	<.0001

Source	DF	Type III SS	Mean Square	F Value	Pr > F
GROUP	1	76.915924	76.915924	0.73	0.3971
BIRTHWT	1	5495.980626	5495.980626	52.18	<.0001

Least Squares Means

GROUP	ADOLESWT LSMEAN	H0: LSMean1 = LSMean2 Pr > |t|
A	50.8365889	0.3971
B	48.2514111	

a. Using the output, what are the numerical values of $SS'_{e(y)}$, $SS'_{a(y)}$, $F = MS'_{a(y)}/MS'_{e(y)}$, adj $\bar{y}_A$, adj $\bar{y}_B$

b. What is the covariate in this study?

    c. Why is the analysis of covariance used in this study?

    d. Is there any evidence that the mean weight of the adolescents who were born to mothers under 15 years old is different from the mean weight of adolescents who were born to older mothers? Why or why not?

    e. Why is the P value for the test of HO: LSMean1 = LSMean2 the same as the P value for the TYPE III analysis of covariance test of the groups?

REVIEW EXERCISES

Decide whether each of the following statements is true or false. Correct each false statement.

13.1. The model $y_{ij} = \mu + \alpha_i + \beta_j + \varepsilon_{ij}$ would apply to covariance analysis if $\bar{x}_{..} = 0$.

13.2. Covariance techniques require that unadjusted y_{ij} as well as adjusted y_{ij} have homogeneous variance.

13.3. The analysis-of-covariance techniques in this chapter are appropriate whether the α_i are fixed or random.

13.4. Analysis-of-covariance techniques are appropriate whether the x_{ij} are fixed or random.

13.5. Analysis-of-covariance techniques are appropriate even though H_0: $\beta_1 = \beta_2 = \beta_3$ is rejected.

13.6. Analysis-of-covariance techniques may be appropriate even though H_0: $\beta = 0$ is not rejected.

13.7. Analysis-of-covariance techniques are appropriate even though $\sum_j (y_{ij} - \hat{y}_{ij})^2 / (n_i - 2)$ is significantly different from group to group.

13.8. Analysis-of-covariance techniques are appropriate even though $\sum_j (x_{ij} - \bar{x}_{i.})^2$ is significantly different from group to group.

13.9. The model for a one-way analysis of covariance is $y_{ij} = \mu + \alpha_i + \beta(x_{ij} - \bar{x}_{..}) + \varepsilon_{ij}$.

13.10. For a valid analysis of covariance, both the x variable and the y variable must be normally distributed.

13.11. For a valid analysis of covariance, both the x variable and the y variable must be random.

13.12. Accepting the hypothesis that the common slope $\beta = 0$ means that there is no relationship between x and y.

13.13. When the hypothesis of parallel regression lines is rejected, it becomes meaningless to discuss differences among adjusted averages that have been based on a common slope.

13.14. Because $\sum_i \sum_j (y_{ij} - \hat{y}_{ij})^2 \leq \sum_i \sum_j (y_{ij} - \bar{y}_{i.})^2$, the adjusted within-group sum of squares can never be greater than the unadjusted within-group sum of squares.

13.15. It is possible that an ANOVA on the unadjusted group averages can yield a significant F test for treatments, but in a similar test after adjustment for the x variable by covariance techniques, group differences may be nonsignificant.

13.16. Analysis of covariance may be used to increase the precision in an experiment even if the regression lines are horizontal.

13.17. The adjusted group means all lie on a vertical line at the overall x average.

13.18. Analysis of covariance can only be applied to 3 treatment groups.

13.19. The pooled estimate of the slope is found by averaging the estimates of the individual slopes.

13.20. Analysis of covariance requires that two different variables which are linearly related are measured on sampling units from 2 or more treatment groups.

SELECTED READINGS

Cochran, W. G. (1957). Analysis of covariance: Its nature and uses. *Biometrics*, **13**, 261–281.

Cox, D. R. (1957). The use of a concomitant variable in selecting an experimental design. *Biometrika*, **44**, 150–158.

Gourlay, N. (1953). Covariance analysis and its application in psychological research. *British Journal of Statistical Psychology*, **6**, 25–34.

Huitema, B. E. (1980). *The Analysis of Covariance and Alternatives*. Wiley, New York.

Snedecor, G. W., and W. G. Cochran (1973). *Statistical Methods*, 6th ed. Iowa State University Press, Ames.

Winer, B. J. (1971). *Statistical Principles in Experimental Design*, 2nd ed. McGraw-Hill, New York.

14 Multiple Regression and Correlation

In Chapter 9 where we were interested in an independent variable x and a dependent variable y we discussed simple linear regression and correlation. In this chapter we generalize the discussion and speak of k independent variables $x_1, x_2, \ldots, x_k$ and a dependent variable y. For the sake of completeness, the computational procedures are demonstrated, but using a computer is the only practical means of performing a multiple regression analysis on a data set of even moderate size. Consequently, greatest stress is placed on interpretation of the computer output for multiple regression analysis. Curvilinear regression is discussed as generalizations of either simple regression or multiple regression.

14.1. MATRIX PROCEDURES

In simple linear regression, we assume that x and y are linearly related, and we use the model $y = \alpha + \beta x + \varepsilon$ to express this relationship. The first step in regression analysis is to estimate α and β in the model. Least-squares procedures are employed, and the least-squares estimates a and b are found by solving two simultaneous equations. The solutions are

$$b = \frac{\sum (x - \bar{x})(y - \bar{y})}{\sum (x - \bar{x})^2} = \frac{S_{xy}}{S_{xx}} \quad \text{and}$$

$$a = \bar{y} - b\bar{x}$$

In a similar way, multiple regression involves use of a model of the form

$$y = \alpha + \beta_1 x_1 + \beta_2 x_2 + \cdots + \beta_k x_k + \varepsilon$$

and the estimates $a, b_1, b_2, \ldots, b_k$ of the α and β's are the least-squares solutions to several simultaneous linear equations.

If there are only two independent variables x_1 and x_2, we can visualize the least-squares procedure as fitting a plane to a set of n data points (x_1, x_2, y) in such a way that $\sum (y - \hat{y})^2$, the sum of the squared deviations of the actual y's from the predicted values, is minimized (Figure 14.1). This is analogous to the fitting of the least-squares trend line in simple linear regression. (If there are more than two independent variables, then the least-squares procedure fits a *hyperplane*, the generalization of a plane in more than three dimensions, to the points.) It is possible to use the equation of the plane for prediction if the plane is not parallel to the x_1, x_2 plane.

Statistics for Research, *Third Edition*, Edited by Shirley Dowdy, Stanley Weardon, and Daniel Chilko. ISBN 0-471-26735-X © 2004 John Wiley & Sons, Inc.

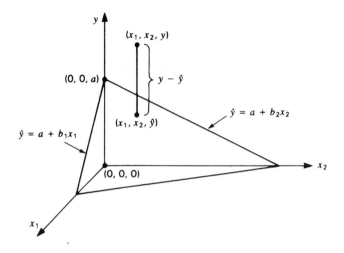

FIGURE 14.1. The least-squares plane $y = a + b_1x + b_2x$.

In this section we develop a computational technique that aids in solving systems of linear equations and also yields some additional information which is necessary for inference related to multiple regression. The computation is straightforward but still tedious for large data sets and for measurement variables containing a large number of digits. We illustrate it with a small data set consisting of variables measured as small integer values. Such data would be unrealistic for most studies involving multiple regression analysis, but they are suitable for demonstrating the computational techniques which are employed. This will dispel some of the mystery which many experience upon first examination of the computer output for multiple regression analysis.

Suppose our data set consists of the age (x_1) in years, weight (x_2) in kilograms, and systolic blood pressure (y) in millimeters of mercury for a random sample of $n = 7$ West Indian women:

Individual	Age (x_1)	Weight (x_2)	Systolic Pressure (y)
A	34	45	108
B	43	44	129
C	49	56	126
D	58	57	149
E	64	65	168
F	73	63	161
G	78	55	174

We want to know whether we can detect a significant linear relationship between the age of a woman and her systolic blood pressure and similarly to determine whether there is a linear relationship between her weight and her pressure. However, we do not perform two simple linear regression analyses, one of pressure on weight and a second of pressure on age, because the results could be misleading if there is also a linear relationship between their ages and their weights. Therefore, we solve a set of linear equations for a, b_1, and b_2 which will take into account any possible linear relationship, termed *collinearity*, between the two independent variables x_1 and x_2. This system of simultaneous linear equations arises from minimizing $\sum(y - \hat{y})^2 = \sum[y - (a + b_1x_1 + b_2x_2)]^2$. The three equations are

$$a = \bar{y} - b_1\bar{x}_1 - b_2\bar{x}_2$$

$$b_1 \sum(x_1 - \bar{x}_1)^2 + b_2 \sum(x_1 - \bar{x}_1)(x_2 - \bar{x}_2) = \sum(x_1 - \bar{x}_1)(y - \bar{y})$$

$$b_1 \sum(x_2 - \bar{x}_2)(x_1 - \bar{x}_1) + b_2 \sum(x_2 - \bar{x}_2)^2 = \sum(x_2 - \bar{x}_2)(y - \bar{y})$$

To set up the equations for solution, we must compute the sums of squares and the sums of products as we have done before when dealing with analyses that involved more than one continuous variable. We first compute

$$\sum x_1 = 399 \qquad \sum x_2 = 385 \qquad \sum y = 1{,}015 \qquad \sum x_1x_2 = 22{,}521$$

$$\sum x_1^2 = 24{,}279 \quad \sum x_2^2 = 21{,}565 \quad \sum y^2 = 150{,}803 \quad \sum x_1y = 60{,}112$$

$$\bar{x}_1 = 57 \qquad\qquad \bar{x}_2 = 55 \qquad\qquad \bar{y} = 145 \qquad\qquad \sum x_2y = 56{,}718$$

and find

$$S_{11} = \sum(x_1 - \bar{x}_1)^2 = \sum x_1^2 - \frac{\left(\sum x_1\right)^2}{n} = 1536$$

$$S_{12} = S_{21} = \sum(x_1 - \bar{x}_1)(x_2 - \bar{x}_2) = \sum x_1x_2 - \frac{\sum x_1 \sum x_2}{n} = 576$$

$$S_{22} = \sum(x_2 - \bar{x}_2)^2 = \sum x_2^2 - \frac{\left(\sum x_2\right)^2}{n} = 390$$

$$S_{1y} = \sum(x_1 - \bar{x}_1)(y - \bar{y}) = \sum x_1y - \frac{\sum x_1 \sum y}{n} = 2257$$

$$S_{2y} = \sum(x_2 - \bar{x}_2)(y - \bar{y}) = \sum x_2y - \frac{\sum x_2 \sum y}{n} = 893$$

This gives the three equations

$$a = 145 - 57b_1 - 55b_2$$
$$1536b_1 + 576b_2 = 2257$$
$$576b_1 + 390b_2 = 893$$

The last two equations can be easily solved using the matrix approach which is to be demonstrated here, and then b_1 and b_2 can be used in the first equation to find a.

In algebra, we learn how to solve two simultaneous equations in two unknowns when such solutions exist. Here, we have the equations

$$1536b_1 + 576b_2 = 2257$$
$$576b_1 + 390b_2 = 893$$

To solve this system of equations, we may multiply or divide any equation by a nonzero constant and we may add or subtract a multiple of one equation from another. We repeat these operations until we obtain an equivalent system of equations of the form

$$1b_1 + 0b_2 = d_1$$
$$0b_1 + 1b_2 = d_2$$

from which we can read the solutions $b_1 = d_1$ and $b_2 = d_2$.

This sequence of operations can be carried out by using a simple matrix approach. A *matrix* is a rectangular array of numbers. For example,

$$\mathbf{X} = \begin{bmatrix} S_{11} & S_{12} \\ S_{21} & S_{22} \end{bmatrix} = \begin{bmatrix} 1536 & 576 \\ 576 & 390 \end{bmatrix}$$

is the matrix of coefficients of the system of equations which we wish to solve, and

$$\mathbf{Y} = \begin{bmatrix} S_{1y} \\ S_{2y} \end{bmatrix} = \begin{bmatrix} 2257 \\ 893 \end{bmatrix}$$

is the matrix of the constants in the two equations. The solution is obtained by starting with

$$[\mathbf{X} \mid \mathbf{Y}] = \begin{bmatrix} 1536 & 576 & 2257 \\ 576 & 390 & 893 \end{bmatrix}$$

an augmented matrix made up of the matrix of coefficients on the left and the matrix of constants on the right. The following steps are the algebraic operations necessary to solve the set of equations:

STEP 1. For the appropriate operation to transform the first coefficient in row one to 1, divide all elements in row one by $S_{11} = 1536$:

$$\begin{bmatrix} 1 & 0.375 & 1.469401 \\ 576 & 390.000 & 893.000000 \end{bmatrix}$$

STEP 2. To transform the first element in row two to 0, multiply the first row by $S_{21} = 576$ and subtract the product from the second row:

$$\begin{bmatrix} 1 & 0.375 & 1.469401 \\ 0 & 174.000 & 46.625024 \end{bmatrix}$$

STEP 3. To obtain a 1 for the second number in row two, divide the second row by $S_{22} - S_{12}^2/S_{11} = 174$:

$$\begin{bmatrix} 1 & 0.375 & 1.469401 \\ 0 & 1.000 & 0.267960 \end{bmatrix}$$

STEP 4. To transform the second element in row one to 0, multiply the second row by $S_{12}/S_{11} = 0.375$ and subtract the product from the first row:

$$\begin{bmatrix} 1 & 0 & | & 1.368916 \\ 0 & 1 & | & 0.267960 \end{bmatrix}$$

Remembering that the above values are the coefficients and constants for a pair of simultaneous equations, we can see that we have obtained the solutions

$$1b_1 + 0b_2 = 1.368916$$
$$0b_1 + 1b_2 = 0.267960$$

To relate these numbers back to the problem concerning the relationships among age, weight, and systolic blood pressure in West Indian women, we have found that, after adjusting for the collinearity between age and weight, on the average systolic pressure increases 1.368916 mm Hg with each increase of one year of age, and it increases 0.267960 mm Hg with each increase of 1 kg in weight. Solutions to the simultaneous equations are given here to six decimal places, even though such a level is far beyond the precision with which blood pressure is usually measured. Instead, six decimal places were carried throughout the computations in order to reduce possibly serious consequences of rounding error. For the same reason, most computer analyses use *double-precision* arithmetic. This expression originally meant that twice as many decimal places were used in computations than provided in the printout; it now implies that the number of decimal digits used is the maximum allowable in the program. In discussing the results, however, we would further round these solutions to an even more sensible number of decimal places such as $b_1 = 1.369$ mm Hg/year and $b_2 = 0.268$ mm Hg/kg.

The shorthand of matrix algebra is quite convenient and is appearing more frequently in various areas of research literature. To promote familiarity with the notation, we will use it to review what was done in the solution of the system of equations. The original matrix form was

$$[\mathbf{X} \mid \mathbf{Y}] = \begin{bmatrix} S_{11} & S_{12} & | & S_{1y} \\ S_{21} & S_{22} & | & S_{2y} \end{bmatrix}$$

representing

$$S_{11}b_1 + S_{12}b_2 = S_{1y}$$
$$S_{21}b_1 + S_{22}b_2 = S_{2y}$$

and it was transformed into

$$[\mathbf{I} \mid \mathbf{B}] = \begin{bmatrix} 1 & 0 & | & b_1 \\ 0 & 1 & | & b_2 \end{bmatrix}$$

in which

$$I = \begin{bmatrix} 1 & 0 \\ 0 & 1 \end{bmatrix}$$

is known as the *identity matrix* and

$$B = \begin{bmatrix} b_1 \\ b_2 \end{bmatrix}$$

is the matrix of solutions.

Although this matrix procedure gives the solutions desired, its usefulness in multiple regression analysis can be improved. As will be seen in the discussion of statistical inference, computation of the standard errors of b_1 and b_2 require elements of the inverse of the matrix $\mathbf{X}$, the sums of squares and products which are the coefficients b_1 and b_2 in the simultaneous equations. The *inverse* can be thought of as the "memory" of the operations which were performed in the solution of the equations, and it can be obtained in a straightforward manner when we augment the beginning matrix with the identity matrix

$$I = \begin{bmatrix} 1 & 0 \\ 0 & 1 \end{bmatrix}$$

as follows:

$$[\mathbf{X} \mid \mathbf{Y} \mid \mathbf{I}] = \begin{bmatrix} S_{11} & S_{12} & S_{1y} & 1 & 0 \\ S_{21} & S_{22} & S_{2y} & 0 & 1 \end{bmatrix}$$

If the same row operations are applied to this form, it is changed into

$$\begin{bmatrix} 1 & 0 & b_1 & p_{11} & p_{12} \\ 0 & 1 & b_2 & p_{21} & p_{22} \end{bmatrix}$$

in which

$$\mathbf{P} = \begin{bmatrix} p_{11} & p_{12} \\ p_{21} & p_{22} \end{bmatrix} = \mathbf{X}^{-1}$$

is the inverse of the matrix of coefficients, that is,

$$\mathbf{PX} = \mathbf{X}^{-1}\mathbf{X} = \begin{bmatrix} p_{11} & p_{12} \\ p_{21} & p_{22} \end{bmatrix}\begin{bmatrix} S_{11} & S_{12} \\ S_{21} & S_{22} \end{bmatrix} = \begin{bmatrix} 1 & 0 \\ 0 & 1 \end{bmatrix} = \mathbf{I}$$

To demonstrate how to obtain the inverse, we perform the same row operations on the augmented matrix:

$$[\mathbf{X} \mid \mathbf{Y} \mid \mathbf{I}] = \begin{bmatrix} 1536 & 576 & 2257 & 1 & 0 \\ 576 & 390 & 893 & 0 & 1 \end{bmatrix}$$

divide row | *one by 1536*

$$\begin{bmatrix} 1 & 0.375 & 1.469401 & 0.000651 & 0 \\ 576 & 390.000 & 893.000000 & 0.000000 & 1 \end{bmatrix}$$

subtract 576 times row | *one from row two*

$$\begin{bmatrix} 1 & 0.375 & 1.469401 & 0.000651 & 0 \\ 0 & 174.000 & 46.625024 & -0.375000 & 1 \end{bmatrix}$$

divide row | *two by 174*

$$\begin{bmatrix} 1 & 0.375 & 1.469401 & 0.000651 & 0.000000 \\ 0 & 1.000 & 0.267960 & -0.002155 & 0.005747 \end{bmatrix}$$

subtract 0.375 times | *row two from row one*

$$\begin{bmatrix} 1 & 0 & 1.368916 & 0.001459 & -0.002155 \\ 0 & 1 & 0.267960 & -0.002155 & 0.005747 \end{bmatrix} = [\mathbf{I} \mid \mathbf{B} \mid \mathbf{P}]$$

The matrix on the right,

$$\mathbf{P} = \mathbf{X}^{-1} = \begin{bmatrix} 0.001459 & -0.002155 \\ -0.002155 & 0.005747 \end{bmatrix}$$

is the inverse $\mathbf{X}^{-1}$ of the sum of squares and products matrix

$$\mathbf{X} = \begin{bmatrix} 1536 & 576 \\ 576 & 390 \end{bmatrix}$$

This can be verified by using the definition of matrix multiplication:

$$\begin{bmatrix} p_{11} & p_{12} \\ p_{21} & p_{22} \end{bmatrix} \begin{bmatrix} S_{11} & S_{12} \\ S_{21} & S_{22} \end{bmatrix} = \begin{bmatrix} p_{11}S_{11} + p_{12}S_{21} & p_{11}S_{12} + p_{12}S_{22} \\ p_{21}S_{11} + p_{22}S_{21} & p_{21}S_{12} + p_{22}S_{22} \end{bmatrix}$$

Thus

$$\mathbf{X}^{-1}\mathbf{X} = \begin{bmatrix} 0.001459 & -0.002155 \\ -0.002155 & 0.005747 \end{bmatrix} \begin{bmatrix} 1536 & 576 \\ 576 & 390 \end{bmatrix}$$

$$= \begin{bmatrix} (0.001459)1536 + (-0.002155)576 & (0.001459)576 + (-0.002155)390 \\ (-0.002155)1536 + (0.005747)576 & (-0.002155)576 + (0.005747)390 \end{bmatrix}$$

$$= \begin{bmatrix} 1.00 & 0.00 \\ 0.00 & 1.00 \end{bmatrix}$$

Only two decimal places are reported here because that is the extent of the accuracy of the computations despite the fact that six decimal places were carried while performing them. We want to stress again the need for carrying a large number of decimal places in multiple regression analysis and indeed, if at all possible, the need to use a reliable computer routine for multiple regression.

EXERCISES

14.1.1. Multiply the following matrices:

a. $\begin{bmatrix} 2 & 5 \\ 5 & 4 \end{bmatrix} \begin{bmatrix} 4 & 1 \\ 1 & 5 \end{bmatrix}$

b. $\begin{bmatrix} -2 & 4 \\ 7 & 3 \end{bmatrix} \begin{bmatrix} 9 & -3 \\ -3 & 1 \end{bmatrix}$

c. $\begin{bmatrix} 4 & -2 \\ -2 & 7 \end{bmatrix} \begin{bmatrix} 10 \\ 12 \end{bmatrix}$

d. $\begin{bmatrix} 6 & 7 & -1 \\ 7 & -2 & 5 \\ -1 & 5 & 4 \end{bmatrix} \begin{bmatrix} 3 \\ -2 \\ 5 \end{bmatrix}$

14.1.2.   **a.** Solve the following system of equations using row operations:

$$4b_1 + 3b_2 = 10$$
$$3b_1 + 5b_2 = 16$$

b. Find the inverse $\mathbf{X}^{-1}$ of the matrix of coefficients

$$\mathbf{X} = \begin{bmatrix} 4 & 3 \\ 3 & 5 \end{bmatrix}.$$

c. Show that $\mathbf{X}^{-1}\mathbf{X} = \mathbf{I}$.

d. Show that $\mathbf{X}^{-1}\mathbf{X} = \mathbf{B}$ where $\mathbf{Y} = \begin{bmatrix} 10 \\ 16 \end{bmatrix}$ and $\mathbf{B}$ is the matrix of solutions.

14.1.3.   Find the inverse of $\begin{bmatrix} 1 & 2 \\ 2 & 5 \end{bmatrix}$.

14.1.4.   Simple linear regression can be approached using matrices. Using the example of employee training in Section 9.1,

Hours of instruction, x:	1	2	3	4	5
Units per hour, y:	5	4	6	8	7

find the estimates of the y intercept and slope as solutions of the systems of normal equations

$$na + b\sum x = \sum y$$
$$a\sum x + b\sum x^2 = \sum xy$$

Compare your answers with the results in Chapter 9.

14.2. ANOVA PROCEDURES FOR MULTIPLE REGRESSION AND CORRELATION

Using the data set involving the age (x_1), weight (x_2), and systolic blood pressure (y) of $n = 7$ women, we have already illustrated the least-squares procedure for obtaining $b_1 = 1.368916$ and $b_2 = 0.267960$. The intercept is estimated by

$$a = \bar{y} - b_1\bar{x}_1 - b_2\bar{x}_2 = 145 - 1.368916(57) - 0.267960(55) = 52.233988$$

Thus the least-squares regression plane is

$$\hat{y} = a + b_1x_1 + b_2x_2 = 52.233988 + 1.368916x_1 + 0.267960x_2$$

To determine whether the plane is parallel to the x_1, x_2 plane, we test H_0: $\beta_1 = \beta_2 = 0$ (parallel) against H_a: $\beta_1 \neq 0$ or $\beta_2 \neq 0$ (not parallel). As in simple regression, this test requires the variance of data points from the regression plane,

$$s_{y.x}^2 = \frac{\sum(y - \hat{y})^2}{n - k - 1}$$

in which n is the number of data points and k is the number of independent variables. Owing to space limitations, only three decimal places will be carried in the prediction equation of $\hat{y} = 52.234 + 1.369x_1 + 0.268x_2$ used to show how to compute this directly:

x_1	x_2	y	$\hat{y} = 52.234 + 1.369x_1 + 0.268x_2$	$y - \hat{y}$	$(y - \hat{y})^2$
34	45	108	$52.234 + 1.369(34) + 0.268(45) = 110.840$	−2.840	8.066
43	44	129	$52.234 + 1.369(43) + 0.268(44) = 122.893$	6.107	37.295
49	56	126	$52.234 + 1.369(49) + 0.268(56) = 134.323$	−8.323	69.272
58	57	149	$52.234 + 1.369(58) + 0.268(57) = 146.912$	2.088	4.360
64	65	168	$52.234 + 1.369(64) + 0.268(65) = 157.270$	10.730	115.133
73	63	161	$52.234 + 1.369(73) + 0.268(63) = 169.055$	−8.055	64.883
78	55	174	$52.234 + 1.369(78) + 0.268(55) = 173.756$	0.244	0.060
					299.069

$$s_{y.x}^2 = \frac{\sum(y - \hat{y})^2}{n - k - 1} = \frac{299.069}{7 - 2 - 1} = 74.767$$

Or we can use a more convenient computational procedure,

$$\sum(y - \hat{y})^2 = S_{yy} - b_1S_{1y} - b_2S_{2y}$$

in which

$$S_{yy} = \sum(y - \bar{y})^2 = \sum y^2 - \frac{\sum y^2}{n} = 3628$$

Thus

$$s_{y.x}^2 = \frac{\sum (y - \hat{y})^2}{n - k - 1} = \frac{3628 - 1.368916(2257) - 0.267960(893)}{7 - 2 - 1}$$

$$= \frac{3628 - 3328.932}{7 - 2 - 1} = \frac{299.068}{4} = 74.767$$

The test of H_0: $\beta_1 = \beta_2 = 0$ is an F test and can be set up in the form of an ANOVA table.

Source	df	SS	MS	F
Due to regression	$k = 2$	$b_1 S_{1y} + b_2 S_{2y} = 3328.932$	1664.466	22.262
Deviations	$n - k - 1 = 4$	$3628 - 3328.932 = 299.068$	74.767	
Total	$n - 1 = 6$	$S_{yy} = 3628.000$		

Since $F_{0.05,2,4} = 6.944$, the computed F is significant, and we can conclude that the regression plane is not parallel to the x_1, x_2 plane but instead is significantly "tilted" because $\beta_1 \neq 0$ or $\beta_2 \neq 0$. We conclude that there is a linear relationship between systolic pressure (y) and age (x_1) or systolic pressure and weight (x_2) or possibly systolic pressure with both independent variables.

Although the F test for H_0: $\beta_1 = \beta_2 = 0$ provides a test of the significance of the regression of the dependent variable on the independent variables, the reliability of the regression equation is very commonly measured by the *multiple correlation coefficient*. The multiple correlation coefficient $R_{y\hat{y}}$ or R can be thought of as the correlation between the observed y's and the $\hat{y}$'s predicted by the regression equation. It can be computed in much the same way as the correlation coefficient was computed for bivariate data:

$$R = \frac{\sum (y - \bar{y})(\hat{y} - \bar{y})}{\sqrt{\sum (y - \bar{y})^2 (\hat{y} - \bar{y})^2}}$$

Unlike the situation for simple correlation, however, $0 \leq R \leq 1$, because it would be impossible to have a negative correlation between the observed and the least-squares predicted values.

The square of the multiple correlation coefficient R^2 can be interpreted as the proportion of the variability that has been accounted for by the regression equation and R^2 is between 0 and 1. If the equation fits the data well, R^2 is close to 1; if the linear model is a poor fit, R^2 will be close to 0.

The formula given above for R is usually cumbersome computationally; instead, R^2 can be computed directly using the formula

$$R^2 = \frac{\sum b_i S_{iy}}{S_{yy}}$$

Then R can be found by taking the positive square root of R^2.

As in the case of simple linear regression and correlation, different assumptions are used when deriving multiple regression and multiple correlation procedures. Multiple regression assumes that the x's are fixed and predetermined by the investigator, the relationship is linear, and the ε's are IND$(0, \sigma^2)$. Multiple correlation assumes that the x's are random and that y, $x_1, \ldots, x_k$ have a multivariate normal distribution.

As a result of these assumptions, all the procedures we discuss in this chapter may be applied to situations that fit the correlation model. If a research situation fits the regression model and has fixed x's, then correlation statistics such as R and R^2 may be calculated, but inference should not be made from these statistics.

If the correlation model is being used, R^2 may be tested. To test the significance of the multiple correlation coefficient, we use hypotheses

$$H_0: \mathbf{P}^2 = 0 \qquad \text{against} \qquad H_a: \mathbf{P}^2 > 0$$

in which $\mathbf{P}$ (the uppercase Greek letter rho) is the true population multiple correlation coefficient. The test statistic is

$$F = \frac{R^2/k}{(1 - R^2)/(n - k - 1)}$$

with $v_1 = k$ and $v_2 = n - k - 1$ degrees of freedom.

For our data set

$$R^2 = \frac{\sum b_i S_{iy}}{S_{yy}} = \frac{3328.932}{3628.000} = 0.917567$$

and

$$F = \frac{R^2/k}{(1 - R^2)/(n - k - 1)} = \frac{0.917567/2}{0.082433/4} = 22.262$$

Except for rounding error, this F test will give the same numerical results as the one used for testing $H_0: \beta_1 = \beta_2 = 0$.

To summarize the results of our use of multiple regression and correlation to examine the linear relationships of systolic pressure with age and weight, we can conclude that there is a significant relationship, and there is good agreement between the observed values and values obtained from a linear prediction equation based on the two independent variables (age and weight). However, in cases such as this where $n - k - 1$ is small, it is possible for a large value of R^2 to result from only moderate linear association between the dependent and independent variables. Thus the physician would very likely want to use a larger sample to confirm these results. Also, it would be helpful to learn whether both independent variables are needed in the prediction equation or whether a simple linear equation, using only one of the independent variables, would be almost as reliable as the one using both x_1 and x_2.

EXERCISES

14.2.1. Given the following sums of squares and cross-products for 27 data points (x_1, x_2, y)

$$S_{11} = 10 \quad S_{22} = 41 \quad S_{yy} = 50$$
$$S_{1y} = 4 \quad S_{2y} = 2 \quad S_{12} = 20$$

a. Complete the augmented matrix of sums of squares and cross-products and the final matrix after row operations:

$$\mathbf{X} \qquad\qquad \mathbf{X}^{-1}$$

$$\left[\begin{array}{cc|c|cc} - & - & - & 1 & 0 \\ - & - & - & 0 & 1 \end{array}\right] \rightarrow \left[\begin{array}{cc|c|cc} - & - & - & 4.1 & -2.0 \\ - & - & - & -2.0 & 1.0 \end{array}\right]$$

(It is not necessary to do the row operations.)

b. Complete the ANOVA table for multiple regression:

Source	df	SS	MS	F	$F_{0.05}$
Regression					
Deviations					

14.2.2. When the age y of a grazing animal is unknown, it can be estimated from the extent of tooth wear x_1 and the amount of gray hair x_2 on the animal's muzzle. In an effort to evaluate and refine this procedure, a random sample of horses of known ages is measured on indices developed to determine tooth wear and graying. The following information is derived:

Augmented Sum of Squares
and Cross-Products Matrix

$$\begin{bmatrix} 64.00 & -39.20 & | -20.00 & | 1 & 0 \\ -39.20 & 49.00 & | 19.39 & | 0 & 1 \end{bmatrix}$$
$$\rightarrow \begin{bmatrix} 1 & 0 & | -0.1375 & | 0.0306 & 0.0245 \\ 0 & 1 & | 0.2856 & | 0.0245 & 0.0400 \end{bmatrix}$$

a. Complete the ANOVA table:

Source	df	SS	MS
Regression	—	8.29	4.145
Deviations	—	16.71	1.671

b. What percentage of the variation in the horses' ages can be explained on the basis of tooth condition and graying?

c. If a multiple regression prediction equation is used, will it explain a significant portion of the variability of age? Why or why not?

d. Do you think the prediction equation would be very useful in estimating the ages of horses when their ages are not known? Why or why not?

14.2.3. In studies of the effect of acid rain on the biomass in freshwater lakes, biologists have found that biomass decreases as acid concentration increases. If the lakes have sources of phosphorous, however, biomass increases with an increase in the amount of phosphorus available. In an effort to make a more thorough study, researchers take water samples from 18 randomly selected lakes and measure the acidity x_1, available phosphorus x_2, and population density y of a certain species of algal plant. The following statistics are computed:

$$\bar{y} = 1{,}400 \qquad S_{yy} = 14{,}400 \qquad S_{12} = 900$$
$$\bar{x}_1 = 2{,}100 \qquad S_{11} = 1{,}600 \qquad S_{1y} = -3{,}000$$
$$\bar{x}_2 = 760 \qquad S_{22} = 3{,}600 \qquad S_{2y} = 2{,}100$$

$$p_{11} = 0.000727 \qquad b_1 = -2.563$$
$$p_{12} = -0.000182 \qquad b_2 = 1.224$$
$$p_{22} = 0.000323 \qquad s_{y.x}^2 = 276$$

a. Compute R^2.

b. Test R^2 for significance.

c. Test $\beta_1 = \beta_2 = 0$.

d. What is the equation of the least-squares plane?

e. If acidity is increased one unit and phosphorus held fixed, what is the effect on population density?

14.2.4. Francis Galton, who gave regression its name, thought everything could be measured, even the power of prayer. Studies of anxiety among the terminally ill now give reason for wanting to measure it. Believing anxiety is due to feeling a lack of control over one's condition, a hospice for the terminally ill conducts a study with the permission of those residing there. For a week each self-administers his or her painkiller up to the maximum prescribed. Since none uses the maximum, the exact amount in milliliters taken is measured. Also daily, the chaplain asks each if he or she would like to pray with him and he records the time in minutes. At the end of the week, residents are given an anxiety scale consisting of a 20-cm. straight line on a piece of paper, and each is asked to make a cross-mark on the line according to the amount of anxiety felt; the farther to the right the mark, the greater the anxiety. The length to the mark is measured with a ruler. Multiple regression is used to analyze the data with distance to the mark (anxiety) the dependent variable and predictor variables are amount of medicine taken and time in prayer. Use the following ANOVA printout to answer the questions.

Rsquare		0.6487		
Average		10.61		
MS Error		9.329		
N		18		
		ANOVA		
Source	**df**	**MS**	**F-Test**	**P-value**
Groups	2	129.174	13.8472	0.0004
Error	15	9.329		

a. How many residents were involved in the study?

b. From data in the ANOVA compute the numerical values of

    i. $b_1 S_{1y} + b_2 S_{2y}$

    ii. $\sum (y - \hat{y})^2$

c. Do prayer and/or self-administration of medication explain a significant portion of the variability in the anxiety among residents? Explain.

d. The fact that none of the patients used maximum medication is thought also to be an expression of control over one's condition. How would you determine why they didn't use all their pain-killing medication? *Hint:* Good researchers use common sense as well as statistics.

14.3. INFERENCE ABOUT EFFECTS OF INDEPENDENT VARIABLES

In an analysis of variance, if the F test is significant, the investigator will perform further tests or compute confidence intervals to pinpoint the specific differences. Similarly, in multiple regression, if H_0: $\beta_1 = \beta_2 = 0$ is rejected, the investigator will want to know which of the x variables contributes to this overall significance. Most commonly this is done by either performing tests of hypotheses on the individual partial regression coefficients (b_i) or placing confidence intervals on them. To use either procedure, however, it is first necessary to compute the standard error of each partial regression coefficient.

To explain the computation of the standard error of a partial regression coefficient, it may be helpful to recall the case of simple regression, where the standard error of the regression coefficient is

$$\text{s.e.}(b) = \frac{s_{y.x}}{\sqrt{S_{xx}}} = \sqrt{\frac{1}{S_{xx}} s_{y.x}^2}$$

We can show how this value would be obtained if we used matrix procedures with simple linear regression. Although the original matrix contains only one row, we can use the same form,

$$[\mathbf{X} \mid \mathbf{Y} \mid \mathbf{I}] = [S_{xx} \mid S_{xy} \mid 1]$$

and we would invert this matrix by dividing all terms in it by S_{xx} to obtain the final form,

$$[\mathbf{I} \mid \mathbf{B} \mid \mathbf{X}^{-1}] = \left[1 \,\middle|\, b = \frac{S_{xy}}{S_{xy}} \,\middle|\, p = \frac{1}{S_{xx}}\right]$$

Thus we can see that the standard error of the simple regression coefficient is the square root of the product of two terms, the variance from the regression line ($s_{y.x}^2 = \text{MS}_e$) and the element of the inverse of the matrix of coefficients ($1/S_{xx}$). In a similar manner, the standard error of any partial regression coefficient (b_i) in multiple regression is the square root of the product of two terms, the variance from the regression plane ($s_{y.x}^2$) and the appropriate element (p_{ii}) from the inverse of the matrix of coefficients:

$$\text{s.e.}(b_i) = \sqrt{p_{ii} s_{y.x}^2}$$

Once the standard error of the partial regression coefficient is obtained, we use it in the same fashion that has become familiar for performing a t test or for setting a confidence interval:

Test of hypothesis H_0: $\beta_i = 0$

$$t = \frac{\text{Estimate} - \text{Hypothesized value}}{\text{Standard error of the estimator}} = \frac{b_i - 0}{\sqrt{p_{ii}s_{y.x}^2}}$$

which is a t test with $v = n - k - 1$ degrees of freedom.

Setting a confidence interval for β_i

$$CI_{1-\alpha}: \text{Estimate} \pm t_{\alpha/2,y} \text{ (Standard error of estimator)}$$

or

$$b_i \pm t_{\alpha/2, n-k-1}\sqrt{p_{ii}s_{y.x}^2}$$

Using the same data we used throughout our discussion of multiple regression, we demonstrate these procedures in an example.

Example 14.1. Inference About Partial Regression Coefficients

Among people living in the United States, both age and weight are known to have positive linear associations with systolic blood pressure. However, the numerical values of the partial regression coefficients are not the same from region to region or from one ethnic group to another. Most physicians are familiar with the situation in North America and anticipate finding positive linear associations in another geographical region and culture such as the West Indies but likely would be unable to predict whether the b_i would be greater or lesser than those found in the United States.

From the analyses already conducted on the data obtained from seven West Indian women, the original augmented matrix of sums of squares and products is

$$[\mathbf{X} \mid \mathbf{Y} \mid \mathbf{I}] = \begin{bmatrix} S_{11} = 1536 & S_{12} = 576 & S_{1y} = 2257 & 1 & 0 \\ S_{21} = 576 & S_{22} = 390 & S_{2y} = 893 & 0 & 1 \end{bmatrix}$$

along with its inverse

$$[\mathbf{I} \mid \mathbf{B} \mid \mathbf{X}^{-1}] = \begin{bmatrix} 1 & 0 & b_1 = 1.368916 & p_{11} = 0.001459 & p_{12} = -0.002155 \\ 0 & 1 & b_2 = 0.267960 & p_{21} = -0.002155 & p_{22} = 0.005747 \end{bmatrix}$$

and the ANOVA table used in testing H_0: $\beta_1 = \beta_2 = 0$:

Source	df	SS	MS	F
Due to regression	$k = 2$	$b_1 S_{1y} + b_2 S_{2y} = 3328.932$	1664.466	22.262
Deviations	$n - k - 1 = 4$	$3628 - 3328.932 = 299.068$	$MS_e = 74.767$	

As already noted, $F > F_{05;k,n-k-1}$; thus there is a significant linear regression of systolic blood pressure (y) on age (x_1) or on weight (x_2) or on both age and weight. The physician knows it is possible to construct a linear equation for predicting systolic blood pressure but does not know whether the reliability of the equation depends on x_1, x_2, or both these independent variables.

The estimated partial regression coefficients b_1 and b_2 can be interpreted as partial slopes. The coefficient $b_1 = 1.368916$ indicates that when age (x_1) increases by one year and weight (x_2) is held constant, on the average systolic pressure increases by 1.368916 mm Hg. Similarly for b_2. However, one must be cautious about directly comparing b_1 and b_2; the first is measured in millimeters per year and the second in millimeters per kilogram. Because of the difference in units of measurement, the fact that b_1 is more than four times greater than b_2 does not mean that x_1 is more important in the prediction equation than is x_2. Also, if x_1 and x_2 are completely independent (unrelated to each other), the partial regression coefficients would be the same as the simple regression coefficients, which would be computed if y were regressed on x_1 and x_2 one at a time. However, age and weight are frequently interrelated, and in multiple regression one can usually expect to find such collinearity among the independent variables. While an x variable can be held fixed in the statistical sense, it may not be possible to do so in the real world. Thus it may be impossible to set up a factorial experiment in which there is every combination of the numerical values of x_1 and x_2, but by using multiple regression analysis, one can still examine the linear effect on y of each x_i independent of the other x variables in the model.

The contribution of each x to the model is determined by testing the partial regression coefficients separately. Because positive relations between y and both x variables have been found in studies conducted in the United States, the physician chooses a one-sided alternative hypothesis: $H_0: \beta_1 = 0$ against $H_a: \beta_1 > 0$ is tested with

$$t = \frac{b_1 - \beta_{1_0}}{\sqrt{p_{11} s_{y.x}^2}} = \frac{1.368916 - 0}{\sqrt{0.001459(74.767)}} = 4.145$$

with $v = n - k - 1 = 4$ degrees of freedom. In the above equation, β_{1_0} is the value of β_1 specified in the null hypothesis; β_{1_0} could be some value other than zero, and in later studies, the physician might want to compare the regression lines obtained from his sample of West Indian women to the values which have been found for other populations. The value p_{11} is the element in the first row and column of $\mathbf{X}^{-1}$, the inverse of the matrix of sums of squares and products found in the process of solving for b_1 and b_2.

Similarly, $H_0: \beta_2 = 0$ against $H_a: \beta_2 > 0$ is tested with

$$t = \frac{b_2 - \beta_{2_0}}{\sqrt{p_{22} s_{y.x}^2}} = \frac{0.267960 - 0}{\sqrt{0.005747(74.767)}} = 0.409$$

For either test, the null hypothesis is rejected if $t \geq t_{0.05,4} = 2.132$. The physician rejects $\beta_1 = 0$ but does not reject $\beta_2 = 0$. He concludes that among these women age (x_1) is significantly associated with systolic blood pressure, but perhaps because of an unrealistically small sample, he is unable to detect any significant effect due to weight (x_2). If the physician wants a prediction equation, so that a woman's actual blood pressure can be compared to that expected for her age and weight, the statistical significance of b_1 indicates that age should be in the prediction equation, but this analysis provides no statistical justification for including weight in the equation.

There are equivalent F tests for testing H_0: $\beta_i = 0$ against H_a: $\beta_i \neq 0$, and some computer programs may provide these F tests in their printout rather than the t tests just examined. Because a t value with v degrees of freedom which is squared is equivalent to an F with 1 and v degrees of freedom, that is, $t_v^2 = F_{1,v}$, the F test is

$$ F = t^2 = \frac{b_i^2}{p_{ii}s_{y.x}^2} = \frac{b_i^2}{p_{ii}MS_e} $$

A computer printout with F tests might appear as

Rsquare	0.9176
Average	145.00
MS Error	74.77
N	7

ANOVA

Source	df	MS	F-Test	P-value
Regression	2	1664.47	22.2620	0.0068
Error	4	74.77		

Term	Coefficient	SS	F-Test	P-value
Age	1.3689	1284.1907	17.1759	0.0143
Weight	0.2680	12.4936	0.1671	0.7076

There is a significant linear relationship between age and systolic blood pressure, and among these women, on the average, systolic blood pressure increases 1.369 mm Hg with each year increase in age. However, because this is only an estimate based on data obtained from just 7 women, the physician needs to set a confidence interval to determine for the entire population how small or large may be the increase in systolic pressure per year of age. A central confidence interval is obtained as follows:

$$ CI_{1-\alpha}: b_i \pm t_{\alpha/2, n-k-1}\sqrt{p_{ii}s_{y.x}^2} $$

Thus the 95% confidence interval for β_1 would be

$$ CI_{0.95}: b_1 \pm t_{0.025,4}\sqrt{p_{11}MS_e} $$

and with the appropriate numerical values replacing their symbols

$$ CI_{0.95}: 1.369 \pm 2.776\sqrt{0.001459(74.767)} $$
$$ 1.369 \pm 2.776(0.330) $$
$$ 1.369 \pm 0.917 $$

This confidence interval is quite wide and very likely would be of limited direct use. However, the expected linear relationship between age and systolic blood pressure has been confirmed in this population of West Indian women, and the physician can proceed with a study involving more women, and he can anticipate obtaining a prediction equation which will be useful in clinical practice.

Procedure. Inference about Individual Partial Regression Coefficients

In making statistical inference about an estimate, it is also necessary to compute the estimated standard error of the estimate. In this case, the estimate is the partial regression coefficient and its standard error is $\sqrt{p_{ii}s_{y \cdot x}^2}$, which is the same as $\sqrt{p_{ii}MS_e}$. We use the estimate and its standard error to perform a t test,

$$t = \frac{\text{Estimate} - \text{Hypothesized value}}{\text{Standard error of the estimator}}$$

or set a confidence interval for a parameter,

$$\text{Estimate} \pm t_{\alpha/2,\nu} \text{ (Standard error)}$$

Test of Hypothesis for Partial Regression Coefficient β_i

$$H_0: \beta_i = \beta_{i_0} \quad \text{against} \quad H_a: \beta_i \neq \beta_{i_0}$$

is tested with

$$t = \frac{b_i - \beta_{i_0}}{\sqrt{p_{ii}s_{y \cdot x}^2}}$$

with $\nu = n - k - 1$ and in which p_{ii} is the ith diagonal element of $\mathbf{X}^{-1}$, the inverse of the matrix of sums of squares and products. The test of $H_0: \beta_i \neq \beta_{i_0}$ against $H_a: \beta_i \neq \beta_{i_0}$ can also be carried out using

$$F = \frac{b_i^2}{p_{ii}s_{y \cdot x}^2}$$

with $\nu_1 = 1$ and $\nu_2 = n - k - 1$; it is equivalent to the above t test.

Confidence Intervals for Partial Regression Coefficients

$$\text{CI}_{1-\alpha}: b_i \pm t_{\alpha/2,n-k-1}\sqrt{p_{ii}s_{y \cdot x}^2}$$

Other Inference about Partial Regression Coefficients

In addition to tests of hypothesis and confidence intervals for individual β_i, other types of inference are possible within a multiple regression analysis. For example, if two or more of the x_i have the same units of measurement, there could be reason for comparing the average increase y per unit increase in these x_i by testing the equality of two partial regression coefficients or by setting confidence intervals for the difference between $\beta_i - \beta_j$. In either case, the estimated standard error for the difference between two regression coefficients will be

$$\sqrt{(p_{ii} + p_{jj} - 2p_{ij})s_{y \cdot x}^2}$$

so we can test H_0: $\beta_i = \beta_j$ against H_a: $\beta_i \neq \beta_j$ with

$$t = \frac{b_i - b_j}{\sqrt{(p_{ii} + p_{jj} - 2p_{ij})s_{x.y}^2}}$$

or set a confidence interval for their difference with

$$\text{CI}_{1-\alpha}: (b_i - b_j) \pm t_{\alpha/2, n-k-1}\sqrt{(p_{ii} + p_{jj} - 2p_{ij})s_{y.x}^2}$$

The term $-2p_{ij}$ in the standard error is due to the possible linear relationship between x_i and x_j.

It is also possible to make tests of hypotheses or find confidence intervals for the estimates obtained using the fitted equation

$$\hat{y} = a + b_1x_1 + \cdots + b_kx_k$$

Given the specific values $x_1 = x_1{}^*, x_2 = x_2{}^*, \ldots, x_k = x_k{}^*$, the standard error of the estimate is

$$\text{s.e.}(\hat{y}) = \sqrt{s_{y.x}^2\left[\frac{1}{n} + \sum_i\sum_j p_{ij}(x_i^* - \bar{x}_i)(x_j^* - \bar{x}_j)\right]}$$

For example, if we want a 95% confidence interval for the mean systolic blood pressure of all West Indian women whose age is $x_1{}^* = 45$ years and whose weight is $x_2{}^* = 50$ kg, the estimate is

$$\hat{y} = 52.234 + 1.369(45) + 0.268(50) = 127.239$$

and the 95% confidence interval for the value which this estimates, $E(y|x_1 = 45, x_2 = 50)$ is $\text{CI}_{0.95}: \hat{y} \pm t_{0.025,4}(\text{s.e.})$, where

$$\text{s.e.}(\hat{y}) = \sqrt{s_{y.x}^2\left[\frac{1}{n} + p_{11}(x_1^* - \bar{x}_1)^2 + 2p_{12}(x_1^* - \bar{x}_1)(x_2^* - \bar{x}_2) + p_{22}(x_2^* - \bar{x}_2)^2\right]}$$

So the confidence interval is

$$127.239 \pm 277[(\text{s.e.}(\hat{y})]$$

where

$$\text{s.e.}(\hat{y}) =$$

$$\sqrt{74.767\left[\frac{1}{7} + 0.001459(45-57)^2 + 3(-0.002155)(45-57)(50-55) + 0.005747(50-55)^2\right]}$$

or

$$127.239 \pm 11.711$$

that is,

$$115.528 \le E(y \mid x_1 = 45, x_2 = 50) \le 138.950$$

If an individual y is predicted, the point estimate is $\hat{y}$ and the prediction interval is

$$\text{PI}_{1-\alpha}: \hat{y} \pm t_{\alpha/2, n-k-1} \sqrt{S_{y \cdot x}^2 \left[1 + \frac{1}{n} + \sum_i \sum_j p_{ij}(x_i^* - \bar{x}_i)(x_j^* - \bar{x}_j)\right]}$$

Because the complexity of these standard errors increases with additional independent variables, we do not want to include in the model x variables that provide little or no additional information about the y variable. In a later section, we show how to simplify the prediction equation by eliminating those x variables that contribute little to the reliability of the prediction.

EXERCISES

14.3.1. In Exercise 14.2.3 on lake biomass:
   a. Place a 95% confidence interval on β_1.
   b. Place a 90% confidence interval on β_2.
   c. Test $\beta_1 = 0$ and $\beta_2 = 0$ separately and interpret the results.
   d. Estimate the mean population density of the algae in a lake with an acidity measurement of 2000 and a phosphorous measurement of 860. Place a 95% confidence interval on this estimate.
   e. Place a 95% prediction interval on the estimate in part d.

14.3.2. Using the example in Exercise 14.2.1, show that the F statistic to test $H_0: \beta_1 = \beta_2 = 0$ can be computed from the multiple correlation coefficient

$$F = \frac{R^2/k}{(1 - R^2)/(n - k - 1)}$$

14.3.3. In Exercise 14.2.2 on grazing animals:
   a. What are the estimates of β_1 and β_2?
   b. Place a 95% confidence interval on each of the regression coefficients.

14.3.4. In his original study of regression, Francis Galton computed the simple regression of adult sons' heights (y) on their fathers' heights (x_1) and in another equation on their mothers' heights (x_2). Suppose he had been able to use multiple regression and had obtained the following (fictional) ANOVA printout. Use the printout data to answer the questions.

Rsquare	0.3325
Average	70.25
MS Error	43.796
N	27

ANOVA

Source	df	MS	F-Test	P-value
Regression	2	327.511	7.4781	0.0030
Error	24	43.796		

Term	Coefficient	SS	F-Test	P-value
Mother	0.3896	339.8492	7.7598	0.0103
Father	0.3504	265.2569	6.0567	0.0214

a. What fraction of the variability among height of sons can be attributed to inheritance or other familial factors?

b. Show how to compute the F used to test:

  i. $H_0: \beta_M = \beta_F = 0$ (the subscript M indicates mother and F father)

  ii. $H_0: \beta_F = 0$

c. What would be the predicted adult height of their newborn son if:

  i. The mother is average height for women and the father is 6 inches taller than the average height for men.

  ii. The mother is 6 inches taller than the average height for women and the father is the average height for men.

d. Assuming there is no change in average height between generations, if the mother is the average height for women, why will the son's height be predicted to be nearer to average male height than is his father's height? (Galton called this "regression toward the mean.")

14.4. COMPUTER USAGE

Multiple Regression
In the SAS System multiple regression is programmed similarly to simple linear regression, as can be seen in the following example. World Health Organization physicians have noted unusually large incidences of hypertension (high blood pressure) in certain communities in the Antilles Islands. A physician at a clinic on one of these islands uses data from a random sample of 30 of his women patients to examine some of the factors which may be related to their blood pressure. Among other data available, he has the age in years, weight in kilograms, and systolic blood pressure in millimeters of mercury for each woman in the sample.

A SAS data set is formed, all simple correlation coefficients are computed, and multiple regression is performed using the following SAS program:

```
DATA PATIENTS;
  INPUT AGE WT SYSTOLIC @@;
  CARDS;
21   67   116   30   53   122   72   64   212   46   49   135
48   47   131   28   44   123   19   63    96   26   55   113
21   59   111   49   43   134   46   69   164   33   56   123
38   43   141   60   44   160   42   48   128   64   63   171
76   48   176   20   63   139   71   60   177   69   49   185
30   49   110   53   52   157   47   64   173   63   50   162
```

```
26    48    108    22    58    122    59    49    154    48    50    139
27    60    132    21    68    128
;

PROC CORR;

PROC REG;

   MODEL SYSTOLIC = AGE WT;
```

The output from **PROC CORR** is as follows.

```
                        The SAS System
                        The CORR Procedure

      3 Variables:        AGE            WT           SYSTOLIC

                       Simple Statistics

Variable    N      Mean      Std Dev    Sum    Minimum     Maximum

AGE         30   42.50000   18.10839   1275   19.00000    76.00000
WT          30   54.50000    8.05049   1635   43.00000    69.00000
SYSTOLIC    30  141.40000   27.25309   4242   96.00000   212.00000


         Pearson Correlation Coefficients, N = 30

              Prob > |r| under H0: Rho = 0

                       AGE            WT          SYSTOLIC
AGE                  1.00000      -0.24304        0.87376
                                   0.1956         <.0001


WT                  -0.24304       1.00000        0.09351
                     0.1956                       0.6231


SYSTOLIC             0.87376       0.09351        1.00000
                     <.0001        0.6231
```

   In the output, descriptive statistics are given for each variable. This is followed by a square array containing the sample correlation coefficient r for each pair of variables. The probability that the sample correlation coefficient is greater than the absolute value of r if the population correlation coefficient ρ is equal to zero is given under each r value. This probability is the P value which can be used to test whether the population correlation coefficient is zero.

   PROG REG is used for multiple regression. The model statement is of the form $y = x1\ x2$, where y is the dependent variable and $x1$ and $x2$ are two independent variables. In this example SYSTOLIC is the dependent variable and AGE and WT are the independent variables. The output is as follows:

```
                        The SAS System

                        The REG Procedure
                        Model: MODEL1
                   Dependent Variable: SYSTOLIC
```

Analysis of Variance

Source	DF	Sum of Squares	Mean Square	F Value	Pr > F
Model	2	18586	9292.89261	84.96	<.0001
Error	27	2953.41479	109.38573		
Corrected Total	29	21539			

Root MSE	10.45876	R-Square	0.8629
Dependent Mean	141.40000	Adj R-Sq	0.8527
Coeff Var	7.39658		

Parameter Estimates

Variable	DF	Parameter Estimate	Standard Error	t Value	Pr > |t|
Intercept	1	20.48318	15.50504	1.32	0.1976
AGE	1	1.43391	0.11057	12.97	<.0001
WT	1	1.10047	0.24870	4.42	0.0001

The significance of the multiple regression model is tested with the F Value and its corresponding P value (Pr > F). In this case $F = 84.96$ with $P < 0.0001$, so this model is a good predictor of systolic blood pressure. The R − Square of 0.8629 indicates that 86.29% of the variability in systolic blood pressure is explained by this model.

The Adj R-Sq, the adjusted R^2, is a version of R^2 that has been adjusted for degrees of freedom, that is, for the number of independent variables in the model. The equation for Adj R-Sq is

$$R_{adj}^2 = 1 - \frac{(1 - R^2)(n - 1)}{n - k - 1}$$

Since R^2 will always increase when additional independent or *regressor variables* are added to the model, this statistic makes it possible to compare models which contain different numbers of independent variables.

The estimate of the constant (Intercept) and the partial regression coefficients follow. The standard error of each of the estimates is the same as that used in the denominator of the t test discussed in the previous section. The output contains the computed t (t Value) and its corresponding P value (Pr > \|t\|).

The SAS program can be modified to provide output which can be used to examine the residuals as discussed in Section 9.2:

```
PROC REG DATA = PATIENTS;
   MODEL SYSTOLIC = AGE WT/NOPRINT;
   OUTPUT OUT = GRAPHS P = PRED_Y R = RESID;
PROC PLOT DATA = GRAPHS;
   PLOT RESID*PRED_Y/VREF = 0;
   PLOT RESID*AGE/VREF = 0;
   PLOT RESID*WT/VREF = 0;
```

In this program the regular output from PROG REG is suppressed by using the option NOPRINT in the MODEL line. The OUTPUT line directs the output to a data file named

GRAPHS (or any other file name we would designate) and in that file the predicted *y* values (P) will be called PRED_Y (or any other name we designate) while the residuals (R) will be called RESID (or any other name we designate).

PROC PLOT is then applied to the data in GRAPHS and the following three graphs are printed. The option VREF = 0 in the PLOT statements will cause a horizontal reference line to be printed on the graphs at zero on the vertical scale:

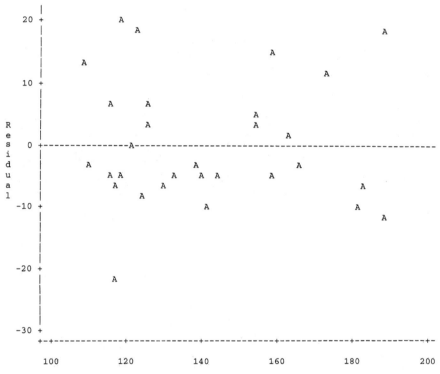

Predicted Value of SYSTOLIC

If the multiple regression model is good for prediction, predicted values can be computed for the independent values in the data set as well as for other values of the independent variables. For example, if an estimate of systolic blood pressure is desired for a woman of age 31 and weight 55 kg, the following SAS program can be used:

```
DATA NEW;
  INPUT AGE WT SYSTOLIC;
  CARDS;
31 55.
;
```

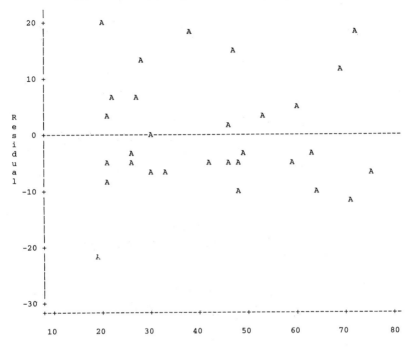

The SAS System

Plot of RESID*AGE.   Legend: A = 1 obs, B = 2 obs, etc.

```
DATA BOTH;
  SET PATIENTS NEW;
PROC REG DATA = BOTH;
  MODEL SYSTOLIC = AGE WT/CLM CLI ;
```

The output follows.

The SAS System

The REG Procedure
Model : MODEL1
Dependent Variable: SYSTOLIC
Analysis of Variance

Source	DF	Sum of Squares	Mean Square	F Value	Pr > F
Model	2	18586	9292.89261	84.96	<.0001
Error	27	2953.41479	109.38573		
Corrected Total	29	21539			

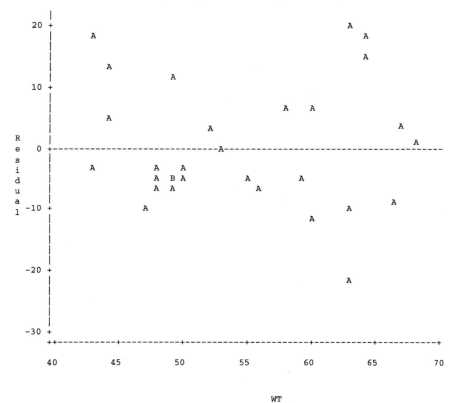

The SAS System

Plot of RESID*WT.   Legend: A = 1 obs, B = 2 obs, etc.

	Root MSE	10.45876	R-Square	0.8629
	Dependent Mean	141.40000	Adj R-Sq	0.8527
	Coeff Var	7.39658		

Parameter Estimates

Variable	DF	Parameter Estimate	Standard Error	t Value	Pr > |t|
Intercept	1	20.48318	15.50504	1.32	0.1976
AGE	1	1.43391	0.11057	12.97	<.0001
WT	1	1.10047	0.24870	4.42	0.0001

The REG Procedure
Model: MODEL1
Dependent Variable: SYSTOLIC

Output Statistics

Obs	Dep Var SYSTOLIC	Predicted Value	Std Error Mean Predict	95% CL Mean	
1	116.0000	124.3269	3.9204	116.2829	132.3709
2	122.0000	121.8255	2.4385	116.8221	126.8288
3	212.0000	194.1547	4.8593	184.1842	204.1253
4	135.0000	140.3661	2.3259	135.5938	145.1384
5	131.0000	141.0329	2.6351	135.6261	146.4398
6	123.0000	109.0534	3.8821	101.0879	117.0188
7	96.0000	117.0572	3.4923	109.8916	124.2229
8	113.0000	118.2908	2.6229	112.9090	123.6725
9	111.0000	115.5231	3.0424	109.2806	121.7657
10	134.0000	138.0650	3.3680	131.1543	144.9756
11	164.0000	162.3755	4.1808	153.7973	170.9538
12	123.0000	129.4286	2.1675	124.9812	133.8760
13	141.0000	122.2920	3.5729	114.9610	129.6229
14	160.0000	154.9384	3.4283	147.9041	161.9727
15	128.0000	133.5300	2.5112	128.3775	138.6825
16	171.0000	181.5830	4.0260	173.3223	189.8437
17	176.0000	182.2828	4.1314	173.8059	190.7597
18	139.0000	118.4911	3.4275	111.4585	125.5237
19	177.0000	188.3189	4.1883	179.7252	196.9127
20	185.0000	173.3459	3.4863	166.1927	180.4991
21	110.0000	117.4236	2.8890	111.4958	123.3513
22	157.0000	153.7048	2.2427	149.1032	158.3065
23	173.0000	158.3071	3.1698	151.8033	164.8109
24	162.0000	165.8430	2.9670	159.7551	171.9308
25	108.0000	110.5875	3.3198	103.7757	117.3992
26	122.0000	115.8566	2.9296	109.8456	121.8675
27	154.0000	159.0069	2.7627	153.3383	164.6754
28	139.0000	144.3344	2.2221	139.7750	148.8937
29	132.0000	125.2270	2.7046	119.6777	130.7764
30	128.0000	125.4274	4.0854	117.0449	133.8099
31	.	125.4603	2.2807	120.7807	130.1399

The REG Procedure
Model: MODEL1
Dependent Variable: SYSTOLIC

Output Statistics

Obs	95% CL Predict		Residual
1	101.4092	147.2446	−8.3269
2	99.7903	143.8606	0.1745

3	170.4920	217.8175	17.8453
4	118.3822	162.3499	−5.3661
5	118.9027	163.1632	−10.0329
6	86.1631	131.9436	13.9466
7	94.4329	139.6816	−21.0572
8	96.1666	140.4149	−5.2908
9	93.1740	137.8723	−4.5231
10	115.5201	160.6098	−4.0650
11	139.2649	185.4862	1.6245
12	107.5130	151.3442	−6.4286
13	99.6147	144.9692	18.7080
14	132.3553	177.5215	5.0616
15	111.4605	155.5995	−5.5300
16	158.5884	204.5777	−10.5830
17	159.2096	205.3560	−6.2828
18	95.9086	141.0737	20.5089
19	165.2026	211.4353	−11.3189
20	150.7255	195.9663	11.6541
21	95.1603	139.6868	−7.4236
22	131.7574	175.6523	3.2952
23	135.8835	180.7306	14.6929
24	143.5365	188.1494	−3.8430
25	88.0727	133.1022	−2.5875
26	93.5710	138.1421	6.1434
27	136.8112	181.2025	−5.0069
28	122.3957	166.2730	−5.3344
29	103.0615	147.3926	6.7730
30	102.3887	148.4661	2.5726
31	103.4964	147.4242	

The REG Procedure
Model: MODEL1
Dependent Variable: SYSTOLIC

Sum of Residuals	0
Sum of Squared Residuals	2953.41479
Predicted Residual SS (PRESS)	3844.53556

14.5. MODEL FITTING

The object of model fitting is to obtain the simplest model that will adequately fit the data for prediction purposes. There may be many independent or regressor variables (x_i) which could logically be included in the model. However, some may be difficult or expensive to obtain, and certainly, as we noted in the previous section, they will increase the complexity of the standard errors of the estimates when they are included in the model. Thus, to be included in the model, a regressor variable should contribute significantly to the accuracy of estimation. This section will examine the process of choosing among many possible independent variables, a suitable set to be retained in the model.

We have already discussed procedures for testing the significance of each b_i in the model, but it is possible for an independent variable to have a significant linear relationship with the y variable and still not be especially useful for prediction purposes, so criteria other than statistical significance are needed in the model-fitting process. Briefly stated, for an x_i to be included in the model, it must simultaneously increase the sum of squares due to regression (SSR $= \sum b_i S_{iy}$) and reduce the mean-square error (MS$_e$) for the model which is chosen. As a consequence of the use of least-squares procedures, except when $b_i = 0$, both SSR $= \sum b_i S_{iy}$ and $R^2 = $ SSR/S$_{yy}$ will increase with the addition of another x_i to the prediction equation (see Figure 14.2). However, the behavior of MS$_e$ is more complex. The mean-square error for any given model is computed as

$$MS_e = \frac{S_{yy} - SSR}{n - k - 1}$$

Thus, when an additional regressor variable is included in the model, it increases SSR to make the numerator smaller, but at the same time, it increases the numerical value of k by one unit, causing the denominator also to be smaller. Hence, if the new variable does not explain very much of the variability in y, the decrease in the numerical value of the numerator of the above equation (S$_{yy}$ − SSR) may be proportionally less than the decrease in the numerical value of the denominator $(n - k - 1)$. Then, as a consequence, the error mean square (MS$_e$) for the model will be greater with the additional regressor variable than it would have been without it (see Figure 14.3).

In model fitting, as a new regressor is added to or deleted from the model, the experimenter must monitor the relative changes in R^2 and MS$_e$. Referring again to Figures 14.2 and 14.3, the ideal model is the one with the set of k predictor variables which occurs at the "knee" of the R^2 curve, the point at which a new predictor variable will not appreciably increase the numerical value of R^2. Similarly, it is that set which produces the minimum MS$_e$ in Figure 14.3. However, there is no guarantee that the same set of independent variables will provide the optimum value on each of the respective curves. In an attempt to manage this problem, in the output of SAS model-fitting programs, there is a statistic that takes into account the relative changes in k, SSR, and MS$_e$. It is *Mallow's C_p statistic*, which will appear in SAS

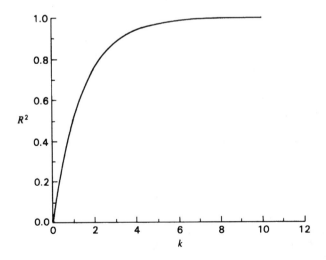

FIGURE 14.2. R^2 as a function of the number of independent variables.

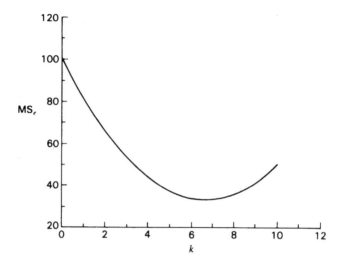

FIGURE 14.3. Mean-square error as a function of the number of independent variables.

output as C(p) and is obtained from the equation

$$C_p = \frac{S_{yy} - SSR}{\text{Full model } MS_e} - n + 2p$$

where p is the number of estimates of parameters in the prediction equation, including the estimate of the intercept a. The equation indicates that as p increases C_p will also increase unless there is an offsetting increase in SSR. If there is an x variable in the model which does not contribute much to prediction, it will increase the value of p but not greatly affect the numerical value of SSR and consequently will cause a larger value of C_p. Thus, when comparing different prediction equations, or models, we choose that which has the smallest numerical value of C_p. In some computer programs, the adjusted coefficient of determination (R_{adj}^2) is used in similar fashion to gauge whether the increase in SSR warrants the expense of increasing k. However, to keep the discussion from becoming too protracted, only the C_p statistic will be demonstrated here.

There are many computer programs for model fitting, but most tend to follow one of two approaches. Some begin with the *full model*, an equation containing all the regressor variables involved in the study, and then delete those which contribute little to prediction. Another approach is to begin with a prediction equation containing only one independent variable and then to continue to add other x variables so long as they improve the predictive ability of the equation. Both approaches require considerable computation and properly should be thought of as computer routines. To remove any mystery about what is being done by the computer, we will use both procedures on the small data set concerning systolic blood pressure of West Indian women.

First we will examine *backward elimination*, a step-down procedure in which the investigator begins with a full model containing all possible regressor variables, and the x_i are eliminated one by one as it is determined that they contribute little to the model. When we first performed the multiple regression analysis, we found that systolic blood pressure has a significant linear relationship with age (x_1) but not with weight (x_2). That in itself provides evidence that, at least for this limited data set, weight does not contribute to the prediction of systolic pressure and should be eliminated from the model. However, let us also examine the information provided by R^2, MS_e, and C_p:

	Full Model (x_1 and x_2)	Model with x_1 Alone
$R^2 = \dfrac{SSR}{S_{yy}}$	$\dfrac{3328.93}{3628.00} = 0.918$	$\dfrac{3316.44}{3628.00} = 0.914$
$MS_e = \dfrac{S_{yy} - SSR}{n - k - 1}$	$\dfrac{3628 - 3328.93}{7 - 2 - 1} = 74.767$	$\dfrac{3628 - 3316.44}{7 - 1 - 1} = 62.312$
$C_p = \dfrac{S_{yy} - SSR}{\text{Full Model } MS_e} - n + 2p$	$\dfrac{299.068}{74.767} - 7 + 2(3) = 3.00$	$\dfrac{311.562}{74.767} - 7 + 2(2) = 1.17$

We note that R^2 is larger for the full model than it is for the model containing x_1 alone, but the increase is only 0.004. When we recall that $100(R^2)$ = percentage of S_{yy} explained by the model, we can see that the full model explains only 0.4% more of the variability in y than does a model containing x_1 alone. Thus in this situation there is no advantage in using the model with age and weight as regressor variables when such a model is so little better than that containing only age as a regressor. This conclusion is further substantiated by examination of the numerical values of MS_e. When k is the number of regressors in the model, $MS_e = SS_e/(n - k - 1)$ may be smaller for a model containing only a few x_i than it is for the full model, and that is indeed the case for this data set; for the full model $MS_e = 74.767$, but it is only 62.312 for the model containing just x_1. For prediction purposes, we generally choose the model with the smallest MS_e, hence that containing x_1 alone.

Mallow's C_p statistic will be discussed only briefly, but recall that it takes into account the number of regressor variables in the model under consideration. When we examine the equation for this statistic, we can see that for the full model C_p will always be equal to $p = k + 1$, but in situations such as ours, where SSR is so nearly the same for a model in which $k = 1$ as it is in the full model with $k = 2$, C_p will be smaller for the model containing only age as a regressor. In general, we want to choose a model for which $C_p < p$. Once again, this would lead us to choose the model containing x_1 alone.

There is a cautionary note to be made with respect to the use of the C_p statistic. Remember in Section 14.1 there was discussion of the linear model and the assumption that the ε's are IND$(0, \sigma^2)$. When the C_p statistic is computed, the variance of the ε's is assumed to be well estimated by the full model MS_e; symbolically we express this as $E(\text{Full model } MS_e) = \sigma^2$. However, we have already noted that full model MS_e can be too large if there are many useless predictor variables in the model, and in such a situation the full model MS_e is a biased overestimate of σ^2. If the denominator in the equation for computing C_p is a seriously inflated overestimate of σ^2, then the relative sizes of the C_p values of two different models may not adequately reflect the real magnitude of the difference in their respective usefulness in prediction.

Example 14.2. Model Building by Backward Elimination

The phantom midge, genus *Chaoborus*, resembles the mosquito in appearance but not in bloodsucking behavior. Swarms of adult chaoborids are a familiar sight along the shoreline of lakes and other bodies of fresh water, but a great portion of the life cycle is spent in the water in the larval stage. The larva burrows into the sediment at the bottom of a lake or pond and remains there during the daylight hours. At night it migrates vertically toward the surface to feed on the fauna in the plankton layer. The larva is itself prey for larger animals and consequently has an important role in the food chain of freshwater fish.

Man-made lakes and other water impoundments seem to create good habitats for chaoborids, so much so that they can become a nuisance. They seem to be little affected by the brackish nature of such water; the reduced oxygen content may even be favorable for an increase in population density. The steep banks and greater depths of man-made impoundments also seem to favor the genus.

To learn more about the contribution of various environmental factors to the habitat of *Chaoborus* larvae, a team of biologists make a study of a recreational lake that was created by damming a small stream. The lake has a surface area of approximately 20 hectares, and to obtain random samples from it, a grid was superimposed on a map of the lake and 30 random sampling points are taken on the grid. By means of surveying equipment, these sampling points are located on the lake. The following variables are measured at each sampling point:

x_1: The depth of the lake at the sampling point, measured to the nearest decimeter (recorded in meters).

x_2: The brackishness (conductivity) of the water, measured from a sample taken at the bottom (recorded in mhos per decimeter).

x_3: The dissolved oxygen (milligrams per liter) in the water sampled from the lake bottom.

y: The number of *Chaoborus* larvae collected in a grab sample of the sediment at the sampling point. The sampling device collected sediment from an area of approximately 225 cm^2 of lake bottom.

A SAS data set is created as follows:

```
DATA LARVAE;

    INPUT MIDGES DEPTH BRACK OXY;

    CARDS;
```

35	8.4	8.0	1.0
10	2.0	6.5	8.5
9	3.5	6.2	6.5
30	10.4	5.0	1.5
20	6.5	6.5	7.5
23	6.2	7.3	4.5
28	12.4	6.4	4.0
8	7.0	6.0	10.0
29	5.8	6.1	3.0
4	3.0	5.4	11.0
18	6.0	7.3	4.5
14	5.5	6.6	5.5
32	9.0	6.5	2.5
6	1.1	5.8	7.0
25	4.3	7.8	3.3
19	9.7	6.7	9.1
39	11.6	4.9	1.2

2	2.6	6.6	13.1
22	2.9	7.4	1.3
26	3.4	6.6	3.0
6	5.8	7.7	10.3
27	3.6	6.2	1.3
12	6.0	5.1	6.8
23	8.0	5.1	5.3
19	4.4	7.1	3.2
29	8.7	6.5	4.4
20	3.0	5.3	6.2
36	12.1	6.8	2.2
24	9.3	7.6	5.2
26	11.0	5.6	2.2

;

In the SAS System backward elimination is performed by the following program:

```
PROC REG DATA = LARVAE;
  MODEL MIDGES = DEPTH BRACK OXY/METHOD = BACKWARD;
```

The output is

The SAS System

The REG Procedure
Model: MODEL1
Dependent Variable: MIDGES

Backward Elimination: Step 0

All Variables Entered: R-Square $= 0.8747$ and $C(p) = 4.0000$
Analysis of Variance

Source	DF	Sum of Squares	Mean Square	F Value	Pr > F
Model	3	2557.76659	852.58886	60.48	<.0001
Error	26	366.53341	14.09744		
Corrected Total	29	2924.30000			

Variable	Parameter Estimate	Standard Error	Type II SS	F Value	Pr > F
Intercept	22.10575	5.98047	192.61068	13.66	0.0010
DEPTH	1.20575	0.23583	368.50263	26.14	<.0001
BRACK	0.33781	0.80394	2.48916	0.18	0.6778
OXY	−2.19334	0.23340	1244.93069	88.31	<.0001

(*continued*)

Variable	Parameter Estimate	Standard Error	Type II SS	F Value	Pr > F
Intercept	22.10575	5.98047	192.61068	13.66	0.0010
DEPTH	1.20575	0.23583	368.50263	26.14	<.0001
BRACK	0.33781	0.80394	2.48916	0.18	0.6778
OXY	−2.19334	0.23340	1244.93069	88.31	<.0001

Bounds on condition number: 1.1983, 10.236

Backward Elimination: Step 1

Variable BRACK Removed: R-Square = 0.8738 and C(p) = 2.1766

Analysis of Variance

Source	DF	Sum of Squares	Mean Square	F Value	Pr > F
Model	2	2555.27743	1277.63872	93.48	<.0001
Error	27	369.02257	13.66750		
Corrected Total	29	2924.30000			

Variable	Parameter Estimate	Standard Error	Type II SS	F Value	Pr > F
Intercept	24.41452	2.32524	1506.77478	110.25	<.0001
DEPTH	1.19268	0.23018	366.94032	26.85	<.0001
OXY	−2.20414	0.22842	1272.64766	93.11	<.0001

Bounds on condition number: 1.1775, 4.7098

All variables left in the model are significant at the 0.1000 level.

Summary of Backward Elimination

Step	Variable Removed	Number Vars In	Partial R-Square	Model R-Square	C(p)	F Value	Pr > F
1	BRACK	2	0.0009	0.8738	2.1766	0.18	0.6778

When the SAS printout is examined, it is seen that there is a Step 0 and a Step 1.

Step 0: the analysis of the full model

This is identified as Step 0 because no regressors have been eliminated, in other words, the full model. There is a test of H_0: $\beta_1 = \beta_2 = \beta_3 = 0$, the hypothesis that the regression hyperplane is nonsignficant. The test of this hypothesis is given in the F test for Model, where the computed value $F = 60.48$ has a $P < 0.0001$. For any conventional α level, the null hypothesis is rejected, so it is obvious that if the prediction equation is based on all three regressor variables, it is significant. For the full model, $R^2 = 0.8747$; hence depth (x_1), conductivity (x_2), and oxygen (x_3) together can account for 87.47% of the variability in midge larval density (y). However, this F test does not indicate whether all of the regressor variables are needed for a prediction equation. Instead, it is necessary to examine the tests of significance for the individual partial regression coefficients in order to determine their relative importance in explaining *Chaoborus* larval density.

Hypothesis	Coefficient (b_i)	Standard Error (s.e.$_i$)	$F = (b_i/\text{s.e.}_i)^2$	P value
$\beta_1 = 0$	$b_1 = 1.2058$	s.e.$_1$ = 0.2358	26.14	<0.0001
$\beta_2 = 0$	$b_2 = 0.3378$	s.e.$_2$ = 0.8039	0.18	0.6778
$\beta_3 = 0$	$b_3 = -2.1933$	s.e.$_3$ = 0.2334	88.31	<0.0001

From these tests, it is seen that x_2 (conductivity or brackishness) adds no significant predictive ability to a multiple regression equation which already contains x_1 (depth) and x_3 (oxygen). Thus it can be dropped from the equation. However, when this is done in Step 1, new values will be computed for b_1 and b_3. These coefficients will be different because, once x_2 has been excluded, none of the least-squares computations will take into account the covariability between x_1 and x_2 or that between x_3 and x_2.

The Type II sums of squares given in this analysis are sometimes called *partial sums of squares*, meaning the added variability explained by adding a regressor to a model which already contains the other $k - 1$ regressors. Thus the Type II SS for conductivity (BRACK) is the additional variability explained by adding x_2 to a model which already contains x_1 and x_3. Similarly, the Type II SS for oxygen (OXY) provides the additional variability explained by adding x_3 to a model which already contains x_1 and x_2. These sums of squares confirm that very little additional variability in larval density is explained by adding BRACK to a model already containing DEPTH and OXY.

Step 1: the analysis of the model with one regressor eliminated
The first information provided in this step is identification of the variable which has been removed, and the rest of the printout consists of a multiple regression analysis on those variables which are retained. Without x_2 (brackishness), the regression plane is still significant; in fact the computed value of F is even larger than it was for the full model. The larger value of F can be explained by the fact that, when compared to the full model, the reduced model shows a numerical value of SSR which is almost the same as it was for the full model, along with a smaller k and a smaller MS$_e$:

	Full Model (x_1, x_2, and x_3)	Reduced Model (x_1 and x_3)
Regression SS (SSR)	2557.7666	2555.2774
Mean-square error (MS$_e$)	14.0974	13.6675
Number of regressors (k)	3	2
$F = \dfrac{\text{SSR}/k}{\text{MS}_e}$	60.48	93.48

There are other comparisons that can be made between the two models. When the values of C_p are computed, for the full model it is the anticipated value of $k + 1 = 4.00$, but for the reduced model, it decreases to 2.18. Furthermore, for the reduced model, $R^2 = 0.8738$ is just slightly smaller than it was for the full model; the difference occurs only in the third decimal place. The last information given in the printout (Summary) is the difference in the two numerical values of R^2 along with a test to see whether there is a significant reduction in R^2 as a consequence of reducing the model:

$$F = \frac{(R_f^2 - R_r^2)/(k_f - k_r)}{(1 - R_f^2)/(n - k_f - 1)} = \frac{0.8747 - 0.8738}{(1 - 0.8747)/26} = 0.18$$

where the subscripts f and r represent the *full* and *reduced* model, respectively.

Based on the information provided in Step 1, all indicators provide evidence that the reduced model is superior to the full model. The numerical values of MS_e and C_p are smaller for the reduced model than for the full one; SSR and R^2 are little changed from their corresponding values in Step 0, and there is a test of significance showing that when x_2 is eliminated from the model, the decrease in R^2 is not significant. Hence the biologists learn they can explain larval density quite effectively without having to use measurements on water conductivity, or brackishness.

The next question to be addressed is whether other regressors can also be eliminated from the model, and the answer is provided in the tests of hypotheses for the partial regression coefficients which are provided in Step 1. The F tests for b_1 (the regression of midge density on depth) and b_3 (the regression of density on oxygen) are both significant at the 0.0001 level. Because the two x variables remaining in the model are significant, neither can be eliminated. Hence the computer routine automatically stops at this point. If a second variable could have been eliminated, there would have been a Step 2 in the printout, and the process would continue to Step 3 and so on until all remaining regressor variables are significant. In the end, the model to be chosen for explaining larval density is that which uses the coefficients given in the last step of the routine. In this example those would be the intercept of 24.41452 and the partial regression coefficients of 1.19268 and -2.20414 for depth and oxygen, respectively. Because the variable which was originally designated as x_2 is no longer in the model, after rounding to fewer decimal places, the prediction equation can be reported as

$$y = 24.415 + 1.193x_1 - 2.204x_2$$

where x_1 represents depth and x_2 now represents oxygen.

The signs of the partial regression coefficients are important. The positive relationship between x_1 and y indicates that larval population density increases with depth of the lake when oxygen content is held constant, whereas there is a negative association between larval count and oxygen content of the water when depth is held constant. Prediction equations based on depth and oxygen content would be valid, but only for the one lake studied. The important ecological information obtained from the study on this lake is the knowledge that oxygen and depth explain a great deal of the variability in *Chaoborus* population density. These are variables that should be included in any future studies involving other lakes. Also, if the biologists decide to conduct experiments to regulate chaoborid larval population density, they have already identified oxygen and depth as two variables which can be used as treatment effects in a factorial experiment.

The second computer routine is that of *stepwise regression*, a process of addition in which the model is built by adding one regressor variable at a time and measuring its contribution to the model. To demonstrate this process on the small, $n = 7$, data set involving blood pressure

(y), age (x_1), and weight (x_1) in Section 14.1, we would first compute the simple correlation coefficients between y and the regressor variables:

$$r_{y1} = \frac{S_{1y}}{\sqrt{S_{11}S_{yy}}} = \frac{2257}{\sqrt{1536(3628)}} = 0.956 \quad \text{and}$$

$$r_{y2} = \frac{S_{2y}}{\sqrt{S_{22}S_{yy}}} = \frac{893}{\sqrt{390(3628)}} = 0.751$$

Then, because r_{y1} is the larger of the correlation coefficients, x_1 would be the first regressor variable to enter the model. We would test its significance using simple regression techniques, and after finding it to be significant, we would then move on to the next variable to try in the model. Because there are only two independent variables in this data set, the next to enter has to be x_2, but if there were other x_i, we would have to compute the *partial correlation coefficient* between y and each x_i, independent of x_1,

$$r_{yi.1} = \frac{r_{yi} - r_{y1}r_{1i}}{\sqrt{(1 - r_{yi}^2)(1 - r_{1i}^2)}}$$

and choose as the second regressor variable that x_i which yields the largest partial correlation coefficient.

After the second regressor variable (x_2) is chosen, a multiple regression analysis is performed and the significance of each of the partial regression coefficients is tested. We have already done this and noted the significant regression on age but not on weight. Thus, based on the results of this multiple regression analysis, x_1 is to be retained in the model but not x_2.

When there are many independent variables to be evaluated in model fitting, the stepwise procedure will continue at each stage to add an x variable to the model and then test it for significance along with all of the others which were kept in the model at earlier stages. If they are significant, they remain in the model; otherwise they are removed. Thus it is possible that a specific x variable will enter the model at one stage of the process only to be removed at a later stage. This can be explained by an example in which there are three possible regressor variables, x_1, x_2, and x_3, but because of collinearity among them, x_1 is little more than a linear function of x_2 and x_3. It is quite possible that x_1 would be the first to enter the model because it indirectly contains information about x_2 and x_3. However, at later stages, when x_2 and x_3 enter the model, x_1 no longer makes any additional contribution to the prediction of the y variable, so it can then be removed from the model. To summarize the consequences of such collinearity, we can say that when x_2 and x_3 are not known, y can be predicted on the basis of x_1 because it is closely related to x_2 and x_3, but if x_2 and x_3 are known, they are more useful in prediction than x_1.

As with the case of the backward procedure, an x variable may have a significant linear relationship with y yet still be of little use in a prediction equation. Thus statistics such as MS_e, R^2, R_{adj}^2, and C_p may be used in addition to the tests of significance in the ultimate choice of a model. The only difference is the order in which they are computed. Thus, if we review all these statistics under the stepwise procedure, we note once again that, on all accounts, for the small data set the model containing age alone is superior to the one containing both age and weight. When there are only a few independent variables in a data set, it is not uncommon to arrive at the same model irrespective of whether the backward or stepwise procedure is used.

However, this will not necessarily be the case when there are many potential regressor variables from which to choose.

Let us note once again that both the backward and the stepwise procedures should be thought of as computer routines. Although the computations were demonstrated on a small data set, we did so only for the purpose of showing the computations on which the procedures are based. Computer routines which perform these procedures are readily available, so it is more important to interpret the results than it is to know how to do the arithmetic. So we will now use the foregoing discussion along with the analyzed data from the study of midge larval population density to explain how to read the computer printout for the stepwise procedure and use it for the purpose of model building.

Example 14.3.   Stepwise Method for Model Building

The research problem is the same as we discussed in Example 14.2: Biologists want to know if measurements on water depth (x_1), conductivity (x_2), and oxygen content (x_3) at a site in a lake can be used to predict the number of *Chaoborus* larval midges to be found in the sediment at the bottom of the lake at the same site.

In the SAS System the stepwise method is performed by the following program:

```
PROC REG DATA = LARVAE;
MODEL MIDGES = DEPTH BRACK OXY/METHOD = STEPWISE;
```

The output is

The SAS System

The REG Procedure
Model: MODEL1
Dependent Variable: MIDGES

Stepwise Selection: Step 1

Variable OXY Entered: R-Square = 0.7483 and C(p) = 26.2054

Analysis of Variance

Source	DF	Sum of Squares	Mean Square	F Value	Pr > F
Model	1	2188.33712	2188.33712	83.26	<.0001
Error	28	735.96288	26.28439		
Corrected Total	29	2924.30000			

Variable	Parameter Estimate	Standard Error	Type II SS	F Value	Pr > F
Intercept	34.47083	1.77592	9902.74206	376.75	<.0001
OXY	−2.66360	0.29192	2188.33712	83.26	<.0001

Bounds on condition number: 1, 1

Stepwise Selection: Step 2

Variable DEPTH Entered: R-Square = 0.8738 and C(p) = 2.1766

Analysis of Variance

Source	DF	Sum of Squares	Mean Square	F Value	Pr > F
Model	2	2555.27743	1277.63872	93.48	<.0001
Error	27	369.02257	13.66750		
Corrected Total	29	2924.30000			

Variable	Parameter Estimate	Standard Error	Type II SS	F Value	Pr > F
Intercept	24.41452	2.32524	1506.77478	110.25	<.0001
DEPTH	1.19268	0.23018	366.94032	26.85	<.0001
OXY	−2.20414	0.22842	1272.64766	93.11	<.0001

Bounds on condition number: 1.1775, 4.7098

All variables left in the model are significant at the 0.1500 level. No other variable met the 0.1500 significance level for entry into the model.

Summary of Stepwise Selection

Step	Variable Entered	Variable Removed	Number Vars In	Partial R-Square	Model R-Square	C(p)	F Value	Pr > F
1	OXY		1	0.7483	0.7483	26.2054	83.26	<.0001
2	DEPTH		2	0.1255	0.8738	2.1766	26.85	<.0001

As was the case with the output for the backward elimination procedure, the computer output is divided into steps, with each step identified by the number of regressor variables added (or deleted in the case of backward elimination). Thus Step 1 begins with simple linear regression analysis using the independent variable with the strongest correlation with the dependent variable; Step 2 is a multiple regression analysis with two x variables, and so on as new variables are introduced into the model.

Step 1: a prediction equation containing only one regressor
The single best independent variable to be used to predict larval count is the one which is entered into the model first, and that is oxygen. Of the three independent variables, this is the one with the greatest simple correlation coefficient with midge larval density (y). While the simple correlation coefficient is not given, its square is identified in $R^2 = 0.7483$, meaning that 74.83% of the variability of larval density from site to site can be attributed to differences in oxygen content of the water. Because only one regressor variable is under consideration in Step 1, the test of the significance of the simple linear regression of larval count on oxygen is given both in the F test for Model and for the variable OXY; hence the numerical value for both F tests is 83.26, which is highly significant ($P < 0.0001$). Thus it is obvious that the

oxygen content of the water has a very important effect on the number of larvae at a site. However, the computed value of Mallow's statistic is $C_p = 26.2054$, which is very much larger than $p = k + 1$, thereby indicating that the model can be improved by the addition of other regressors. This is done in Step 2.

Step 2: a prediction equation containing two regressors

The printout for this step indicates that, in a model already containing oxygen, the second most useful regressor variable is depth. This was determined by holding oxygen level constant and finding the partial correlation coefficient between y (larval count) and the other x variables (depth and brackishness, respectively). Although the partial correlation coefficients are not part of the printout, that involving depth was the larger; hence that variable was selected to come into the model at Step 2.

The improvement in the model which is due to the addition of depth as a regressor can be seen by comparing the analyses for Steps 1 and 2. The numerical value of C_p drops dramatically from 26.21 in Step 1 to 2.18 in the second analysis, and one of the gauges of a useful model is for it to have a C_p value less than $k + 1$. Furthermore, the addition of depth to the control causes MS_e to decrease from 26.28 in Step 1 to 13.67. This also indicates that the model in Step 2 is superior to that containing only oxygen as a regressor. As a final measure of the improvement in the model, it is seen that the coefficient of determination for Step 2 ($R_{II}^2 = 0.8738$) is greater than that for Step 1 ($R_I^2 = 0.7483$). The additional variability accounted for by depth is $0.8738 - 0.7483 = 12.55\%$. As evidenced in the summary of the stepwise procedure, this difference in two R^2 values is significant. Thus, with respect to the percentage of variability explained, the two-variable model is significantly better than that containing only oxygen as a regressor.

The final action in each analysis of the stepwise procedure is to make a test of the individual H_0: $\beta_i = 0$. This is to determine whether variables added in prior steps are still useful for prediction purposes after the addition of the new variable. The F values for oxygen and depth are 93.11 and 26.85, respectively, and both have P values of less than 0.0001. Hence both are significant and should be kept in the model.

After Step 2 is completed, a new partial correlation coefficient is computed, that between larval density and conductivity, with both oxygen and depth held constant. If this partial correlation was significant, there would be a Step 3 in which the third regressor would be introduced. However, because it is not significant, the computer routine automatically stops after Step 2.

The model chosen by stepwise regression is that which uses oxygen and depth as regressors, the same two variables which were chosen by backward elimination. Furthermore, the numerical values of a and the b coefficients are the same for the two procedures. However, the b_i are reversed in order because the stepwise procedure brought oxygen into the model first. This indicates that, if the prediction of larval density should be based on only one regressor, that variable should be oxygen, since it explains the most variability in larval density ($R_I^2 = 0.7483$). If a two-variable model is to be used, it should contain both oxygen and depth, for these two together can explain significantly more variability ($R_{II}^2 = 0.8738$) than does oxygen alone. However, if all three predictor variables are used, the R^2 for the full model may be greater, but it will not be significantly greater.

In deciding which of the two procedures to use, the choice is arbitrary and largely a matter of personal preference. Some researchers use backward elimination because they want to see how much variability is explained by all the independent variables they included in their study, that is, the full model. They are less satisfied with the stepwise procedure because it provides information only about those regressors which are significant when added to the model. An opposite opinion is held by those who prefer the stepwise procedure because they want to know how much variability is explained by the single best predictor variable, and they find the backward procedure limiting because it stops the elimination process with the first significant x variable. However, when one is using computer routines, once the data have been entered, it is quite easy to perform more than one analysis. By using several different options in multiple regression analysis, one can usually obtain all the information desired.

EXERCISES

14.5.1. Using the data and analyses for the *Chaoborus* larvae study in this section:
   a. Compute the respective numerical values of R^2_{adj} for:
      i. The model containing oxygen and depth as regressors
      ii. The full model containing all the independent variable
   b. If R^2_{adj} is used as the criterion for selecting the prediction equation in this study, which model will be chosen? Explain your answer.

14.5.2. In a study of factors which contribute to successful farming, a random sample was taken of farms of similar size and farming operations. Then for each farm and farmer records were obtained on the following variables:

Education: the number of years of formal education of the farmer

Experience: the number of years of farming experience

Age: the age, in years, of the farmer

Profit: the profit, in dollars, of the previous 12 months of operation

The data were analyzed by stepwise regression and the following results obtained:

The SAS System

The REG Procedure
Model: MODEL1
Dependent Variable: PROFIT

Stepwise Selection: Step 1

Variable AGE Entered: R-Square $= 0.9865$ and C(p) $= 114.80$

Analysis of Variance

Source	DF	Sum of Squares	Mean Square	F Value	Pr > F
Model	1	2843978654.7533	2843978654.7533	4130.88	<.0001
Error	44	30292610.5728	688468.4221		
Corrected Total	45	2874271265.3261			

Variable	Parameter Estimate	Standard Error	Type II SS	F Value	Pr > F
Intercept	−309.1045				
AGE	592.2699	9.2151	2843978654.7533	4130.88	<.0001

Stepwise Selection: Step 2

Variable EXP Entered: R-Square = 0.9918 and C (p) = 82.78

Analysis of Variance

Source	DF	Sum of Squares	Mean Square	F Value	Pr > F
Model	2	2850550540.5368	1425275270.2684	2583.68	<.0001
Error	43	23720724.7893	551644.7625		
Corrected					
Total	45	2874271265.3261			

Variable	Parameter Estimate	Standard Error	Type II SS	F Value	Pr > F
Intercept	5275.8158				
EXP	197.1492	57.1189	6571885.7835	11.91	0.0013
AGE	358.2077	68.3133	15167663.2354	27.50	0.0001

Stepwise Selection: Step 3

Variable ED Entered: R-Square = 0.9972 and C (p) = 4.0000

Analysis of Variance

Source	DF	Sum of Squares	Mean Square	F Value	Pr > F
Model	3	2866157182.1712	955385727.3904	4945.25	<.0001
Error	43	8114083.1549	193192.4561		
Corrected					
Total	45	2874271265.3261			

Variable	Parameter Estimate	Standard Error	Type II SS	F Value	Pr > F
Intercept	4880.0870				
ED	632.9172	70.4186	15606641.6344	80.78	<.0001
EXP	649.9674	60.6696	22173310.6004	114.77	<.0001
AGE	−51.0535	60.8911	135810.5613	0.70	0.4065

Stepwise Selection: Step 4

Variable AGE Removed: R-Square = 0.9971 and C (p) = 2.7000

Analysis of Variance

Source	DF	Sum of Squares	Mean Square	F Value	Pr > F
Model	2	2866021371.6098	1433010685.8049	7469.12	<.0001
Error	43	8249893.7163	191857.9934		
Corrected					
Total	45	2874271265.3261			

Variable	Parameter Estimate	Standard Error	Type II SS	F Value	Pr > F
Intercept	4362.9729				
ED	588.7656	46.5906	30638494.3084	159.69	<.0001
EXP	599.7007	9.2677	803342471.0076	4187.17	<.0001

All variables left in the model are significant at the 0.1500 level.
No other variable met the 0.1500 significance level for entry into the
model.

Summary of Stepwise Selection

Step	Variable Entered	Variable Removed	Number Vars In	Partial R-Square	Model R-Square	C(p)	F Value	Pr > F
1	AGE		1	0.9859	0.9859	114.80	4130.88	<.0001
2	EXP		2	0.0023	0.9917	82.78	11.91	0.0013
3	ED		3	0.0054	0.9972	4.00	80.78	<.0001
4		AGE	2	0.0000	0.9971	2.70	0.70	0.4065

a. How would the C_p value for Step 3 be known in advance?

b. If only one regressor variable is to be used to predict farm profit, which variable would you choose? Explain the reason for your choice.

c. If the prediction of farm profit is to be based on two regressor variables, which variables would you choose? Explain the reason for your choice.

d. In comparing different prediction equations, what fraction of S_{yy} is explained by:

  i. Adding experience to an equation which already contains age?

  ii. Adding age to an equation which already contains experience and education?

  iii. Adding education to an equation which already contains experience and age?

e. Compute the numerical value of R^2_{adj} for Step 4.

f. Based on the results of this analysis:

  i. Tell which prediction equation should be used to predict farm profit.

  ii. Use that equation to predict the profit for a farm operated by a 35-year-old farmer who has a 12th-grade education and 16 years experience in farming.

14.5.3. Prairie chickens, a species of grouse that was once abundant throughout the Great Plains, are now found primarily in a few counties in Kansas and Nebraska. To learn more about their habitat, a random sample is taken of pastures in areas in which these birds live. Data are recorded on each pasture, and then bird dogs are used to flush the prairie chickens so that the number in the pasture can be recorded. Thus the following data are recorded for each pasture in the sample:

Acres: the size of the pasture recorded in acres.

Field: the type of pasture, whether original prairie grass or improved. Note that this is recorded on the nominal scale, but for analytical purposes, a *dummy variable* can be created by giving a code of $x_{2j} = 0$ to pasture containing original grass and $x_{2j} = 1$ for improved pastures. (If there are more than two classification variables, the coding becomes more complicated.)

Distance: the distance, in yards, from the field to nearest occupied house.

Birds: the number of prairie chickens flushed from the pasture.

The SAS System

The REG Procedure

Model: MODEL1
Dependent Variable: BIRDS

Backward Elimination: Step 0

All Variables Entered: R-Square = 0.8810 and C(p) = 4.0000

Analysis of Variance

Source	DF	Sum of Squares	Mean Square	F Value	Pr > F
Model	3	346.65356	115.55119	44.44	<.0001
Error	18	46.80099	2.60005		
Corrected Total	21	393.45454			

Variable	Parameter Estimate	Standard Error	Type II SS	F Value	Pr > F
Intercept	−3.74744	1.70233	12.59983	4.85	0.0410
ACRES	0.02059	0.00180	339.45813	130.56	<.0001
FIELD	0.85594	0.81630	2.85868	1.10	0.3083
DISTANCE	0.00182	0.00131	5.02160	1.93	0.1816

Bounds on condition number: 1.3979, 11.3875

Backward Elimination: Step 1

Variable FIELD Removed: R-Square = 0.8738 and C(p) = 3.0995

Analysis of Variance

Source	DF	Sum of Squares	Mean Square	F Value	Pr > F
Model	2	343.79488	171.89744	65.77	<.0001
Error	19	49.65967	2.61367		
Corrected Total	21	393.45454			

Variable	Parameter Estimate	Standard Error	Type II SS	F Value	Pr > F
Intercept	−2.64358	1.34128	10.15309	3.88	0.0635
ACRES	0.02066	0.00180	342.57002	131.07	<.0001
DISTANCE	0.00109	0.00111	2.50978	0.96	0.3394

Bounds on condition number: 1.0007, 4.0027

Backward Elimination: Step 2

Variable DISTANCE Removed: R-Square = 0.8674 and C(p) = 2.0647

Analysis of Variance

Source	DF	Sum of Squares	Mean Square	F Value	Pr > F
Model	1	341.28509	341.28509	130.84	<.0001
Error	20	52.16945	2.60847		
Corrected Total	21	393.45454			

Variable	Parameter Estimate	Standard Error	Type II SS	F Value	Pr > F
Intercept	−1.52314	0.70050	12.33257	4.73	0.0419
ACRES	0.02062	0.00180	341.28509	130.84	<.0001

Bounds on condition number: 1.0007, 1.0000

```
All variables left in the model are significant at the 0.1000 level.
```

Summary of Backward Elimination

Step	Variable Removed	Number Vars In	Partial R-Squar	Model R-Squar	C(p)	F Value	Pr > F
1	FIELD	2	0.0073	0.8738	3.0995	1.10	0.3083
2	DISTANCE	1	0.0064	0.8674	2.0647	0.96	0.3394

a. For Step 0, analysis of the full model, show how to compute R^2 and R^2_{adj}.

b. In building a prediction model, what fraction of S_{yy} is explained by adding:

   i. The distance to the nearest house (x_3) to a model already containing the acreage of the field (x_1)?

   ii. The distance to the nearest house (x_3) to a model already containing the acreage (x_1) and type (x_2) of the field?

c. Show how to compute the C_p value for a model which contains acreage (x_1) as the only regressor.

d. Based on the results of this analysis:

   i. Tell which prediction equation should be used to predict the number of prairie chickens in a pasture.

   ii. Make a test of significance to determine whether this model has an R^2 which is significantly smaller than that for the full model.

14.6. LOGARITHMIC TRANSFORMATIONS

There is a tendency among those who use linear regression techniques to drop the term "linear" when they speak and write about the relationship between variables x and y. Also, most researchers wisely seek the simplest solution first and test for a linear association before looking for a more complex relationship between the variables. Thus there is the danger of implying that all relationships are linear or that least-squares techniques are not appropriate for nonlinear relationships.

The problem in testing for more complex relationships is knowing what sort of relationship we should test. If the relationship is not linear, there are an infinite number of other possible relationships in which y is a function of x. In this section and the next one, we examine some functions of x that are curves rather than straight lines. We assume as before that there will be deviations from the trend line, that these deviations are normally distributed, and that the deviations have the same variance for all x values.

We look at two techniques for nonlinear functions: logarithmic transformations and polynomial regression. Log transformations are discussed in this section and polynomial regression in Section 14.7.

If there is a log-linearizable relationship between x and y, then we can obtain a straight line by transforming x to logs, y to logs, or both x and y to logs. Each of these procedures rectifies (straightens out) a different sort of relationship. The three types of relationships are shown in Figure 14.4 along with the logarithmic transformations to be used.

The type of logarithmic transformation to use may be determined in several ways. The nature of the two variables may indicate it, such as the exponential growth rate of single-cell organisms or investment strategy when earnings are reinvested. Sometimes there may be an

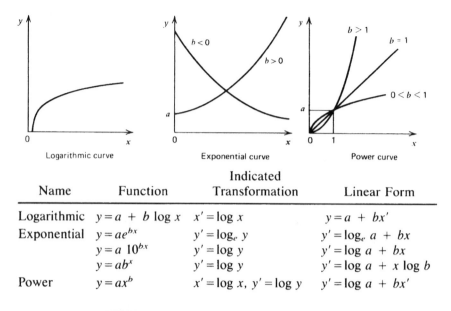

Name	Function	Indicated Transformation	Linear Form
Logarithmic	$y = a + b \log x$	$x' = \log x$	$y = a + bx'$
Exponential	$y = ae^{bx}$	$y' = \log_e y$	$y' = \log_e a + bx$
	$y = a\,10^{bx}$	$y' = \log y$	$y' = \log a + bx$
	$y = ab^x$	$y' = \log y$	$y' = \log a + x \log b$
Power	$y = ax^b$	$x' = \log x, \; y' = \log y$	$y' = \log a + bx'$

FIGURE 14.4. Log-linearizable function ($a > 0$, $x > 0$).

absolute upper or lower bound to the y variable, and this asymptotic value is approached experimentally. Frequently, the research literature in the area reveals that earlier experimenters have successfully used a logarithmic transformation, and one can anticipate that such a procedure will serve again. Finally, the experimenter may choose to plot the data points on semilog graph paper or on log-log graph paper to see whether a certain transformation appears to work. It is worth remembering, however, that the experimental α level is affected when one uses a "try it and see how it works" approach to data analysis. If one has a truly independent set of x and y variables, it may still be possible to find a seemingly significant relationship if enough different transformations are tried and the best fit is chosen for statistical analysis.

Example 14.4.   Log Transformation of the Independent Variable

Research workers in nuclear medicine have been interested in establishing cytogenetic dose–response relationships for various levels of radioactivity. Early work depended on evaluating cytogenetic lesions in tissue cultures of lymphocytes from individuals accidentally exposed to nuclear radiation and from those undergoing radiation therapy. Now, procedures are available that make it possible to establish dose–response curves for human lymphocytes that are exposed *in vitro* (outside the body). Blood can be drawn from healthy individuals and the white cells collected, exposed to the appropriate dose, and placed in tissue-culture solution. Cell division is arrested at a stage when the chromosomes are clearly distinguishable and can be examined for radiation damage.

In the biological sciences associated with medicine, the logarithmic transformation of dosage is so common that consulting statisticians almost anticipate using it. Thus, when data are obtained, the statistician has it plotted on graph paper that has vertical rulings on an arithmetic scale (to plot the y variable) and horizontal ruling on a logarithmic scale (for the x variable) or a computer package can be used to plot y against $\log x$. If his suspicions about log dose–response are confirmed, he will proceed with the sort of analysis demonstrated below. (Specific activity, dosage, is measured in nanocuries per milliliter, nCi/mL.)

Specific Activity	Log of Activity x	Dicentric Chromosomes y
40	1.6021	2
40	1.6021	4
40	1.6021	5
80	1.9031	9
80	1.9031	6
80	1.9031	16
160	2.2041	14
160	2.2041	19
160	2.2041	23
320	2.5051	35
320	2.5051	32
320	2.5051	26
Total	24.6432	191

$$\sum xy = 433.0231 \qquad \sum y^2 = 4429.00$$

$$\frac{\sum x \sum y}{n} = \underline{392.2376} \qquad \frac{\left(\sum y\right)^2}{n} = \underline{3040.08}$$

$$S_{xy} = 40.7855 \qquad S_{yy} = 1388.92$$

$$\sum x^2 = 51.9663 \qquad b = \frac{40.7855}{1.3590} = 30.011$$

$$\frac{\left(\sum x\right)^2}{n} = \underline{50.6073}$$

$$S_{xx} = 1.3590$$

The variance from the trend line is obtained in the same manner as it was for simple regression:

$$s_{y.x}^2 = \frac{\sum (y - \hat{y})^2}{n - 2} = \frac{S_{yy} - S_{xy}^2/S_{xx}}{n - 2}$$

$$= \frac{1388.92 - (40.7855)^2/1.3590}{10}$$

$$= 16.49$$

and the test significance for H_0: $\beta = 0$ against H_a: $\beta > 0$ is

$$t = \frac{b - 0}{\sqrt{s_{y \cdot x}^2 / S_{xx}}} = \frac{30.011}{\sqrt{16.49/1.359}}$$

$$= \frac{30.011}{3.483}$$

$$= 8.616$$

When compared with $t_{0.05,\ 10} = 1.812$, the trend is found to be significant. The coefficient of determination is

$$r^2 = \frac{S_{xy}^2 / S_{xx}}{S_{yy}} = \frac{1224.03}{1388.92} = 0.881$$

which is a relatively large value, indicating a reasonably good fit that could be useful in predicting the chromosomal transmutations that result from specific levels of radioactivity. Additional studies would be necessary to determine the association between *in vivo* (within-the-body) chromosomal changes and those obtained by this procedure. However, the experimenter should feel encouraged by this experiment, for it indicates a useful technique in the study of genetic damage caused by exposure to radioactive substances.

In the SAS System the analysis is carried out by the following program:

```
DATA DOSE;
   INPUT ACT CHROMO;
   L_ACT = LOG10 (ACT);
CARDS;
   40        2
   40        4
   40        5
   80        9
   80        6
   80       16
  160       14
  160       19
  160       23
  320       35
  320       32
  320       26
   ;
PROC PLOT;
   PLOT CHROMO * L_ACT;
PROC REG;
   MODEL CHROMO = L_ACT;
```

The output follows.

The SAS System

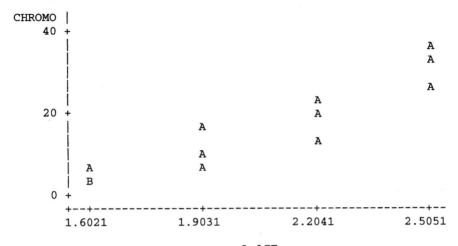

```
 Plot of CHROMO*L_ACT.  Legend: A = 1 obs, B = 2 obs, etc.

CHROMO |
    40 +
       |                                                        A
       |                                                        A
       |
       |                                                        A
       |                                     A
    20 +                                     A
       |                      A
       |                                     A
       |                      A
       |       A              A
       |       B
     0 +
       +--+---------------+---------------+---------------+--
       1.6021         1.9031          2.2041          2.5051

                             L_ACT
```

The REG Procedure
Model: MODEL1
Dependent Variable: CHROMO

Analysis of Variance

Source	DF	Sum of Squares	Mean Square	F Value	Pr > F
Model	1	1224.01667	1224.01667	74.23	<.0001
Error	10	164.90000	16.49000		
Corrected Total	11	1388.91667			

Root MSE	4.06079	R-Square	0.8813
Dependent Mean	15.91667	Adj R-Sq	0.8694
Coeff Var	25.51280		

Parameter Estimates

Variable	DF	Parameter Estimate	Standard Error	t Value	Pr > |t|
Intercept	1	−45.70808	7.24815	−6.31	<.0001
L_ACT	1	30.00808	3.48301	8.62	<.0001

Similar techniques can be used for exponential relationships. An example follows in which the independent variable *y* is the transformation of a measurement variable.

Example 14.5. Log Transformation of the Dependent Variable

The use of insecticides is a benefit but also a source of concern to the fruit industry. Insecticides protect the fruit from insect damage, but they are also toxic compounds that can be ingested by human beings. There are federally set tolerances on the amount of insecticides that fresh fruit and fruit pulp can contain, and fruit is carefully washed to meet those tolerances. Consequently, fruit processors are eager to gain as much information as they can about the deposition of insecticides and how they can be removed.

Insecticides are applied topically by spraying the fruit trees, so if the skin of the fruit has not been broken, all of the insecticide lies on the surface. Consequently, the larger the fruit, the more insecticide is deposited on it. To study the relationship between the size of peaches and the amount of insecticide retained on them, a horticulturist sprays an orchard according to USDA recommendations and, after the fruit is harvested, takes a random sample of 10 peaches and measures their diameter x. She then washes each peach with a constant volume of detergent solution and makes a chemical determination of the amount of insecticide u in the solution after cleaning. Because she expects the amount of insecticide u to be an exponential function of diameter, $u = a10^{bx}$, she transforms the measurements on the u variable to common logarithms:

Peach	Diameter (cm) x	Insecticide (ppm) u	$\log u = y$
1	6.0	0.5	$-0.3010^{\dagger}$
2	7.0	6.4	0.8062
3	6.6	1.0	0.0000
4	5.8	0.2	-0.6990
5	6.8	5.5	0.7404
6	7.4	14.2	1.1523
7	7.2	8.2	0.9138
8	5.4	0.1	-1.0000
9	5.6	0.3	-0.5229
10	6.2	0.6	-0.2218
Total	64.0		0.8680

The log transformation of the independent variable occurs prior to any analytical computations. After the independent variable has been adjusted to the logarithmic scale, the arithmetic is the same as for any other simple regression analysis and consequently need not be demonstrated here. However, it will be useful to examine the computer analysis for *log y regression*, meaning regression with the y variable on the log scale.

The program in the SAS System is as follows:

```
DATA FRUIT;
    INPUT X U;
    Y = LOG10 (U);
    CARDS;
6.0        0.5
7.0        6.4
6.6        1.0
```

$^{\dagger}\log(0.5) = \log(5) - \log(10)$.

```
5.8          0.2
6.8          5.5
7.4         14.2
7.2          8.2
5.4          0.1
5.6          0.3
6.2          0.6
;
```

```
PROC PLOT;
  PLOT Y * X;
PROC REG;
  MODEL Y = X;
```

and the output is

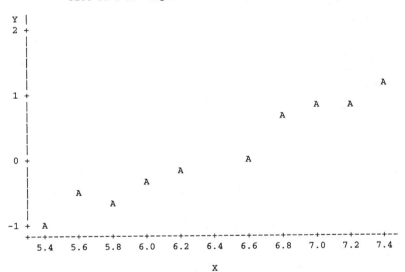

Plot of Y*X. Legend: A = 1 obs, B = 2 obs, etc.

```
The SAS System
```

The SAS System

The REG Procedure
Model: MODEL1
Dependent Variable: Y

Analysis of Variance

Source	DF	Sum of Squares	Mean Square	F Value	Pr > F
Model	1	4.94736	4.94736	164.91	<.0001
Error	8	0.24000	0.03000		

```
Corrected Total      9     5.18736
```

Root MSE	0.17320	R-Square	0.9537	
Dependent Mean	0.08679	Adj R-Sq	0.9480	
Coeff Var	199.56326			

Parameter Estimates

Variable	DF	Parameter Estimate	Standard Error	t Value	Pr > |t|
Intercept	1	−6.69962	0.53129	−12.61	<.0001
X	1	1.06038	0.08257	12.84	<.0001

As can be seen from the computer printout, the test of significance for the relationship H_0: $\beta = 0$ against H_a: $\beta > 0$ produces the test statistic $t = 12.84$ with a $P < 0.0001$, and the coefficient of determination for this data set is found to be $r^2 = 0.9537$. Thus, if a logarithmic relationship is used, the diameter of a peach in this orchard can be used as a very reliable indicator of the amount of insecticide that has been deposited on its surface. This information may have some bearing on the thoroughness with which different-sized peaches should be washed prior to marketing.

A similar technique can be used for exponential functions of the form $y = ae^{bx}$. In this case, $\log_e y$ is used for the transformation. If desired, common logarithms base 10 may be used and then converted to natural logarithms by the relationship

$$\log_e y = 2.303 \log y$$

If the function is of the form $y = ax^b$, then it can be linearized by transforming the variables to $\log y$ and $\log x$.

Consulting statisticians are frequently asked by economists to assist in the analysis of data that involve the regression of $\log y$ on $\log x$. The economists refer to the equations that are obtained as Cobb–Douglas functions. In other fields of research, there are also associations of the form

$$y = ax^b$$

but in economics they have been used with sufficient frequency to have gained a special designation. An example of their use would be a situation in which y is a measure of production in a certain industry and x is a measure of labor. Thus an economist could take a random sample of, say, bottling plants, gain access to their records, and find the regression of log(cases of soda) on log(man-hours). With this procedure, it is not uncommon to see multiple regression techniques employed as well. Thus the function becomes

$$y = ax_1^{b_1} x_2^{b_2}$$

Such a study might involve log(production) as a function of log(labor) and log(capital invested).

Having already demonstrated $\log x$ regression and $\log y$ regression, it seems unnecessary to give a numerical example of this procedure as well. However, it might be worthwhile to review the assumptions that are made in regression analysis. Irrespective of the units on the x and y axes, for the diagram, it is assumed that

1. the relationship is linear for the units of x and y used,

2. y has a normal distribution, and

3. y has the same variance throughout the range of x in the study.

Thus, if y is measured on the log scale, it implies that the log of the original units of measurement—log(cases of soda) in the Cobb–Douglas example—has a normal distribution with the same variance from the trend line irrespective of the number of workers involved. If the researcher is uncertain whether these assumptions should be made, then preliminary data should be obtained and used to investigate their distribution under the transformation. The arithmetic can be performed and numerical values obtained whether or not the assumptions hold true, but probability statements and inference are meaningless if the assumptions are not valid.

EXERCISES

14.6.1. Dicentric chromosomes result from the fusion of parts of two shattered chromosomes to form a single large chromosome. When dicentric chromosomes are formed, there are other chromosome fragments which are not reassembled and are eventually lost from the karyotype (chromosome composition). In the example demonstrating curvilinear regression rectified by log x, the dicentric chromosomes are used as the y variable; suppose that in a similar experiment chromosome fragments are also counted and the following results obtained:

Specific activity	x:	40	80	160	320
Fragments	y:	10, 12	14, 20	22, 34	42, 70

a. Complete the regression of y on log x and test it for significance.

b. In studies of this sort, the variances sometimes increase proportionally. Is there cause for concern about that possibility in these data? What might the experimenters do to determine whether or not variances are homogeneous irrespective of dosage?

c. Compute r and r^2.

d. Compute the expected number of chromosome fragments for 100 nCi/mL specific activity. Place a 95% CI on the estimate.

In the log y transformation example in this section concerning insecticide residue, estimate a in the function $u = a10^{bx}$.

To study the efficiency of microwave cooking in sterilizing meat, a food scientist takes a random sample of nine sausage links, and by means of a hypodermic needle she inoculates each with the same volume of a nutrient broth containing a heavy suspension of salmonella. She then cooks each link for a different length of time in a microwave oven set for a constant temperature. The contents of the sausages are then mixed with an agar solution and poured into petri dishes. The dishes are placed in an incubator. After 18 hours of incubation, the number of salmonella colonies per dish are counted. The results are

Time cooked in microwave (min)	x:	0	2	4	6	8	10	12	14	16
Number of salmonella colonies	y:	740	410	210	100	45	25	10	6	4

a. Graph the data. What type of function seems to model the relationship of y to x?

b. Make a $\log_e$ transformation on the y variable, graph the transformed data, and compute:

    i. The regression coefficient

    ii. The correlation coefficient

    iii. The coefficient of determination

c. In testing a hypothesis about the slope of the regression line:

    i. Why would the food scientist use a one-sided alternative?

    ii. Why would she reject the null hypothesis for $\alpha = 0.05$?

d. Based on the results of this study:

    i. What is the expected number of colonies to develop in sausage cooked 15 minutes?

    ii. Place a 95% CI for the value estimated above.

    iii. How long should sausage be cooked in the microwave oven in order to produce an expected salmonella survival of zero?

A learning model used in experimental psychology is $T_i = ab^i$, in which T_i is the time it takes to perform a task on the ith occasion. Since $\log T_i = \log a + i \log b$, this relationship is log linearizable. An experiment is performed which is believed to follow this model:

i:	1	2	3	4	5	6	7
T_i (min):	27	17	11	7	5	3	2

Compute a and b.

14.7. POLYNOMIAL REGRESSION

Multiple regression procedures can be used to analyze for polynomial regression. A number of geometrical curves involve selected powers of x. For example, the quadratic curve (parabola) can be written

$$y = a + b_1 x + b_2 x^2$$

and the cubic curve can be written

$$y = a + b_1 x + b_2 x^2 + b_3 x^3$$

In general, there are as many maximum points (extrema) on the curve as 1 less than the highest power of x in the model (Figure 14.5). It is possible to discuss quartic, quintic, and even more complex curves, but most experimenters find it difficult to explain curves with more than two maximum or minimum points. Thus we discuss only the quadratic and cubic curves.

A quadratic curve is utilized by agronomists when they study the effect of fertilizer. Agronomists know that there is a diminishing return from the use of more than a certain amount of fertilizer. In soil that is deficient in nitrogen, yield of crop increases with additional applications of nitrogen fertilizer, but it is possible to apply more nitrogen than the crop can use. In fact, too much fertilizer can damage and even kill the crop. Thus it is important to

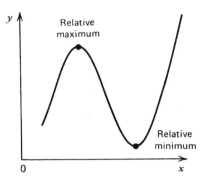

FIGURE 14.5. Polynomial functions of x.

identify the range of safe application and to exclude applications beyond the maximum, the point of diminishing return.

To find the maximum point, agronomists set up experimental plots and use fertilizers in a series of applications. This series should extend through the supposed safe range and even into the range that is thought to be dangerous. The data can then be analyzed for a quadratic trend. A specific example follows.

Example 14.6. Quadratic Regression

The Jerusalem artichoke, *Helianthus tuberosus*, resembles the sunflower, but as its scientific name implies, it produces tubers. The polysaccharide stored in the *Helianthus* tubers is inulin, which cannot be converted into sugars as can the starch stored in many tubers and roots. But it can be fermented to produce alcohol. The plant has the added advantage of being able to grow on relatively poor soil; consequently, it does not compete for the farmland used to grow beets, cane, corn, sorghum, and other sources of sugar and carbohydrates. Thus the Jerusalem artichoke has potential as a source of the polysaccharides needed to produce alcohol for use in industry, transportation, and beverages. However, the plant has been grown mainly as a flower, a curiosity, or a cover plant, and little is known about its culture as a cash crop. To gain information about the response to fertilizer for this species, an agronomist plants Jerusalem artichoke on 12 hillside plots and randomly assigns three hillsides to each of four fertilizer regimens (0, 4, 8, and 12 hundredweight per acre). Yield, measured in hundredweight inulin per acre, is given below

Fertilizer x	Yield y	Necessary Computations	
0	35.0	$\Sigma x = 72$	$\Sigma y = 468.0$
0	38.7	$\Sigma x^2 = 672$	$\Sigma xy = 2839.2$
0	33.1	$\Sigma x^3 = 6912$	$\Sigma x^2 y = 26{,}169.6$
4	42.6	$\Sigma x^4 = 75{,}264$	$\Sigma x^3 y = 267{,}072.0$
4	40.5	$\Sigma x^5 = 847{,}872$	$\Sigma y^2 = 18{,}373.38$
4	43.8	$\Sigma x^6 = 9{,}756{,}672$	$n = 12$

Fertilizer x	Yield y	Necessary Computations
8	41.0	
8	42.1	$S_{11} = \Sigma x^2 - \Sigma(x)^2/n$
8	36.9	$S_{12} = \Sigma x x^2 - (\Sigma x)(\Sigma x^2)/n = \Sigma x^3 - (\Sigma x)(\Sigma x^2)/n$
12	36.1	$S_{13} = \Sigma x x^3 - (\Sigma x)(\Sigma x^3)/n = \Sigma x^4 - (\Sigma x)(\Sigma x^3)/n$
12	40.8	$S_{22} = \Sigma x^2 x^2 - (\Sigma x^2)(\Sigma x^2)/n = \Sigma x^4 - (\Sigma x^2)^2/n$
12	37.4	$S_{23} = \Sigma x^2 x^3 - (\Sigma x^2)(\Sigma x^3)/n = \Sigma x^5 - (\Sigma x^2)(\Sigma x^3)/n$
		$S_{33} = \Sigma x^3 x^3 - (\Sigma x^3)(\Sigma x^3)/n = \Sigma x^6 - (\Sigma x^3)^2/n$
		$S_{1y} = \Sigma x y - (\Sigma x)(\Sigma y)/n$
		$S_{2y} = \Sigma x^2 y - (\Sigma x^2)(\Sigma y)/n$
		$S_{3y} = \Sigma x^3 y - (\Sigma x^3)(\Sigma y)/n$
		$S_{yy} = \Sigma y^2 - (\Sigma y)^2/n$

Given are the sums of the x variable raised to different powers, along with the summations of their cross-products with the y variable. Also given are the computational equations for the corrected sums of squares and cross-products (S_{ij}) which are needed for the simultaneous equations which must be solved. However, these are intended only as evidence of the size of the numerical values which must be dealt with and the amount of computation involved. Except for quite small samples and relatively small numerical values of x, using a computer routine is the most sensible method of performing polynomial regression analysis.

The computational procedures are the same as for other multiple regression analyses; the only difference is that the x_i are not different measurement variables, but instead they are different powers of the same measurement variable. It is common to perform polynomial regression analysis in a stepwise fashion, starting with simple linear regression as the first model for fitting the data,

$$\text{Linear model: } \hat{y} = a + bx$$

and then advancing to the next level of complexity, the second-degree polynomial, by adding x^2 to the model,

$$\text{Quadratic model: } \hat{y} = a + b_1 x + b_2 x^2$$

and, if desired, one can continue to increase the complexity of the model simply by including in it the next power of the x variable. In our case a third-degree polynomial is obtained with

$$\text{Cubic model: } \hat{y} = a + b_1 x + b_2 x^2 + b_3 x^3$$

It is frequently advised that one include one more level of complexity than that expected for the actual curvilinear relationship between the dependent and independent variables. When this is done, it provides a measure of the "lack of fit" of the model to the data. Thus, if the agronomist is expecting a quadratic response, he designs the experiment with four levels of fertilizer so that there will be enough points on the x axis to fit a cubic curve. If there were only three different values of x, a quadratic would be forced through the three means, and there would be no opportunity to examine the extent to which the curve of interest fails to fit the data.

In the SAS System the program and output are as follows:

```
DATA TUBERS;
  INPUT X Y;
  CARDS;
  0              35.0
  0              38.7
  0              33.1
  4              42.6
  4              40.5
  4              43.5
  8              41.0
  8              42.1
  8              36.9
  12             36.1
  12             40.8
  12             37.4
PROC GLM;
  MODEL Y = X/SS1;
PROC GLM;
  MODEL Y = X   X*X/SS1;
PROC GLM;
  MODEL Y = X   X*X*X*X/SS1;
```

The SAS System

The GLM Procedure

Number of observations   12

Dependent Variable: Y

Source	DF	Sum of Squares	Mean Square	F Value	Pr > F
Model	1	4.0560000	4.0560000	0.35	0.5696
Error	10	117.3240000	11.7324000		
Corrected Total	11	121.3800000			

R-Square	Coeff Var	Root MSE	Y Mean
0.033416	8.782716	3.425259	39.00000

Source	DF	Type I SS	Mean Square	F Value	Pr > F
X	1	4.05600000	4.05600000	0.35	0.5696

Parameter	Estimate	Standard Error	t Value	Pr > |t|
Intercept	38.22000000	1.65455734	23.10	<.0001
X	0.13000000	0.22109953	0.59	0.5696

The GLM Procedure

Number of observations   12

Dependent Variable: Y

Source	DF	Sum of Squares	Mean Square	F Value	Pr > F
Model	2	59.5260000	29.7630000	4.33	0.0481
Error	9	61.8540000	6.8726667		
Corrected Total	11	121.3800000			

R-Square	Coeff Var	Root MSE	Y Mean
0.490410	6.721993	2.621577	39.00000

Source	DF	Type I SS	Mean Square	F Value	Pr > F
X	1	4.05600000	4.05600000	0.59	0.4620
X*X	1	55.47000000	55.47000000	8.07	0.0194

Parameter	Estimate	Standard Error	t Value	Pr > |t|
Intercept	36.07000000	1.47524386	24.45	<.0001
X	1.74250000	0.59227727	2.94	0.0164
X*X	−0.13437500	0.04729901	−2.84	0.0194

The GLM Procedure

Number of observations   12

Dependent Variable: Y

Source	DF	Sum of Squares	Mean Square	F Value	Pr > F
Model	3	72.7800000	24.2600000	3.99	0.0521
Error	8	48.6000000	6.0750000		
Corrected Total	11	121.3800000			

R-Square	Coeff Var	Root MSE	Y Mean
0.599605	6.319876	2.464752	39.00000

Source	DF	Type I SS	Mean Square	F Value	Pr > F
X	1	4.05600000	4.05600000	0.67	0.4375
X*X	1	55.47000000	55.47000000	9.13	0.0165
X*X*X	1	13.25400000	13.25400000	2.18	0.1779

Parameter	Estimate	Standard Error	t Value	Pr > |t|
Intercept	35.60000000	1.42302495	25.02	<.0001
X	3.58333333	1.36502060	2.63	0.0304
X*X	−0.57500000	0.30160702	−1.91	0.0930
X*X*X	0.02447917	0.01657282	1.48	0.1779

From the results of the computer analyses the following information about how well the data are fit by each type of curve can be extracted:

Curve	Model	F Test for Model	R^2	MS_e
Linear	$38.22 + 0.13x$	0.35	0.0334	11.7324
Quadratic	$36.07 + 1.74x - 0.13x^2$	4.33	0.4904	6.8727
Cubic	$35.60 + 3.58x - 0.58x^2 + 0.02x^3$	3.99	0.5996	6.0750

It can readily be seen that the linear model is ineffective in explaining the response to the different levels of fertilizer. The F test for this model is nonsignificant; R^2 is very small, and the MS_e is the largest for the three models. When the quadratic curve is fit to the data, the F test is significant; R^2 increases greatly, and the MS_e for this model is almost half the value of that for the linear model. Thus the criteria which are used here for comparison all indicate that the quadratic response curve is a superior fit to that for the linear. However, the decision is not so clear for the cubic response curve where R^2 increases, but MS_e is changed little from that for the quadratic, and the numerical value of the F test actually decreases. To understand what is happening, the agronomist remembers that when x^3 is added to the model, the degrees of freedom associated with the model increase to $k = 3$, and those associated with the MS_e decrease to $n - k - 1 = 8$. Thus the increase in R^2 for the cubic curve does not justify the additional degree of freedom associated with it. For example, from previous sections the F test for a model is

$$F = \frac{SSR/k}{MS_e} = \frac{R^2 S_{yy}/k}{MS_e}$$

In this study, $S_{yy} = 121.38$, so this value along with knowledge of the values of R^2 and MS_e for the quadratic and cubic models, respectively, can be used to obtain F tests for each model as well as for the improvement in fit provided by the cubic as compared to that for the quadratic.

Quadratic model:

$$F = \frac{R_Q^2 S_{yy}/k_Q}{\text{Quad.MS}_e} = \frac{0.4904(121.38)/2}{6.8727} = 4.33$$

Cubic model:

$$F = \frac{R_C^2 S_{yy}/k_C}{\text{Cubic MS}_e} = \frac{0.5996(121.38)/3}{6.0750} = 3.99$$

Improvement of cubic vs +quadratic:

$$F = \frac{(R_C^2 - R_Q^2)S_{yy}/(k_C - k_Q)}{\text{Cubic MS}_e} = \frac{0.1092(121.38)/1}{6.0750} = 2.18$$

The third F test can be thought of as a test of "lack of fit," or the extent to which the quadratic curve fails to fit the $\bar{y}_i$ found at the four different points on the x axis. This test is not significant, meaning the cubic curve does not fit the data significantly better than does the quadratic curve. Hence the agronomist has even further evidence that the response of Jerusalem artichoke yield to fertilizer can best be described by a quadratic curve.

With many of the available computer routines, it is not necessary to perform the hand computations shown for the third F value above. To obtain a test for the improvement of one polynomial model over another, each successive power of x is brought into the model, one at a time, and then the improvement in the model is tested in much the same fashion as was done above. The instruction for this process in the SAS System is found in the model statement. When instructions were given for the analysis of the cubic model, the model statement was

$$\text{MODEL Y} = \text{X} \quad \text{X} * \text{X} \quad \text{X} * \text{X} * \text{X} / \text{SS1};$$

The above statement gave instructions to bring x into the model and test the variability explained by it alone, then to bring x^2 into the equation and test the improvement in the equation due to this second term, and finally to bring x^3 into the equation and test the improvement due to the third term. The printout for the third analysis gives the following information:

SOURCE	DF	Type I SS	F Value
X (Linear trend alone)	1	$R_L^2 S_{yy} = 4.056$	0.67
X*X (Quadratic after linear)	1	$(R_Q^2 - R_L^2)S_{yy} = 55.470$	9.13
X*X*X (Cubic after quadratic)	1	$(R_C^2 - R_Q^2)S_{yy} = 13.254$	2.18

Among the F tests in this analysis, the only one which is significant ($P = 0.0165$) is that for the improvement of a quadratic model as compared to a linear model. Once again, the quadratic model is the one which should be chosen for describing the response of yield to increased levels of fertilizer.

The plant scientist had intended to apply fertilizer rates beyond the point of diminishing return, and it has been confirmed that the response curve can be better described by a parabola than by a straight line or a cubic curve with its two extrema. The quadratic curve selected to model the response in yield to different levels of fertilizer is found to be

$$\hat{y} = 36.0700 + 1.7425x - 0.1344x^2$$

The maximum, or point of diminishing return, can be found by setting the first derivative of y with respect to x equal to zero. Thus the maximum y is at

$$x_m = \frac{-b_1}{2b_2} = \frac{-1.7425}{2(-0.1344)} = 6.48$$

as illustrated in Figure 14.6. The implication from this experiment is that when fertilizer is applied to Jerusalem artichokes at a rate greater than 6.48 hundredweight per acre, there is not likely to be any further increase in yield. In fact, the results of this experiment indicate that yield would begin to decrease with the application of a greater amount of fertilizer.

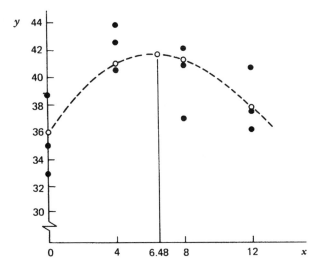

FIGURE 14.6. The maximum of the model in Example 14.6.

We have seen how polynomial regression can be used to fit a linear, quadratic, or cubic response. A third extremum in the regression line can be obtained by using x, x^2, x^3, and x^4 and fitting a quartic curve with $k = 4$ degrees of freedom. Polynomial regression is an extremely useful technique, but as with the other statistical techniques we have discussed, there are also limitations, cautions, and assumptions to be considered before drawing inference from these procedures. Here are some of the things the research worker should consider before using polynomial regression:

1. Not all curves with a single extremum are parabolas, and similarly polynomial curves may not provide the best fit for more complex curves. The polynomial curves have symmetrical features which make them unsuitable for fitting data that follow a nonsymmetrical trend. It is always useful to gather preliminary data, plot it, and then discuss with a statistician or mathematician what function may provide the best fit of y.

2. The number of different values of x is more important than the number of data points in polynomial regression. In the example where inulin yield was fitted to fertilizer applications, there were 12 data points but only $a = 4$ different values of x. The best possible fit (the maximum R^2) is obtained when $k = a - 1$, so it is a waste of time and effort to try to fit a very complex polynomial curve to data for which there are only a few different x values.

3. In polynomial regression, $x_k = x^k$, or as we saw in the Jerusalem artichoke example, $x_2 = x^2$. Because of this, if the x's are greater than 1, S_{22} will be larger than S_{11}, and if we use $x_3 = x^3$ and $x_4 = x^4$, then S_{33} and S_{44} will be still larger. A great disparity in the size of the S_{ii} makes it difficult to invert the sum of squares and cross-products matrix accurately.

4. As always, it is necessary to make the assumption that the deviations from the trend line are normally distributed with the same variance all along the segment of the line for which inference will be made.

A technique called *orthogonal polynomials* further addresses some of the concerns given here and shows how good experimental design can permit easy tests of significance for higher order polynomial regression. We conclude this chapter with a discussion of orthogonal polynomials.

It might first be useful to review Section 10.4 on orthogonal contrasts, since very similar techniques are demonstrated here. If the x's are equally spaced and there are a constant number of observations n at each x, then one can use tabulated orthogonal polynomials to determine which kind of polynomial curve best fits the data. This is usually done in conjunction with an ANOVA in which each value of x is considered an experimental group.

The procedure can be demonstrated with the data obtained from the Jerusalem artichoke experiment, for the x's are equally spaced, that is, there is a 4-hundredweight interval between adjacent levels of fertilizer, and there are $n = 3$ yields obtained for each level of fertilizer. The data can be grouped for an ANOVA as follows:

	0 cwt	4 cwt	8 cwt	12 cwt
	35.0	42.6	41.0	36.1
	38.7	40.5	42.1	40.8
	33.1	43.8	36.9	37.4
$\sum y_i = T_i$	106.8	126.9	120.0	114.3

$$T = 18{,}373.38 \text{ (uncorrected total sum of squares)}$$
$$A = 18{,}324.78 \text{ (uncorrected group sum of squares)}$$
$$CF = 18{,}252.000$$

Source	df	SS	MS	F	$F_{0.05;3,8}$
Levels	3	72.78	24.26	4.00	4.066
Error	8	48.60	6.07		
Total		121.38			

The coefficients to be used for computing the contributions of x, x^2, and x^3 to the model can be obtained from Table A.19 (see Appendix) for $a = 4$ levels. These are used to compute the three sums of squares which partition the sum of squares for levels as follows:

Degree Polynomial		Level: 0 T_i: 106.8	4 126.9	8 120.0	12 114.3	$\sum a_i T_i$	$\sum a_i^2$	$\left(\sum a_i T_i\right)^2 / n\sum a_i^2$
Linear	a_{Li}:	-3	-1	$+1$	$+3$	15.6	20	4.056
Quadratic	a_{Qi}:	$+1$	-1	-1	$+1$	-25.8	4	55.470
Cubic	a_{Ci}:	-1	$+3$	-3	$+1$	28.2	20	13.254
								72.780

The ANOVA table can be expanded to take into account these three orthogonal sums of squares, each with 1 degree of freedom.

Source	df	SS	MS	F	$F_{0.05;1,8}$
Levels	3	72.78			
Linear	1	4.056	4.056	0.668	5.318
Quadratic	1	55.470	55.470	9.138	5.318
Cubic	1	13.254	13.254	2.183	5.318
Error	8	48.60	6.07		

When we compare the three sums of squares computed here with the results of the third SAS analysis of the Jerusalem artichoke yields, we can see how the resulting sums of squares correspond identically:

Orthogonal Coefficients	Polynomial Regression	SS
Linear	$R_L^2 S_{yy}$	4.056
Quadratic	$(R_Q^2 - R_L^2) S_{yy}$	55.470
Cubic	$(R_C^2 - R_Q^2) S_{yy}$	13.254

Thus we can evaluate the nature of the response in y to increasing levels of x, either by using polynomial regression or by using ANOVA techniques which are then followed by use of orthogonal polynomials to obtain a sums of squares each with 1 degree of freedom. When the levels of x are equally spaced and the number of observations (n) at each level is the same, there may be some convenience in using the ANOVA and orthogonal polynomials, but under other circumstances, it is usually found to be easier to use polynomial regression.

The orthogonal polynomial coefficients are given in Table A.19 in the Appendix for various levels of a, and they can be used as shown here provided, as has been pointed out, the $a - 1$ levels are equally spaced and n is the same at all levels. The coefficients can be obtained in a fashion similar to that used in covariance to obtain one variable adjusted to another. Thus, to obtain the coefficients for the quadratic polynomial, the variable $x_2 = x^2$ must be adjusted for x and the resulting values coded so that they will sum to zero. Fortunately, the advent of good computer programs such as SAS has made these simple but tedious arithmetic procedures unnecessary.

EXERCISES

14.7.1. An experiment similar to that studying the yield of inulin in Jerusalem artichoke is performed with sugar beets. The yield is measured in cwt of sugar:

x:	0	4	8	12
y:	34.5, 37.9, 31.4	39.2, 39.8, 43.4	45.1, 40.3, 43.0	43.2, 38.8, 43.4

a. Find the numerical values of b_1 and b_2 and test them for significance.

b. Does a quadratic curve fit the data significantly ($\alpha = 0.05$) better than a straight line? On what computations do you base your answers?

c. Find the maximum response of yield as a function of fertilizer.

d. The x values are deliberately kept the same in this problem as they were in the numerical example. This is to provide a computational guide for those who chose to invert the sum of squares and cross-products matrix. How can one use the results of the numerical example to perform the analysis without having to invert the matrix?

Biologists are studying the effect of temperature on the germination of seed from cold-resistant trees. Seeds of Korean ash (*Fraxinus chinensis*) are collected and kept in dry storage for eight months, when 14 groups of 100 seeds each are established through random sampling. For seven groups, the pericarp (a plant ovary part which serves as a seed covering) is picked away from the seeds; for the other seven groups, it is left intact. Each group of seeds is placed in a separate flat containing vermiculite, and two flats, one of each kind of seed treatment, are assigned at random to each of seven temperature chambers. The numbers of seed germinating for each temperature and seed treatment are given below.

Seed Treatment	Temperature (°C)						
	5	10	15	20	25	30	35
With pericarp	4	5	9	31	58	75	77
Without pericarp	3	4	9	18	36	65	96

Computational hint for hand calculation: Because the settings of the temperature chambers are in multiples of 5, these observations can be easily coded by dividing by 5. This simplifies the arithmetic when powers of x are employed.

a. Test for curvilinear regression by using x, x^2, and x^3 in multiple regression:

   i. For germination of seed with pericarp

   ii. For germination of seed with the pericarp removed

b. The simple linear trend of germination on temperature (uncoded data) is 2.914 seeds/degree. The regression coefficient for the coded data is $b_1 = 14.571$. What is the effect on the simple linear regression coefficient of dividing the x values by 5? What is the effect on the other coefficients if multiple regression is performed?

c. To determine how complex the model must be to explain Korean ash germination under different conditions, give the percentage of variability explained by each model below:

   i. Germination y as a simple linear function of temperature x for (1) seed with pericarp and (2) seed without pericarp

   ii. Germination as a quadratic function of temperature for (1) seed with pericarp and (2) seed without

   iii. Germination as a cubic function of temperature for (1) seed with pericarp and (2) seed without

d. Is there evidence that the relationship to temperature is significantly different for the two seed treatments?

e. Using the information gained here, could the biologists properly use the techniques covered in the section on covariance to adjust similar data for different temperatures in order to compare the two seed treatments at a common temperature?

The yield of sugar from beets has been studied, and there is interest in determining the response of yield to the amounts of fertilizer applied.
The data are

	Fertilizer (cwt)			
	0	4	8	12
	34.5	39.2	45.1	43.2
	37.9	39.8	40.3	38.8
	31.4	43.4	43.0	43.4
Total	103.8	122.4	128.4	125.4

a. Perform the ANOVA and test for differences among levels of fertilizer.

b. Test for linear, quadratic, and cubic trends.

c. Is there evidence that the range of applications of fertilizer encompasses the point of diminishing return?

d. These data were analyzed in the sugar beet experiment of Exercise 14.7.1 using x and x^2 as independent variables in multiple regression. Compare results from the two techniques.

14.8. LOGISTIC REGRESSION

One or more independent variables can also be used to predict a dependent variable that is nominal rather than numerical. The procedure is called *logistic regression*. Although more than one regressor variable can be used, we will demonstrate it for data recorded for x, y pairs, and even then calculations are extensive. To perform the calculations one must only know how to find first and second derivatives and know the matrix procedures of Section 14.1. However, many iterations are often needed to obtain estimates of the parameters, and the repetitiveness is tedious. We demonstrate the calculations only to dispel mathematical mystery, but logistic regression should be thought of as a procedure always performed by a statistical computer package.

Logistic regression is used when there is a continuous variable such as hours of study on the night prior to an exam, and we want to see if it has a predictable effect on the discrete exam grade. If the grade was numerical we would use regression techniques, but if it is nominal, such as fail or pass, least squares techniques are not appropriate. Even if we make $y = 0$ for a failure and $y = 1$ for success, the assumptions of linear regression are still not met because the y variable is binomial, hence there will not be a common variance. To solve the experimental problem we base it on how hours of study improves the probability of a pass, or how hours of study change the odds of a passing grade. Then we use logistic regression with $x =$ hours of study and $y = \log_e$ (odds).

It is usually helpful to demonstrate a new procedure on a small sample, but to do so with logistic regression would increase the number of iterations. Logistic regression requires a computer program, but it also requires large data sets. So we will begin the discussion with a large data set in which x = diameter of a thoracic aortic aneurysm and y = odds it will rupture within 5 years.

An aortic aneurysm is a marked dilation of a particular portion of the aorta in either the thoracic or abdominal portion. Such aneurysms have a 5-year mortality which is nearly 75%. One-third to one-half of these deaths result from rupture of the aneurysm. Surgical repair constitutes the only effective treatment, but treatment decisions need to balance the complications of the dilated aneurysm with the complications resulting from the surgery itself.

A group of physicians has collected information from new patients for several years. One item is the initial aneurysm size determined by radiology. Another item is whether the aneurysm ruptured. A summary of this information is given in the following table:

Initial Aneurysm Size	Number of Ruptures	Number of Patients	Proportion of Ruptures
3.5–3.9 cm	0	33	0.0000
4.0–4.9 cm	3	133	0.0226
5.0–5.9 cm	4	78	0.0513
6.0 cm or more	6	60	0.1000

These investigators want to predict the rupture outcome. The outcome is a dichotomous variable, essentially a yes or a no. They want an equation that will predict the proportion of yes outcomes or, equivalently, estimate the probability that a patient's aneurysm ruptures. They cannot use an ordinary linear regression equation because it might predict proportions less than zero or greater than 1, which would be meaningless. Also, it is reasonable to conjecture that the probability of rupture is virtually zero until some threshold aneurysm size is reached. The probability of rupture increases as the aneurysm size increases until some size is reached beyond which the probability of rupture is virtually 1. It is reasonable to relate the probability of rupture to aneurysm size by an S-shaped function.

Instead of using the proportion, they use the log of the odds of the proportion as the dependent variable. This is called the *logit* of the proportion:

$$\text{logit}(\pi) = \log_e \left(\frac{\pi}{1 - \pi} \right)$$

If the proportion π is zero, the logit is minus infinity. If the proportion π is 1, the logit is plus infinity. For yes or no dichotomous variables, the logit is

$$\log_e[P(\text{yes})] - \log_e[P(\text{no})]$$

This implies that if we change the focus from the occurrence of an event to the nonoccurrence of that event, the magnitude remains the same but the sign changes.

Model

If the investigators assume the relationship between aneurysm size and the logit is linear, they might use the predictive model

$$\log_e \left(\frac{\pi}{1 - \pi} \right) = \alpha + \beta x$$

where x is the predictor variable, α and β are unknown parameters, and π is the proportion. The procedure that relates a quantitative independent variable to the probability of the outcome of a dichotomous dependent variable is called *logistic regression*.

There is no error term in the logistic regression model because the predicted value is not a yes or no. It is the probability distribution of a yes (or no). For example, if the equation predicts a 90% chance of rupture, we wouldn't say it erred if the outcome was no. Instead, as a way of evaluating the utility of the prediction equation, we might sum the negative logarithms of the predicted probabilities of the events that actually occurred. So, if $\hat{\pi}$ is the predicted probability of rupture, we would "score" the predictions by assigning $-\log_e(\hat{\pi})$ if the patient's aneurysm ruptures and $-\log_e(1 - \hat{\pi})$ if the patient's aneurysm doesn't rupture. A perfect prediction would come up with a $\hat{\pi}$ of 1 when the patient's aneurysm ruptures and a $\hat{\pi}$ of 0 when the patient's aneurysm doesn't rupture. In either case the score is zero. A predicted probability of 0 for an event that occurs means the score is plus infinity. The smaller the sum of the scores, the better the prediction. The sum of the scores is

$$\sum_i^n [-y_i \log_e(\hat{\pi}) - (1 - y_i) \log_e(1 - \hat{\pi})]$$

if we code rupture as $y_i = 1$ and no rupture as $y_i = 0$.

Maximum-likelihood estimation

The inverse logit of the model expresses the probability for each outcome. Solving for π in the logit model produces

$$\pi(x) = \frac{1}{1 + e^{-(\alpha + \beta x)}}$$

The estimates of α and β are found so as to maximize the *likelihood*. In Section 3.3 we observed that the sample proportion is the maximum-likelihood estimator of the binomial parameter π. Such estimators find values of parameters that make the outcome observed more likely than it would be with any other value. Likelihood means the probability has been evaluated as a function of the parameters with the data fixed. The calculation of the likelihood estimators is simplified by two shortcuts:

a. The joint probability of all the observations is the product of the probability function for each observation.
b. Maximizing the log of the likelihood produces the same result as maximizing the likelihood. The log likelihood is the sum of the logarithms of the probabilities. Finding the maximum-likelihood estimators is the same as minimizing the negative sum of logs of the probabilities attributed to the response levels that actually occurred for each observation.

The estimates a and b in the case of simple linear regression are, in fact, maximum-likelihood estimators of α and β because minimizing the negative sum of the logs of the probabilities produces the same function as the least-squares method.

Log-likelihood equations

If we code each response y_i as 0 or 1 and let x_i represent the corresponding aneurysm size, the contribution to an observation to the likelihood is

$$\pi(x_i)^{y_i}(1 - \pi(x_i))^{1-y_i}$$

Since the observations are independent, the likelihood of all the observations is the product of each contribution

$$l(\alpha, \beta) = \prod_{i=1}^{n} \pi(x_i)^{y_i}[1 - \pi(x_i)]^{1-y_i}$$

and the log likelihood is

$$L(\alpha, \beta) = \sum_{i}^{n} \{y_i \log_e[\pi(x_i)] + (1 - y_i) \log_e[1 - \pi(x_i)]\}$$

To find the values of α and β that maximize $L(\alpha, \beta)$, we differentiate $L(\alpha, \beta)$ with respect to α and β and set the resulting equations to zero. These likelihood equations are

$$\sum [y_i - \pi(x_i)] = 0$$

and

$$\sum x_i[y_i - \pi(x_i)] = 0$$

In linear regression the derivatives of the sum of squared deviations with respect to α and β produce equations that are linear with respect to α and β and are easy to solve. For logistic regression $L(\alpha, \beta)$ is nonlinear in α and β and the solutions of the likelihood equations need special methods. One such method is the iterative Newton–Raphson procedure. This procedure requires the second derivatives of the log likelihood with respect to α and β. The second derivative with respect to α is

$$\frac{\delta^2 L(\alpha, \beta)}{\delta \alpha^2} = - \sum \pi(x_i)[1 - \pi(x_i)]$$

The derivative with respect to α and β is

$$\frac{\delta^2 L(\alpha, \beta)}{\delta \alpha \delta \beta} = - \sum x_i \pi(x_i)[1 - \pi(x_i)]$$

The second derivative with respect to β is

$$\frac{\delta^2 L(\alpha, \beta)}{\delta \beta^2} = - \sum x_i^2 \pi(x_i)[1 - \pi(x_i)]$$

The procedure starts with initial values for α and β. It calculates the log likelihood and evaluates the likelihood equations and the second derivatives. It uses the results of the product of the inverse of the second derivative matrix and the likelihood functions to calculate adjustments for α and β. Using matrix notation, the adjustment is

$$
\begin{bmatrix} \hat{\alpha} \\ \hat{\beta} \end{bmatrix}^{new} = \begin{bmatrix} \hat{\alpha} \\ \hat{\beta} \end{bmatrix}^{old} + \begin{bmatrix} -\dfrac{\delta^2 L(\alpha, \beta)}{\delta \alpha^2} & -\dfrac{\delta^2 L(\alpha, \beta)}{\delta \alpha \delta \beta} \\ -\dfrac{\delta^2 L(\alpha, \beta)}{\delta \alpha \delta \beta} & -\dfrac{\delta^2 L(\alpha, \beta)}{\delta \beta^2} \end{bmatrix}^{-1} \begin{bmatrix} \sum [y_i - \pi(x_i)] = 0 \\ \sum x_i[y_i - \pi(x_i)] = 0 \end{bmatrix}
$$

The procedure repeats the calculations until the changes in the likelihood are small.

Test of hypothesis

Of primary interest in logistic regression is to learn if there is a log-linear increase in the odds ratio as the x variable (size of aneurysm) increases. The null and alternative hypotheses can be stated in symbols as:

$$H_0: \beta = 0 \quad \text{and} \quad H_a: \beta \neq 0$$

And in words as:

H_0: *Odds of rupture do not change with aneurysm size*

and

H_a: *Odds of rupture change with aneurysm size*

Likelihood also can be used to perform tests of the hypothesis in the following way:

a. Find the likelihood with no constraints on the parameters.

b. Find the likelihood with the parameters constrained by the null hypothesis.

Two times the difference between the log likelihoods,

$$2[\log \text{likelihood(unconstrained)} - \log \text{likelihood(constrained)}]$$

has an approximate chi-square distribution. These tests are called *likelihood ratio chi squares*.

In the iterative procedure described above we will start the procedure by constraining the estimate of β to be zero. We use the overall proportion of ruptures to obtain a starting value for the estimate of α. We calculate the log likelihood and use the likelihood equations to calculate adjustments to the estimates of α and β. We remove the constraint on the estimate of β and recalculate until the changes in the log likelihood fall below some criterion. Two times the difference between the last log likelihood and the initial log likelihood is a chi-square statistic with 1 degree of freedom. We can use it to test the hypothesis that β is equal to zero.

In simple linear regression, the test that β is equal to 0 requires the assumption that the errors are normally distributed. With that assumption the test statistic has a t distribution regardless of the sample size. In logistic regression the test statistic has an approximate chi-square distribution. The approximation improves with larger sample sizes.

A statistically equivalent test is the *Wald test*, which may provide a different P value than the likelihood chi-square, but will almost always lead to the same decision about the null hypothesis. This test is performed by dividing the maximum likelihood estimate of the parameter by its standard error. Under the null hypothesis that the parameter is 0, this ratio has a standard normal distribution.

Confidence intervals for the parameters

The basis for constructing confidence intervals for the parameters is the Wald test. For example, confidence intervals for the slope and intercept are based on the respective Wald tests. The $100(1 - \alpha)\%$ confidence interval for β is

$$\hat{\beta} \pm z_{1-\alpha/2}\text{s.e.}(\hat{\beta})$$

and for α is

$$\hat{\alpha} \pm z_{1-\alpha/2}\text{s.e.}(\hat{\alpha})$$

The standard errors of the estimates are obtained from the square root of the diagonal elements of the inverse of the matrix of second derivatives.

Calculations

To perform a logistic regression of the aneurysm-rupture information presented above, we choose to use the midpoint of size intervals as the value for the independent variable and code the rupture information into 0's and 1's and construct two columns of counts, y_1 the number of ruptures and y_0 the number that did not rupture.

x	y_1	y_0
3.75	0	33
4.5	3	130
5.5	4	74
6.25	6	54
Total	13	291

Step 1 (Null model, $\beta = 0$) $\hat{\pi} = 13/(13 + 291) = 13/304 = 0.04276$ $\hat{\alpha} = \log_e(13/291)$ $= -3.1084$ $\hat{\beta} = 0$

x	y_1	y_0	$\hat{\pi}(x)$	$-2\log_e L$
3.75	0	33	0.0428	2.8845
4.5	3	130	0.0428	30.2756
5.5	4	74	0.0428	31.6849
6.25	6	54	0.0428	42.5450
Total	13	291		107.39

$$-2\log_e L = -2y_1 \log_e[\hat{\pi}(x)] - 2y_0 \log_e[1 - \hat{\pi}(x)]$$

Parameter estimates adjustment:

$$\begin{bmatrix} \hat{\alpha} \\ \hat{\beta} \end{bmatrix}^{new} = \begin{bmatrix} -3.1084 \\ 0 \end{bmatrix}^{old} + \begin{bmatrix} 12.441 & 62.4762 \\ 62.4762 & 321.768 \end{bmatrix}^{-1} \begin{bmatrix} 0 \\ 7.7327 \end{bmatrix}$$

$$\begin{bmatrix} \hat{\alpha} \\ \hat{\beta} \end{bmatrix}^{new} = \begin{bmatrix} -3.1084 \\ 0 \end{bmatrix}^{old} + \begin{bmatrix} 3.1913 & -0.6196 \\ -0.6196 & 0.1234 \end{bmatrix} \begin{bmatrix} 0 \\ 7.7327 \end{bmatrix}$$

$$\begin{bmatrix} -7.8999 \\ 0.9544 \end{bmatrix}^{new} = \begin{bmatrix} -3.1084 \\ 0 \end{bmatrix}^{old} + \begin{bmatrix} -4.7915 \\ 0.9544 \end{bmatrix}$$

Step 2 $\hat{\alpha} = -7.8999$ $\hat{\beta} = 0.9554$

x	y_1	y_0	$\hat{\pi}(x)$	$-2 \log_e L$
3.75	0	33	0.013	0.8712
4.5	3	130	0.0265	28.7651
5.5	4	74	0.0659	31.8478
6.25	6	54	0.1262	39.4082
Total	13	291		100.8922

Parameter estimates adjustment:

$$\begin{bmatrix} \hat{\alpha} \\ \hat{\beta} \end{bmatrix}^{new} = \begin{bmatrix} -7.8999 \\ 0.9544 \end{bmatrix}^{old} + \begin{bmatrix} 15.2739 & 84.7948 \\ 84.7948 & 479.1635 \end{bmatrix}^{-1} \begin{bmatrix} -3.6674 \\ -20.0726 \end{bmatrix}$$

$$\begin{bmatrix} \hat{\alpha} \\ \hat{\beta} \end{bmatrix}^{new} = \begin{bmatrix} -7.8999 \\ 0.9544 \end{bmatrix}^{old} + \begin{bmatrix} 3.7279 & -0.6597 \\ -0.6597 & 0.1188 \end{bmatrix} \begin{bmatrix} -3.6674 \\ -20.0726 \end{bmatrix}$$

$$\begin{bmatrix} -8.3296 \\ 0.9885 \end{bmatrix}^{new} = \begin{bmatrix} -7.8999 \\ 0.9544 \end{bmatrix}^{old} + \begin{bmatrix} -0.4296 \\ 0.0341 \end{bmatrix}$$

Step 3 $\hat{\alpha} = -8.3296$ $\hat{\beta} = 0.9885$

x	y_1	y_0	$\hat{\pi}(x)$	$-2 \log_e L$
3.75	0	33	0.0097	0.6454
4.50	3	130	0.0202	28.7179
5.50	4	74	0.0525	31.5570
6.25	6	54	0.1042	39.0215
Total	13	291		99.9418

Parameter estimates adjustment

$$\begin{bmatrix} \hat{\alpha} \\ \hat{\beta} \end{bmatrix}^{new} = \begin{bmatrix} -8.3296 \\ 0.9885 \end{bmatrix}^{old} + \begin{bmatrix} 12.4338 & 69.3968 \\ 69.3968 & 393.9999 \end{bmatrix}^{-1} \begin{bmatrix} -0.3580 \\ -1.9087 \end{bmatrix}$$

$$\begin{bmatrix} \hat{\alpha} \\ \hat{\beta} \end{bmatrix}^{new} = \begin{bmatrix} -8.3296 \\ 0.9885 \end{bmatrix}^{old} + \begin{bmatrix} 4.7466 & -0.8360 \\ -0.8360 & 0.1498 \end{bmatrix} \begin{bmatrix} -0.3580 \\ -1.9087 \end{bmatrix}$$

$$\begin{bmatrix} -8.4331 \\ 1.0019 \end{bmatrix}^{new} = \begin{bmatrix} -8.3296 \\ 0.9885 \end{bmatrix}^{old} + \begin{bmatrix} -0.1035 \\ 0.0134 \end{bmatrix}$$

Step 4 $\hat{\alpha} = -8.4331$ $\hat{\beta} = 1.0019$

x	y_1	y_0	$\hat{\pi}(x)$	$-2\log_e L$
3.75	0	33	0.0092	0.6121
4.50	3	130	0.0194	28.7498
5.50	4	74	0.0510	31.5547
6.25	6	54	0.1024	39.0137
Total	13	291		99.9302

Parameter estimates adjustment

$$\begin{bmatrix} \hat{\alpha} \\ \hat{\beta} \end{bmatrix}^{new} = \begin{bmatrix} -8.4331 \\ 1.0091 \end{bmatrix}^{old} + \begin{bmatrix} 12.1204 & 67.7427 \\ 67.7427 & 385.0812 \end{bmatrix}^{-1} \begin{bmatrix} -0.0052 \\ -0.0260 \end{bmatrix}$$

$$\begin{bmatrix} \hat{\alpha} \\ \hat{\beta} \end{bmatrix}^{new} = \begin{bmatrix} -8.4331 \\ 1.0019 \end{bmatrix}^{old} + \begin{bmatrix} 4.9215 & -0.8658 \\ -0.8658 & 0.1549 \end{bmatrix} \begin{bmatrix} -0.0052 \\ -0.0260 \end{bmatrix}$$

$$\begin{bmatrix} -8.4360 \\ 1.0024 \end{bmatrix}^{new} = \begin{bmatrix} -8.4331 \\ 1.0019 \end{bmatrix}^{old} + \begin{bmatrix} -0.0029 \\ 0.0005 \end{bmatrix}$$

Step 5 $\hat{\alpha} = -8.4360$ $\hat{\beta} = 1.0024$

x	y_1	y_0	$\hat{\pi}(x)$	$-2\log_e L$
3.75	0	33	0.0092	0.6113
4.5	3	130	0.0194	28.7506
5.5	4	74	0.0510	31.5547
6.25	6	54	0.1024	39.0136
Total	13	291		99.9302

Parameter estimates adjustment:

$$\begin{bmatrix} \hat{\alpha} \\ \hat{\beta} \end{bmatrix}^{new} = \begin{bmatrix} -8.4360 \\ 1.0024 \end{bmatrix}^{old} + \begin{bmatrix} 12.1156 & 67.7190 \\ 67.7190 & 384.9607 \end{bmatrix}^{-1} \begin{bmatrix} -0.0000 \\ -0.0000 \end{bmatrix}$$

$$\begin{bmatrix} \hat{\alpha} \\ \hat{\beta} \end{bmatrix}^{new} = \begin{bmatrix} -8.4360 \\ 1.0024 \end{bmatrix}^{old} + \begin{bmatrix} 4.9251 & -0.8664 \\ -0.8664 & 0.1550 \end{bmatrix} \begin{bmatrix} -0.0000 \\ -0.0000 \end{bmatrix}$$

$$\begin{bmatrix} -8.4360 \\ 1.0024 \end{bmatrix}^{new} = \begin{bmatrix} -8.4360 \\ 1.0024 \end{bmatrix}^{old} + \begin{bmatrix} -0.0000 \\ 0.0000 \end{bmatrix}$$

At the end of this step the changes in the chi-square $[-2(\log \text{likelihood})]$ value is 0 to four decimal places. The adjustments to the parameter estimates are also 0 to four decimal places. The procedure has converged to a solution.

We wish to test the logistic regression equation for significance. To do this we use 2(initial log-likelihood − final log-likelihood) = 2(107.3900−99.9302) = 7.4549. If the null hypothesis is true, this statistic has a chi-square distribution with 1 degree of freedom. For the 0.05 level of significance, the critical value is 3.8416, hence the model is significant and it is confirmed that an increase in size of an aneurysm significantly increases the odds it will rupture.

The estimate of α is $\hat{\alpha} = -8.3460$ and the estimate of β is $\hat{\beta} = 1.0024$. The estimated standard error of $\hat{\alpha}$ is 2.219. The estimated standard error of $\hat{\beta}$ is 0.3937.

A 95% confidence interval for β is

$$1.0024 \pm 1.96(0.3937)$$

$$1.0024 \pm 0.7716$$

$$0.2308 - 1.774$$

Odds Ratio

Because the logistic regression equation predicts the log odds, the coefficient β represents the difference between two logs which is the same as a log of an odds ratio. The inverse of the coefficient, the odds ratio, is the factor by which the odds will be multiplied for a unit increase in x. Therefore a 1-cm increase in the aneurysm is a $e^{1.0024} = 2.72$-fold increase in the odds for rupture.

Computer Usage

Most statistical software will perform the computations necessary for logistic regression. The following SAS program can be used to create a SAS data set and perform a logistic regression for the aneurysm ruptures:

```
Data;
input size rupture count;
cards;
3.75  1   0
3.75  0   33
4.5   1   3
4.5   0   130
5.5   1   4
5.5   0   74
6.25  1   6
6.25  0   54
proc logistic;
freq count;
model rupture(event = '1') = size;
```

The output follows.

The SAS System

The LOGISTIC Procedure

Model Information

Data Set	WORK.DATA1
Response Variable	rupture
Number of Response Levels	2
Number of Observations	7
Frequency Variable	count
Sum of Frequencies	304
Model	binary logit
Optimization Technique	Fisher's scoring

Response Profile

Ordered Value	rupture	Total Frequency
1	0	291
2	1	13

Probability modeled is rupture = 1.

NOTE: 1 observation having zero frequency or weight was excluded since it does not contribute to the analysis.

Model Convergence Status

Convergence criterion (GCONV = 1E-8) satisfied.

Model Fit Statistics

Criterion	Intercept Only	Intercept and Covariates
AIC	109.390	103.930
SC	113.107	111.364
−2 Log L	107.390	99.930

Testing Global Null Hypothesis: BETA = 0

Test	Chi-Square	DF	Pr > ChiSq
Likelihood Ratio	7.4598	1	0.0063
Score	7.3800	1	0.0066
Wald	6.4821	1	0.0109

The LOGISTIC Procedure

Analysis of Maximum Likelihood Estimates

Parameter	DF	Estimate	Standard Error	Wald Chi-Square	Pr > ChiSq
Intercept	1	−8.4360	2.2192	14.4500	0.0001
size	1	1.0024	0.3937	6.4821	0.0109

```
                      Odds Ratio Estimates

                   Point                95% Wald
    Effect         Estimate         Confidence Limits
    size            2.725           1.260      5.894
```

The $-2 \log_e L$ is 107.390 for the intercept-only model and 99.930 for the intercept-and-covariates model. The likelihood ratio chi-square is 7.4598. Observe that the estimates are -8.4360 and 1.0024. In addition, the odds ratio estimate is 2.725.

EXERCISES

14.8.1. A serum thought to be effective in preventing colds is given to 300 persons. Their records for one year are compared with those of 200 untreated persons with the following results:

Group	No Colds	Colds
Treated	145	155
Untreated	80	120

   a. What is the estimate of the odds ratio?
   b. Find a 95% confidence interval for the odds ratio.
   c. Is 1 in the confidence interval? Interpret.
   d. Compare these results with the results of Exercise 7.5.8.

Hint: The odds ratio can be computed by the SAS logistic procedure by coding the data. Colds and Untreated are coded 1. No Colds and Treated are coded 0. (SAS gives two-sided confidence intervals for odds ratios, but experimenters usually know the direction of the trend if it exists and use one-sided confidence intervals.) For dichotomous variables the relationship between the regression coefficient β and the odds ratio ϕ is

$$\phi = e^{\beta}$$

Confidence intervals for ϕ can be obtained from the Wald confidence intervals of β by transforming the endpoints.

Some of the SAS output follows:

```
              Analysis of Maximum Likelihood Estimates

                          Standard        Wald
Parameter    DF    Estimate    Error    Chi-Square   Pr > ChiSq
Intercept    1     -0.4055    0.1443     7.8913        0.0050
Group        1      0.3388    0.1849     3.3576        0.0669
```

```
                    Odds Ratio Estimates

                    Point              95% Wald
        Effect      Estimate      Confidence Limits
        Group        1.403        0.977      2.016
```

14.8.2. One of the strategies employed in American football is to "control the ball," to maintain possession of the ball for long periods of time hoping to score points or at least denying the opponents opportunities to score. Suppose there are data for a team on length of time it held the ball in games with no tied scores or overtime play, and the results are

Median Time Ball Controlled	Games Won	Games Lost
20	10	15
30	25	20
40	45	5

Some logistic regression computer output follows:

```
                    The LOGISTIC Procedure

        Analysis of Maximum Likelihood Estimates

                                Standard      Wald
Parameter      DF    Estimate     Error    Chi-Square    Pr > ChiSq

Intercept       1     -3.2884     0.9196     12.7877       0.0003
time_
controlled      1      0.1283     0.0296     18.7530      <.0001


                    Odds Ratio Estimates

                        Point           95% Wald
            Effect      Estimate    Confidence Limits
    time_controlled      1.137       1.073     1.205
```

a. Is there evidence that the idea of ball control is a valid strategy? That is, are the odds of winning related to the length of time the team "controlled the ball"? Explain.

b. What would be the odds of a win for a team that controls the ball for 40 minutes?

14.8.3. To determine why his tea was sometimes bitter, Francis Galton designed a teapot with a thermometer so he could maintain the heat between $180°$ and $190°$F. Using a balance to weight the tea, he was able to use the same amount of tea for each brewing. Then, while holding temperature and amount of tea constant, he was able to examine the effect of time the tea was allowed to remain in the hot water. After each brewing he recorded whether or not the tea was bitter. He repeated the experiment for "numerous days," varying only the time the tea remained in the hot water. Suppose his results were

Time Tea Remained in Hot Water (min)	Number of Pots of Tea Made	Number of Pots of Bitter Tea
8	40	5
9	40	25
10	40	35

The SAS System

The LOGISTIC Procedure

Analysis of Maximum Likelihood Estimates

Parameter	DF	Estimate	Standard Error	Wald Chi-Square	Pr > ChiSq
Intercept	1	-15.7142	3.0974	25.7391	<.0001
time	1	1.7849	0.3450	26.7634	<.0001

Odds Ratio Estimates

Effect	Point Estimate	95% Wald Confidence Limits	
time	5.959	3.030	11.718

a. What are the hypotheses that would be of interest to Galton? Test these hypotheses.

b. What is the increase in the odds of bitter tea as the time of brewing increases by 1 minute? Is that increase in odds significant? Explain.

c. Galton reported that it is critical that tea not be brewed for more than 8 minutes. Is there statistically significant evidence to support this claim?

REVIEW EXERCISES

Decide whether each of the following is true or false. If a statement is false, explain why.

14.1. Multiple regression techniques require that all x variables have the same variance.

14.2. If surface area of an animal seems to be a function of its weight raised to a power, a logarithmic transformation on the area is indicated before a regression analysis.

14.3. All F tests of coefficients in a multiple regression analysis have one and $n - k$ degrees of freedom associated with them.

14.4. The experimenter may be as interested in determining which variables are nonsignificant as in determining those which are related to the dependent variable.

14.5. The test of significance of the multiple regression coefficient R is against a one-sided alternative.

14.6. When comparing different multiple regression models, the one with the largest C_p statistic is the best fit.

14.7. In a regression of y on x_1 and x_2, it is possible to use the least-squares plane for prediction if it is perpendicular to the y axis.

14.8. The total variability in y can be split into two nonoverlapping parts: the portion explained by regression and the unexplained portion.

14.9. The multiple correlation coefficient R is never negative.

14.10. Multiple regression and multiple correlation analysis require the same assumptions.

14.11. It is possible for R^2 to equal 0.90 and the regression equation may be the wrong model for the data.

14.12. The partial regression coefficients are unit free.

14.13. The partial regression coefficients are always in the same units.

14.14. Standardized partial regression coefficients are unit free.

14.15. Partial correlation coefficients can never be negative.

14.16. Backward elimination and stepwise regression always lead to the same model.

14.17. The log transformations are used to simplify the computations involved in regression.

14.18. Polynomial regression is multiple regression with $x_i = x^i$.

14.19. The model $\hat{y} = \alpha + \beta_1 x + \beta_2 x^2$ will always fit a data set better than $\hat{y} = \alpha + \beta_1 x$ because it contains a term with a higher power of x.

14.20. In logistic regression, the independent variable is measured on the categorical (nominal) scale and the dependent variable on the measurement scale.

SELECTED READINGS

Anderson, R. L., D. M. Allen, and F. B. Cady (1972). Selection of predictor variables in linear multiple regression. In *Statistical Papers in Honor of George W. Snedecor*, edited by T. A. Bancroft. Iowa State University Press, Ames.

Andrews, D. F. (1971). A note on the selection of data transformations. *Biometrika*, **58**, 249–254.

Bradley, R. A., and S. S. Srivastava (1979). Correlation in polynomial regression. *American Statistician*, **33**, 11–14.

Brogden, H. E. (1946). On the interpretation of the correlation coefficient as a measure of predictive efficiency. *Journal of Educational Psychology*, **37**, 65–76.

Chatterjee, S., and B. Price (1991). *Regression Analysis by Example*, 2nd. ed. Wiley, New York.

Cochran, W. G. (1970). Some effects of errors of measurement on multiple correlation. *Journal of the American Statistical Association*, **65**, 22–34.

Cramer, E. M. (1972). Significant tests and tests of models in multiple regression. *American Statistician*, **26** (Oct.), 26–30.

Crocker, D. C. (1972). Some interpretations of the multiple correlation coefficient. *American Statistician*, **26** (Apr.), 31–33.

Draper, N., and H. Smith (1998). *Applied Regression Analysis*, 3rd ed. Wiley, New York.

Ellenberg, J. H. (1976). Testing for a single outlier from a general linear regression. *Biometrics*, **32**, 637–645.

Furnival, G. M. (1971). All possible regressions with less computation. *Technometrics*, **13**, 403–408.

Garside, M. J. (1965). The best subset in multiple regression analysis. *Applied Statistics*, **14**, 196–200.

Gorman, J. W., and R. J. Toman (1966). Selection of variables for fitting equations to data. *Technometrics*, **8**, 27–51.

Heren, D. A. (1968). A note on log-linear regression. *Journal of the American Statistical Association*, **63**, 1034–1038.

Hosmer, D. W., and S. Lemeshow (2000). *Applied Logistic Regression*, 2nd. ed. Wiley, New York.

Hill, R. C., G. G. Judge, and T. B. Fomby (1978). On testing the adequacy of a regression model. *Technometrics*, **20**, 491–494.

Hill, R. C., G. G. Judge, and T. B. Fomby (1980). Is the regression equation adequate?—A reply. *Technometrics*, **22**, 127–128.

Hocking, R. R., and R. N. Leslie (1967). Selection of the best sub-set in regression analysis. *Technometrics*, **9**, 531–540.

Kramer, K. H. (1963). Tables for constructing confidence limits on the multiple correlation coefficient. *Journal of the American Statistical Association*, **58**, 1082–1085.

Lindley, D. V. (1968). The choice of variables in multiple regression. *Journal of the Royal Statistical Society*, Series B, **30**, 31–66.

Montgomery, D. C., and E. A. Peck (1982). *Introduction to Linear Regression Analysis*. Wiley, New York.

Mullet, G. M. (1972). A graphical illustration of simple (total) and partial regression. *American Statistician*, **26** (Dec.), 25–27.

Ramsey, F. L., and D. W. Shafer (1997). *The Statistical Sleuth: A Course in Methods of Data Analysis*. Duxbury Press, Belmont, California.

Robson, D. S. (1959). A simple method for constructing orthogonal polynomials when the independent variable is unequally spaced. *Biometrics*, **15**, 187–191.

Searle, S. R. (1971). *Linear Models*. Wiley, New York.

Suich, R., and G. C. Deringer (1977). Is the regression equation adequate?—One criterion. *Technometrics*, **19**, 213–216.

Suich, R., and G. C. Deringer (1980). Is the regression equation adequate—A further note. *Technometrics*, **22**, 125–126.

Weisberg, S. (1985). *Applied Linear Regression*, 2nd ed. Wiley, New York.

Weiss, N. S. (1970). A graphical representation of the relationships between multiple regression and multiple correlation. *American Statistician*, **24** (Apr.), 25–29.

Wilkie, D. (1965). Complete set of leading coefficients, $\lambda(r, n)$, for orthogonal polynomials up to $n = 26$. *Technometrics*, **7**, 644–648.

Appendix of Useful Tables

Statistics for Research, *Third Edition*, Edited by Shirley Dowdy, Stanley Weardon, and Daniel Chilko.
ISBN 0-471-26735-X © 2004 John Wiley & Sons, Inc.

TABLE A.1. 2500 RANDOM DIGITS

These computer produced pseudorandom digits may be read in any direction: vertical, up or down; horizontal, left-to-right or right-to-left; or along any diagonal, up or down. Single digits or groups of any size may be read; the five-digit groupings are only for ease of reading and should be ignored when reading the table. Care should be taken not to use the same portion of the table repeatedly, especially for the same experiment. This can be accomplished by using a random start (see Section 2.2) or by starting at one corner of the table and striking out the digits as they are used so that each portion of the table is used only once.

	1-5	6-10	11-15	16-20	21-25	26-30	31-35	36-40	41-45	46-50
1	38742	24201	25580	18631	30563	11548	08022	62261	74563	54597
2	01448	28091	45285	81470	09829	49377	88809	59780	46891	29447
3	34768	23715	37836	17206	26527	21554	62118	78918	30845	78748
4	89533	67552	74970	68065	50599	85529	20588	59726	84051	44388
5	74163	13487	64602	07271	03530	88954	66174	68319	25323	05476
6	92837	06594	01664	43011	27981	81256	75467	28245	29149	70357
7	69008	55983	22496	55337	74159	11283	13316	27479	63079	34060
8	92404	00156	38141	06269	51599	11371	24120	88150	99649	54740
9	45369	68854	67952	06245	32056	67900	84670	50098	29179	47904
10	16929	17418	70611	53752	39997	53621	67393	24891	53738	77251
11	95400	57951	64492	52389	86037	52586	42206	74681	82599	24606
12	36981	75140	26771	67681	54042	26121	70479	50295	43593	08220
13	37705	05124	60924	24374	99850	12414	13982	83219	26396	93876
14	67830	54660	89150	92919	90913	49560	49845	98239	78807	87479
15	32789	25115	44030	86301	61900	17173	34870	37043	40625	17954
16	60127	17491	59011	37625	03435	77178	08520	49910	34898	34345
17	17115	42174	81592	04300	68875	30353	48630	86132	55173	05788
18	27760	36661	85617	06242	09725	10642	44142	29625	49415	98360
19	04494	95805	16053	37126	54750	12617	09310	94021	38471	57427
20	34753	89545	33847	78318	41551	18705	64107	18200	56834	74584
21	63319	12471	56242	06344	94606	89207	26550	93261	17931	79259
22	98802	54600	92170	51425	74130	10301	08763	56046	00093	03793
23	82661	67501	01368	91079	54810	68160	11860	84288	27053	00917
24	99251	10088	48345	72786	81066	54353	17546	31595	77246	40514
25	72756	52088	29291	46169	14636	26380	35201	07490	28845	02341
26	96723	05193	38941	33288	13923	46860	12385	94973	43259	85010
27	96169	16158	24345	78561	46611	66869	17678	38209	24023	56259
28	96678	41518	88402	17882	79991	00083	29337	39994	06328	06476
29	97329	58496	55229	90839	93840	67032	77411	57137	06172	11036
30	38143	94319	58015	71878	42332	28120	80481	41745	68085	88776

	1–5	6–10	11–15	16–20	21–25	26–30	31–35	36–40	41–45	46–50
32	98898	39140	50371	20646	07782	63276	66375	88305	77405	74749
33	04406	76609	46544	55985	72507	98678	48840	16601	44598	50487
34	55997	34203	29784	12914	37942	86041	48431	11784	28492	28049
35	95911	19810	65733	05412	18498	79393	37322	75911	92047	61599
36	67151	13303	12466	08918	27140	22886	61210	67131	52278	95829
37	59368	23548	60681	09171	18170	62627	48209	62135	44727	12937
38	75670	78997	76059	83474	15744	71892	52740	22930	92624	93036
39	94444	45866	42304	85506	26762	24841	47226	34746	90302	70785
40	73516	82157	24805	75928	02150	84557	12930	63123	11922	76960
41	89059	45446	56541	62549	21737	78963	30917	37046	81184	83397
42	94958	71785	47469	29362	91492	80902	80586	66162	74551	87221
43	21739	80710	61346	04257	09821	17188	80855	76589	36971	41982
44	93859	78783	46343	03715	12473	48553	02762	45114	75502	42382
45	14263	52552	17964	20078	82454	35167	35631	81815	18879	93676
46	22894	01894	47934	54594	43739	51301	22511	39456	51031	58121
47	29316	85620	09294	67074	77403	82789	22212	52358	69310	57604
48	31889	40095	98007	15605	93206	86857	29784	63937	83545	50407
49	60096	11744	74086	65948	37934	35941	25731	30787	68848	14320
50	42450	70020	43245	05233	21149	85898	73527	55648	65388	55211

TABLE A.2. FACTORIALS

$n!$	n
1	1
2	2
6	3
24	4
120	5
720	6
5,040	7
40,320	8
362,880	9
3,628,800	10
39,916,800	11
479,001,600	12
6,227,020,800	13
87,178,291,200	14
1,307,674,368,000	15
20,922,789,888,000	16
355,687,428,096,000	17
6,402,373,705,728,000	18
121,645,100,408,832,000	19
2,432,902,008,176,640,000	20
51,090,942,171,709,440,000	21
1,124,000,727,777,607,680,000	22
25,852,016,738,884,976,640,000	23
620,448,401,733,239,439,360,000	24
15,511,210,043,330,985,984,000,000	25

$n! = 1 \cdot 2 \cdot 3 \cdot \ldots \cdot n$ $0! = 1$ by definition.

TABLE A.3. BINOMIAL COEFFICIENTS $\binom{n}{y}$

n \ y	0	1	2	3	4	5	6	7	8	9	10	11	12
1	1	1											
2	1	2	1										
3	1	3	3	1									
4	1	4	6	4	1								
5	1	5	10	10	5	1							
6	1	6	15	20	15	6	1						
7	1	7	21	35	35	21	7	1					
8	1	8	28	56	70	56	28	8	1				
9	1	9	36	84	126	126	84	36	9	1			
10	1	10	45	120	210	252	210	120	45	10	1		
11	1	11	55	165	330	462	462	330	165	55	11	1	
12	1	12	66	220	495	792	924	792	495	220	66	12	1
13	1	13	78	286	715	1287	1716	1716	1287	715	286	78	13
14	1	14	91	364	1001	2002	3003	3432	3003	2002	1001	364	91
15	1	15	105	455	1365	3003	5005	6435	6435	5005	3003	1365	455
16	1	16	120	560	1820	4368	8008	11440	12870	11440	8008	4368	1820
17	1	17	136	680	2380	6188	12376	19448	24310	24310	19448	12376	6188
18	1	18	153	816	3060	8568	18564	31824	43758	48620	43758	31824	18564
19	1	19	171	969	3876	11628	27132	50388	75582	92378	92378	75582	50388
20	1	20	190	1140	4845	15504	38760	77520	125970	167960	184756	167960	125970
21	1	21	210	1330	5985	20349	54264	116280	203490	293930	352716	352716	293930
22	1	22	231	1540	7315	26334	74613	170544	319770	497420	646646	705432	646646
23	1	23	253	1771	8855	33649	100947	245157	490314	817190	1144066	1352078	1352078
24	1	24	276	2024	10626	42504	134596	346104	735471	1307504	1961256	2496144	2704156
25	1	25	300	2300	12650	53130	177100	480700	1081575	2042975	3268760	4457400	5200300

$\binom{n}{y} = \dfrac{n!}{y!(n-y)!}$ Use $\binom{n}{y} = \binom{n}{n-y}$ for $y > 12$.

TABLE A.4a. BINOMIAL DISTRIBUTIONS, $n = 20$

y \ π	.05	.10	.15	.20	.25	.30	.35	.40	.45	.50	.55	.60	.65	.70	.75	.80	.85	.90	.95	y
0	.358	.122	.039	.012	.003	.001	.000	.000	.000	.000	.000	.000	.000	.000	.000	.000	.000	.000	.000	0
1	.377	.270	.137	.058	.021	.007	.002	.000	.000	.000	.000	.000	.000	.000	.000	.000	.000	.000	.000	1
2	.189	.285	.229	.137	.067	.028	.010	.003	.001	.000	.000	.000	.000	.000	.000	.000	.000	.000	.000	2
3	.060	.190	.243	.205	.134	.072	.032	.012	.004	.001	.000	.000	.000	.000	.000	.000	.000	.000	.000	3
4	.013	.090	.182	.218	.190	.130	.074	.035	.014	.005	.001	.000	.000	.000	.000	.000	.000	.000	.000	4
5	.002	.032	.103	.175	.202	.179	.127	.075	.036	.015	.005	.001	.000	.000	.000	.000	.000	.000	.000	5
6	.000	.009	.045	.109	.169	.192	.171	.124	.075	.037	.015	.005	.001	.000	.000	.000	.000	.000	.000	6
7	.000	.002	.016	.055	.112	.164	.184	.166	.122	.074	.037	.015	.004	.001	.000	.000	.000	.000	.000	7
8	.000	.000	.005	.022	.061	.114	.161	.180	.162	.120	.073	.035	.014	.004	.001	.000	.000	.000	.000	8
9	.000	.000	.001	.007	.027	.065	.116	.160	.177	.160	.119	.071	.034	.012	.003	.000	.000	.000	.000	9
10	.000	.000	.000	.002	.010	.031	.069	.117	.159	.176	.159	.117	.069	.031	.010	.002	.000	.000	.000	10
11	.000	.000	.000	.000	.003	.012	.034	.071	.119	.160	.177	.160	.116	.065	.027	.007	.001	.000	.000	11
12	.000	.000	.000	.000	.001	.004	.014	.035	.073	.120	.162	.180	.161	.114	.061	.022	.005	.000	.000	12
13	.000	.000	.000	.000	.000	.001	.004	.015	.037	.074	.122	.166	.184	.164	.112	.055	.016	.002	.000	13
14	.000	.000	.000	.000	.000	.000	.001	.005	.015	.037	.075	.124	.171	.192	.169	.109	.045	.009	.000	14
15	.000	.000	.000	.000	.000	.000	.000	.001	.005	.015	.036	.075	.127	.179	.202	.175	.103	.032	.002	15
16	.000	.000	.000	.000	.000	.000	.000	.000	.001	.005	.014	.035	.074	.130	.190	.218	.182	.090	.013	16
17	.000	.000	.000	.000	.000	.000	.000	.000	.000	.001	.004	.012	.032	.072	.134	.205	.243	.190	.060	17
18	.000	.000	.000	.000	.000	.000	.000	.000	.000	.000	.001	.003	.010	.028	.067	.137	.229	.285	.189	18
19	.000	.000	.000	.000	.000	.000	.000	.000	.000	.000	.000	.000	.002	.007	.021	.058	.137	.270	.377	19
20	.000	.000	.000	.000	.000	.000	.000	.000	.000	.000	.000	.000	.000	.001	.003	.012	.039	.122	.358	20

Step boundaries give approximate 90% confidence intervals for π. (See Section 3.3.).

TABLE A.4b. BINOMIAL DISTRIBUTIONS, $n = 25$

y \ π	.05	.10	.15	.20	.25	.30	.35	.40	.45	.50	.55	.60	.65	.70	.75	.80	.85	.90	.95	y
0	.277	.072	.017	.004	.001	.000	.000	.000	.000	.000	.000	.000	.000	.000	.000	.000	.000	.000	.000	0
1	.365	.199	.076	.024	.006	.001	.000	.000	.000	.000	.000	.000	.000	.000	.000	.000	.000	.000	.000	1
2	.231	.266	.161	.071	.025	.007	.002	.000	.000	.000	.000	.000	.000	.000	.000	.000	.000	.000	.000	2
3	.093	.226	.217	.136	.064	.024	.008	.002	.000	.000	.000	.000	.000	.000	.000	.000	.000	.000	.000	3
4	.027	.138	.211	.187	.118	.057	.022	.007	.002	.000	.000	.000	.000	.000	.000	.000	.000	.000	.000	4
5	.006	.065	.156	.196	.165	.103	.051	.020	.006	.002	.000	.000	.000	.000	.000	.000	.000	.000	.000	5
6	.001	.024	.092	.163	.183	.147	.091	.044	.017	.005	.001	.000	.000	.000	.000	.000	.000	.000	.000	6
7	.000	.007	.044	.111	.165	.171	.133	.080	.038	.014	.004	.001	.000	.000	.000	.000	.000	.000	.000	7
8	.000	.002	.017	.062	.124	.165	.161	.120	.070	.032	.012	.003	.001	.000	.000	.000	.000	.000	.000	8
9	.000	.000	.006	.029	.078	.134	.163	.151	.108	.061	.027	.009	.002	.000	.000	.000	.000	.000	.000	9
10	.000	.000	.002	.012	.042	.092	.141	.161	.142	.097	.052	.021	.006	.001	.000	.000	.000	.000	.000	10
11	.000	.000	.000	.004	.019	.054	.103	.147	.158	.133	.087	.043	.016	.004	.001	.000	.000	.000	.000	11
12	.000	.000	.000	.001	.007	.027	.065	.114	.151	.155	.124	.076	.035	.011	.002	.000	.000	.000	.000	12
13	.000	.000	.000	.000	.002	.011	.035	.076	.124	.155	.151	.114	.065	.027	.007	.001	.000	.000	.000	13
14	.000	.000	.000	.000	.001	.004	.016	.043	.087	.133	.158	.147	.103	.054	.019	.004	.000	.000	.000	14
15	.000	.000	.000	.000	.000	.001	.006	.021	.052	.097	.142	.161	.141	.092	.042	.012	.002	.000	.000	15
16	.000	.000	.000	.000	.000	.000	.002	.009	.027	.061	.108	.151	.163	.134	.078	.029	.006	.000	.000	16
17	.000	.000	.000	.000	.000	.000	.001	.003	.012	.032	.070	.120	.161	.165	.124	.062	.017	.002	.000	17
18	.000	.000	.000	.000	.000	.000	.000	.001	.004	.014	.038	.080	.133	.171	.165	.111	.044	.007	.000	18
19	.000	.000	.000	.000	.000	.000	.000	.000	.001	.005	.017	.044	.091	.147	.183	.163	.092	.024	.001	19
20	.000	.000	.000	.000	.000	.000	.000	.000	.000	.002	.006	.020	.051	.103	.165	.196	.156	.065	.006	20
21	.000	.000	.000	.000	.000	.000	.000	.000	.000	.000	.002	.007	.022	.057	.118	.187	.211	.138	.027	21
22	.000	.000	.000	.000	.000	.000	.000	.000	.000	.000	.000	.002	.008	.024	.064	.136	.217	.226	.093	22
23	.000	.000	.000	.000	.000	.000	.000	.000	.000	.000	.000	.000	.002	.007	.025	.071	.161	.266	.231	23
24	.000	.000	.000	.000	.000	.000	.000	.000	.000	.000	.000	.000	.000	.001	.006	.024	.076	.199	.365	24
25	.000	.000	.000	.000	.000	.000	.000	.000	.000	.000	.000	.000	.000	.000	.001	.004	.017	.072	.277	25

Step boundaries give approximate 90% confidence intervals for π. (See Section 3.3.)

517

TABLE A.5a THROUGH A.5e CONFIDENCE INTERVALS ON THE BINOMIAL PARAMETER π

Each of the following tables gives central confidence intervals at the $\alpha = 0.10$, $\alpha = 0.05$, and $\alpha = 0.01$ levels. ($L =$ lower confidence limit; $U =$ upper confidence limit.) For sample sizes $n = 25$ and $n = 50$ (Tables A.5a and A.5b), if y cases of the outcome of interest occur in the sample, $CI_{1-\alpha}$: $L \leq \pi \leq U$ is found by referring to row y and reading L and U under the appropriate α level.

Example: If $\alpha = 0.10$, $n = 50$, and $y = 31$,

$$CI_{0.90}: \ -0.494 \leq \pi \leq 0.735$$

For $n = 100$, the procedure is the same except that if $y > 50$, row $100 - y$ must be used to find the confidence interval and $L = 1 - U$ (of row $100 - y$) and $U = 1 - L$ (of row $100 - y$).

Example: If $\alpha = 0.01$, $y = 75$, and $n = 100$, then $100 - y = 25$ and

$$CI_{0.90}: \ 1 - 0.377 \leq \pi \leq 1 - 0.148$$

and

$$CI_{0.99}: \ 0.623 \leq \pi \leq 0.852$$

For $n = 250$ and $n = 500$ (Tables A.5d and A.5e), the confidence interval is found using y/n.

Example: If $\alpha = 0.05$, $y = 100$, and $n = 250$, then $y/n = 100/250 = 0.40$ and

$$CI_{.95}: \ 0.339 \leq \pi \leq 0.464$$

If $y/n > 0.50$, L is $1 - U$ (of row $1 - y/n$) and $U = 1 - L$ (of row $1 - y/n$).

Linear interpolation can be used with these tables for sample sizes intermediate to the ones given in the tables. Linear interpolation can also be used if y/n is intermediate to those values listed in Tables A.5d and A.5e.

The confidence intervals in these tables were derived with the use of the formulas given on page 960 of the *Handbook of Mathematical Functions With Formulas, Graphs, and Mathematical Tables*, edited by M. Abramowitz and I. A. Stegun, U.S. Department of Commercse, National Bureau of Standards, Applied Mathematics Series 55, 1964.

TABLE A.5a. CONFIDENCE INTERVALS ON THE BINOMIAL PARAMETER π

sample size $n = 25$

y	$\alpha = 0.10$ L	$\alpha = 0.10$ U	$\alpha = 0.05$ L	$\alpha = 0.05$ U	$\alpha = 0.01$ L	$\alpha = 0.01$ U
0	0.000	0.113	0.000	0.137	0.000	0.191
1	0.002	0.176	0.001	0.204	0.000	0.261
2	0.014	0.231	0.010	0.260	0.004	0.321
3	0.034	0.282	0.025	0.312	0.014	0.374
4	0.057	0.330	0.045	0.361	0.028	0.424
5	0.082	0.375	0.068	0.407	0.046	0.470
6	0.110	0.420	0.094	0.451	0.066	0.514
7	0.139	0.462	0.121	0.494	0.089	0.556
8	0.170	0.504	0.150	0.535	0.113	0.596
9	0.202	0.544	0.180	0.575	0.140	0.633
10	0.236	0.583	0.211	0.613	0.167	0.670
11	0.270	0.621	0.244	0.651	0.198	0.705
12	0.305	0.659	0.278	0.687	0.228	0.740
13	0.341	0.695	0.313	0.722	0.260	0.772
14	0.379	0.730	0.349	0.756	0.295	0.802
15	0.417	0.764	0.387	0.789	0.330	0.833
16	0.456	0.798	0.425	0.820	0.367	0.860
17	0.496	0.830	0.465	0.850	0.404	0.887
18	0.538	0.861	0.506	0.879	0.444	0.911
19	0.580	0.890	0.549	0.906	0.486	0.934
20	0.625	0.918	0.593	0.932	0.530	0.954
21	0.670	0.943	0.639	0.955	0.576	0.972
22	0.718	0.966	0.688	0.975	0.626	0.986
23	0.769	0.986	0.740	0.990	0.679	0.996
24	0.824	0.998	0.796	0.999	0.739	1.000
25	0.887	1.000	0.863	1.000	0.809	1.000

TABLE A.5b. CONFIDENCE INTERVALS ON THE BINOMIAL PARAMETER π

sample size $n = 50$

y	$\alpha = 0.10$ L	$\alpha = 0.10$ U	$\alpha = 0.05$ L	$\alpha = 0.05$ U	$\alpha = 0.01$ L	$\alpha = 0.01$ U
0	0.000	0.058	0.000	0.071	0.000	0.101
1	0.001	0.091	0.001	0.107	0.000	0.140
2	0.007	0.121	0.005	0.137	0.002	0.172
3	0.017	0.148	0.013	0.165	0.007	0.203
4	0.028	0.174	0.022	0.192	0.014	0.231
5	0.040	0.199	0.033	0.218	0.022	0.258
6	0.054	0.223	0.045	0.243	0.032	0.284
7	0.068	0.247	0.058	0.267	0.043	0.309
8	0.082	0.270	0.072	0.291	0.054	0.333
9	0.097	0.293	0.086	0.314	0.066	0.358
10	0.113	0.316	0.100	0.337	0.078	0.380
11	0.129	0.338	0.115	0.360	0.092	0.403
12	0.145	0.360	0.131	0.382	0.106	0.426
13	0.161	0.381	0.146	0.403	0.120	0.447
14	0.178	0.403	0.162	0.425	0.134	0.469
15	0.195	0.424	0.179	0.446	0.149	0.490
16	0.212	0.445	0.195	0.467	0.164	0.511
17	0.230	0.465	0.212	0.488	0.180	0.531
18	0.247	0.486	0.229	0.508	0.196	0.552
19	0.265	0.506	0.246	0.528	0.212	0.571
20	0.283	0.526	0.264	0.548	0.229	0.591
21	0.301	0.546	0.282	0.568	0.246	0.610
22	0.320	0.566	0.300	0.587	0.262	0.629
23	0.339	0.585	0.318	0.607	0.280	0.648
24	0.357	0.605	0.337	0.626	0.298	0.666
25	0.376	0.624	0.355	0.645	0.315	0.685
26	0.395	0.643	0.374	0.663	0.334	0.702
27	0.415	0.661	0.393	0.682	0.352	0.720
28	0.434	0.680	0.413	0.700	0.371	0.738
29	0.454	0.699	0.432	0.718	0.390	0.754
30	0.474	0.717	0.452	0.736	0.409	0.771
31	0.494	0.735	0.472	0.754	0.429	0.788
32	0.514	0.753	0.492	0.771	0.448	0.804
33	0.535	0.770	0.512	0.788	0.469	0.820
34	0.555	0.788	0.533	0.805	0.489	0.836
35	0.576	0.805	0.554	0.821	0.510	0.851
36	0.597	0.822	0.575	0.838	0.531	0.866

TABLE A.5b. Continued

			sample size $n = 50$			
	$\alpha = 0.10$		$\alpha = 0.05$		$\alpha = 0.01$	
y	L	U	L	U	L	U
37	0.619	0.839	0.597	0.854	0.553	0.880
38	0.640	0.855	0.618	0.869	0.574	0.894
39	0.662	0.871	0.640	0.885	0.597	0.908
40	0.684	0.887	0.663	0.900	0.620	0.922
41	0.707	0.903	0.686	0.914	0.642	0.934
42	0.730	0.918	0.709	0.928	0.667	0.946
43	0.753	0.932	0.733	0.942	0.691	0.957
44	0.777	0.946	0.757	0.955	0.716	0.968
45	0.801	0.960	0.782	0.967	0.742	0.978
46	0.826	0.972	0.808	0.978	0.769	0.986
47	0.852	0.983	0.835	0.987	0.797	0.993
48	0.879	0.993	0.863	0.995	0.828	0.998
49	0.909	0.999	0.893	0.999	0.860	1.000
50	0.942	1.000	0.929	1.000	0.899	1.000

TABLE A.5c. CONFIDENCE INTERVALS ON THE BINOMIAL PARAMETER π

			sample size $n = 100$			
	$\alpha = 0.10$		$\alpha = 0.05$		$\alpha = 0.01$	
y	L	U	L	U	L	U
0	0.000	0.029	0.000	0.036	0.000	0.052
1	0.001	0.047	0.000	0.054	0.000	0.072
2	0.004	0.062	0.002	0.070	0.001	0.089
3	0.008	0.076	0.006	0.085	0.003	0.106
4	0.014	0.089	0.011	0.099	0.007	0.121
5	0.020	0.102	0.016	0.113	0.011	0.135
6	0.026	0.115	0.022	0.126	0.016	0.149
7	0.033	0.127	0.029	0.139	0.021	0.163
8	0.040	0.140	0.035	0.152	0.026	0.176
9	0.048	0.152	0.042	0.164	0.032	0.189
10	0.055	0.164	0.049	0.176	0.038	0.202
11	0.063	0.175	0.056	0.188	0.044	0.215
12	0.071	0.187	0.064	0.200	0.051	0.227

(*Table continued*)

| | sample size $n = 100$ | | | | | |
| | $\alpha = 0.10$ | | $\alpha = 0.05$ | | $\alpha = 0.01$ | |
y	L	U	L	U	L	U
13	0.079	0.199	0.071	0.212	0.058	0.239
14	0.087	0.210	0.079	0.224	0.064	0.251
15	0.095	0.222	0.086	0.235	0.072	0.263
16	0.103	0.233	0.094	0.247	0.079	0.275
17	0.111	0.244	0.102	0.258	0.086	0.287
18	0.120	0.255	0.110	0.269	0.093	0.298
19	0.128	0.266	0.118	0.281	0.101	0.310
20	0.137	0.277	0.127	0.292	0.108	0.321
21	0.145	0.288	0.135	0.303	0.116	0.332
22	0.154	0.299	0.143	0.314	0.124	0.344
23	0.163	0.310	0.152	0.325	0.132	0.355
24	0.171	0.321	0.160	0.336	0.140	0.366
25	0.180	0.331	0.169	0.347	0.148	0.377
26	0.189	0.342	0.177	0.357	0.156	0.388
27	0.198	0.353	0.186	0.368	0.164	0.398
28	0.207	0.363	0.195	0.379	0.172	0.409
29	0.216	0.374	0.204	0.389	0.181	0.420
30	0.225	0.384	0.212	0.400	0.189	0.431
31	0.234	0.395	0.221	0.410	0.198	0.441
32	0.243	0.405	0.230	0.421	0.206	0.452
33	0.252	0.415	0.239	0.431	0.215	0.462
34	0.261	0.426	0.248	0.442	0.223	0.473
35	0.271	0.436	0.257	0.452	0.232	0.483
36	0.280	0.446	0.266	0.462	0.240	0.493
37	0.289	0.457	0.276	0.472	0.250	0.503
38	0.299	0.467	0.285	0.483	0.259	0.514
39	0.308	0.477	0.294	0.493	0.267	0.523
40	0.318	0.487	0.303	0.503	0.276	0.533
41	0.327	0.497	0.313	0.513	0.286	0.544
42	0.336	0.507	0.322	0.523	0.294	0.554
43	0.346	0.517	0.331	0.533	0.303	0.563
44	0.356	0.527	0.341	0.543	0.313	0.573
45	0.365	0.537	0.350	0.553	0.322	0.583
46	0.375	0.547	0.360	0.563	0.331	0.593
47	0.384	0.557	0.369	0.572	0.341	0.603
48	0.394	0.567	0.379	0.582	0.350	0.612
49	0.404	0.577	0.389	0.592	0.359	0.622
50	0.414	0.586	0.398	0.602	0.369	0.631

TABLE A.5d. CONFIDENCE INTERVALS ON THE BINOMIAL PARAMETER π

sample size $n = 250$

y/n	$\alpha = 0.10$		$\alpha = 0.05$		$\alpha = 0.01$	
	L	U	L	U	L	U
0.00	0.000	0.012	0.000	0.015	0.000	0.021
0.02	0.008	0.042	0.007	0.046	0.004	0.056
0.04	0.022	0.067	0.019	0.072	0.015	0.084
0.06	0.037	0.091	0.034	0.097	0.028	0.110
0.08	0.054	0.114	0.050	0.121	0.042	0.134
0.10	0.070	0.137	0.066	0.144	0.057	0.159
0.12	0.088	0.159	0.082	0.167	0.073	0.182
0.14	0.105	0.181	0.099	0.189	0.089	0.205
0.16	0.123	0.203	0.117	0.211	0.105	0.228
0.18	0.141	0.225	0.134	0.233	0.122	0.250
0.20	0.159	0.246	0.152	0.255	0.139	0.273
0.22	0.178	0.267	0.170	0.277	0.156	0.295
0.24	0.196	0.289	0.188	0.298	0.174	0.316
0.26	0.215	0.310	0.207	0.319	0.192	0.338
0.28	0.233	0.331	0.225	0.340	0.210	0.359
0.30	0.252	0.351	0.244	0.361	0.228	0.380
0.32	0.271	0.372	0.263	0.382	0.246	0.401
0.34	0.290	0.393	0.281	0.402	0.264	0.422
0.36	0.310	0.413	0.300	0.423	0.283	0.442
0.38	0.329	0.433	0.320	0.443	0.302	0.463
0.40	0.348	0.454	0.339	0.464	0.321	0.483
0.42	0.368	0.474	0.358	0.484	0.340	0.503
0.44	0.387	0.494	0.377	0.504	0.359	0.523
0.46	0.407	0.514	0.397	0.524	0.378	0.543
0.48	0.426	0.534	0.417	0.544	0.398	0.563
0.50	0.446	0.554	0.436	0.564	0.417	0.583

TABLE A.5e. CONFIDENCE INTERVALS ON THE BINOMIAL PARAMETER π

sample size $n = 500$

y/n	$\alpha = 0.10$ L	$\alpha = 0.10$ U	$\alpha = 0.05$ L	$\alpha = 0.05$ U	$\alpha = 0.01$ L	$\alpha = 0.01$ U
0.00	0.000	0.006	0.000	0.007	0.000	0.011
0.01	0.004	0.021	0.003	0.023	0.002	0.028
0.02	0.011	0.034	0.010	0.036	0.007	0.042
0.03	0.019	0.046	0.017	0.049	0.014	0.056
0.04	0.027	0.058	0.025	0.061	0.021	0.068
0.05	0.035	0.069	0.033	0.073	0.028	0.081
0.06	0.044	0.081	0.041	0.085	0.036	0.093
0.07	0.052	0.092	0.049	0.096	0.044	0.105
0.08	0.061	0.103	0.058	0.107	0.052	0.116
0.09	0.070	0.114	0.066	0.119	0.060	0.128
0.10	0.079	0.125	0.075	0.130	0.068	0.139
0.11	0.088	0.136	0.084	0.141	0.077	0.151
0.12	0.097	0.147	0.093	0.152	0.085	0.162
0.13	0.106	0.157	0.102	0.163	0.094	0.173
0.14	0.115	0.168	0.111	0.174	0.103	0.184
0.15	0.124	0.179	0.120	0.184	0.111	0.196
0.16	0.134	0.189	0.129	0.195	0.120	0.207
0.17	0.143	0.200	0.138	0.206	0.129	0.217
0.18	0.152	0.211	0.147	0.217	0.138	0.228
0.19	0.162	0.221	0.157	0.227	0.147	0.239
0.20	0.171	0.232	0.166	0.238	0.156	0.250
0.21	0.180	0.242	0.175	0.248	0.165	0.261
0.22	0.190	0.253	0.184	0.259	0.174	0.271
0.23	0.199	0.263	0.194	0.269	0.183	0.282
0.24	0.209	0.274	0.203	0.280	0.192	0.292
0.25	0.218	0.284	0.213	0.290	0.202	0.303
0.26	0.228	0.294	0.222	0.301	0.211	0.314
0.27	0.237	0.305	0.232	0.311	0.220	0.324
0.28	0.247	0.315	0.241	0.322	0.230	0.335
0.29	0.257	0.325	0.251	0.332	0.239	0.345
0.30	0.266	0.336	0.260	0.342	0.248	0.356
0.31	0.276	0.346	0.270	0.353	0.258	0.366
0.32	0.286	0.356	0.279	0.363	0.267	0.376
0.33	0.295	0.366	0.289	0.373	0.277	0.387
0.34	0.305	0.376	0.299	0.383	0.286	0.397
0.35	0.315	0.387	0.308	0.394	0.296	0.407

TABLE A.5e. Continued

<table>
<tr><td colspan="7" align="center">sample size $n = 500$</td></tr>
<tr><td rowspan="2">y/n</td><td colspan="2" align="center">$\alpha = 0.10$</td><td colspan="2" align="center">$\alpha = 0.05$</td><td colspan="2" align="center">$\alpha = 0.01$</td></tr>
<tr><td>L</td><td>U</td><td>L</td><td>U</td><td>L</td><td>U</td></tr>
<tr><td>0.36</td><td>0.324</td><td>0.397</td><td>0.318</td><td>0.404</td><td>0.305</td><td>0.417</td></tr>
<tr><td>0.37</td><td>0.334</td><td>0.407</td><td>0.328</td><td>0.414</td><td>0.315</td><td>0.428</td></tr>
<tr><td>0.38</td><td>0.344</td><td>0.417</td><td>0.337</td><td>0.424</td><td>0.325</td><td>0.438</td></tr>
<tr><td>0.39</td><td>0.354</td><td>0.427</td><td>0.347</td><td>0.434</td><td>0.334</td><td>0.448</td></tr>
<tr><td>0.40</td><td>0.363</td><td>0.437</td><td>0.357</td><td>0.444</td><td>0.344</td><td>0.458</td></tr>
<tr><td>0.41</td><td>0.373</td><td>0.448</td><td>0.367</td><td>0.455</td><td>0.353</td><td>0.468</td></tr>
<tr><td>0.42</td><td>0.383</td><td>0.458</td><td>0.376</td><td>0.465</td><td>0.363</td><td>0.478</td></tr>
<tr><td>0.43</td><td>0.393</td><td>0.468</td><td>0.386</td><td>0.475</td><td>0.373</td><td>0.489</td></tr>
<tr><td>0.44</td><td>0.403</td><td>0.478</td><td>0.396</td><td>0.485</td><td>0.383</td><td>0.498</td></tr>
<tr><td>0.45</td><td>0.413</td><td>0.488</td><td>0.406</td><td>0.495</td><td>0.392</td><td>0.509</td></tr>
<tr><td>0.46</td><td>0.423</td><td>0.498</td><td>0.416</td><td>0.505</td><td>0.402</td><td>0.519</td></tr>
<tr><td>0.47</td><td>0.432</td><td>0.508</td><td>0.426</td><td>0.515</td><td>0.412</td><td>0.529</td></tr>
<tr><td>0.48</td><td>0.442</td><td>0.518</td><td>0.435</td><td>0.525</td><td>0.422</td><td>0.539</td></tr>
<tr><td>0.49</td><td>0.452</td><td>0.528</td><td>0.445</td><td>0.535</td><td>0.432</td><td>0.548</td></tr>
<tr><td>0.50</td><td>0.462</td><td>0.538</td><td>0.455</td><td>0.545</td><td>0.442</td><td>0.558</td></tr>
</table>

TABLE A.6. VALUES OF $e^{-\lambda}$

λ	.00	.05	.10	.15	.20	.25	.30	.35	.40	.45
0.00	1.000000	.951229	.904837	.860708	.818731	.778801	.740818	.704688	.670320	.637628
1.00	.367879	.349938	.332871	.316637	.301194	.286505	.272532	.259240	.246597	.234570
2.00	.135335	.128735	.122456	.116484	.110803	.105399	.100259	.095369	.090718	.086294
3.00	.049787	.047359	.045049	.042852	.040762	.038774	.036883	.035084	.033373	.031746
4.00	.018316	.017422	.016573	.015764	.014996	.014264	.013569	.012907	.012277	.011679
5.00	.006738	.006409	.006097	.005799	.005517	.005248	.004992	.004748	.004517	.004296
6.00	.002479	.002358	.002243	.002133	.002029	.001930	.001836	.001747	.001662	.001581
7.00	.000912	.000867	.000825	.000785	.000747	.000710	.000676	.000643	.000611	.000581
8.00	.000335	.000319	.000304	.000289	.000275	.000261	.000249	.000236	.000225	.000214
9.00	.000123	.000117	.000112	.000106	.000101	.000096	.000091	.000087	.000083	.000079
10.00	.000045	.000043	.000041	.000039	.000037	.000035	.000034	.000032	.000030	.000029
11.00	.000017	.000016	.000015	.000014	.000014	.000013	.000012	.000012	.000011	.000011
12.00	.000006	.000006	.000006	.000005	.000005	.000005	.000005	.000004	.000004	.000004
13.00	.000002	.000002	.000002	.000002	.000002	.000002	.000002	.000002	.000002	.000001
14.00	.000001	.000001	.000001	.000001	.000001	.000001	.000001	.000001	.000001	.000001

λ	.50	.55	.60	.65	.70	.75	.80	.85	.90	.95
0.00	.606531	.576950	.548812	.522046	.496585	.472367	.449329	.427415	.406570	.386741
1.00	.223130	.212248	.201897	.192050	.182684	.173774	.165299	.157237	.149569	.142274
2.00	.082085	.078082	.074274	.070651	.067206	.063928	.060810	.057844	.055023	.052340
3.00	.030197	.028725	.027324	.025991	.024724	.023518	.022371	.021280	.020242	.019255
4.00	.011109	.010567	.010052	.009562	.009095	.008652	.008230	.007828	.007447	.007083

TABLE A.6. Continued

λ	.50	.55	.60	.65	.70	.75	.80	.85	.90	.95
5.00	.004087	.003887	.003698	.003518	.003346	.003183	.003028	.002880	.002739	.002606
6.00	.001503	.001430	.001360	.001294	.001231	.001171	.001114	.001059	.001008	.000959
7.00	.000553	.000526	.000500	.000476	.000453	.000431	.000410	.000390	.000371	.000353
8.00	.000203	.000194	.000184	.000175	.000167	.000158	.000151	.000143	.000136	.000130
9.00	.000075	.000071	.000068	.000064	.000061	.000058	.000055	.000053	.000050	.000048
10.00	.000028	.000026	.000025	.000024	.000023	.000021	.000020	.000019	.000018	.000018
11.00	.000010	.000010	.000009	.000009	.000008	.000008	.000008	.000007	.000007	.000006
12.00	.000004	.000004	.000003	.000003	.000003	.000003	.000003	.000003	.000002	.000002
13.00	.000001	.000001	.000001	.000001	.000001	.000001	.000001	.000001	.000001	.000001
14.00	.000001	.000000	.000000	.000000	.000000	.000000	.000000	.000000	.000000	.000000

TABLE A.7. POISSON DISTRIBUTIONS

y \ λ	0.05	0.10	0.20	0.30	0.40	0.50	0.60
0	.9512	.9048	.8187	.7408	.6703	.6065	.5488
1	.0476	.0905	.1637	.2222	.2681	.3033	.3293
2	.0012	.0045	.0164	.0333	.0536	.0758	.0988
3	.0000	.0002	.0011	.0033	.0072	.0126	.0198
4	.0000	.0000	.0001	.0003	.0007	.0016	.0030
5	.0000	.0000	.0000	.0000	.0001	.0002	.0004
6	.0000	.0000	.0000	.0000	.0000	.0000	.0000

y \ λ	0.70	0.80	0.90	1.00	1.20	1.40	1.60
0	.4966	.4493	.4066	.3679	.3012	.2466	.2019
1	.3476	.3595	.3659	.3679	.3614	.3452	.3230
2	.1217	.1438	.1647	.1839	.2169	.2417	.2584
3	.0284	.0383	.0494	.0613	.0867	.1128	.1378
4	.0050	.0077	.0111	.0153	.0260	.0395	.0551
5	.0007	.0012	.0020	.0031	.0062	.0111	.0176
6	.0001	.0002	.0003	.0005	.0012	.0026	.0047
7	.0000	.0000	.0000	.0001	.0002	.0005	.0011
8	.0000	.0000	.0000	.0000	.0000	.0001	.0002
9	.0000	.0000	.0000	.0000	.0000	.0000	.0000

y \ λ	1.80	2.00	2.20	2.40	2.60	2.80	3.00
0	.1653	.1353	.1108	.0907	.0743	.0608	.0498
1	.2975	.2707	.2438	.2177	.1931	.1703	.1494
2	.2678	.2707	.2681	.2613	.2510	.2384	.2240
3	.1607	.1804	.1966	.2090	.2176	.2225	.2240
4	.0723	.0902	.1082	.1254	.1414	.1557	.1680
5	.0260	.0361	.0476	.0602	.0735	.0872	.1008
6	.0078	.0120	.0174	.0241	.0319	.0407	.0504
7	.0020	.0034	.0055	.0083	.0118	.0163	.0216
8	.0005	.0009	.0015	.0025	.0038	.0057	.0081
9	.0001	.0002	.0004	.0007	.0011	.0018	.0027
10	.0000	.0000	.0001	.0002	.0003	.0005	.0008
11	.0000	.0000	.0000	.0000	.0001	.0001	.0002
12	.0000	.0000	.0000	.0000	.0000	.0000	.0001
13	.0000	.0000	.0000	.0000	.0000	.0000	.0000

y \ λ	3.50	4.00	4.50	5.00	5.50	6.00	6.50
0	.0302	.0183	.0111	.0067	.0041	.0025	.0015
1	.1057	.0733	.0500	.0337	.0225	.0149	.0098

TABLE A.7. Continued

y\λ	3.50	4.00	4.50	5.00	5.50	6.00	6.50
2	.1850	.1465	.1125	.0842	.0618	.0446	.0318
3	.2158	.1954	.1687	.1404	.1133	.0892	.0688
4	.1888	.1954	.1898	.1755	.1558	.1339	.1118
5	.1322	.1563	.1708	.1755	.1714	.1606	.1454
6	.0771	.1042	.1281	.1462	.1571	.1606	.1575
7	.0385	.0595	.0824	.1044	.1234	.1377	.1462
8	.0169	.0298	.0463	.0653	.0849	.1033	.1188
9	.0066	.0132	.0232	.0363	.0519	.0688	.0858
10	.0023	.0053	.0104	.0181	.0285	.0413	.0558
11	.0007	.0019	.0043	.0082	.0143	.0225	.0330
12	.0002	.0006	.0016	.0034	.0065	.0113	.0179
13	.0001	.0002	.0006	.0013	.0028	.0052	.0089
14	.0000	.0001	.0002	.0005	.0011	.0022	.0041
15	.0000	.0000	.0001	.0002	.0004	.0009	.0018
16	.0000	.0000	.0000	.0000	.0001	.0003	.0007
17	.0000	.0000	.0000	.0000	.0000	.0001	.0003
18	.0000	.0000	.0000	.0000	.0000	.0000	.0001
19	.0000	.0000	.0000	.0000	.0000	.0000	.0000

y\λ	7.00	8.00	9.00	10.00	11.00	12.00	13.00
0	.0009	.0003	.0001	.0000	.0000	.0000	.0000
1	.0064	.0027	.0011	.0005	.0002	.0001	.0000
2	.0223	.0107	.0050	.0023	.0010	.0004	.0002
3	.0521	.0286	.0150	.0076	.0037	.0018	.0008
4	.0912	.0573	.0337	.0189	.0102	.0053	.0027
5	.1277	.0916	.0607	.0378	.0224	.0127	.0070
6	.1490	.1221	.0911	.0631	.0411	.0255	.0152
7	.1490	.1396	.1171	.0901	.0646	.0437	.0281
8	.1304	.1396	.1318	.1126	.0888	.0655	.0457
9	.1014	.1241	.1318	.1251	.1085	.0874	.0661
10	.0710	.0993	.1186	.1251	.1194	.1048	.0859
11	.0452	.0722	.0970	.1137	.1194	.1144	.1015
12	.0263	.0481	.0728	.0948	.1094	.1144	.1099
13	.0142	.0296	.0504	.0729	.0926	.1056	.1099
14	.0071	.0169	.0324	.0521	.0728	.0905	.1021
15	.0033	.0090	.0194	.0347	.0534	.0724	.0885
16	.0014	.0045	.0109	.0217	.0367	.0543	.0719
17	.0006	.0021	.0058	.0128	.0237	.0383	.0550
18	.0002	.0009	.0029	.0071	.0145	.0255	.0397
19	.0001	.0004	.0014	.0037	.0084	.0161	.0272
20	.0000	.0002	.0006	.0019	.0046	.0097	.0177

(Table continued)

TABLE A.7. Continued

y\\λ	7.00	8.00	9.00	10.00	11.00	12.00	13.00
21	.0000	.0001	.0003	.0009	.0024	.0055	.0109
22	.0000	.0000	.0001	.0004	.0012	.0030	.0065
23	.0000	.0000	.0000	.0002	.0006	.0016	.0037
24	.0000	.0000	.0000	.0001	.0003	.0008	.0020
25	.0000	.0000	.0000	.0000	.0001	.0004	.0010
26	.0000	.0000	.0000	.0000	.0000	.0002	.0005
27	.0000	.0000	.0000	.0000	.0000	.0001	.0002
28	.0000	.0000	.0000	.0000	.0000	.0000	.0001
29	.0000	.0000	.0000	.0000	.0000	.0000	.0001
30	.0000	.0000	.0000	.0000	.0000	.0000	.0000

TABLE A.8. CENTRAL POISSON CONFIDENCE INTERVALS

	$1 - \alpha = 0.80$		$1 - \alpha = 0.90$		$1 - \alpha = 0.95$	
y	L	U	L	U	L	U
0	0.0000	2.3026	0.0000	2.9957	0.0000	3.6889
1	0.1054	3.8897	0.0513	4.7439	0.0253	5.5716
2	0.5318	5.3223	0.3554	6.2958	0.2422	7.2247
3	1.1021	6.6808	0.8177	7.7537	0.6187	8.7673
4	1.7448	7.9936	1.3663	9.1535	1.0899	10.2416
5	2.4326	9.2747	1.9701	10.5130	1.6235	11.6683
6	3.1519	10.5321	2.6130	11.8424	2.2019	13.0595
7	3.8948	11.7709	3.2853	13.1481	2.8144	14.4227
8	4.6561	12.9947	3.9808	14.4346	3.4538	15.7632
9	5.4325	14.2060	4.6952	15.7052	4.1154	17.0848
10	6.2213	15.4066	5.4254	16.9622	4.7954	18.3904
11	7.0207	16.5981	6.1690	18.2075	5.4912	19.6820
12	7.8293	17.7816	6.9242	19.4426	6.2006	20.9616
13	8.6459	18.9580	7.6896	20.6686	6.9220	22.2304
14	9.4696	20.1280	8.4639	21.8865	7.6539	23.4896
15	10.2996	21.2924	9.2463	23.0971	8.3954	24.7402
16	11.1353	22.4516	10.0360	24.3012	9.1454	25.9830
17	11.9761	23.6061	10.8321	25.4992	9.9031	27.2186
18	12.8216	24.7563	11.6343	26.6918	10.6679	28.4478
19	13.6715	25.9025	12.4420	27.8792	11.4392	29.6709
20	14.5253	27.0451	13.2547	29.0620	12.2165	30.8884

TABLE A.9. CRITICAL CHI-SQUARE VALUES

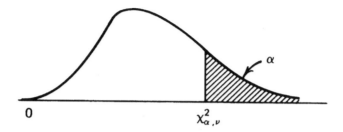

$$P(\chi^2 > \chi^2_{\alpha,\nu}) = P(\chi^2 > \text{tabular value}) = \alpha$$

Examples:

1. $P(\chi^2 > \chi^2_{0.025,5}) = P(\chi^2 > 12.833) = 0.025$
2. $P(\chi^2 > \chi^2_{0.995,10}) = P(\chi^2 > 2.156) = 0.995$

TABLE A.9. CRITICAL CHI-SQUARE VALUES

ν $\backslash$ α	0.995	0.990	0.975	0.950	0.050	0.025	0.010	0.005
1	0.000	0.000	0.001	0.004	3.841	5.024	6.635	7.879
2	0.010	0.020	0.051	0.103	5.991	7.378	9.210	10.597
3	0.072	0.115	0.216	0.352	7.815	9.348	11.345	12.838
4	0.207	0.297	0.484	0.711	9.488	11.143	13.277	14.860
5	0.412	0.554	0.831	1.145	11.070	12.833	15.086	16.750
6	0.676	0.872	1.237	1.635	12.592	14.449	16.812	18.548
7	0.989	1.239	1.690	2.167	14.067	16.013	18.475	20.278
8	1.344	1.646	2.180	2.733	15.507	17.535	20.090	21.955
9	1.735	2.088	2.700	3.325	16.919	19.023	21.666	23.589
10	2.156	2.558	3.247	3.940	18.307	20.483	23.209	25.188
11	2.603	3.053	3.816	4.575	19.675	21.920	24.725	26.757
12	3.074	3.571	4.404	5.226	21.026	23.337	26.217	28.300
13	3.565	4.107	5.009	5.892	22.362	24.736	27.688	29.819
14	4.075	4.660	5.629	6.571	23.685	26.119	29.141	31.319
15	4.601	5.229	6.262	7.261	24.996	27.488	30.578	32.801
16	5.142	5.812	6.908	7.962	26.296	28.845	32.000	34.267
17	5.697	6.408	7.564	8.672	27.587	30.191	33.409	35.718
18	6.265	7.015	8.231	9.390	28.869	31.526	34.805	37.156
19	6.844	7.633	8.907	10.117	30.144	32.852	36.191	38.582
20	7.434	8.260	9.591	10.851	31.410	34.170	37.566	39.997

α ν	0.995	0.990	0.975	0.950	0.050	0.025	0.010	0.005
21	8.034	8.897	10.283	11.591	32.671	35.479	38.932	41.401
22	8.643	9.542	10.982	12.338	33.924	36.781	40.289	42.796
23	9.260	10.196	11.689	13.091	35.172	38.076	41.638	44.181
24	9.886	10.856	12.401	13.848	36.415	39.364	42.980	45.559
25	10.520	11.524	13.120	14.611	37.652	40.646	44.314	46.928
26	11.160	12.198	13.844	15.379	38.885	41.923	45.642	48.290
27	11.808	12.879	14.573	16.151	40.113	43.195	46.963	49.645
28	12.461	13.565	15.308	16.928	41.337	44.461	48.278	50.993
29	13.121	14.256	16.047	17.708	42.557	45.722	49.588	52.336
30	13.787	14.953	16.791	18.493	43.773	46.979	50.892	53.672
32	15.134	16.362	18.291	20.072	46.194	49.480	53.486	56.328
34	16.501	17.789	19.806	21.664	48.602	51.966	56.061	58.964
36	17.887	19.233	21.336	23.269	50.998	54.437	58.619	61.581
38	19.289	20.691	22.878	24.884	53.384	56.896	61.162	64.181
40	20.707	22.164	24.433	26.509	55.758	59.342	63.691	66.766
42	22.138	23.650	25.999	28.144	58.124	61.777	66.206	69.336
44	23.584	25.148	27.575	29.787	60.481	64.201	68.710	71.893
46	25.041	26.657	29.160	31.439	62.830	66.617	71.201	74.437
48	26.511	28.177	30.755	33.098	65.171	69.023	73.683	76.969
50	27.991	29.707	32.357	34.764	67.505	71.420	76.154	79.490
60	35.534	37.485	40.482	43.188	79.082	83.298	88.379	91.952
70	43.275	45.442	48.758	51.739	90.531	95.023	100.425	104.215
80	51.172	53.540	57.153	60.391	101.879	106.629	112.329	116.321
90	59.196	61.754	65.647	69.126	113.145	118.136	124.116	128.299
100	67.328	70.065	74.222	77.929	124.342	129.561	135.807	140.169

TABLE A.10. THE STANDARD NORMAL DISTRIBUTION

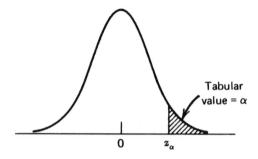

Values α in the body of the table are the probability that z is greater than the positive value z_α given in the margins.

Example:

$$P(z > 1.54) = 0.062$$

or

$$z_{0.062} = 1.54$$

For negative z values, the probability of a greater value can be found using the symmetry of the distribution.

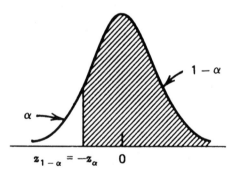

$$P(z > z_\alpha) = 1 - \alpha = P(z > z_{1-\alpha})$$

Example:

$$P(x > -1.54) = 1 - 0.062 = 0.938$$

or

$$z_{0.938} = -1.54$$

534

TABLE A.10. THE STANDARD NORMAL DISTRIBUTION

z	.00	.01	.02	.03	.04	.05	.06	.07	.08	.09
0.00	.500	.496	.492	.488	.484	.480	.476	.472	.468	.464
0.10	.460	.456	.452	.448	.444	.440	.436	.433	.429	.425
0.20	.421	.417	.413	.409	.405	.401	.397	.394	.390	.386
0.30	.382	.378	.374	.371	.367	.363	.359	.356	.352	.348
0.40	.345	.341	.337	.334	.330	.326	.323	.319	.316	.312
0.50	.309	.305	.302	.298	.295	.291	.288	.284	.281	.278
0.60	.274	.271	.268	.264	.261	.258	.255	.251	.248	.245
0.70	.242	.239	.236	.233	.230	.227	.224	.221	.218	.215
0.80	.212	.209	.206	.203	.200	.198	.195	.192	.189	.187
0.90	.184	.181	.179	.176	.174	.171	.169	.166	.164	.161
1.00	.159	.156	.154	.152	.149	.147	.145	.142	.140	.138
1.10	.136	.133	.131	.129	.127	.125	.123	.121	.119	.117
1.20	.115	.113	.111	.109	.107	.106	.104	.102	.100	.099
1.30	.097	.095	.093	.092	.090	.089	.087	.085	.084	.082
1.40	.081	.079	.078	.076	.075	.074	.072	.071	.069	.068
1.50	.067	.066	.064	.063	.062	.061	.059	.058	.057	.056
1.60	.055	.054	.053	.052	.051	.049	.048	.047	.046	.046
1.70	.045	.044	.043	.042	.041	.040	.039	.038	.038	.037
1.80	.036	.035	.034	.034	.033	.032	.031	.031	.030	.029
1.90	.029	.028	.027	.027	.026	.026	.025	.024	.024	.023
2.00	.023	.022	.022	.021	.021	.020	.020	.019	.019	.018
2.10	.018	.017	.017	.017	.016	.016	.015	.015	.015	.014
2.20	.014	.014	.013	.013	.013	.012	.012	.012	.011	.011
2.30	.011	.010	.010	.010	.010	.009	.009	.009	.009	.008
2.40	.008	.008	.008	.008	.007	.007	.007	.007	.007	.006
2.50	.006	.006	.006	.006	.006	.005	.005	.005	.005	.005
2.60	.005	.005	.004	.004	.004	.004	.004	.004	.004	.004
2.70	.003	.003	.003	.003	.003	.003	.003	.003	.003	.003
2.80	.003	.002	.002	.002	.002	.002	.002	.002	.002	.002
2.90	.002	.002	.002	.002	.002	.002	.002	.001	.001	.001
3.00	.001	.001	.001	.001	.001	.001	.001	.001	.001	.001

TABLE A.11. CRITICAL *t* VALUES

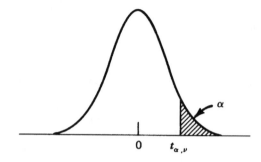

$$P(t > t_{\alpha,v}) = P(t > \text{tabular value}) = \alpha$$

Example:

$$P(t > t_{0.05,10}) = P(t > 1.812) = 0.05$$

Symmetry is used to find negative *t* values.
    Example:

$$t_{0.95,10} = -t_{0.05,10} = -1.812$$

The last row of the *t* table gives critical *z* values, that is,

$$t_{\alpha,\infty} = z_{\alpha}$$

TABLE A.11. CRITICAL *t* VALUES

v \ α	0.100	0.050	0.025	0.010	0.005
1	3.078	6.314	12.706	31.821	63.657
2	1.886	2.920	4.303	6.965	9.925
3	1.638	2.353	3.182	4.541	5.841
4	1.533	2.132	2.776	3.747	4.604
5	1.476	2.015	2.571	3.365	4.032
6	1.440	1.943	2.447	3.143	3.707
7	1.415	1.895	2.365	2.998	3.499
8	1.397	1.860	2.306	2.896	3.355
9	1.383	1.833	2.262	2.821	3.250
10	1.372	1.812	2.228	2.764	3.169

TABLE A.11. Continued

α	0.100	0.050	0.025	0.010	0.005
ν					
11	1.363	1.796	2.201	2.718	3.106
12	1.356	1.782	2.179	2.681	3.055
13	1.350	1.771	2.160	2.650	3.012
14	1.345	1.761	2.145	2.624	2.977
15	1.341	1.753	2.131	2.602	2.947
16	1.337	1.746	2.120	2.583	2.921
17	1.333	1.740	2.110	2.567	2.898
18	1.330	1.734	2.101	2.552	2.878
19	1.328	1.729	2.093	2.539	2.861
20	1.325	1.725	2.086	2.528	2.845
21	1.323	1.721	2.080	2.518	2.831
22	1.321	1.717	2.074	2.508	2.819
23	1.319	1.714	2.069	2.500	2.807
24	1.318	1.711	2.064	2.492	2.797
25	1.316	1.708	2.060	2.485	2.787
26	1.315	1.706	2.056	2.479	2.779
27	1.314	1.703	2.052	2.473	2.771
28	1.313	1.701	2.048	2.467	2.763
29	1.311	1.699	2.045	2.462	2.756
30	1.310	1.697	2.042	2.457	2.750
40	1.303	1.684	2.021	2.423	2.704
60	1.296	1.671	2.000	2.390	2.660
120	1.289	1.658	1.980	2.358	2.617
INF	1.282	1.645	1.960	2.326	2.576

TABLE A12a THROUGH A12x. CRITICAL F VALUES

v_1 = numerator degrees of freedom

v_2 = denominator degrees of freedom

$$P(F > F_{\alpha,v_1,v_2}) = \alpha$$

Example:

$$F_{0.025,2,4} = 10.649$$

For lower critical F values, use the relationship

$$F_{1-\alpha,v_1,v_2} = \frac{1}{F_{a,v_2,v_1}}$$

Example:

$$F_{0.995,10,8} = \frac{1}{F_{0.005,8,10}} = \frac{1}{6.116} = 0.1635$$

Table for a Given Pair of Degrees of Freedom

		Numerator Degrees of Freedom					
		1–5	6–10	11–15	16–20	21–25	26–30
Denominator	1–10	A.12a	A.12b	A.12c	A.12d	A.12e	A.12f
Degrees of	11–20	A.12g	A.12h	A.12i	A.12j	A.12k	A.12l
Freedom	21–30	A.12m	A.12n	A.12o	A.12p	A.12q	A.12r
	40–200	A.12s	A.12t	A.12u	A.12v	A.12w	A.12x

TABLE A.12a. CRITICAL *F* VALUES

Denominator ν	α	Numerator ν				
		1	2	3	4	5
1	0.050	161.448	199.500	215.707	224.583	230.162
	0.025	647.790	799.500	864.163	899.583	921.848
	0.010	4052.194	4999.506	5403.355	5624.584	5763.660
	0.005	16210.873	19999.499	21614.726	22499.596	23055.762
	0.001	405293.184	499996.121	540378.670	562498.442	576406.763
2	0.050	18.513	19.000	19.164	19.247	19.296
	0.025	38.506	39.000	39.165	39.248	39.298
	0.010	98.503	99.000	99.166	99.249	99.299
	0.005	198.501	199.000	199.166	199.250	199.300
	0.001	998.505	998.991	999.168	999.257	999.302
3	0.050	10.128	9.552	9.277	9.117	9.013
	0.025	17.443	16.044	15.439	15.101	14.885
	0.010	34.116	30.817	29.457	28.710	28.237
	0.005	55.552	49.799	47.467	46.195	45.392
	0.001	167.030	148.501	141.109	137.099	134.581
4	0.050	7.709	6.944	6.591	6.388	6.256
	0.025	12.218	10.649	9.979	9.605	9.364
	0.010	21.198	18.000	16.694	15.977	15.522
	0.005	31.333	26.284	24.259	23.155	22.456
	0.001	74.137	61.245	56.177	53.436	51.711
5	0.050	6.608	5.786	5.409	5.192	5.050
	0.025	10.007	8.434	7.764	7.388	7.146
	0.010	16.258	13.274	12.060	11.392	10.967
	0.005	22.785	18.314	16.530	15.556	14.940
	0.001	47.181	37.122	33.203	31.085	29.753
6	0.050	5.987	5.143	4.757	4.534	4.387
	0.025	8.813	7.260	6.599	6.227	5.988
	0.010	13.745	10.925	9.780	9.148	8.746
	0.005	18.635	14.544	12.917	12.028	11.464
	0.001	35.508	27.000	23.703	21.924	20.803
7	0.050	5.591	4.737	4.347	4.120	3.972
	0.025	8.073	6.542	5.890	5.523	5.285
	0.010	12.246	9.547	8.451	7.847	7.460
	0.005	16.236	12.404	10.882	10.050	9.522
	0.001	29.245	21.689	18.772	17.198	16.206
8	0.050	5.318	4.459	4.066	3.838	3.687
	0.025	7.571	6.059	5.416	5.053	4.817
	0.010	11.259	8.649	7.591	7.006	6.632
	0.005	14.688	11.042	9.596	8.805	8.302
	0.001	25.415	18.494	15.829	14.392	13.485

(*Table continued*)

TABLE A.12a. Continued

Denominator ν	α	Numerator ν				
		1	2	3	4	5
9	0.050	5.117	4.256	3.863	3.633	3.482
	0.025	7.209	5.715	5.078	4.718	4.484
	0.010	10.561	8.022	6.992	6.422	6.057
	0.005	13.614	10.107	8.717	7.956	7.471
	0.001	22.857	16.387	13.902	12.560	11.714
10	0.050	4.965	4.103	3.708	3.478	3.326
	0.025	6.937	5.456	4.826	4.468	4.236
	0.010	10.044	7.559	6.552	5.994	5.636
	0.005	12.826	9.427	8.081	7.343	6.872
	0.001	21.040	14.905	12.553	11.283	10.481

TABLE A.12b. CRITICAL F VALUES

Denominator ν	α	Numerator ν				
		6	7	8	9	10
1	0.050	233.986	236.768	238.883	240.543	241.882
	0.025	937.110	948.218	956.656	963.285	968.627
	0.010	5858.981	5928.349	5981.073	6022.471	6055.850
	0.005	23437.141	23714.565	23925.451	24091.033	24224.533
	0.001	585927.903	592864.102	598136.821	602279.789	605630.027
2	0.050	19.330	19.353	19.371	19.385	19.396
	0.025	39.331	39.355	39.373	39.387	39.398
	0.010	99.333	99.356	99.374	99.388	99.399
	0.005	199.333	199.357	199.375	199.388	199.399
	0.001	999.329	999.360	999.376	999.387	999.409
3	0.050	8.941	8.887	8.845	8.812	8.786
	0.025	14.735	14.624	14.540	14.473	14.419
	0.010	27.911	27.672	27.489	27.345	27.229
	0.005	44.838	44.434	44.126	43.882	43.686
	0.001	132.848	131.584	130.619	129.860	129.247
4	0.050	6.163	6.094	6.041	5.999	5.964
	0.025	9.197	9.074	8.980	8.905	8.844
	0.010	15.207	14.976	14.799	14.659	14.546
	0.005	21.975	21.622	21.352	21.139	20.967
	0.001	50.525	49.658	48.996	48.474	48.053
5	0.050	4.950	4.876	4.818	4.772	4.735
	0.025	6.978	6.853	6.757	6.681	6.619
	0.010	10.672	10.456	10.289	10.158	10.051
	0.005	14.513	14.200	13.961	13.772	13.618

Denominator v	α	Numerator v				
		6	7	8	9	10
	0.001	28.834	28.163	27.649	27.245	26.916
6	0.050	4.284	4.207	4.147	4.099	4.060
	0.025	5.820	5.695	5.600	5.523	5.461
	0.010	8.466	8.260	8.102	7.976	7.874
	0.005	11.073	10.786	10.566	10.391	10.250
	0.001	20.030	19.463	19.030	18.688	18.411
7	0.050	3.866	3.787	3.726	3.677	3.637
	0.025	5.119	4.995	4.899	4.823	4.761
	0.010	7.191	6.993	6.840	6.719	6.620
	0.005	9.155	8.885	8.678	8.514	8.380
	0.001	15.521	15.019	14.634	14.330	14.083
8	0.050	3.581	3.500	3.438	3.388	3.347
	0.025	4.652	4.529	4.433	4.357	4.295
	0.010	6.371	6.178	6.029	5.911	5.814
	0.005	7.952	7.694	7.496	7.339	7.211
	0.001	12.858	12.398	12.046	11.767	11.540
9	0.050	3.374	3.293	3.230	3.179	3.137
	0.025	4.320	4.197	4.102	4.026	3.964
	0.010	5.802	5.613	5.467	5.351	5.257
	0.005	7.134	6.885	6.693	6.541	6.417
	0.001	11.128	10.698	10.368	10.107	9.894
10	0.050	3.217	3.135	3.072	3.020	2.978
	0.025	4.072	3.950	3.855	3.779	3.717
	0.010	5.386	5.200	5.057	4.942	4.849
	0.005	6.545	6.302	6.116	5.968	5.847
	0.001	9.926	9.517	9.204	8.956	8.754

TABLE A.12c. CRITICAL F VALUES

Denominator v	α	Numerator v				
		11	12	13	14	15
1	0.050	242.984	243.906	244.690	245.364	245.950
	0.025	973.025	976.709	979.837	982.527	984.866
	0.010	6083.321	6106.329	6125.853	6142.674	6157.294
	0.005	24334.361	24426.333	24504.525	24571.721	24630.203
	0.001	608357.024	610674.243	612614.192	614311.903	615752.317
2	0.050	19.405	19.413	19.419	19.424	19.429
	0.025	39.407	39.415	39.421	39.426	39.431

(*Table continued*)

Denominator ν	α	Numerator ν				
		11	12	13	14	15
	0.010	99.408	99.416	99.422	99.428	99.432
	0.005	199.408	199.416	199.423	199.428	199.433
	0.001	999.412	999.421	999.422	999.437	999.426
3	0.050	8.763	8.745	8.729	8.715	8.703
	0.025	14.374	14.337	14.304	14.277	14.253
	0.010	27.133	27.052	26.983	26.924	26.872
	0.005	43.524	43.387	43.272	43.172	43.085
	0.001	128.742	128.317	127.957	127.645	127.376
4	0.050	5.936	5.912	5.891	5.873	5.858
	0.025	8.794	8.751	8.715	8.684	8.657
	0.010	14.452	14.374	14.307	14.249	14.198
	0.005	20.824	20.705	20.603	20.515	20.438
	0.001	47.704	47.412	47.163	46.948	46.761
5	0.050	4.704	4.678	4.655	4.636	4.619
	0.025	6.568	6.525	6.488	6.456	6.428
	0.010	9.963	9.888	9.825	9.770	9.722
	0.005	13.491	13.384	13.293	13.215	13.146
	0.001	26.646	26.418	26.224	26.057	25.911
6	0.050	4.027	4.000	3.976	3.956	3.938
	0.025	5.410	5.366	5.329	5.297	5.269
	0.010	7.790	7.718	7.657	7.605	7.559
	0.005	10.133	10.034	9.950	9.877	9.814
	0.001	18.182	17.989	17.824	17.682	17.559
7	0.050	3.603	3.575	3.550	3.529	3.511
	0.025	4.709	4.666	4.628	4.596	4.568
	0.010	6.538	6.469	6.410	6.359	6.314
	0.005	8.270	8.176	8.097	8.028	7.968
	0.001	13.879	13.707	13.561	13.434	13.324
8	0.050	3.313	3.284	3.259	3.237	3.218
	0.025	4.243	4.200	4.162	4.130	4.101
	0.010	5.734	5.667	5.609	5.559	5.515
	0.005	7.104	7.015	6.938	6.872	6.814
	0.001	11.352	11.194	11.060	10.943	10.841
9	0.050	3.102	3.073	3.048	3.025	3.006
	0.025	3.912	3.868	3.831	3.798	3.769
	0.010	5.178	5.111	5.055	5.005	4.962
	0.005	6.314	6.227	6.153	6.089	6.032
	0.001	9.718	9.570	9.443	9.334	9.238
10	0.050	2.943	2.913	2.887	2.865	2.845
	0.025	3.665	3.621	3.583	3.550	3.522
	0.010	4.772	4.706	4.650	4.601	4.558
	0.005	5.746	5.661	5.589	5.526	5.471
	0.001	8.586	8.445	8.324	8.220	8.129

TABLE A.12d. CRITICAL *F* VALUES

Denominator ν	α	Numerator ν				
		16	17	18	19	20
1	0.050	246.464	246.918	247.323	247.686	248.013
	0.025	986.919	988.733	990.350	991.797	993.102
	0.010	6170.090	6181.436	6191.527	6200.577	6208.737
	0.005	24681.450	24726.829	24767.214	24803.335	24835.957
	0.001	617053.889	618188.763	619195.633	620086.602	620918.989
2	0.050	19.433	19.437	19.440	19.443	19.446
	0.025	39.435	39.439	39.442	39.445	39.448
	0.010	99.437	99.440	99.444	99.447	99.449
	0.005	199.437	199.441	199.444	199.447	199.449
	0.001	999.428	999.436	999.440	999.441	999.443
3	0.050	8.692	8.683	8.675	8.667	8.660
	0.025	14.232	14.213	14.196	14.181	14.167
	0.010	26.827	26.787	26.751	26.719	26.690
	0.005	43.008	42.941	42.880	42.826	42.778
	0.001	127.136	126.927	126.738	126.572	126.418
4	0.050	5.844	5.832	5.821	5.811	5.803
	0.025	8.633	8.611	8.592	8.575	8.560
	0.010	14.154	14.115	14.080	14.048	14.020
	0.005	20.371	20.311	20.258	20.210	20.167
	0.001	46.597	46.451	46.322	46.205	46.100
5	0.050	4.604	4.590	4.579	4.568	4.558
	0.025	6.403	6.381	6.362	6.344	6.329
	0.010	9.680	9.643	9.610	9.580	9.553
	0.005	13.086	13.033	12.985	12.942	12.903
	0.001	25.783	25.669	25.568	25.477	25.395
6	0.050	3.922	3.908	3.896	3.884	3.874
	0.025	5.244	5.222	5.202	5.184	5.168
	0.010	7.519	7.483	7.451	7.422	7.396
	0.005	9.758	9.709	9.664	9.625	9.589
	0.001	17.450	17.353	17.267	17.190	17.120
7	0.050	3.494	3.480	3.467	3.455	3.445
	0.025	4.543	4.521	4.501	4.483	4.467
	0.010	6.275	6.240	6.209	6.181	6.155
	0.005	7.915	7.868	7.826	7.788	7.754
	0.001	13.227	13.140	13.063	12.994	12.932
8	0.050	3.202	3.187	3.173	3.161	3.150
	0.025	4.076	4.054	4.034	4.016	3.999
	0.010	5.477	5.442	5.412	5.384	5.359
	0.005	6.763	6.718	6.678	6.641	6.608
	0.001	10.752	10.672	10.601	10.537	10.480
9	0.050	2.989	2.974	2.960	2.948	2.936
	0.025	3.744	3.722	3.701	3.683	3.667
	0.010	4.924	4.890	4.860	4.833	4.808

(*Table continued*)

TABLE A.12d. Continued

Denominator ν	α	Numerator ν				
		16	17	18	19	20
	0.005	5.983	5.939	5.899	5.864	5.832
	0.001	9.154	9.079	9.012	8.952	8.898
10	0.050	2.828	2.812	2.798	2.785	2.774
	0.025	3.496	3.474	3.453	3.435	3.419
	0.010	4.520	4.487	4.457	4.430	4.405
	0.005	5.422	5.379	5.340	5.305	5.274
	0.001	8.048	7.977	7.913	7.856	7.804

TABLE A.12e. CRITICAL *F* VALUES

Denominator ν	α	Numerator ν				
		21	22	23	24	25
1	0.050	248.309	248.579	248.826	249.052	249.260
	0.025	994.286	995.363	996.346	997.249	998.081
	0.010	6216.126	6222.855	6228.993	6234.629	6239.826
	0.005	24865.611	24892.464	24916.926	24939.664	24960.416
	0.001	621653.353	622320.075	622924.674	623495.668	624013.102
2	0.050	19.448	19.450	19.452	19.454	19.456
	0.025	39.450	39.452	39.454	39.456	39.458
	0.010	99.452	99.454	99.456	99.457	99.459
	0.005	199.452	199.454	199.456	199.458	199.460
	0.001	999.452	999.452	999.456	999.456	999.460
3	0.050	8.654	8.648	8.643	8.639	8.634
	0.025	14.155	14.144	14.134	14.124	14.115
	0.010	26.664	26.640	26.618	26.598	26.579
	0.005	42.733	42.693	42.656	42.622	42.591
	0.001	126.281	126.155	126.041	125.935	125.840
4	0.050	5.795	5.787	5.781	5.774	5.769
	0.025	8.546	8.533	8.522	8.511	8.501
	0.010	13.994	13.970	13.949	13.929	13.911
	0.005	20.128	20.093	20.060	20.030	20.002
	0.001	46.005	45.918	45.839	45.766	45.699
5	0.050	4.549	4.541	4.534	4.527	4.521
	0.025	6.314	6.301	6.289	6.278	6.268
	0.010	9.528	9.506	9.485	9.466	9.449
	0.005	12.868	12.836	12.807	12.780	12.755
	0.001	25.320	25.252	25.190	25.133	25.080
6	0.050	3.865	3.856	3.849	3.841	3.835
	0.025	5.154	5.141	5.128	5.117	5.107

TABLE A.12e. Continued

Denominator ν	α	Numerator ν				
		21	22	23	24	25
	0.010	7.372	7.351	7.331	7.313	7.296
	0.005	9.556	9.526	9.499	9.474	9.451
	0.001	17.057	16.999	16.946	16.897	16.853
7	0.050	3.435	3.426	3.418	3.410	3.404
	0.025	4.452	4.439	4.426	4.415	4.405
	0.010	6.132	6.111	6.092	6.074	6.058
	0.005	7.723	7.695	7.669	7.645	7.623
	0.001	12.875	12.823	12.776	12.732	12.692
8	0.050	3.140	3.131	3.123	3.115	3.108
	0.025	3.985	3.971	3.959	3.947	3.937
	0.010	5.336	5.316	5.297	5.279	5.263
	0.005	6.578	6.551	6.526	6.503	6.482
	0.001	10.427	10.379	10.336	10.295	10.258
9	0.050	2.926	2.917	2.908	2.900	2.893
	0.025	3.652	3.638	3.626	3.614	3.604
	0.010	4.786	4.765	4.746	4.729	4.713
	0.005	5.803	5.776	5.752	5.729	5.708
	0.001	8.848	8.803	8.762	8.724	8.689
10	0.050	2.764	2.754	2.745	2.737	2.730
	0.025	3.403	3.390	3.377	3.365	3.355
	0.010	4.383	4.363	4.344	4.327	4.311
	0.005	5.245	5.219	5.195	5.173	5.153
	0.001	7.757	7.713	7.674	7.638	7.604

TABLE A.12f. CRITICAL F VALUES

Denominator ν	α	Numerator ν				
		26	27	28	29	30
1	0.050	249.453	249.631	249.797	249.951	250.095
	0.025	998.849	999.561	1000.222	1000.839	1001.414
	0.010	6244.624	6249.061	6253.195	6257.053	6260.644
	0.005	24979.489	24997.314	25013.859	25029.224	25043.644
	0.001	624504.229	624947.959	625346.713	625750.603	626089.462
2	0.050	19.457	19.459	19.460	19.461	19.462
	0.025	39.459	39.461	39.462	39.463	39.465
	0.010	99.461	99.462	99.464	99.465	99.466
	0.005	199.461	199.462	199.464	199.465	199.466
	0.001	999.456	999.462	999.464	999.466	999.474

(Table continued)

TABLE A.12f. Continued

Denominator		Numerator v				
v	α	26	27	28	29	30
3	0.050	8.630	8.626	8.623	8.620	8.617
	0.025	14.107	14.100	14.093	14.087	14.081
	0.010	26.562	26.546	26.531	26.517	26.505
	0.005	42.562	42.536	42.511	42.487	42.466
	0.001	125.749	125.666	125.587	125.517	125.448
4	0.050	5.763	5.759	5.754	5.750	5.746
	0.025	8.492	8.483	8.475	8.468	8.461
	0.010	13.894	13.878	13.864	13.850	13.838
	0.005	19.977	19.953	19.931	19.911	19.891
	0.001	45.637	45.579	45.525	45.475	45.428
5	0.050	4.515	4.510	4.505	4.500	4.496
	0.025	6.258	6.250	6.242	6.234	6.227
	0.010	9.433	9.418	9.404	9.391	9.379
	0.005	12.732	12.711	12.691	12.673	12.656
	0.001	25.032	24.987	24.944	24.906	24.869
6	0.050	3.829	3.823	3.818	3.813	3.808
	0.025	5.097	5.088	5.080	5.072	5.065
	0.010	7.280	7.266	7.253	7.240	7.229
	0.005	9.430	9.410	9.392	9.374	9.358
	0.001	16.811	16.773	16.737	16.703	16.672
7	0.050	3.397	3.391	3.386	3.381	3.376
	0.025	4.395	4.386	4.378	4.370	4.362
	0.010	6.043	6.029	6.016	6.003	5.992
	0.005	7.603	7.584	7.566	7.550	7.534
	0.001	12.655	12.620	12.588	12.558	12.530
8	0.050	3.102	3.095	3.090	3.084	3.079
	0.025	3.927	3.918	3.909	3.901	3.894
	0.010	5.248	5.234	5.221	5.209	5.198
	0.005	6.462	6.444	6.427	6.411	6.396
	0.001	10.224	10.192	10.162	10.135	10.109
9	0.050	2.886	2.880	2.874	2.869	2.864
	0.025	3.594	3.584	3.576	3.568	3.560
	0.010	4.698	4.685	4.672	4.660	4.649
	0.005	5.689	5.671	5.655	5.639	5.625
	0.001	8.656	8.626	8.598	8.572	8.548
10	0.050	2.723	2.716	2.710	2.705	2.700
	0.025	3.345	3.335	3.327	3.319	3.311
	0.010	4.296	4.283	4.270	4.258	4.247
	0.005	5.134	5.116	5.100	5.085	5.071
	0.001	7.573	7.544	7.517	7.492	7.469

TABLE A.12g. CRITICAL *F* VALUES

Denominator ν	α	Numerator ν				
		1	2	3	4	5
11	0.050	4.844	3.982	3.587	3.357	3.204
	0.025	6.724	5.256	4.630	4.275	4.044
	0.010	9.646	7.206	6.217	5.668	5.316
	0.005	12.226	8.912	7.600	6.881	6.422
	0.001	19.687	13.812	11.561	10.346	9.578
12	0.050	4.747	3.885	3.490	3.259	3.106
	0.025	6.554	5.096	4.474	4.121	3.891
	0.010	9.330	6.927	5.953	5.412	5.064
	0.005	11.754	8.510	7.226	6.521	6.071
	0.001	18.643	12.974	10.804	9.633	8.892
13	0.050	4.667	3.806	3.411	3.179	3.025
	0.025	6.414	4.965	4.347	3.996	3.767
	0.010	9.074	6.701	5.739	5.205	4.862
	0.005	11.374	8.186	6.926	6.233	5.791
	0.001	17.815	12.313	10.209	9.073	8.354
14	0.050	4.600	3.739	3.344	3.112	2.958
	0.025	6.298	4.857	4.242	3.892	3.663
	0.010	8.862	6.515	5.564	5.035	4.695
	0.005	11.060	7.922	6.680	5.998	5.562
	0.001	17.143	11.779	9.729	8.622	7.922
15	0.050	4.543	3.682	3.287	3.056	2.901
	0.025	6.200	4.765	4.153	3.804	3.576
	0.010	8.683	6.359	5.417	4.893	4.556
	0.005	10.798	7.701	6.476	5.803	5.372
	0.001	16.587	11.339	9.335	8.253	7.567
16	0.050	4.494	3.634	3.239	3.007	2.852
	0.025	6.115	4.687	4.077	3.729	3.502
	0.010	8.531	6.226	5.292	4.773	4.437
	0.005	10.575	7.514	6.303	5.638	5.212
	0.001	16.120	10.971	9.006	7.944	7.272
17	0.050	4.451	3.592	3.197	2.965	2.810
	0.025	6.042	4.619	4.011	3.665	3.438
	0.010	8.400	6.112	5.185	4.669	4.336
	0.005	10.384	7.354	6.156	5.497	5.075
	0.001	15.722	10.658	8.727	7.683	7.022
18	0.050	4.414	3.555	3.160	2.928	2.773
	0.025	5.978	4.560	3.954	3.608	3.382
	0.010	8.285	6.013	5.092	4.579	4.248
	0.005	10.218	7.215	6.028	5.375	4.956
	0.001	15.379	10.390	8.487	7.459	6.808

(Table continued)

TABLE A.12g. Continued

Denominator ν	α	Numerator ν				
		1	2	3	4	5
19	0.050	4.381	3.522	3.127	2.895	2.740
	0.025	5.922	4.508	3.903	3.559	3.333
	0.010	8.185	5.926	5.010	4.500	4.171
	0.005	10.073	7.093	5.916	5.268	4.853
	0.001	15.081	10.157	8.280	7.265	6.622
20	0.050	4.351	3.493	3.098	2.866	2.711
	0.025	5.871	4.461	3.859	3.515	3.289
	0.010	8.096	5.849	4.938	4.431	4.103
	0.005	9.944	6.986	5.818	5.174	4.762
	0.001	14.819	9.953	8.098	7.096	6.461

TABLE A.12h. CRITICAL F VALUES

Denominator ν	α	Numerator ν				
		6	7	8	9	10
11	0.050	3.095	3.012	2.948	2.896	2.854
	0.025	3.881	3.759	3.664	3.588	3.526
	0.010	5.069	4.886	4.744	4.632	4.539
	0.005	6.102	5.865	5.682	5.537	5.418
	0.001	9.047	8.655	8.355	8.116	7.922
12	0.050	2.996	2.913	2.849	2.796	2.753
	0.025	3.728	3.607	3.512	3.436	3.374
	0.010	4.821	4.640	4.499	4.388	4.296
	0.005	5.757	5.525	5.345	5.202	5.085
	0.001	8.379	8.001	7.710	7.480	7.292
13	0.050	2.915	2.832	2.767	2.714	2.671
	0.025	3.604	3.483	3.388	3.312	3.250
	0.010	4.620	4.441	4.302	4.191	4.100
	0.005	5.482	5.253	5.076	4.935	4.820
	0.001	7.856	7.489	7.206	6.982	6.799
14	0.050	2.848	2.764	2.699	2.646	2.602
	0.025	3.501	3.380	3.285	3.209	3.147
	0.010	4.456	4.278	4.140	4.030	3.939
	0.005	5.257	5.031	4.857	4.717	4.603
	0.001	7.436	7.077	6.802	6.583	6.404
15	0.050	2.790	2.707	2.641	2.588	2.544
	0.025	3.415	3.293	3.199	3.123	3.060
	0.010	4.318	4.142	4.004	3.895	3.805

Denominator ν	α	Numerator ν				
		6	7	8	9	10
	0.005	5.071	4.847	4.674	4.536	4.424
	0.001	7.092	6.741	6.471	6.256	6.081
16	0.050	2.741	2.657	2.591	2.538	2.494
	0.025	3.341	3.219	3.125	3.049	2.986
	0.010	4.202	4.026	3.890	3.780	3.691
	0.005	4.913	4.692	4.521	4.384	4.272
	0.001	6.805	6.460	6.195	5.984	5.812
17	0.050	2.699	2.614	2.548	2.494	2.450
	0.025	3.277	3.156	3.061	2.985	2.922
	0.010	4.102	3.927	3.791	3.682	3.593
	0.005	4.779	4.559	4.389	4.254	4.142
	0.001	6.562	6.223	5.962	5.754	5.584
18	0.050	2.661	2.577	2.510	2.456	2.412
	0.025	3.221	3.100	3.005	2.929	2.866
	0.010	4.015	3.841	3.705	3.597	3.508
	0.005	4.663	4.445	4.276	4.141	4.030
	0.001	6.355	6.021	5.763	5.558	5.390
19	0.050	2.628	2.544	2.477	2.423	2.378
	0.025	3.172	3.051	2.956	2.880	2.817
	0.010	3.939	3.765	3.631	3.523	3.434
	0.005	4.561	4.345	4.177	4.043	3.933
	0.001	6.175	5.845	5.590	5.388	5.222
20	0.050	2.599	2.514	2.447	2.393	2.348
	0.025	3.128	3.007	2.913	2.837	2.774
	0.010	3.871	3.699	3.564	3.457	3.368
	0.005	4.472	4.257	4.090	3.956	3.847
	0.001	6.019	5.692	5.440	5.239	5.075

TABLE A.12i. CRITICAL *F* VALUES

Denominator ν	α	Numerator ν				
		11	12	13	14	15
11	0.050	2.818	2.788	2.761	2.739	2.719
	0.025	3.474	3.430	3.392	3.359	3.330
	0.010	4.462	4.397	4.342	4.293	4.251
	0.005	5.320	5.236	5.165	5.103	5.049
	0.001	7.761	7.626	7.509	7.409	7.321
12	0.050	2.717	2.687	2.660	2.637	2.617
	0.025	3.321	3.277	3.239	3.206	3.177

(Table continued)

TABLE A.12i. Continued

Denominator ν	α	Numerator ν				
		11	12	13	14	15
	0.010	4.220	4.155	4.100	4.052	4.010
	0.005	4.988	4.906	4.836	4.775	4.721
	0.001	7.136	7.005	6.892	6.794	6.709
13	0.050	2.635	2.604	2.577	2.554	2.533
	0.025	3.197	3.153	3.115	3.082	3.053
	0.010	4.025	3.960	3.905	3.857	3.815
	0.005	4.724	4.643	4.573	4.513	4.460
	0.001	6.647	6.519	6.409	6.314	6.231
14	0.050	2.565	2.534	2.507	2.484	2.463
	0.025	3.095	3.050	3.012	2.979	2.949
	0.010	3.864	3.800	3.745	3.698	3.656
	0.005	4.508	4.428	4.359	4.299	4.247
	0.001	6.256	6.130	6.023	5.930	5.848
15	0.050	2.507	2.475	2.448	2.424	2.403
	0.025	3.008	2.963	2.925	2.891	2.862
	0.010	3.730	3.666	3.612	3.564	3.522
	0.005	4.329	4.250	4.181	4.122	4.070
	0.001	5.935	5.812	5.707	5.615	5.535
16	0.050	2.456	2.425	2.397	2.373	2.352
	0.025	2.934	2.889	2.851	2.817	2.788
	0.010	3.616	3.553	3.498	3.451	3.409
	0.005	4.179	4.099	4.031	3.972	3.920
	0.001	5.668	5.547	5.443	5.353	5.274
17	0.050	2.413	2.381	2.353	2.329	2.308
	0.025	2.870	2.825	2.786	2.753	2.723
	0.010	3.519	3.455	3.401	3.353	3.312
	0.005	4.050	3.971	3.903	3.844	3.793
	0.001	5.443	5.324	5.221	5.132	5.054
18	0.050	2.374	2.342	2.314	2.290	2.269
	0.025	2.814	2.769	2.730	2.696	2.667
	0.010	3.434	3.371	3.316	3.269	3.227
	0.005	3.938	3.860	3.793	3.734	3.683
	0.001	5.250	5.132	5.031	4.943	4.866
19	0.050	2.340	2.308	2.280	2.256	2.234
	0.025	2.765	2.720	2.681	2.647	2.617
	0.010	3.360	3.297	3.242	3.195	3.153
	0.005	3.841	3.763	3.696	3.638	3.587
	0.001	5.084	4.967	4.867	4.780	4.704
20	0.050	2.310	2.278	2.250	2.225	2.203
	0.025	2.721	2.676	2.637	2.603	2.573
	0.010	3.294	3.231	3.177	3.130	3.088
	0.005	3.756	3.678	3.611	3.553	3.502
	0.001	4.939	4.823	4.724	4.637	4.562

TABLE A.12j. CRITICAL *F* VALUES

Denominator ν	α	Numerator ν				
		16	17	18	19	20
11	0.050	2.701	2.685	2.671	2.658	2.646
	0.025	3.304	3.282	3.261	3.243	3.226
	0.010	4.213	4.180	4.150	4.123	4.099
	0.005	5.001	4.959	4.921	4.886	4.855
	0.001	7.244	7.175	7.113	7.058	7.008
12	0.050	2.599	2.583	2.568	2.555	2.544
	0.025	3.152	3.129	3.108	3.090	3.073
	0.010	3.972	3.939	3.909	3.883	3.858
	0.005	4.674	4.632	4.595	4.561	4.530
	0.001	6.634	6.567	6.507	6.454	6.405
13	0.050	2.515	2.499	2.484	2.471	2.459
	0.025	3.027	3.004	2.983	2.965	2.948
	0.010	3.778	3.745	3.716	3.689	3.665
	0.005	4.413	4.372	4.334	4.301	4.270
	0.001	6.158	6.093	6.034	5.982	5.934
14	0.050	2.445	2.428	2.413	2.400	2.388
	0.025	2.923	2.900	2.879	2.861	2.844
	0.010	3.619	3.586	3.556	3.529	3.505
	0.005	4.200	4.159	4.122	4.089	4.059
	0.001	5.776	5.712	5.655	5.604	5.557
15	0.050	2.385	2.368	2.353	2.340	2.328
	0.025	2.836	2.813	2.792	2.773	2.756
	0.010	3.485	3.452	3.423	3.396	3.372
	0.005	4.024	3.983	3.946	3.913	3.883
	0.001	5.464	5.402	5.345	5.294	5.248
16	0.050	2.333	2.317	2.302	2.288	2.276
	0.025	2.761	2.738	2.717	2.698	2.681
	0.010	3.372	3.339	3.310	3.283	3.259
	0.005	3.875	3.834	3.797	3.764	3.734
	0.001	5.205	5.143	5.087	5.037	4.992
17	0.050	2.289	2.272	2.257	2.243	2.230
	0.025	2.697	2.673	2.652	2.633	2.616
	0.010	3.275	3.242	3.212	3.186	3.162
	0.005	3.747	3.707	3.670	3.637	3.607
	0.001	4.986	4.924	4.869	4.820	4.775
18	0.050	2.250	2.233	2.217	2.203	2.191
	0.025	2.640	2.617	2.596	2.576	2.559
	0.010	3.190	3.158	3.128	3.101	3.077
	0.005	3.637	3.597	3.560	3.527	3.498
	0.001	4.798	4.738	4.683	4.634	4.590

(Table continued)

TABLE A.12j. Continued

Denominator ν	α	Numerator ν				
		16	17	18	19	20
19	0.050	2.215	2.198	2.182	2.168	2.155
	0.025	2.591	2.567	2.546	2.526	2.509
	0.010	3.116	3.084	3.054	3.027	3.003
	0.005	3.541	3.501	3.465	3.432	3.402
	0.001	4.636	4.576	4.522	4.474	4.430
20	0.050	2.184	2.167	2.151	2.137	2.124
	0.025	2.547	2.523	2.501	2.482	2.464
	0.010	3.051	3.018	2.989	2.962	2.938
	0.005	3.457	3.416	3.380	3.347	3.318
	0.001	4.495	4.435	4.382	4.334	4.290

TABLE A.12k. CRITICAL F VALUES

Denominator ν	α	Numerator ν				
		21	22	23	24	25
11	0.050	2.636	2.626	2.617	2.609	2.601
	0.025	3.211	3.197	3.184	3.173	3.162
	0.010	4.077	4.057	4.038	4.021	4.005
	0.005	4.827	4.801	4.778	4.756	4.736
	0.001	6.962	6.920	6.882	6.847	6.815
12	0.050	2.533	2.523	2.514	2.505	2.498
	0.025	3.057	3.043	3.031	3.019	3.008
	0.010	3.836	3.816	3.798	3.780	3.765
	0.005	4.502	4.476	4.453	4.431	4.412
	0.001	6.361	6.320	6.283	6.249	6.217
13	0.050	2.448	2.438	2.429	2.420	2.412
	0.025	2.932	2.918	2.905	2.893	2.882
	0.010	3.643	3.622	3.604	3.587	3.571
	0.005	4.243	4.217	4.194	4.173	4.153
	0.001	5.891	5.851	5.815	5.781	5.751
14	0.050	2.377	2.367	2.357	2.349	2.341
	0.025	2.828	2.814	2.801	2.789	2.778
	0.010	3.483	3.463	3.444	3.427	3.412
	0.005	4.031	4.006	3.983	3.961	3.942
	0.001	5.514	5.475	5.440	5.407	5.377
15	0.050	2.316	2.306	2.297	2.288	2.280
	0.025	2.740	2.726	2.713	2.701	2.689
	0.010	3.350	3.330	3.311	3.294	3.278

Denominator ν	α	Numerator ν				
		21	22	23	24	25
	0.005	3.855	3.830	3.807	3.786	3.766
	0.001	5.207	5.168	5.133	5.101	5.071
16	0.050	2.264	2.254	2.244	2.235	2.227
	0.025	2.665	2.651	2.637	2.625	2.614
	0.010	3.237	3.216	3.198	3.181	3.165
	0.005	3.707	3.682	3.659	3.638	3.618
	0.001	4.951	4.913	4.878	4.846	4.817
17	0.050	2.219	2.208	2.199	2.190	2.181
	0.025	2.600	2.585	2.572	2.560	2.548
	0.010	3.139	3.119	3.101	3.084	3.068
	0.005	3.580	3.555	3.532	3.511	3.492
	0.001	4.734	4.697	4.663	4.631	4.602
18	0.050	2.179	2.168	2.159	2.150	2.141
	0.025	2.543	2.529	2.515	2.503	2.491
	0.010	3.055	3.035	3.016	2.999	2.983
	0.005	3.471	3.446	3.423	3.402	3.382
	0.001	4.549	4.512	4.478	4.447	4.418
19	0.050	2.144	2.133	2.123	2.114	2.106
	0.025	2.493	2.478	2.465	2.452	2.441
	0.010	2.981	2.961	2.942	2.925	2.909
	0.005	3.375	3.350	3.327	3.306	3.287
	0.001	4.390	4.353	4.319	4.288	4.259
20	0.050	2.112	2.102	2.092	2.082	2.074
	0.025	2.448	2.434	2.420	2.408	2.396
	0.010	2.916	2.895	2.877	2.859	2.843
	0.005	3.291	3.266	3.243	3.222	3.203
	0.001	4.250	4.214	4.180	4.149	4.121

TABLE A.12l. CRITICAL F VALUES

Denominator ν	α	Numerator ν				
		26	27	28	29	30
11	0.050	2.594	2.588	2.582	2.576	2.570
	0.025	3.152	3.142	3.133	3.125	3.118
	0.010	3.990	3.977	3.964	3.952	3.941
	0.005	4.717	4.700	4.684	4.668	4.654
	0.001	6.785	6.757	6.731	6.707	6.684
12	0.050	2.491	2.484	2.478	2.472	2.466
	0.025	2.998	2.988	2.979	2.971	2.963

(*Table continued*)

Denominator ν	α	Numerator ν				
		26	27	28	29	30
	0.010	3.750	3.736	3.724	3.712	3.701
	0.005	4.393	4.376	4.360	4.345	4.331
	0.001	6.188	6.161	6.136	6.112	6.090
13	0.050	2.405	2.398	2.392	2.386	2.380
	0.025	2.872	2.862	2.853	2.845	2.837
	0.010	3.556	3.543	3.530	3.518	3.507
	0.005	4.134	4.117	4.101	4.087	4.073
	0.001	5.722	5.695	5.671	5.647	5.626
14	0.050	2.333	2.326	2.320	2.314	2.308
	0.025	2.767	2.758	2.749	2.740	2.732
	0.010	3.397	3.383	3.371	3.359	3.348
	0.005	3.923	3.906	3.891	3.876	3.862
	0.001	5.349	5.323	5.298	5.275	5.254
15	0.050	2.272	2.265	2.259	2.253	2.247
	0.025	2.679	2.669	2.660	2.652	2.644
	0.010	3.264	3.250	3.237	3.225	3.214
	0.005	3.748	3.731	3.715	3.701	3.687
	0.001	5.043	5.018	4.994	4.971	4.950
16	0.050	2.220	2.212	2.206	2.200	2.194
	0.025	2.603	2.594	2.584	2.576	2.568
	0.010	3.150	3.137	3.124	3.112	3.101
	0.005	3.600	3.583	3.567	3.553	3.539
	0.001	4.789	4.764	4.740	4.718	4.697
17	0.050	2.174	2.167	2.160	2.154	2.148
	0.025	2.538	2.528	2.519	2.510	2.502
	0.010	3.053	3.039	3.026	3.014	3.003
	0.005	3.473	3.457	3.441	3.426	3.412
	0.001	4.575	4.550	4.526	4.504	4.484
18	0.050	2.134	2.126	2.119	2.113	2.107
	0.025	2.481	2.471	2.461	2.453	2.445
	0.010	2.968	2.955	2.942	2.930	2.919
	0.005	3.364	3.347	3.332	3.317	3.303
	0.001	4.391	4.366	4.343	4.321	4.301
19	0.050	2.098	2.090	2.084	2.077	2.071
	0.025	2.430	2.420	2.411	2.402	2.394
	0.010	2.894	2.880	2.868	2.855	2.844
	0.005	3.269	3.252	3.236	3.221	3.208
	0.001	4.233	4.208	4.185	4.163	4.143
20	0.050	2.066	2.059	2.052	2.045	2.039
	0.025	2.385	2.375	2.366	2.357	2.349
	0.010	2.829	2.815	2.802	2.790	2.778
	0.005	3.184	3.168	3.152	3.137	3.123
	0.001	4.094	4.070	4.047	4.025	4.005

TABLE A.12m. CRITICAL *F* VALUES

Denominator ν	α	Numerator ν				
		1	2	3	4	5
21	0.050	4.325	3.467	3.072	2.840	2.685
	0.025	5.827	4.420	3.819	3.475	3.250
	0.010	8.017	5.780	4.874	4.369	4.042
	0.005	9.830	6.891	5.730	5.091	4.681
	0.001	14.587	9.772	7.938	6.947	6.318
22	0.050	4.301	3.443	3.049	2.817	2.661
	0.025	5.786	4.383	3.783	3.440	3.215
	0.010	7.945	5.719	4.817	4.313	3.988
	0.005	9.727	6.806	5.652	5.017	4.609
	0.001	14.380	9.612	7.796	6.814	6.191
23	0.050	4.279	3.422	3.028	2.796	2.640
	0.025	5.750	4.349	3.750	3.408	3.183
	0.010	7.881	5.664	4.765	4.264	3.939
	0.005	9.635	6.730	5.582	4.950	4.544
	0.001	14.195	9.469	7.669	6.696	6.078
24	0.050	4.260	3.403	3.009	2.776	2.621
	0.025	5.717	4.319	3.721	3.379	3.155
	0.010	7.823	5.614	4.718	4.218	3.895
	0.005	9.551	6.661	5.519	4.890	4.486
	0.001	14.028	9.339	7.554	6.589	5.977
25	0.050	4.242	3.385	2.991	2.759	2.603
	0.025	5.686	4.291	3.694	3.353	3.129
	0.010	7.770	5.568	4.675	4.177	3.855
	0.005	9.475	6.598	5.462	4.835	4.433
	0.001	13.877	9.223	7.451	6.493	5.885
26	0.050	4.225	3.369	2.975	2.743	2.587
	0.025	5.659	4.265	3.670	3.329	3.105
	0.010	7.721	5.526	4.637	4.140	3.818
	0.005	9.406	6.541	5.409	4.785	4.384
	0.001	13.739	9.116	7.357	6.406	5.802
27	0.050	4.210	3.354	2.960	2.728	2.572
	0.025	5.633	4.242	3.647	3.307	3.083
	0.010	7.677	5.488	4.601	4.106	3.785
	0.005	9.342	6.489	5.361	4.740	4.340
	0.001	13.613	9.019	7.272	6.326	5.726
28	0.050	4.196	3.340	2.947	2.714	2.558
	0.025	5.610	4.221	3.626	3.286	3.063
	0.010	7.636	5.453	4.568	4.074	3.754
	0.005	9.284	6.440	5.317	4.698	4.300
	0.001	13.498	8.931	7.193	6.253	5.656
29	0.050	4.183	3.328	2.934	2.701	2.545
	0.025	5.588	4.201	3.607	3.267	3.044

(Table continued)

TABLE A.12m. Continued

Denominator ν	α	Numerator ν				
		1	2	3	4	5
	0.010	7.598	5.420	4.538	4.045	3.725
	0.005	9.230	6.396	5.276	4.659	4.262
	0.001	13.391	8.849	7.121	6.186	5.593
30	0.050	4.171	3.316	2.922	2.690	2.534
	0.025	5.568	4.182	3.589	3.250	3.026
	0.010	7.562	5.390	4.510	4.018	3.699
	0.005	9.180	6.355	5.239	4.623	4.228
	0.001	13.293	8.773	7.054	6.125	5.534

TABLE A.12n. CRITICAL F VALUES

Denominator ν	α	Numerator ν				
		6	7	8	9	10
21	0.050	2.573	2.488	2.420	2.366	2.321
	0.025	3.090	2.969	2.874	2.798	2.735
	0.010	3.812	3.640	3.506	3.398	3.310
	0.005	4.393	4.179	4.013	3.880	3.771
	0.001	5.881	5.557	5.308	5.109	4.946
22	0.050	2.549	2.464	2.397	2.342	2.297
	0.025	3.055	2.934	2.839	2.763	2.700
	0.010	3.758	3.587	3.453	3.346	3.258
	0.005	4.322	4.109	3.944	3.812	3.703
	0.001	5.758	5.438	5.190	4.993	4.832
23	0.050	2.528	2.442	2.375	2.320	2.275
	0.025	3.023	2.902	2.808	2.731	2.668
	0.010	3.710	3.539	3.406	3.299	3.211
	0.005	4.259	4.047	3.882	3.750	3.642
	0.001	5.649	5.331	5.085	4.890	4.730
24	0.050	2.508	2.423	2.355	2.300	2.255
	0.025	2.995	2.874	2.779	2.703	2.640
	0.010	3.667	3.496	3.363	3.256	3.168
	0.005	4.202	3.991	3.826	3.695	3.587
	0.001	5.550	5.235	4.991	4.797	4.638
25	0.050	2.490	2.405	2.337	2.282	2.236
	0.025	2.969	2.848	2.753	2.677	2.613
	0.010	3.627	3.457	3.324	3.217	3.129
	0.005	4.150	3.939	3.776	3.645	3.537
	0.001	5.462	5.148	4.906	4.713	4.555

TABLE A.12n. Continued

Denominator		Numerator v				
v	α	6	7	8	9	10
26	0.050	2.474	2.388	2.321	2.265	2.220
	0.025	2.945	2.824	2.729	2.653	2.590
	0.010	3.591	3.421	3.288	3.182	3.094
	0.005	4.103	3.893	3.730	3.599	3.492
	0.001	5.381	5.070	4.829	4.637	4.480
27	0.050	2.459	2.373	2.305	2.250	2.204
	0.025	2.923	2.802	2.707	2.631	2.568
	0.010	3.558	3.388	3.256	3.149	3.062
	0.005	4.059	3.850	3.687	3.557	3.450
	0.001	5.308	4.998	4.759	4.568	4.412
28	0.050	2.445	2.359	2.291	2.236	2.190
	0.025	2.903	2.782	2.687	2.611	2.547
	0.010	3.528	3.358	3.226	3.120	3.032
	0.005	4.020	3.811	3.649	3.519	3.412
	0.001	5.241	4.933	4.695	4.505	4.349
29	0.050	2.432	2.346	2.278	2.223	2.177
	0.025	2.884	2.763	2.669	2.592	2.529
	0.010	3.499	3.330	3.198	3.092	3.005
	0.005	3.983	3.775	3.613	3.483	3.377
	0.001	5.179	4.873	4.636	4.447	4.292
30	0.050	2.421	2.334	2.266	2.211	2.165
	0.025	2.867	2.746	2.651	2.575	2.511
	0.010	3.473	3.304	3.173	3.067	2.979
	0.005	3.949	3.742	3.580	3.450	3.344
	0.001	5.122	4.817	4.581	4.393	4.239

TABLE A.12o. CRITICAL F VALUES

Denominator		Numerator v				
v	α	11	12	13	14	15
21	0.050	2.283	2.250	2.222	2.197	2.176
	0.025	2.682	2.637	2.598	2.564	2.534
	0.010	3.236	3.173	3.119	3.072	3.030
	0.005	3.680	3.602	3.536	3.478	3.427
	0.001	4.811	4.696	4.597	4.512	4.437
22	0.050	2.259	2.226	2.198	2.173	2.151
	0.025	2.647	2.602	2.563	2.528	2.498
	0.010	3.184	3.121	3.067	3.019	2.978

(Table continued)

TABLE A.12o. Continued

Denominator ν	α	Numerator ν				
		11	12	13	14	15
	0.005	3.612	3.535	3.469	3.411	3.360
	0.001	4.697	4.583	4.486	4.401	4.326
23	0.050	2.236	2.204	2.175	2.150	2.128
	0.025	2.615	2.570	2.531	2.497	2.466
	0.010	3.137	3.074	3.020	2.973	2.931
	0.005	3.551	3.475	3.408	3.351	3.300
	0.001	4.596	4.483	4.386	4.301	4.227
24	0.050	2.216	2.183	2.155	2.130	2.108
	0.025	2.586	2.541	2.502	2.468	2.437
	0.010	3.094	3.032	2.977	2.930	2.889
	0.005	3.497	3.420	3.354	3.296	3.246
	0.001	4.505	4.393	4.296	4.212	4.139
25	0.050	2.198	2.165	2.136	2.111	2.089
	0.025	2.560	2.515	2.476	2.441	2.411
	0.010	3.056	2.993	2.939	2.892	2.850
	0.005	3.447	3.370	3.304	3.247	3.196
	0.001	4.423	4.312	4.216	4.132	4.059
26	0.050	2.181	2.148	2.119	2.094	2.072
	0.025	2.536	2.491	2.451	2.417	2.387
	0.010	3.021	2.958	2.904	2.857	2.815
	0.005	3.402	3.325	3.259	3.202	3.151
	0.001	4.349	4.238	4.142	4.059	3.986
27	0.050	2.166	2.132	2.103	2.078	2.056
	0.025	2.514	2.469	2.429	2.395	2.364
	0.010	2.988	2.926	2.871	2.824	2.783
	0.005	3.360	3.284	3.218	3.161	3.110
	0.001	4.281	4.171	4.075	3.993	3.920
28	0.050	2.151	2.118	2.089	2.064	2.041
	0.025	2.494	2.448	2.409	2.374	2.344
	0.010	2.959	2.896	2.842	2.795	2.753
	0.005	3.322	3.246	3.180	3.123	3.073
	0.001	4.219	4.109	4.014	3.932	3.859
29	0.050	2.138	2.104	2.075	2.050	2.027
	0.025	2.475	2.430	2.390	2.355	2.325
	0.010	2.931	2.868	2.814	2.767	2.726
	0.005	3.287	3.211	3.145	3.088	3.038
	0.001	4.162	4.053	3.958	3.876	3.804
30	0.050	2.126	2.092	2.063	2.037	2.015
	0.025	2.458	2.412	2.372	2.338	2.307
	0.010	2.906	2.843	2.789	2.742	2.700
	0.005	3.255	3.179	3.113	3.056	3.006
	0.001	4.110	4.001	3.907	3.825	3.753

TABLE A.12p. CRITICAL *F* VALUES

Denominator ν	α	Numerator ν				
		16	17	18	19	20
21	0.050	2.156	2.139	2.123	2.109	2.096
	0.025	2.507	2.483	2.462	2.442	2.425
	0.010	2.993	2.960	2.931	2.904	2.880
	0.005	3.382	3.342	3.305	3.273	3.243
	0.001	4.371	4.311	4.258	4.210	4.167
22	0.050	2.131	2.114	2.098	2.084	2.071
	0.025	2.472	2.448	2.426	2.407	2.389
	0.010	2.941	2.908	2.879	2.852	2.827
	0.005	3.315	3.275	3.239	3.206	3.176
	0.001	4.260	4.201	4.149	4.101	4.058
23	0.050	2.109	2.091	2.075	2.061	2.048
	0.025	2.440	2.416	2.394	2.374	2.357
	0.010	2.894	2.861	2.832	2.805	2.781
	0.005	3.255	3.215	3.179	3.146	3.116
	0.001	4.162	4.103	4.051	4.004	3.961
24	0.050	2.088	2.070	2.054	2.040	2.027
	0.025	2.411	2.386	2.365	2.345	2.327
	0.010	2.852	2.819	2.789	2.762	2.738
	0.005	3.201	3.161	3.125	3.092	3.062
	0.001	4.074	4.015	3.963	3.916	3.873
25	0.050	2.069	2.051	2.035	2.021	2.007
	0.025	2.384	2.360	2.338	2.318	2.300
	0.010	2.813	2.780	2.751	2.724	2.699
	0.005	3.151	3.111	3.075	3.043	3.013
	0.001	3.994	3.936	3.884	3.837	3.794
26	0.050	2.052	2.034	2.018	2.003	1.990
	0.025	2.360	2.335	2.314	2.294	2.276
	0.010	2.778	2.745	2.715	2.688	2.664
	0.005	3.107	3.067	3.031	2.998	2.968
	0.001	3.921	3.864	3.812	3.765	3.723
27	0.050	2.036	2.018	2.002	1.987	1.974
	0.025	2.337	2.313	2.291	2.271	2.253
	0.010	2.746	2.713	2.683	2.656	2.632
	0.005	3.066	3.026	2.990	2.957	2.928
	0.001	3.856	3.798	3.747	3.700	3.658
28	0.050	2.021	2.003	1.987	1.972	1.959
	0.025	2.317	2.292	2.270	2.251	2.232
	0.010	2.716	2.683	2.653	2.626	2.602
	0.005	3.028	2.988	2.952	2.919	2.890
	0.001	3.795	3.738	3.687	3.640	3.598

(Table continued)

TABLE A.12p. Continued

Denominator ν	α	Numerator ν				
		16	17	18	19	20
29	0.050	2.007	1.989	1.973	1.958	1.945
	0.025	2.298	2.273	2.251	2.231	2.213
	0.010	2.689	2.656	2.626	2.599	2.574
	0.005	2.993	2.953	2.917	2.885	2.855
	0.001	3.740	3.683	3.632	3.585	3.543
30	0.050	1.995	1.976	1.960	1.945	1.932
	0.025	2.280	2.255	2.233	2.213	2.195
	0.010	2.663	2.630	2.600	2.573	2.549
	0.005	2.961	2.921	2.885	2.853	2.823
	0.001	3.689	3.632	3.581	3.535	3.493

TABLE A.12q. CRITICAL F VALUES

Denominator ν	α	Numerator ν				
		21	22	23	24	25
21	0.050	2.084	2.073	2.063	2.054	2.045
	0.025	2.409	2.394	2.380	2.368	2.356
	0.010	2.857	2.837	2.818	2.801	2.785
	0.005	3.216	3.191	3.168	3.147	3.128
	0.001	4.127	4.091	4.058	4.027	3.999
22	0.050	2.059	2.048	2.038	2.028	2.020
	0.025	2.373	2.358	2.344	2.331	2.320
	0.010	2.805	2.785	2.766	2.749	2.733
	0.005	3.149	3.125	3.102	3.081	3.061
	0.001	4.019	3.983	3.949	3.919	3.891
23	0.050	2.036	2.025	2.014	2.005	1.996
	0.025	2.340	2.325	2.312	2.299	2.287
	0.010	2.758	2.738	2.719	2.702	2.686
	0.005	3.089	3.065	3.042	3.021	3.001
	0.001	3.921	3.886	3.853	3.822	3.794
24	0.050	2.015	2.003	1.993	1.984	1.975
	0.025	2.311	2.296	2.282	2.269	2.257
	0.010	2.716	2.695	2.676	2.659	2.643
	0.005	3.035	3.011	2.988	2.967	2.947
	0.001	3.834	3.799	3.766	3.735	3.707
25	0.050	1.995	1.984	1.974	1.964	1.955
	0.025	2.284	2.269	2.255	2.242	2.230
	0.010	2.677	2.657	2.638	2.620	2.604

TABLE A.12p. Continued

Denominator ν	α	Numerator ν				
		21	22	23	24	25
	0.005	2.986	2.961	2.939	2.918	2.898
	0.001	3.756	3.720	3.687	3.657	3.629
26	0.050	1.978	1.966	1.956	1.946	1.938
	0.025	2.259	2.244	2.230	2.217	2.205
	0.010	2.642	2.621	2.602	2.585	2.569
	0.005	2.941	2.917	2.894	2.873	2.853
	0.001	3.684	3.649	3.616	3.586	3.558
27	0.050	1.961	1.950	1.940	1.930	1.921
	0.025	2.237	2.222	2.208	2.195	2.183
	0.010	2.609	2.589	2.570	2.552	2.536
	0.005	2.900	2.876	2.853	2.832	2.812
	0.001	3.619	3.584	3.551	3.521	3.493
28	0.050	1.946	1.935	1.924	1.915	1.906
	0.025	2.216	2.201	2.187	2.174	2.161
	0.010	2.579	2.559	2.540	2.522	2.506
	0.005	2.863	2.838	2.815	2.794	2.775
	0.001	3.560	3.524	3.492	3.462	3.434
29	0.050	1.932	1.921	1.910	1.901	1.891
	0.025	2.196	2.181	2.167	2.154	2.142
	0.010	2.552	2.531	2.512	2.495	2.478
	0.005	2.828	2.803	2.780	2.759	2.740
	0.001	3.505	3.470	3.437	3.407	3.380
30	0.050	1.919	1.908	1.897	1.887	1.878
	0.025	2.178	2.163	2.149	2.136	2.124
	0.010	2.526	2.506	2.487	2.469	2.453
	0.005	2.796	2.771	2.748	2.727	2.708
	0.001	3.454	3.419	3.387	3.357	3.330

TABLE A.12r. CRITICAL F VALUES

Denominator ν	α	Numerator ν				
		26	27	28	29	30
21	0.050	2.037	2.030	2.023	2.016	2.010
	0.025	2.345	2.335	2.325	2.317	2.308
	0.010	2.770	2.756	2.743	2.731	2.720
	0.005	3.110	3.093	3.077	3.063	3.049
	0.001	3.972	3.948	3.925	3.904	3.884
22	0.050	2.012	2.004	1.997	1.990	1.984
	0.025	2.309	2.299	2.289	2.280	2.272
	0.010	2.718	2.704	2.691	2.679	2.667

(Table continued)

| Denominator | | Numerator ν | | | | |
ν	α	26	27	28	29	30
	0.005	3.043	3.026	3.011	2.996	2.982
	0.001	3.864	3.840	3.817	3.796	3.776
23	0.050	1.988	1.981	1.973	1.967	1.961
	0.025	2.276	2.266	2.256	2.247	2.239
	0.010	2.671	2.657	2.644	2.632	2.620
	0.005	2.983	2.966	2.951	2.936	2.922
	0.001	3.768	3.744	3.721	3.700	3.680
24	0.050	1.967	1.959	1.952	1.945	1.939
	0.025	2.246	2.236	2.226	2.217	2.209
	0.010	2.628	2.614	2.601	2.589	2.577
	0.005	2.929	2.912	2.897	2.882	2.868
	0.001	3.681	3.657	3.634	3.613	3.593
25	0.050	1.947	1.939	1.932	1.926	1.919
	0.025	2.219	2.209	2.199	2.190	2.182
	0.010	2.589	2.575	2.562	2.550	2.538
	0.005	2.880	2.863	2.847	2.833	2.819
	0.001	3.603	3.579	3.556	3.535	3.515
26	0.050	1.929	1.921	1.914	1.907	1.901
	0.025	2.194	2.184	2.174	2.165	2.157
	0.010	2.554	2.540	2.526	2.514	2.503
	0.005	2.835	2.818	2.802	2.788	2.774
	0.001	3.532	3.508	3.486	3.464	3.445
27	0.050	1.913	1.905	1.898	1.891	1.884
	0.025	2.171	2.161	2.151	2.142	2.133
	0.010	2.521	2.507	2.494	2.481	2.470
	0.005	2.794	2.777	2.761	2.747	2.733
	0.001	3.467	3.443	3.421	3.400	3.380
28	0.050	1.897	1.889	1.882	1.875	1.869
	0.025	2.150	2.140	2.130	2.121	2.112
	0.010	2.491	2.477	2.464	2.451	2.440
	0.005	2.756	2.739	2.724	2.709	2.695
	0.001	3.408	3.384	3.362	3.341	3.321
29	0.050	1.883	1.875	1.868	1.861	1.854
	0.025	2.131	2.120	2.110	2.101	2.092
	0.010	2.463	2.449	2.436	2.423	2.412
	0.005	2.722	2.705	2.689	2.674	2.660
	0.001	3.354	3.330	3.308	3.287	3.267
30	0.050	1.870	1.862	1.854	1.847	1.841
	0.025	2.112	2.102	2.092	2.083	2.074
	0.010	2.437	2.423	2.410	2.398	2.386
	0.005	2.689	2.672	2.657	2.642	2.628
	0.001	3.304	3.280	3.258	3.237	3.217

TABLE A.12s. CRITICAL *F* VALUES

Denominator ν	α	Numerator ν				
		1	2	3	4	5
40	0.050	4.085	3.232	2.839	2.606	2.449
	0.025	5.424	4.051	3.463	3.126	2.904
	0.010	7.314	5.179	4.313	3.828	3.514
	0.005	8.828	6.066	4.976	4.374	3.986
	0.001	12.609	8.251	6.595	5.698	5.128
45	0.050	4.057	3.204	2.812	2.579	2.422
	0.025	5.377	4.009	3.422	3.086	2.864
	0.010	7.234	5.110	4.249	3.767	3.454
	0.005	8.715	5.974	4.892	4.294	3.909
	0.001	12.392	8.086	6.450	5.564	5.001
50	0.050	4.034	3.183	2.790	2.557	2.400
	0.025	5.340	3.975	3.390	3.054	2.833
	0.010	7.171	5.057	4.199	3.720	3.408
	0.005	8.626	5.902	4.826	4.232	3.849
	0.001	12.222	7.956	6.336	5.459	4.901
60	0.050	4.001	3.150	2.758	2.525	2.368
	0.025	5.286	3.925	3.343	3.008	2.786
	0.010	7.077	4.977	4.126	3.649	3.339
	0.005	8.495	5.795	4.729	4.140	3.760
	0.001	11.973	7.768	6.171	5.307	4.757
70	0.050	3.978	3.128	2.736	2.503	2.346
	0.025	5.247	3.890	3.309	2.975	2.754
	0.010	7.011	4.922	4.074	3.600	3.291
	0.005	8.403	5.720	4.661	4.076	3.698
	0.001	11.799	7.637	6.057	5.201	4.656
80	0.050	3.960	3.111	2.719	2.486	2.329
	0.025	5.218	3.864	3.284	2.950	2.730
	0.010	6.963	4.881	4.036	3.563	3.255
	0.005	8.335	5.665	4.611	4.029	3.652
	0.001	11.671	7.540	5.972	5.123	4.582
90	0.050	3.947	3.098	2.706	2.473	2.316
	0.025	5.196	3.844	3.265	2.932	2.711
	0.010	6.925	4.849	4.007	3.535	3.228
	0.005	8.282	5.623	4.573	3.992	3.617
	0.001	11.573	7.466	5.908	5.064	4.526
100	0.050	3.936	3.087	2.696	2.463	2.305
	0.025	5.179	3.828	3.250	2.917	2.696
	0.010	6.895	4.824	3.984	3.513	3.206

(Table continued)

TABLE A.12s. Continued

Denominator ν	α	Numerator ν				
		1	2	3	4	5
	0.005	8.241	5.589	4.542	3.963	3.589
	0.001	11.495	7.408	5.857	5.017	4.482
150	0.050	3.904	3.056	2.665	2.432	2.274
	0.025	5.126	3.781	3.204	2.872	2.652
	0.010	6.807	4.749	3.915	3.447	3.142
	0.005	8.118	5.490	4.453	3.878	3.508
	0.001	11.267	7.236	5.707	4.879	4.351
200	0.050	3.888	3.041	2.650	2.417	2.259
	0.025	5.100	3.758	3.182	2.850	2.630
	0.010	6.763	4.713	3.881	3.414	3.110
	0.005	8.0507	5.441	4.408	3.837	3.467
	0.001	11.155	7.152	5.634	4.812	4.287

TABLE A.12t. CRITICAL F VALUES

Denominator ν	α	Numerator ν				
		6	7	8	9	10
40	0.050	2.336	2.249	2.180	2.124	2.077
	0.025	2.744	2.624	2.529	2.452	2.388
	0.010	3.291	3.124	2.993	2.888	2.801
	0.005	3.713	3.509	3.350	3.222	3.117
	0.001	4.731	4.436	4.207	4.024	3.874
45	0.050	2.308	2.221	2.152	2.096	2.049
	0.025	2.705	2.584	2.489	2.412	2.348
	0.010	3.232	3.066	2.935	2.830	2.743
	0.005	3.638	3.435	3.276	3.149	3.044
	0.001	4.608	4.316	4.090	3.909	3.760
50	0.050	2.286	2.199	2.130	2.073	2.026
	0.025	2.674	2.553	2.458	2.381	2.317
	0.010	3.186	3.020	2.890	2.785	2.698
	0.005	3.579	3.376	3.219	3.092	2.988
	0.001	4.512	4.222	3.998	3.818	3.671
60	0.050	2.254	2.167	2.097	2.040	1.993
	0.025	2.627	2.507	2.412	2.334	2.270

Denominator ν	α	Numerator ν				
		6	7	8	9	10
	0.010	3.119	2.953	2.823	2.718	2.632
	0.005	3.492	3.291	3.134	3.008	2.904
	0.001	4.372	4.086	3.865	3.687	3.541
70	0.050	2.231	2.143	2.074	2.017	1.969
	0.025	2.595	2.474	2.379	2.302	2.237
	0.010	3.071	2.906	2.777	2.672	2.585
	0.005	3.431	3.232	3.075	2.950	2.846
	0.001	4.275	3.992	3.773	3.596	3.452
80	0.050	2.214	2.126	2.056	1.999	1.951
	0.025	2.571	2.450	2.355	2.277	2.213
	0.010	3.036	2.871	2.742	2.637	2.551
	0.005	3.387	3.188	3.032	2.907	2.803
	0.001	4.204	3.923	3.705	3.530	3.386
90	0.050	2.201	2.113	2.043	1.986	1.938
	0.025	2.552	2.432	2.336	2.259	2.194
	0.010	3.009	2.845	2.715	2.611	2.524
	0.005	3.352	3.154	2.999	2.873	2.770
	0.001	4.150	3.870	3.653	3.479	3.336
100	0.050	2.191	2.103	2.032	1.975	1.927
	0.025	2.537	2.417	2.321	2.244	2.179
	0.010	2.988	2.823	2.694	2.590	2.503
	0.005	3.325	3.127	2.972	2.847	2.744
	0.001	4.107	3.829	3.612	3.439	3.296
150	0.050	2.160	2.071	2.001	1.943	1.894
	0.025	2.494	2.373	2.278	2.200	2.135
	0.010	2.924	2.761	2.632	2.528	2.441
	0.005	3.245	3.048	2.894	2.770	2.667
	0.001	3.981	3.706	3.493	3.321	3.179
200	0.050	2.144	2.056	1.985	1.927	1.878
	0.025	2.472	2.351	2.256	2.178	2.113
	0.010	2.893	2.730	2.601	2.497	2.411
	0.005	3.206	3.010	2.856	2.732	2.629
	0.001	3.920	3.647	3.434	3.264	3.123

TABLE A.12u. CRITICAL *F* VALUES

Denominator ν	α	Numerator ν				
		11	12	13	14	15
40	0.050	2.038	2.003	1.974	1.948	1.924
	0.025	2.334	2.288	2.248	2.213	2.182
	0.010	2.727	2.665	2.611	2.563	2.522
	0.005	3.028	2.953	2.888	2.831	2.781
	0.001	3.749	3.642	3.551	3.471	3.400
45	0.050	2.009	1.974	1.945	1.918	1.895
	0.025	2.294	2.248	2.208	2.172	2.141
	0.010	2.670	2.608	2.553	2.506	2.464
	0.005	2.956	2.881	2.816	2.759	2.709
	0.001	3.636	3.530	3.439	3.360	3.290
50	0.050	1.986	1.952	1.921	1.895	1.871
	0.025	2.263	2.216	2.176	2.140	2.109
	0.010	2.625	2.562	2.508	2.461	2.419
	0.005	2.900	2.825	2.760	2.703	2.653
	0.001	3.548	3.443	3.352	3.273	3.204
60	0.050	1.952	1.917	1.887	1.860	1.836
	0.025	2.216	2.169	2.129	2.093	2.061
	0.010	2.559	2.496	2.442	2.394	2.352
	0.005	2.817	2.742	2.677	2.620	2.570
	0.001	3.419	3.315	3.226	3.147	3.078
70	0.050	1.928	1.893	1.863	1.836	1.812
	0.025	2.183	2.136	2.095	2.059	2.028
	0.010	2.512	2.450	2.395	2.348	2.306
	0.005	2.759	2.684	2.619	2.563	2.513
	0.001	3.330	3.227	3.138	3.060	2.991
80	0.050	1.910	1.875	1.845	1.817	1.793
	0.025	2.158	2.111	2.071	2.035	2.003
	0.010	2.478	2.415	2.361	2.313	2.271
	0.005	2.716	2.641	2.577	2.520	2.470
	0.001	3.265	3.162	3.074	2.996	2.927
90	0.050	1.897	1.861	1.830	1.803	1.779
	0.025	2.140	2.092	2.051	2.015	1.983
	0.010	2.451	2.389	2.334	2.286	2.244
	0.005	2.683	2.608	2.544	2.487	2.437
	0.001	3.215	3.113	3.024	2.947	2.879
100	0.050	1.886	1.850	1.819	1.792	1.768
	0.025	2.124	2.077	2.036	2.000	1.968
	0.010	2.430	2.368	2.313	2.265	2.223

Denominator ν	α	Numerator ν				
		11	12	13	14	15
	0.005	2.657	2.583	2.518	2.461	2.411
	0.001	3.176	3.074	2.986	2.908	2.840
150	0.050	1.853	1.817	1.786	1.758	1.734
	0.025	2.080	2.032	1.991	1.955	1.922
	0.010	2.368	2.305	2.251	2.203	2.160
	0.005	2.580	2.506	2.441	2.385	2.335
	0.001	3.061	2.959	2.872	2.795	2.727
200	0.050	1.837	1.801	1.769	1.742	1.717
	0.025	2.058	2.010	1.969	1.932	1.900
	0.010	2.338	2.275	2.220	2.172	2.129
	0.005	2.543	2.468	2.404	2.347	2.297
	0.001	3.005	2.904	2.816	2.740	2.672

TABLE A.12v. CRITICAL F VALUES

Denominator ν	α	Numerator ν				
		16	17	18	19	20
40	0.050	1.904	1.885	1.868	1.853	1.839
	0.025	2.154	2.129	2.107	2.086	2.068
	0.010	2.484	2.451	2.421	2.394	2.369
	0.005	2.737	2.697	2.661	2.628	2.598
	0.001	3.338	3.282	3.232	3.186	3.145
45	0.050	1.874	1.855	1.838	1.823	1.808
	0.025	2.113	2.088	2.066	2.045	2.026
	0.010	2.427	2.393	2.363	2.336	2.311
	0.005	2.665	2.625	2.589	2.556	2.527
	0.001	3.228	3.172	3.122	3.077	3.036
50	0.050	1.850	1.831	1.814	1.798	1.784
	0.025	2.081	2.056	2.033	2.012	1.993
	0.010	2.382	2.348	2.318	2.290	2.265
	0.005	2.609	2.569	2.533	2.500	2.470
	0.001	3.142	3.086	3.037	2.992	2.951
60	0.050	1.815	1.796	1.778	1.763	1.748
	0.025	2.033	2.008	1.985	1.964	1.944
	0.010	2.315	2.281	2.251	2.223	2.198

(*Table continued*)

TABLE A.12v. Continued

Denominator ν	α	Numerator ν				
		16	17	18	19	20
	0.005	2.526	2.486	2.450	2.417	2.387
	0.001	3.017	2.962	2.912	2.867	2.827
70	0.050	1.790	1.771	1.753	1.737	1.722
	0.025	1.999	1.974	1.950	1.929	1.910
	0.010	2.268	2.234	2.204	2.176	2.150
	0.005	2.468	2.428	2.392	2.359	2.329
	0.001	2.930	2.875	2.826	2.781	2.741
80	0.050	1.772	1.752	1.734	1.718	1.703
	0.025	1.974	1.948	1.925	1.904	1.884
	0.010	2.233	2.199	2.169	2.141	2.115
	0.005	2.425	2.385	2.349	2.316	2.286
	0.001	2.867	2.812	2.763	2.718	2.677
90	0.050	1.757	1.737	1.720	1.703	1.688
	0.025	1.955	1.929	1.905	1.884	1.864
	0.010	2.206	2.172	2.142	2.114	2.088
	0.005	2.393	2.353	2.316	2.283	2.253
	0.001	2.818	2.763	2.714	2.670	2.629
100	0.050	1.746	1.726	1.708	1.691	1.676
	0.025	1.939	1.913	1.890	1.868	1.849
	0.010	2.185	2.151	2.120	2.092	2.067
	0.005	2.367	2.326	2.290	2.257	2.227
	0.001	2.780	2.725	2.676	2.632	2.591
150	0.050	1.711	1.691	1.673	1.656	1.641
	0.025	1.893	1.867	1.843	1.821	1.801
	0.010	2.122	2.088	2.057	2.029	2.003
	0.005	2.290	2.250	2.213	2.180	2.150
	0.001	2.667	2.613	2.564	2.519	2.479
200	0.050	1.694	1.674	1.656	1.639	1.623
	0.025	1.870	1.844	1.820	1.798	1.778
	0.010	2.091	2.057	2.026	1.997	1.971
	0.005	2.252	2.212	2.175	2.142	2.112
	0.001	2.612	2.558	2.509	2.465	2.424

TABLE A.12w. CRITICAL *F* VALUES

Denominator ν	α	Numerator ν				
		21	22	23	24	25
40	0.050	1.826	1.814	1.803	1.793	1.783
	0.025	2.051	2.035	2.020	2.007	1.994
	0.010	2.346	2.325	2.306	2.288	2.271
	0.005	2.571	2.546	2.523	2.502	2.482
	0.001	3.107	3.073	3.041	3.011	2.984
45	0.050	1.795	1.783	1.772	1.762	1.752
	0.025	2.009	1.993	1.978	1.965	1.952
	0.010	2.288	2.267	2.248	2.230	2.213
	0.005	2.499	2.474	2.451	2.430	2.410
	0.001	2.998	2.964	2.932	2.902	2.875
50	0.050	1.771	1.759	1.748	1.737	1.727
	0.025	1.976	1.960	1.945	1.931	1.919
	0.010	2.242	2.221	2.202	2.183	2.167
	0.005	2.443	2.418	2.395	2.373	2.353
	0.001	2.913	2.879	2.847	2.817	2.790
60	0.050	1.735	1.722	1.711	1.700	1.690
	0.025	1.927	1.911	1.896	1.882	1.869
	0.010	2.175	2.153	2.134	2.115	2.098
	0.005	2.360	2.335	2.311	2.290	2.270
	0.001	2.789	2.755	2.723	2.694	2.667
70	0.050	1.709	1.696	1.685	1.674	1.664
	0.025	1.892	1.876	1.861	1.847	1.833
	0.010	2.127	2.106	2.086	2.067	2.050
	0.005	2.302	2.276	2.253	2.231	2.211
	0.001	2.703	2.669	2.637	2.608	2.581
80	0.050	1.689	1.677	1.665	1.654	1.644
	0.025	1.866	1.850	1.835	1.820	1.807
	0.010	2.092	2.070	2.050	2.032	2.015
	0.005	2.259	2.233	2.210	2.188	2.168
	0.001	2.640	2.606	2.574	2.545	2.518
90	0.050	1.675	1.662	1.650	1.639	1.629
	0.025	1.846	1.830	1.814	1.800	1.787
	0.010	2.065	2.043	2.023	2.004	1.987
	0.005	2.226	2.200	2.177	2.155	2.134
	0.001	2.592	2.558	2.526	2.497	2.469
100	0.050	1.663	1.650	1.638	1.627	1.616
	0.025	1.830	1.814	1.798	1.784	1.770
	0.010	2.043	2.021	2.001	1.983	1.965

(*Table continued*)

TABLE A.12w. Continued

Denominator ν	α	Numerator ν				
		21	22	23	24	25
	0.005	2.199	2.174	2.150	2.128	2.108
	0.001	2.554	2.519	2.488	2.458	2.431
150	0.050	1.627	1.614	1.602	1.590	1.580
	0.025	1.783	1.766	1.750	1.736	1.722
	0.010	1.979	1.957	1.937	1.918	1.900
	0.005	2.122	2.096	2.072	2.050	2.030
	0.001	2.442	2.407	2.376	2.346	2.319
200	0.050	1.609	1.596	1.583	1.572	1.561
	0.025	1.759	1.742	1.726	1.712	1.698
	0.010	1.947	1.925	1.905	1.886	1.868
	0.005	2.084	2.058	2.034	2.012	1.991
	0.001	2.387	2.353	2.321	2.292	2.264

TABLE A.12x. CRITICAL *F* VALUES

Denominator ν	α	Numerator ν				
		26	27	28	29	30
40	0.050	1.775	1.766	1.759	1.751	1.744
	0.025	1.983	1.972	1.962	1.952	1.943
	0.010	2.256	2.241	2.228	2.215	2.203
	0.005	2.464	2.447	2.431	2.416	2.401
	0.001	2.958	2.935	2.912	2.892	2.872
45	0.050	1.743	1.735	1.727	1.720	1.713
	0.025	1.940	1.929	1.919	1.909	1.900
	0.010	2.197	2.183	2.169	2.156	2.144
	0.005	2.392	2.374	2.358	2.343	2.329
	0.001	2.850	2.826	2.804	2.783	2.763
50	0.050	1.718	1.710	1.702	1.694	1.687
	0.025	1.907	1.895	1.885	1.875	1.866
	0.010	2.151	2.136	2.123	2.110	2.098
	0.005	2.335	2.317	2.301	2.286	2.272
	0.001	2.765	2.741	2.719	2.698	2.679
60	0.050	1.681	1.672	1.664	1.656	1.649
	0.025	1.857	1.845	1.835	1.825	1.815
	0.010	2.083	2.068	2.054	2.041	2.028

TABLE A.12x. Continued

Denominator ν	α	Numerator ν				
		26	27	28	29	30
	0.005	2.251	2.234	2.217	2.202	2.187
	0.001	2.641	2.617	2.595	2.574	2.555
70	0.050	1.654	1.646	1.637	1.629	1.622
	0.025	1.821	1.810	1.799	1.789	1.779
	0.010	2.034	2.019	2.005	1.992	1.980
	0.005	2.192	2.175	2.158	2.143	2.128
	0.001	2.555	2.532	2.509	2.489	2.469
80	0.050	1.634	1.626	1.617	1.609	1.602
	0.025	1.795	1.783	1.772	1.762	1.752
	0.010	1.999	1.983	1.969	1.956	1.944
	0.005	2.149	2.131	2.115	2.099	2.084
	0.001	2.492	2.468	2.446	2.425	2.406
90	0.050	1.619	1.610	1.601	1.593	1.586
	0.025	1.774	1.763	1.752	1.741	1.731
	0.010	1.971	1.956	1.942	1.928	1.916
	0.005	2.115	2.098	2.081	2.065	2.051
	0.001	2.444	2.420	2.398	2.377	2.357
100	0.050	1.607	1.598	1.589	1.581	1.573
	0.025	1.758	1.746	1.735	1.725	1.715
	0.010	1.949	1.934	1.919	1.906	1.893
	0.005	2.089	2.071	2.054	2.039	2.024
	0.001	2.406	2.382	2.360	2.339	2.319
150	0.050	1.570	1.560	1.552	1.543	1.535
	0.025	1.709	1.697	1.686	1.675	1.665
	0.010	1.884	1.868	1.854	1.840	1.827
	0.005	2.010	1.992	1.975	1.959	1.944
	0.001	2.293	2.270	2.247	2.226	2.206
200	0.050	1.551	1.542	1.533	1.524	1.516
	0.025	1.685	1.673	1.661	1.650	1.640
	0.010	1.851	1.836	1.821	1.807	1.794
	0.005	1.972	1.953	1.936	1.920	1.905
	0.001	2.239	2.215	2.192	2.171	2.151

TABLE A.13a. FISHER'S z TRANSFORMATION

r	0.00	0.01	0.02	0.03	0.04	0.05	0.06	0.07	0.08	0.09
0.0	0.000	0.010	0.020	0.030	0.040	0.050	0.060	0.070	0.080	0.090
0.1	0.100	0.110	0.121	0.131	0.141	0.151	0.161	0.172	0.182	0.192
0.2	0.203	0.213	0.224	0.234	0.245	0.255	0.266	0.277	0.288	0.299
0.3	0.310	0.321	0.332	0.343	0.354	0.365	0.377	0.388	0.400	0.412
0.4	0.424	0.436	0.448	0.460	0.472	0.485	0.497	0.510	0.523	0.536
0.5	0.549	0.563	0.576	0.590	0.604	0.618	0.633	0.648	0.662	0.678
0.6	0.693	0.709	0.725	0.741	0.758	0.775	0.793	0.811	0.829	0.848
0.7	0.867	0.887	0.908	0.929	0.950	0.973	0.996	1.020	1.045	1.071
0.8	1.099	1.127	1.157	1.188	1.221	1.256	1.293	1.333	1.376	1.422

r	0.000	0.001	0.002	0.003	0.004	0.005	0.006	0.007	0.008	0.009
0.90	1.472	1.478	1.483	1.488	1.494	1.499	1.505	1.510	1.516	1.522
0.91	1.528	1.533	1.539	1.545	1.551	1.557	1.564	1.570	1.576	1.583
0.92	1.589	1.596	1.602	1.609	1.616	1.623	1.630	1.637	1.644	1.651
0.93	1.658	1.666	1.673	1.681	1.689	1.697	1.705	1.713	1.721	1.730
0.94	1.738	1.747	1.756	1.764	1.774	1.783	1.792	1.802	1.812	1.822
0.95	1.832	1.842	1.853	1.863	1.874	1.886	1.897	1.909	1.921	1.933
0.96	1.946	1.959	1.972	1.986	2.000	2.014	2.029	2.044	2.060	2.076
0.97	2.092	2.109	2.127	2.146	2.165	2.185	2.205	2.227	2.249	2.273
0.98	2.298	2.323	2.351	2.380	2.410	2.443	2.477	2.515	2.555	2.599
0.99	2.647	2.700	2.759	2.826	2.903	2.994	3.106	3.250	3.453	3.800

Tabular value is $z_r = \log_e \sqrt{(1+r)/(1-r)}$. For example, $z_{0.35} = \log_e \sqrt{(1+0.35)/(1-0.35)} = 0.365$.

TABLE A.13b. INVERSE OF FISHER'S z TRANSFORMATION

z	.00	.01	.02	.03	.04	.05	.06	.07	.08	.09
0.0	.000	.010	.020	.030	.040	.050	.060	.070	.080	.090
0.1	.100	.110	.119	.129	.139	.149	.159	.168	.178	.188
0.2	.197	.207	.217	.226	.235	.245	.254	.264	.273	.282
0.3	.291	.300	.310	.319	.327	.336	.345	.354	.363	.371
0.4	.380	.388	.397	.405	.414	.422	.430	.438	.446	.454
0.5	.462	.470	.478	.485	.493	.501	.508	.515	.523	.530
0.6	.537	.544	.551	.558	.565	.572	.578	.585	.592	.598
0.7	.604	.611	.617	.623	.629	.635	.641	.647	.653	.658

TABLE A.13b. Continued

z	.00	.01	.02	.03	.04	.05	.06	.07	.08	.09
0.8	.664	.670	.675	.680	.686	.691	.696	.701	.706	.711
0.9	.716	.721	.726	.731	.735	.740	.744	.749	.753	.757
1.0	.762	.766	.770	.774	.778	.782	.786	.789	.793	.797
1.1	.800	.804	.808	.811	.814	.818	.821	.824	.827	.831
1.2	.834	.837	.840	.843	.845	.848	.851	.854	.856	.859
1.3	.862	.864	.867	.869	.872	.874	.876	.879	.881	.883
1.4	.885	.887	.890	.892	.894	.896	.898	.900	.901	.903
1.5	.905	.907	.909	.910	.912	.914	.915	.917	.919	.920
1.6	.922	.923	.925	.926	.927	.929	.930	.932	.933	.934
1.7	.935	.937	.938	.939	.940	.941	.943	.944	.945	.946
1.8	.947	.948	.949	.950	.951	.952	.953	.954	.954	.955
1.9	.956	.957	.958	.959	.960	.960	.961	.962	.963	.963
2.0	.964	.965	.965	.966	.967	.967	.968	.969	.969	.970
2.1	.970	.971	.972	.972	.973	.973	.974	.974	.975	.975
2.2	.976	.976	.977	.977	.978	.978	.978	.979	.979	.980
2.3	.980	.980	.981	.981	.982	.982	.982	.983	.983	.983
2.4	.984	.984	.984	.985	.985	.985	.986	.986	.986	.986
2.5	.987	.987	.987	.987	.988	.988	.988	.988	.989	.989
2.6	.989	.989	.989	.990	.990	.990	.990	.990	.991	.991
2.7	.991	.991	.991	.992	.992	.992	.992	.992	.992	.992
2.8	.993	.993	.993	.993	.993	.993	.993	.994	.994	.994
2.9	.994	.994	.994	.994	.994	.995	.995	.995	.995	.995

Tabular value is r. For example, if $z_r = 1.72$, then $r = 0.938$.

TABLE A.14a. CRITICAL VALUES FOR DUNCAN'S NEW MULTIPLE RANGE TEST $\alpha = 0.05$

v \ r	2	3	4	5	6	7	8	9	10
1	17.97	17.97	17.97	17.97	17.97	17.97	17.97	17.97	17.97
2	6.085	6.085	6.085	6.085	6.085	6.085	6.085	6.085	6.085
3	4.501	4.516	4.516	4.516	4.516	4.516	4.516	4.516	4.516
4	3.927	4.013	4.033	4.033	4.033	4.033	4.033	4.033	4.033
5	3.635	3.749	3.797	3.814	3.814	3.814	3.814	3.814	3.814
6	3.461	3.587	3.649	3.680	3.694	3.697	3.697	3.697	3.697
7	3.344	3.477	3.548	3.588	3.611	3.622	3.626	3.626	3.626
8	3.261	3.399	3.475	3.521	3.549	3.566	3.575	3.579	3.579
9	3.199	3.339	3.420	3.470	3.502	3.523	3.536	3.544	3.547
10	3.151	3.293	3.376	3.430	3.465	3.489	3.505	3.516	3.522
11	3.113	3.256	3.342	3.397	3.435	3.462	3.480	3.493	3.501
12	3.082	3.225	3.313	3.370	3.410	3.439	3.459	3.474	3.484
13	3.055	3.200	3.289	3.348	3.389	3.419	3.442	3.458	3.470
14	3.033	3.178	3.268	3.329	3.372	3.403	3.426	3.444	3.457
15	3.014	3.160	3.250	3.312	3.356	3.389	3.413	3.432	3.446
16	2.998	3.144	3.235	3.298	3.343	3.376	3.402	3.422	3.437
17	2.984	3.130	3.222	3.285	3.331	3.366	3.392	3.412	3.429
18	2.971	3.118	3.210	3.274	3.321	3.356	3.383	3.405	3.421
19	2.960	3.107	3.199	3.264	3.311	3.347	3.375	3.397	3.415
20	2.950	3.097	3.190	3.255	3.303	3.339	3.368	3.391	3.409
24	2.919	3.066	3.160	3.226	3.276	3.315	3.345	3.370	3.390
30	2.888	3.035	3.131	3.199	3.250	3.290	3.322	3.349	3.371
40	2.858	3.006	3.102	3.171	3.224	3.266	3.300	3.328	3.352
60	2.829	2.976	3.073	3.143	3.198	3.241	3.277	3.307	3.333
120	2.800	2.947	3.045	3.116	3.172	3.217	3.254	3.287	3.314
INF	2.772	2.918	3.017	3.089	3.146	3.193	3.232	3.265	3.294

v \ r	11	12	13	14	15	16	17	18	19
1	17.97	17.97	17.97	17.97	17.97	17.97	17.97	17.97	17.97
2	6.085	6.085	6.085	6.085	6.085	6.085	6.085	6.085	6.085
3	4.516	4.516	4.516	4.516	4.516	4.516	4.516	4.516	4.516
4	4.033	4.033	4.033	4.033	4.033	4.033	4.033	4.033	4.033
5	3.814	3.814	3.814	3.814	3.814	3.814	3.814	3.814	3.814
6	3.697	3.697	3.697	3.697	3.697	3.697	3.697	3.697	3.697
7	3.626	3.626	3.626	3.626	3.626	3.626	3.626	3.626	3.626
8	3.579	3.579	3.579	3.579	3.579	3.579	3.579	3.579	3.579
9	3.547	3.547	3.547	3.547	3.547	3.547	3.547	3.547	3.547
10	3.525	3.526	3.526	3.526	3.526	3.526	3.526	3.526	3.526

TABLE A.14a. Continued

ν\r	11	12	13	14	15	16	17	18	19
11	3.506	3.509	3.510	3.510	3.510	3.510	3.510	3.510	3.510
12	3.491	3.496	3.498	3.499	3.499	3.499	3.499	3.499	3.499
13	3.478	3.484	3.488	3.490	3.490	3.490	3.490	3.490	3.490
14	3.467	3.474	3.479	3.482	3.484	3.484	3.485	3.485	3.485
15	3.457	3.465	3.471	3.476	3.478	3.480	3.481	3.481	3.481
16	3.449	3.458	3.465	3.470	3.473	3.477	3.478	3.478	3.478
17	3.441	3.451	3.459	3.465	3.469	3.473	3.475	3.476	3.476
18	3.435	3.445	3.454	3.460	3.465	3.470	3.472	3.474	3.474
19	3.429	3.440	3.449	3.456	3.462	3.467	3.470	3.472	3.473
20	3.424	3.436	3.445	3.453	3.459	3.464	3.467	3.470	3.472
24	3.406	3.420	3.432	3.441	3.449	3.456	3.461	3.465	3.469
30	3.389	3.405	3.418	3.430	3.439	3.447	3.454	3.460	3.466
40	3.373	3.390	3.405	3.418	3.429	3.439	3.448	3.456	3.463
60	3.355	3.374	3.391	3.406	3.419	3.431	3.442	3.451	3.460
120	3.337	3.359	3.377	3.394	3.409	3.423	3.435	3.446	3.457
INF	3.320	3.343	3.363	3.382	3.399	3.414	3.428	3.442	3.454

ν\r	20	22	24	26	28	30	32	34	36
1	17.97	17.97	17.97	17.97	17.97	17.97	17.97	17.97	17.97
2	6.085	6.085	6.085	6.085	6.085	6.085	6.085	6.085	6.085
3	4.516	4.516	4.516	4.516	4.516	4.516	4.516	4.516	4.516
4	4.033	4.033	4.033	4.033	4.033	4.033	4.033	4.033	4.033
5	3.814	3.814	3.814	3.814	3.814	3.814	3.814	3.814	3.814
6	3.697	3.697	3.697	3.697	3.697	3.697	3.697	3.697	3.697
7	3.626	3.626	3.626	3.626	3.626	3.626	3.626	3.626	3.626
8	3.579	3.579	3.579	3.579	3.579	3.579	3.579	3.579	3.579
9	3.547	3.547	3.547	3.547	3.547	3.547	3.547	3.547	3.547
10	3.526	3.526	3.526	3.526	3.526	3.526	3.526	3.526	3.526
11	3.510	3.510	3.510	3.510	3.510	3.510	3.510	3.510	3.510
12	3.499	3.499	3.499	3.499	3.499	3.499	3.499	3.499	3.499
13	3.490	3.490	3.490	3.490	3.490	3.490	3.490	3.490	3.490
14	3.485	3.485	3.485	3.485	3.485	3.485	3.485	3.485	3.485
15	3.481	3.481	3.481	3.481	3.481	3.481	3.481	3.481	3.481
16	3.478	3.478	3.478	3.478	3.478	3.478	3.478	3.478	3.478
17	3.476	3.476	3.476	3.476	3.476	3.476	3.476	3.476	3.476
18	3.474	3.474	3.474	3.474	3.474	3.474	3.474	3.474	3.474
19	3.474	3.474	3.474	3.474	3.474	3.474	3.474	3.474	3.474
20	3.473	3.474	3.474	3.474	3.474	3.474	3.474	3.474	3.474
24	3.471	3.475	3.477	3.477	3.477	3.477	3.477	3.477	3.477
30	3.470	3.477	3.481	3.484	3.486	3.486	3.486	3.486	3.486
40	3.469	3.479	3.486	3.492	3.497	3.500	3.503	3.504	3.504
60	3.467	3.481	3.492	3.501	3.509	3.515	3.521	3.525	3.529
120	3.466	3.483	3.498	3.511	3.522	3.532	3.541	3.548	3.555

(Table continued)

INF   3.466   3.486   3.505   3.522   3.536   3.550   3.562   3.574   3.584

TABLE A.14a. Continued

ν \ r	38	40	50	60	70	80	90	100
1	17.97	17.97	17.97	17.97	17.97	17.97	17.97	17.97
2	6.085	6.085	6.085	6.085	6.085	6.085	6.085	6.085
3	4.516	4.516	4.516	4.516	4.516	4.516	4.516	4.516
4	4.033	4.033	4.033	4.033	4.033	4.033	4.033	4.033
5	3.814	3.814	3.814	3.814	3.814	3.814	3.814	3.814
6	3.697	3.697	3.697	3.697	3.697	3.697	3.697	3.697
7	3.626	3.626	3.626	3.626	3.626	3.626	3.626	3.626
8	3.579	3.579	3.579	3.579	3.579	3.579	3.579	3.579
9	3.547	3.547	3.547	3.547	3.547	3.547	3.547	3.547
10	3.526	3.526	3.526	3.526	3.526	3.526	3.526	3.526
11	3.510	3.510	3.510	3.510	3.510	3.510	3.510	3.510
12	3.499	3.499	3.499	3.499	3.499	3.499	3.499	3.499
13	3.490	3.490	3.490	3.490	3.490	3.490	3.490	3.490
14	3.485	3.485	3.485	3.485	3.485	3.485	3.485	3.485
15	3.481	3.481	3.481	3.481	3.481	3.481	3.481	3.481
16	3.478	3.478	3.478	3.478	3.478	3.478	3.478	3.478
17	3.476	3.476	3.476	3.476	3.476	3.476	3.476	3.476
18	3.474	3.474	3.474	3.474	3.474	3.474	3.474	3.474
19	3.474	3.474	3.474	3.474	3.474	3.474	3.474	3.474
20	3.474	3.474	3.474	3.474	3.474	3.474	3.474	3.474
24	3.477	3.477	3.477	3.477	3.477	3.477	3.477	3.477
30	3.486	3.486	3.486	3.486	3.486	3.486	3.486	3.486
40	3.504	3.504	3.504	3.504	3.504	3.504	3.504	3.504
60	3.531	3.534	3.537	3.537	3.537	3.537	3.537	3.537
120	3.561	3.566	3.585	3.596	3.600	3.601	3.601	3.601
INF	3.594	3.603	3.640	3.668	3.690	3.708	3.722	3.735

TABLE A.14b. CRITICAL VALUES FOR DUNCAN'S NEW MULTIPLE RANGE TEST $\alpha = 0.01$

ν \ r	2	3	4	5	6	7	8	9	10
1	90.03	90.03	90.03	90.03	90.03	90.03	90.03	90.03	90.03
2	14.04	14.04	14.04	14.04	14.04	14.04	14.04	14.04	14.04
3	8.261	8.321	8.321	8.321	8.321	8.321	8.321	8.321	8.321
4	6.512	6.677	6.740	6.756	6.756	6.756	6.756	6.756	6.756
5	5.702	5.893	5.989	6.040	6.065	6.074	6.074	6.074	6.074
6	5.243	5.439	5.549	5.614	5.655	5.680	5.694	5.701	5.703
7	4.949	5.145	5.260	5.334	5.383	5.416	5.439	5.454	5.464
8	4.746	4.939	5.057	5.135	5.189	5.227	5.256	5.276	5.291
9	4.596	4.787	4.906	4.986	5.043	5.086	5.118	5.142	5.160
10	4.482	4.671	4.790	4.871	4.931	4.975	5.010	5.037	5.058
11	4.392	4.579	4.697	4.780	4.841	4.887	4.924	4.952	4.975
12	4.320	4.504	4.622	4.706	4.767	4.815	4.852	4.883	4.907
13	4.260	4.442	4.560	4.644	4.706	4.755	4.793	4.824	4.850
14	4.210	4.391	4.508	4.591	4.654	4.704	4.743	4.775	4.802
15	4.168	4.347	4.463	4.547	4.610	4.660	4.700	4.733	4.760
16	4.131	4.309	4.425	4.509	4.572	4.622	4.663	4.696	4.724
17	4.099	4.275	4.391	4.475	4.539	4.589	4.630	4.664	4.693
18	4.071	4.246	4.362	4.445	4.509	4.560	4.601	4.635	4.664
19	4.046	4.220	4.335	4.419	4.483	4.534	4.575	4.610	4.639
20	4.024	4.197	4.312	4.395	4.459	4.510	4.552	4.587	4.617
24	3.956	4.126	4.239	4.322	4.386	4.437	4.480	4.516	4.546
30	3.889	4.056	4.168	4.250	4.314	4.366	4.409	4.445	4.477
40	3.825	3.988	4.098	4.180	4.244	4.296	4.339	4.376	4.408
60	3.762	3.922	4.031	4.111	4.174	4.226	4.270	4.307	4.340
120	3.702	3.858	3.965	4.044	4.107	4.158	4.202	4.239	4.272
INF	3.643	3.796	3.900	3.978	4.040	4.091	4.135	4.172	4.205

ν \ r	11	12	13	14	15	16	17	18	19
1	90.03	90.03	90.03	90.03	90.03	90.03	90.03	90.03	90.03
2	14.04	14.04	14.04	14.04	14.04	14.04	14.04	14.04	14.04
3	8.321	8.321	8.321	8.321	8.321	8.321	8.321	8.321	8.321
4	6.756	6.756	6.756	6.756	6.756	6.756	6.756	6.756	6.756
5	6.074	6.074	6.074	6.074	6.074	6.074	6.074	6.074	6.074
6	5.703	5.703	5.703	5.703	5.703	5.703	5.703	5.703	5.703
7	5.470	5.472	5.472	5.472	5.472	5.472	5.472	5.472	5.472
8	5.302	5.309	5.314	5.316	5.317	5.317	5.317	5.317	5.317
9	5.174	5.185	5.193	5.199	5.203	5.205	5.206	5.206	5.206
10	5.074	5.088	5.098	5.106	5.112	5.117	5.120	5.122	5.124
11	4.994	5.009	5.021	5.031	5.039	5.045	5.050	5.054	5.057
12	4.927	4.944	4.958	4.969	4.978	4.986	4.993	4.998	5.002

(*Table continued*)

TABLE A.14b. Continued

r ν	11	12	13	14	15	16	17	18	19
13	4.872	4.889	4.904	4.917	4.928	4.937	4.944	4.950	4.956
14	4.824	4.843	4.859	4.872	4.884	4.894	4.902	4.910	4.916
15	4.783	4.803	4.820	4.834	4.846	4.857	4.866	4.874	4.881
16	4.748	4.768	4.786	4.800	4.813	4.825	4.835	4.844	4.851
17	4.717	4.738	4.756	4.771	4.785	4.797	4.807	4.816	4.824
18	4.689	4.711	4.729	4.745	4.759	4.772	4.783	4.792	4.801
19	4.665	4.686	4.705	4.722	4.736	4.749	4.761	4.771	4.780
20	4.642	4.664	4.684	4.701	4.716	4.729	4.741	4.751	4.761
24	4.573	4.596	4.616	4.634	4.651	4.665	4.678	4.690	4.700
30	4.504	4.528	4.550	4.569	4.586	4.601	4.615	4.628	4.640
40	4.436	4.461	4.483	4.503	4.521	4.537	4.553	4.566	4.579
60	4.368	4.394	4.417	4.438	4.456	4.474	4.490	4.504	4.518
120	4.301	4.327	4.351	4.372	4.392	4.410	4.426	4.442	4.456
INF	4.235	4.261	4.285	4.307	4.327	4.345	4.363	4.379	4.394

r ν	20	22	24	26	28	30	32	34	36
1	90.03	90.03	90.03	90.03	90.03	90.03	90.03	90.03	90.03
2	14.04	14.04	14.04	14.04	14.04	14.04	14.04	14.04	14.04
3	8.321	8.321	8.321	8.321	8.321	8.321	8.321	8.321	8.321
4	6.756	6.756	6.756	6.756	6.756	6.756	6.756	6.756	6.756
5	6.074	6.074	6.074	6.074	6.074	6.074	6.074	6.074	6.074
6	5.703	5.703	5.703	5.703	5.703	5.703	5.703	5.703	5.703
7	5.472	5.472	5.472	5.472	5.472	5.472	5.472	5.472	5.472
8	5.317	5.317	5.317	5.317	5.317	5.317	5.317	5.317	5.317
9	5.206	5.206	5.206	5.206	5.206	5.206	5.206	5.206	5.206
10	5.124	5.124	5.124	5.124	5.124	5.124	5.124	5.124	5.124
11	5.059	5.061	5.061	5.061	5.061	5.061	5.061	5.061	5.061
12	5.006	5.010	5.011	5.011	5.011	5.011	5.011	5.011	5.011
13	4.960	4.966	4.970	4.972	4.972	4.972	4.972	4.972	4.972
14	4.921	4.929	4.935	4.938	4.940	4.940	4.940	4.940	4.940
15	4.887	4.897	4.904	4.909	4.912	4.914	4.914	4.914	4.914
16	4.858	4.869	4.877	4.883	4.887	4.890	4.892	4.892	4.892
17	4.832	4.844	4.853	4.860	4.865	4.869	4.872	4.873	4.874
18	4.808	4.821	4.832	4.839	4.846	4.850	4.854	4.856	4.857
19	4.788	4.802	4.812	4.821	4.828	4.833	4.838	4.841	4.843
20	4.769	4.784	4.795	4.805	4.813	4.818	4.823	4.827	4.830
24	4.710	4.727	4.741	4.752	4.762	4.770	4.777	4.783	4.788
30	4.650	4.669	4.685	4.699	4.711	4.721	4.730	4.738	4.744
40	4.591	4.611	4.630	4.645	4.659	4.671	4.682	4.692	4.700
60	4.530	4.553	4.573	4.591	4.607	4.620	4.633	4.645	4.655
120	4.469	4.494	4.516	4.535	4.552	4.568	4.583	4.596	4.609
INF	4.408	4.434	4.457	4.478	4.497	4.514	4.530	4.545	4.559

TABLE A.14b. Continued

r / ν	38	40	50	60	70	80	90	100
1	90.03	90.03	90.03	90.03	90.03	90.03	90.03	90.03
2	14.04	14.04	14.04	14.04	14.04	14.04	14.04	14.04
3	8.321	8.321	8.321	8.321	8.321	8.321	8.321	8.321
4	6.756	6.756	6.756	6.756	6.756	6.756	6.756	6.756
5	6.074	6.074	6.074	6.074	6.074	6.074	6.074	6.074
6	5.703	5.703	5.703	5.703	5.703	5.703	5.703	5.703
7	5.472	5.472	5.472	5.472	5.472	5.472	5.472	5.472
8	5.317	5.317	5.317	5.317	5.317	5.317	5.317	5.317
9	5.206	5.206	5.206	5.206	5.206	5.206	5.206	5.206
10	5.124	5.124	5.124	5.124	5.124	5.124	5.124	5.124
11	5.061	5.061	5.061	5.061	5.061	5.061	5.061	5.061
12	5.011	5.011	5.011	5.011	5.011	5.011	5.011	5.011
13	4.972	4.972	4.972	4.972	4.972	4.972	4.972	4.972
14	4.940	4.940	4.940	4.940	4.940	4.940	4.940	4.940
15	4.914	4.914	4.914	4.914	4.914	4.914	4.914	4.914
16	4.892	4.892	4.892	4.892	4.892	4.892	4.892	4.892
17	4.874	4.874	4.874	4.874	4.874	4.874	4.874	4.874
18	4.858	4.858	4.858	4.858	4.858	4.858	4.858	4.858
19	4.844	4.845	4.845	4.845	4.845	4.845	4.845	4.845
20	4.832	4.833	4.833	4.833	4.833	4.833	4.833	4.833
24	4.791	4.794	4.802	4.802	4.802	4.802	4.802	4.802
30	4.750	4.755	4.772	4.777	4.777	4.777	4.777	4.777
40	4.708	4.715	4.740	4.754	4.761	4.764	4.764	4.764
60	4.665	4.673	4.707	4.730	4.745	4.755	4.761	4.765
120	4.619	4.630	4.673	4.703	4.727	4.745	4.759	4.770
INF	4.572	4.584	4.635	4.675	4.707	4.734	4.756	4.776

TABLE A.15a. CRITICAL VALUES FOR THE STUDENTIZED RANGE $\alpha = 0.05$

ν \ r	2	3	4	5	6	7	8	9	10
1	17.97	26.98	32.82	37.08	40.41	43.12	45.40	47.36	49.07
2	6.085	8.331	9.798	10.88	11.74	12.44	13.03	13.54	13.99
3	4.501	5.910	6.825	7.502	8.037	8.478	8.853	9.177	9.462
4	3.927	5.040	5.757	6.287	6.707	7.053	7.347	7.602	7.826
5	3.635	4.602	5.218	5.673	6.033	6.330	6.582	6.802	6.995
6	3.461	4.339	4.896	5.305	5.628	5.895	6.122	6.319	6.493
7	3.344	4.165	4.681	5.060	5.359	5.606	5.815	5.998	6.158
8	3.261	4.041	4.529	4.886	5.167	5.399	5.597	5.767	5.918
9	3.199	3.949	4.415	4.756	5.024	5.244	5.432	5.595	5.739
10	3.151	3.877	4.327	4.654	4.912	5.124	5.305	5.461	5.599
11	3.113	3.820	4.256	4.574	4.823	5.028	5.202	5.353	5.487
12	3.082	3.773	4.199	4.508	4.751	4.950	5.119	5.265	5.395
13	3.055	3.735	4.151	4.453	4.690	4.885	5.049	5.192	5.318
14	3.033	3.702	4.111	4.407	4.639	4.829	4.990	5.131	5.254
15	3.014	3.674	4.076	4.367	4.595	4.782	4.940	5.077	5.198
16	2.998	3.649	4.046	4.333	4.557	4.741	4.897	5.031	5.150
17	2.984	3.628	4.020	4.303	4.524	4.705	4.858	4.991	5.108
18	2.971	3.609	3.997	4.277	4.495	4.673	4.824	4.956	5.071
19	2.960	3.593	3.977	4.253	4.469	4.645	4.794	4.924	5.038
20	2.950	3.578	3.958	4.232	4.445	4.620	4.768	4.896	5.008
24	2.919	3.532	3.901	4.166	4.373	4.541	4.684	4.807	4.915
30	2.888	3.486	3.845	4.102	4.302	4.464	4.602	4.720	4.824
40	2.858	3.442	3.791	4.039	4.232	4.389	4.521	4.635	4.735
60	2.829	3.399	3.737	3.977	4.163	4.314	4.441	4.550	4.646
120	2.800	3.356	3.685	3.917	4.096	4.241	4.363	4.468	4.560
INF	2.772	3.314	3.633	3.858	4.030	4.170	4.286	4.387	4.474

ν \ r	11	12	13	14	15	16	17	18	19
1	50.59	51.96	53.20	54.33	55.36	56.32	57.22	58.04	58.83
2	4.39	14.75	15.08	15.38	15.65	15.91	16.14	16.37	16.57
3	9.717	9.946	10.15	10.35	10.53	10.69	10.84	10.98	11.11
4	8.027	8.208	8.373	8.525	8.664	8.794	8.914	9.028	9.134
5	7.168	7.324	7.466	7.596	7.717	7.828	7.932	8.030	8.122
6	6.649	6.789	6.917	7.034	7.143	7.244	7.338	7.426	7.508
7	6.302	6.431	6.550	6.658	6.759	6.852	6.939	7.020	7.097
8	6.054	6.175	6.287	6.389	6.483	6.571	6.653	6.729	6.802
9	5.867	5.983	6.089	6.186	6.276	6.359	6.437	6.510	6.579
10	5.722	5.833	5.935	6.028	6.114	6.194	6.269	6.339	6.405

TABLE A.15a. Continued

ν \ r	11	12	13	14	15	16	17	18	19
11	5.605	5.713	5.811	5.901	5.984	6.062	6.134	6.202	6.265
12	5.511	5.615	5.710	5.798	5.878	5.953	6.023	6.089	6.151
13	5.431	5.533	5.625	5.711	5.789	5.862	5.931	5.995	6.055
14	5.364	5.463	5.554	5.637	5.714	5.786	5.852	5.915	5.974
15	5.306	5.404	5.493	5.574	5.649	5.720	5.785	5.846	5.904
16	5.256	5.352	5.439	5.520	5.593	5.662	5.727	5.786	5.843
17	5.212	5.307	5.392	5.471	5.544	5.612	5.675	5.734	5.790
18	5.174	5.267	5.352	5.429	5.501	5.568	5.630	5.688	5.743
19	5.140	5.231	5.315	5.391	5.462	5.528	5.589	5.647	5.701
20	5.108	5.199	5.282	5.357	5.427	5.493	5.553	5.610	5.663
24	5.012	5.099	5.179	5.251	5.319	5.381	5.439	5.494	5.545
30	4.917	5.001	5.077	5.147	5.211	5.271	5.327	5.379	5.429
40	4.824	4.904	4.977	5.044	5.106	5.163	5.216	5.266	5.313
60	4.732	4.808	4.878	4.942	5.001	5.056	5.107	5.154	5.199
120	4.641	4.714	4.781	4.842	4.898	4.950	4.998	5.044	5.086
INF	4.552	4.622	4.685	4.743	4.796	4.845	4.891	4.934	4.974

ν \ r	20	22	24	26	28	30	32	34	36
1	59.56	60.91	62.12	63.22	64.23	65.15	66.01	66.81	67.56
2	16.77	17.13	17.45	17.75	18.02	18.27	18.50	18.72	18.92
3	11.24	11.47	11.68	11.87	12.05	12.21	12.36	12.50	12.63
4	9.233	9.418	9.584	9.736	9.875	10.00	10.12	10.23	10.34
5	8.208	8.368	8.512	8.643	8.764	8.875	8.979	9.075	9.165
6	7.587	7.730	7.861	7.979	8.088	8.189	8.283	8.370	8.452
7	7.170	7.303	7.423	7.533	7.634	7.728	7.814	7.895	7.972
8	6.870	6.995	7.109	7.212	7.307	7.395	7.477	7.554	7.625
9	6.644	6.763	6.871	6.970	7.061	7.145	7.222	7.295	7.363
10	6.467	6.582	6.686	6.781	6.868	6.948	7.023	7.093	7.159
11	6.326	6.436	6.536	6.628	6.712	6.790	6.863	6.930	6.994
12	6.209	6.317	6.414	6.503	6.585	6.660	6.731	6.796	6.858
13	6.112	6.217	6.312	6.398	6.478	6.551	6.620	6.684	6.744
14	6.029	6.132	6.224	6.309	6.387	6.459	6.526	6.588	6.647
15	5.958	6.059	6.149	6.233	6.309	6.379	6.445	6.506	6.564
16	5.897	5.995	6.084	6.166	6.241	6.310	6.374	6.434	6.491
17	5.842	5.940	6.027	6.107	6.181	6.249	6.313	6.372	6.427
18	5.794	5.890	5.977	6.055	6.128	6.195	6.258	6.316	6.371
19	5.752	5.846	5.932	6.009	6.081	6.147	6.209	6.267	6.321
20	5.714	5.807	5.891	5.968	6.039	6.104	6.165	6.222	6.275
24	5.594	5.683	5.764	5.838	5.906	5.968	6.027	6.081	6.132
30	5.475	5.561	5.638	5.709	5.774	5.833	5.889	5.941	5.990
40	5.358	5.439	5.513	5.581	5.642	5.700	5.753	5.803	5.849
60	5.241	5.319	5.389	5.453	5.512	5.566	5.617	5.664	5.708
120	5.126	5.200	5.266	5.327	5.382	5.434	5.481	5.526	5.568
INF	5.012	5.081	5.144	5.201	5.253	5.301	5.346	5.388	5.427

(*Table continued*)

581

TABLE A.15a. Continued

ν \ r	38	40	50	60	70	80	90	100
1	68.26	68.92	71.73	73.97	75.82	77.40	78.77	79.98
2	19.11	19.28	20.05	20.66	21.16	21.59	21.96	22.29
3	12.75	12.87	13.36	13.76	14.08	14.36	14.61	14.82
4	10.44	10.53	10.93	11.24	11.51	11.73	11.92	12.09
5	9.250	9.330	9.674	9.949	10.18	10.38	10.54	10.69
6	8.529	8.601	8.913	9.163	9.370	9.548	9.702	9.839
7	8.043	8.110	8.400	8.632	8.824	8.989	9.133	9.261
8	7.693	7.756	8.029	8.248	8.430	8.586	8.722	8.843
9	7.428	7.488	7.749	7.958	8.132	8.281	8.410	8.526
10	7.220	7.279	7.529	7.730	7.897	8.041	8.166	8.276
11	7.053	7.110	7.352	7.546	7.708	7.847	7.968	8.075
12	6.916	6.970	7.205	7.394	7.552	7.687	7.804	7.909
13	6.800	7.854	7.083	7.267	7.421	7.552	7.667	7.769
14	6.702	6.754	6.979	7.159	7.309	7.438	7.550	7.650
15	6.618	6.669	6.888	7.065	7.212	7.339	7.449	7.546
16	6.544	6.594	6.810	6.984	7.128	7.252	7.360	7.457
17	6.479	6.529	6.741	6.912	7.054	7.176	7.283	7.377
18	6.422	6.471	6.680	6.848	6.989	7.109	7.213	7.307
19	6.371	6.419	6.626	6.792	6.930	7.048	7.152	7.244
20	6.325	6.373	6.576	6.740	6.877	6.994	7.097	7.187
24	6.181	6.226	6.421	6.579	6.710	6.822	6.920	7.008
30	6.037	6.080	6.267	6.417	6.543	6.650	6.744	6.827
40	5.893	5.934	6.112	6.255	6.375	6.477	6.566	6.645
60	5.750	5.789	5.958	6.093	6.206	6.303	6.387	6.462
120	5.607	5.644	5.802	5.929	6.035	6.126	6.205	6.275
INF	5.463	5.498	5.646	5.764	5.863	5.947	6.020	6.085

NOTE: Tables A.15a and A.15b are reproduced, with the author's permission, from H. Leon Harter's *Order Statistics and Their Use in Testing and Estimation*, Vol. 1, U.S. Government Printing Office, Washington, D.C., 1970.

TABLE A.15b. CRITICAL VALUES FOR THE STUDENTIZED RANGE $\alpha = 0.01$

v \ r	2	3	4	5	6	7	8	9	10
1	90.03	135.0	164.3	185.6	202.2	215.8	227.2	237.0	245.6
2	14.04	19.02	22.29	24.72	26.63	28.20	29.53	30.68	31.69
3	8.261	10.62	12.17	13.33	14.24	15.00	15.64	16.20	16.69
4	6.512	8.120	9.173	9.958	10.58	11.10	11.55	11.93	12.27
5	5.702	6.976	7.804	8.421	8.913	9.321	9.669	9.972	10.24
6	5.243	6.331	7.033	7.556	7.973	8.318	8.613	8.869	9.097
7	4.949	5.919	6.543	7.005	7.373	7.679	7.939	8.166	8.368
8	4.746	5.635	6.204	6.625	6.960	7.237	7.474	7.681	7.863
9	4.596	5.428	5.957	6.348	6.658	6.915	7.134	7.325	7.495
10	4.482	5.270	5.769	6.136	6.428	6.669	6.875	7.055	7.213
11	4.392	5.146	5.621	5.970	6.247	6.476	6.672	6.842	6.992
12	4.320	5.046	5.502	5.836	6.101	6.321	6.507	6.670	6.814
13	4.260	4.964	5.404	5.727	5.981	6.192	6.372	6.528	6.667
14	4.210	4.895	5.322	5.634	5.881	6.085	6.258	6.409	6.543
15	4.168	4.836	5.252	5.556	5.796	5.994	6.162	6.309	6.439
16	4.131	4.786	5.192	5.489	5.722	5.915	6.079	6.222	6.349
17	4.099	4.742	5.140	5.430	5.659	5.847	6.007	6.147	6.270
18	4.071	4.703	5.094	5.379	5.603	5.788	5.944	6.081	6.201
19	4.046	4.670	5.054	5.334	5.554	5.735	5.889	6.022	6.141
20	4.024	4.639	5.018	5.294	5.510	5.688	5.839	5.970	6.087
24	3.956	4.546	4.907	5.168	5.374	5.542	5.685	5.809	5.919
30	3.889	4.455	4.799	5.048	5.242	5.401	5.536	5.653	5.756
40	3.825	4.367	4.696	4.931	5.114	5.265	5.392	5.502	5.599
60	3.762	4.282	4.595	4.818	4.991	5.133	5.253	5.356	5.447
120	3.702	4.200	4.497	4.709	4.872	5.005	5.118	5.214	5.299
INF	3.643	4.120	4.403	4.603	4.757	4.882	4.987	5.078	5.157

v \ r	11	12	13	14	15	16	17	18	19
1	253.2	260.0	266.2	271.8	277.0	281.8	286.3	290.4	294.3
2	32.59	33.40	34.13	34.81	35.43	36.00	36.53	37.03	37.50
3	17.13	17.53	17.89	18.22	18.52	18.81	19.07	19.32	19.55
4	12.57	12.84	13.09	13.32	13.53	13.73	13.91	14.08	14.24
5	10.48	10.70	10.89	11.08	11.24	11.40	11.55	11.68	11.81
6	9.301	9.485	9.653	9.808	9.951	10.08	10.21	10.32	10.43
7	8.548	8.711	8.860	8.997	9.124	9.242	9.353	9.456	9.554
8	8.027	8.176	8.312	8.436	8.552	8.659	8.760	8.854	8.943
9	7.647	7.784	7.910	8.025	8.132	8.232	8.325	8.412	8.495
10	7.356	7.485	7.603	7.712	7.812	7.906	7.993	8.076	8.153
11	7.128	7.250	7.362	7.465	7.560	7.649	7.732	7.809	7.883
12	6.943	7.060	7.167	7.265	7.356	7.441	7.520	7.594	7.665
13	6.791	6.903	7.006	7.101	7.188	7.269	7.345	7.417	7.485

(*Table continued*)

583

TABLE A.15b. Continued

ν\\r	11	12	13	14	15	16	17	18	19
14	6.664	6.772	6.871	6.962	7.047	7.126	7.199	7.268	7.333
15	6.555	6.660	6.757	6.845	6.927	7.003	7.074	7.142	7.204
16	6.462	6.564	6.658	6.744	6.823	6.898	6.967	7.032	7.093
17	6.381	6.480	6.572	6.656	6.734	6.806	6.873	6.937	6.997
18	6.310	6.407	6.497	6.579	6.655	6.725	6.792	6.854	6.912
19	6.247	6.342	6.430	6.510	6.585	6.654	6.719	6.780	6.837
20	6.191	6.285	6.371	6.450	6.523	6.591	6.654	6.714	6.771
24	6.017	6.106	6.186	6.261	6.330	6.394	6.453	6.510	6.563
30	5.849	5.932	6.008	6.078	6.143	6.203	6.259	6.311	6.361
40	5.686	5.764	5.835	5.900	5.961	6.017	6.069	6.119	6.165
60	5.528	5.601	5.667	5.728	5.785	5.837	5.886	5.931	5.974
120	5.375	5.443	5.505	5.562	5.614	5.662	5.708	5.750	5.790
INF	5.227	5.290	5.348	5.400	5.448	5.493	5.535	5.574	5.611

ν\\r	20	22	24	26	28	30	32	34	36
1	298.0	304.7	310.8	316.3	321.3	326.0	330.3	334.3	338.0
2	37.95	38.76	39.49	40.15	40.76	41.32	41.84	42.33	42.78
3	19.77	20.17	20.53	20.86	21.16	21.44	21.70	21.95	22.17
4	14.40	14.68	14.93	15.16	15.37	15.57	15.75	15.92	16.08
5	11.93	12.16	12.36	12.54	12.71	12.87	13.02	13.15	13.28
6	10.54	10.73	10.91	11.06	11.21	11.34	11.47	11.58	11.69
7	9.646	9.815	9.970	10.11	10.24	10.36	10.47	10.58	10.67
8	9.027	9.182	9.322	9.450	9.569	9.678	9.779	9.874	9.964
9	8.573	8.717	8.847	8.966	9.075	9.177	9.271	9.360	9.443
10	8.226	8.361	8.483	8.595	8.698	8.794	8.883	8.966	9.044
11	7.952	8.080	8.196	8.303	8.400	8.491	8.575	8.654	8.728
12	7.731	7.853	7.964	8.066	8.159	8.246	8.327	8.402	8.473
13	7.548	7.665	7.772	7.870	7.960	8.043	8.121	8.193	8.262
14	7.395	7.508	7.611	7.705	7.792	7.873	7.948	8.018	8.084
15	7.264	7.374	7.474	7.566	7.650	7.728	7.800	7.869	7.932
16	7.152	7.258	7.356	7.445	7.527	7.602	7.673	7.739	7.802
17	7.053	7.158	7.253	7.340	7.420	7.493	7.563	7.627	7.687
18	6.968	7.070	7.163	7.247	7.325	7.398	7.465	7.528	7.587
19	6.891	6.992	7.082	7.166	7.242	7.313	7.379	7.440	7.498
20	6.823	6.922	7.011	7.092	7.168	7.237	7.302	7.362	7.419
24	6.612	6.705	6.789	6.865	6.936	7.001	7.062	7.119	7.173
30	6.407	6.494	6.572	6.644	6.710	6.772	6.828	6.881	6.932
40	6.209	6.289	6.362	6.429	6.490	6.547	6.600	6.650	6.697
60	6.015	6.090	6.158	6.220	6.277	6.330	6.378	6.424	6.467
120	5.827	5.897	5.959	6.016	6.069	6.117	6.162	6.204	6.244
INF	5.645	5.709	5.766	5.818	5.866	5.911	5.952	5.990	6.026

TABLE A.15b. Continued

r / ν	38	40	50	60	70	80	90	100
1	341.5	344.8	358.9	370.1	379.4	387.3	394.1	400.1
2	43.21	43.61	45.33	46.70	47.83	48.80	49.64	50.38
3	22.39	22.59	23.45	24.13	24.71	25.19	25.62	25.99
4	16.23	16.37	16.98	17.46	17.86	18.20	18.50	18.77
5	13.40	13.52	14.00	14.39	14.72	14.99	15.23	15.45
6	11.80	11.90	12.31	12.65	12.92	13.16	13.37	13.55
7	10.77	10.85	11.23	11.52	11.77	11.99	12.17	12.34
8	10.05	10.13	10.47	10.75	10.97	11.17	11.34	11.49
9	9.521	9.594	9.912	10.17	10.38	10.57	10.73	10.87
10	9.117	9.187	9.486	9.726	9.927	10.10	10.25	10.39
11	8.798	8.864	9.148	9.377	9.568	9.732	9.875	10.00
12	8.539	8.603	8.875	9.094	9.277	9.434	9.571	9.693
13	8.326	8.387	8.648	8.859	9.035	9.187	9.318	9.436
14	8.146	8.204	8.457	8.661	8.832	8.978	9.106	9.219
15	7.992	8.049	8.295	8.492	8.658	8.800	8.924	9.035
16	7.860	7.916	8.154	8.347	8.507	8.646	8.767	8.874
17	7.745	7.799	8.031	8.219	8.377	8.511	8.630	8.735
18	7.643	7.696	7.924	8.107	8.261	8.393	8.508	8.611
19	7.553	7.605	7.828	8.008	8.159	8.288	8.401	8.502
20	7.473	7.523	7.742	7.919	8.067	8.194	8.305	8.404
24	7.223	7.270	7.476	7.642	7.780	7.900	8.004	8.097
30	6.978	7.023	7.215	7.370	7.500	7.611	7.709	7.796
40	6.740	6.782	6.960	7.104	7.225	7.328	7.419	7.500
60	6.507	6.546	6.710	6.843	6.954	7.050	7.133	7.207
120	6.281	6.316	6.467	6.588	6.689	6.776	6.852	6.919
INF	6.060	6.092	6.228	6.338	6.429	6.507	6.575	6.636

TABLE A.16. CRITICAL VALUES OF THE RATIO F_{max}

v \ a	2	3	4	5	6	7	8	9	10	11	12
					$\alpha = 0.05$						
2	39.0	87.5	142	202	266	333	403	475	550	626	704
3	15.4	27.8	39.2	50.7	62.0	72.9	83.5	93.9	104	114	124
4	9.60	15.5	20.6	25.2	29.5	33.6	37.5	41.1	44.6	48.0	51.4
5	7.15	10.8	13.7	16.3	18.7	20.8	22.9	24.7	26.5	28.2	29.9
6	5.82	8.38	10.4	12.1	13.7	15.0	16.3	17.5	18.6	19.7	20.7
7	4.99	6.94	8.44	9.70	10.8	11.8	12.7	13.5	14.3	15.1	15.8
8	4.43	6.00	7.18	8.12	9.03	9.78	10.5	11.1	11.7	12.2	12.7
9	4.03	5.34	6.31	7.11	7.80	8.41	8.95	9.45	9.91	10.3	10.7
10	3.72	4.85	5.67	6.34	6.92	7.42	7.87	8.28	8.66	9.01	9.34
12	3.28	4.16	4.79	5.30	5.72	6.09	6.42	6.72	7.00	7.25	7.48
15	2.86	3.54	4.01	4.37	4.68	4.95	5.19	5.40	5.59	5.77	5.93
20	2.46	2.95	3.29	3.54	3.76	3.94	4.10	4.24	4.37	4.49	4.59
30	2.07	2.40	2.61	2.78	2.91	3.02	3.12	3.21	3.29	3.36	3.39
60	1.67	1.85	1.96	2.04	2.11	2.17	2.22	2.26	2.30	2.33	2.36
INF	1.00	1.00	1.00	1.00	1.00	1.00	1.00	1.00	1.00	1.00	1.00
					$\alpha = 0.01$						
2	199	448	729	1036	1362	1705	2063	2432	2813	3204	3605
3	47.5	85	120	151	184	21(6)	24(9)	28(1)	31(0)	33(7)	36(1)
4	23.2	37	49	59	69	79	89	97	106	113	120
5	14.9	22	28	33	38	42	46	50	54	57	60
6	11.1	15.5	19.1	22	25	27	30	32	34	36	37
7	8.89	12.1	14.5	16.5	18.4	20	22	23	24	26	27
8	7.50	9.9	11.7	13.2	14.5	15.8	16.9	17.9	18.9	19.8	21
9	6.54	8.5	9.9	11.1	12.1	13.1	13.9	14.7	15.3	16.0	16.6
10	5.85	7.4	8.6	9.6	10.4	11.1	11.8	12.4	12.9	13.4	13.9
12	4.91	6.1	6.9	7.6	8.2	8.7	9.1	9.5	9.9	10.2	10.6
15	4.07	4.9	5.5	6.0	6.4	6.7	7.1	7.3	7.5	7.8	8.0
20	3.32	3.8	4.3	4.6	4.9	5.1	5.3	5.5	5.6	5.8	5.9
30	2.63	3.0	3.3	3.4	3.6	3.7	3.8	3.9	4.0	4.1	4.2
60	1.96	2.2	2.3	2.4	2.4	2.5	2.5	2.6	2.6	2.7	2.7
INF	1.00	1.0	1.0	1.0	1.0	1.0	1.0	1.0	1.0	1.0	1.0

Reproduced with permission of the Biometrika Trust, from *Biometrika Tables for Statisticians*, Vol. 1, 3rd edition, 1966, edited by E.S. Pearson and H.O. Hartley.

TABLE A.17. LOGS BASE TEN

	.00	.01	.02	.03	.04	.05	.06	.07	.08	.09
1.0	.0000	.0043	.0086	.0128	.0170	.0212	.0253	.0294	.0334	.0374
1.1	.0414	.0453	.0492	.0531	.0569	.0607	.0645	.0682	.0719	.0755
1.2	.0792	.0828	.0864	.0899	.0934	.0969	.1004	.1038	.1072	.1106
1.3	.1139	.1173	.1206	.1239	.1271	.1303	.1335	.1367	.1399	.1430
1.4	.1461	.1492	.1523	.1553	.1584	.1614	.1644	.1673	.1703	.1732
1.5	.1761	.1790	.1818	.1847	.1875	.1903	.1931	.1959	.1987	.2014
1.6	.2041	.2068	.2095	.2122	.2148	.2175	.2201	.2227	.2253	.2279
1.7	.2304	.2330	.2355	.2380	.2405	.2430	.2455	.2480	.2504	.2529
1.8	.2553	.2577	.2601	.2625	.2648	.2672	.2695	.2718	.2742	.2765
1.9	.2788	.2810	.2833	.2856	.2878	.2900	.2923	.2945	.2967	.2989
2.0	.3010	.3032	.3054	.3075	.3096	.3118	.3139	.3160	.3181	.3201
2.1	.3222	.3243	.3263	.3284	.3304	.3324	.3345	.3365	.3385	.3404
2.2	.3424	.3444	.3464	.3483	.3502	.3522	.3541	.3560	.3579	.3598
2.3	.3617	.3636	.3655	.3674	.3692	.3711	.3729	.3747	.3766	.3784
2.4	.3802	.3820	.3838	.3856	.3874	.3892	.3909	.3927	.3945	.3962
2.5	.3979	.3997	.4014	.4031	.4048	.4065	.4082	.4099	.4116	.4133
2.6	.4150	.4166	.4183	.4200	.4216	.4232	.4249	.4265	.4281	.4298
2.7	.4314	.4330	.4346	.4362	.4378	.4393	.4409	.4425	.4440	.4456
2.8	.4472	.4487	.4502	.4518	.4533	.4548	.4564	.4579	.4594	.4609
2.9	.4624	.4639	.4654	.4669	.4683	.4698	.4713	.4728	.4742	.4757
3.0	.4771	.4786	.4800	.4814	.4829	.4843	.4857	.4871	.4886	.4900
3.1	.4914	.4928	.4942	.4955	.4969	.4983	.4997	.5011	.5024	.5038
3.2	.5051	.5065	.5079	.5092	.5105	.5119	.5132	.5145	.5159	.5172
3.3	.5185	.5198	.5211	.5224	.5237	.5250	.5263	.5276	.5289	.5302
3.4	.5315	.5328	.5340	.5353	.5366	.5378	.5391	.5403	.5416	.5428
3.5	.5441	.5453	.5465	.5478	.5490	.5502	.5514	.5527	.5539	.5551
3.6	.5563	.5575	.5587	.5599	.5611	.5623	.5635	.5647	.5658	.5670
3.7	.5682	.5694	.5705	.5717	.5729	.5740	.5752	.5763	.5775	.5786
3.8	.5798	.5809	.5821	.5832	.5843	.5855	.5866	.5877	.5888	.5899
3.9	.5911	.5922	.5933	.5944	.5955	.5966	.5977	.5988	.5999	.6010
4.0	.6021	.6031	.6042	.6053	.6064	.6075	.6085	.6096	.6107	.6117
4.1	.6128	.6138	.6149	.6159	.6170	.6180	.6191	.6201	.6212	.6222
4.2	.6232	.6243	.6253	.6263	.6274	.6284	.6294	.6304	.6314	.6325
4.3	.6335	.6345	.6355	.6365	.6375	.6385	.6395	.6405	.6415	.6425
4.4	.6435	.6444	.6454	.6464	.6474	.6484	.6493	.6503	.6513	.6522
4.5	.6532	.6542	.6551	.6561	.6571	.6580	.6590	.6599	.6609	.6618

(Table continued)

TABLE A.17. Continued

	.00	.01	.02	.03	.04	.05	.06	.07	.08	.09
4.6	.6628	.6637	.6646	.6656	.6665	.6675	.6684	.6693	.6702	.6712
4.7	.6721	.6730	.6739	.6749	.6758	.6767	.6776	.6785	.6794	.6803
4.8	.6812	.6821	.6830	.6839	.6848	.6857	.6866	.6875	.6884	.6893
4.9	.6902	.6911	.6920	.6928	.6937	.6946	.6955	.6964	.6972	.6981
5.0	.6990	.6998	.7007	.7016	.7024	.7033	.7042	.7050	.7059	.7067
5.1	.7076	.7084	.7093	.7101	.7110	.7118	.7126	.7135	.7143	.7152
5.2	.7160	.7168	.7177	.7185	.7193	.7202	.7210	.7218	.7226	.7235
5.3	.7243	.7251	.7259	.7267	.7275	.7284	.7292	.7300	.7308	.7316
5.4	.7324	.7332	.7340	.7348	.7356	.7364	.7372	.7380	.7388	.7396
5.5	.7404	.7412	.7419	.7427	.7435	.7443	.7451	.7459	.7466	.7474
5.6	.7482	.7490	.7497	.7505	.7513	.7520	.7528	.7536	.7543	.7551
5.7	.7559	.7566	.7574	.7582	.7589	.7597	.7604	.7612	.7619	.7627
5.8	.7634	.7642	.7649	.7657	.7664	.7672	.7679	.7686	.7694	.7701
5.9	.7709	.7716	.7723	.7731	.7738	.7745	.7752	.7760	.7767	.7774
6.0	.7782	.7789	.7796	.7803	.7810	.7818	.7825	.7832	.7839	.7846
6.1	.7853	.7860	.7868	.7875	.7882	.7889	.7896	.7903	.7910	.7917
6.2	.7924	.7931	.7938	.7945	.7952	.7959	.7966	.7973	.7980	.7987
6.3	.7993	.8000	.8007	.8014	.8021	.8028	.8035	.8041	.8048	.8055
6.4	.8062	.8069	.8075	.8082	.8089	.8096	.8102	.8109	.8116	.8122
6.5	.8129	.8136	.8142	.8149	.8156	.8162	.8169	.8176	.8182	.8189
6.6	.8195	.8202	.8209	.8215	.8222	.8228	.8235	.8241	.8248	.8254
6.7	.8261	.8267	.8274	.8280	.8287	.8293	.8299	.8306	.8312	.8319
6.8	.8325	.8331	.8338	.8344	.8351	.8357	.8363	.8370	.8376	.8382
6.9	.8388	.8395	.8401	.8407	.8414	.8420	.8426	.8432	.8439	.8445
7.0	.8451	.8457	.8463	.8470	.8476	.8482	.8488	.8494	.8500	.8506
7.1	.8513	.8519	.8525	.8531	.8537	.8543	.8549	.8555	.8561	.8567
7.2	.8573	.8579	.8585	.8591	.8597	.8603	.8609	.8615	.8621	.8627
7.3	.8633	.8639	.8645	.8651	.8657	.8663	.8669	.8675	.8681	.8686
7.4	.8692	.8698	.8704	.8710	.8716	.8722	.8727	.8733	.8739	.8745
7.5	.8751	.8756	.8762	.8768	.8774	.8779	.8785	.8791	.8797	.8802
7.6	.8808	.8814	.8820	.8825	.8831	.8837	.8842	.8848	.8854	.8859
7.7	.8865	.8871	.8876	.8882	.8887	.8893	.8899	.8904	.8910	.8915
7.8	.8921	.8927	.8932	.8938	.8943	.8949	.8954	.8960	.8965	.8971
7.9	.8976	.8982	.8987	.8993	.8998	.9004	.9009	.9015	.9020	.9025
8.0	.9031	.9036	.9042	.9047	.9053	.9058	.9063	.9069	.9074	.9079
8.1	.9085	.9090	.9096	.9101	.9106	.9112	.9117	.9122	.9128	.9133
8.2	.9138	.9143	.9149	.9154	.9159	.9165	.9170	.9175	.9180	.9186
8.3	.9191	.9196	.9201	.9206	.9212	.9217	.9222	.9227	.9232	.9238
8.4	.9243	.9248	.9253	.9258	.9263	.9269	.9274	.9279	.9284	.9289

TABLE A.17. Continued

	.00	.01	.02	.03	.04	.05	.06	.07	.08	.09
8.5	.9294	.9299	.9304	.9309	.9315	.9320	.9325	.9330	.9335	.9340
8.6	.9345	.9350	.9355	.9360	.9365	.9370	.9375	.9380	.9385	.9390
8.7	.9395	.9400	.9405	.9410	.9415	.9420	.9425	.9430	.9435	.9440
8.8	.9445	.9450	.9455	.9460	.9465	.9469	.9474	.9479	.9484	.9489
8.9	.9494	.9499	.9504	.9509	.9513	.9518	.9523	.9528	.9533	.9538
9.0	.9542	.9547	.9552	.9557	.9562	.9566	.9571	.9576	.9581	.9586
9.1	.9590	.9595	.9600	.9605	.9609	.9614	.9619	.9624	.9628	.9633
9.2	.9638	.9643	.9647	.9652	.9657	.9661	.9666	.9671	.9675	.9680
9.3	.9685	.9689	.9694	.9699	.9703	.9708	.9713	.9717	.9722	.9727
9.4	.9731	.9736	.9741	.9745	.9750	.9754	.9759	.9763	.9768	.9773
9.5	.9777	.9782	.9786	.9791	.9795	.9800	.9805	.9809	.9814	.9818
9.6	.9823	.9827	.9832	.9836	.9841	.9845	.9850	.9854	.9859	.9863
9.7	.9868	.9872	.9877	.9881	.9886	.9890	.9894	.9899	.9903	.9908
9.8	.9912	.9917	.9921	.9926	.9930	.9934	.9939	.9943	.9948	.9952
9.9	.9956	.9961	.9965	.9969	.9974	.9978	.9983	.9987	.9991	.9996

TABLE A.18. ANGULAR TRANSFORMATION ARC SIN $\sqrt{\% \times 0.01}$

%	.0	.1	.2	.3	.4	.5	.6	.7	.8	.9
0	0.00	1.81	2.56	3.14	3.63	4.05	4.44	4.80	5.13	5.44
1	5.74	6.02	6.29	6.55	6.80	7.03	7.27	7.49	7.71	7.92
2	8.13	8.33	8.53	8.72	8.91	9.10	9.28	9.46	9.63	9.80
3	9.97	10.14	10.30	10.47	10.63	10.78	10.94	11.09	11.24	11.39
4	11.54	11.68	11.83	11.97	12.11	12.25	12.38	12.52	12.66	12.79
5	12.92	13.05	13.18	13.31	13.44	13.56	13.69	13.81	13.94	14.06
6	14.18	14.30	14.42	14.54	14.65	14.77	14.89	15.00	15.12	15.23
7	15.34	15.45	15.56	15.68	15.79	15.89	16.00	16.11	16.22	16.32
8	16.43	16.54	16.64	16.74	16.85	16.95	17.05	17.15	17.26	17.36
9	17.46	17.56	17.66	17.76	17.85	17.95	18.05	18.15	18.24	18.34
10	18.43	18.53	18.63	18.72	18.81	18.91	19.00	19.09	19.19	19.28
11	19.37	19.46	19.55	19.64	19.73	19.82	19.91	20.00	20.09	20.18
12	20.27	20.36	20.44	20.53	20.62	20.70	20.79	20.88	20.96	21.05
13	21.13	21.22	21.30	21.39	21.47	21.56	21.64	21.72	21.81	21.89
14	21.97	22.06	22.14	22.22	22.30	22.38	22.46	22.54	22.63	22.71
15	22.79	22.87	22.95	23.03	23.11	23.18	23.26	23.34	23.42	23.50
16	23.58	23.66	23.73	23.81	23.89	23.97	24.04	24.12	24.20	24.27
17	24.35	24.43	24.50	24.58	24.65	24.73	24.80	24.88	24.95	25.03
18	25.10	25.18	25.25	25.33	25.40	25.47	25.55	25.62	25.70	25.77
19	25.84	25.91	25.99	26.06	26.13	26.21	26.28	26.35	26.42	26.49
20	26.57	26.64	26.71	26.78	26.85	26.92	26.99	27.06	27.13	27.20
21	27.27	27.34	27.42	27.49	27.56	27.62	27.69	27.76	27.83	27.90
22	27.97	28.04	28.11	28.18	28.25	28.32	28.39	28.45	28.52	28.59
23	28.66	28.73	28.79	28.86	28.93	29.00	29.06	29.13	29.20	29.27
24	29.33	29.40	29.47	29.53	29.60	29.67	29.73	29.80	29.87	29.93
25	30.00	30.07	30.13	30.20	30.26	30.33	30.40	30.46	30.53	30.59
26	30.66	30.72	30.79	30.85	30.92	30.98	31.05	31.11	31.18	31.24
27	31.31	31.37	31.44	31.50	31.56	31.63	31.69	31.76	31.82	31.88
28	31.95	32.01	32.08	32.14	32.20	32.27	32.33	32.39	32.46	32.52
29	32.58	32.65	32.71	32.77	32.83	32.90	32.96	33.02	33.09	33.15
30	33.21	33.27	33.34	33.40	33.46	33.52	33.58	33.65	33.71	33.77
31	33.83	33.90	33.96	34.02	34.08	34.14	34.20	34.27	34.33	34.39
32	34.45	34.51	34.57	34.63	34.70	34.76	34.82	34.88	34.94	35.00
33	35.06	35.12	35.18	35.24	35.30	35.37	35.43	35.49	35.55	35.61
34	35.67	35.73	35.79	35.85	35.91	35.97	36.03	36.09	36.15	36.21

TABLE A.18. Continued

%	.0	.1	.2	.3	.4	.5	.6	.7	.8	.9
35	36.27	36.33	36.39	36.45	36.51	36.57	36.63	36.69	36.75	36.81
36	36.87	36.93	36.99	37.05	37.11	37.17	37.23	37.29	37.35	37.41
37	37.46	37.52	37.58	37.64	37.70	37.76	37.82	37.88	37.94	38.00
38	38.06	38.12	38.17	38.23	38.29	38.35	38.41	38.47	38.53	38.59
39	38.65	38.70	38.76	38.82	38.88	38.94	39.00	39.06	39.11	39.17
40	39.23	39.29	39.35	39.41	39.47	39.52	39.58	39.64	39.70	39.76
41	39.82	39.87	39.93	39.99	40.05	40.11	40.16	40.22	40.28	40.34
42	40.40	40.45	40.51	40.57	40.63	40.69	40.74	40.80	40.86	40.92
43	40.98	41.03	41.09	41.15	41.21	41.27	41.32	41.38	41.44	41.50
44	41.55	41.61	41.67	41.73	41.78	41.84	41.90	41.96	42.02	42.07
45	42.13	42.19	42.25	42.30	42.36	42.42	42.48	42.53	42.59	42.65
46	42.71	42.76	42.82	42.88	42.94	42.99	43.05	43.11	43.17	43.22
47	43.28	43.34	43.39	43.45	43.51	43.57	43.62	43.68	43.74	43.80
48	43.85	43.91	43.97	44.03	44.08	44.14	44.20	44.26	44.31	44.37
49	44.43	44.48	44.54	44.60	44.66	44.71	44.77	44.83	44.89	44.94
50	45.00	45.06	45.11	45.17	45.23	45.29	45.34	45.40	45.46	45.52
51	45.57	45.63	45.69	45.74	45.80	45.86	45.92	45.97	46.03	46.09
52	46.15	46.20	46.26	46.32	46.38	46.43	46.49	46.55	46.61	46.66
53	46.72	46.78	46.83	46.89	46.95	47.01	47.06	47.12	47.18	47.24
54	47.29	47.35	47.41	47.47	47.52	47.58	47.64	47.70	47.75	47.81
55	47.87	47.93	47.98	48.04	48.10	48.16	48.22	48.27	48.33	48.39
56	48.45	48.50	48.56	48.62	48.68	48.73	48.79	48.85	48.91	48.97
57	49.02	49.08	49.14	49.20	49.26	49.31	49.37	49.43	49.49	49.55
58	49.60	49.66	49.72	49.78	49.84	49.89	49.95	50.01	50.07	50.13
59	50.18	50.24	50.30	50.36	50.42	50.48	50.53	50.59	50.65	50.71
60	50.77	50.83	50.89	50.94	51.00	51.06	51.12	51.18	51.24	51.30
61	51.35	51.41	51.47	51.53	51.59	51.65	51.71	51.77	51.83	51.88
62	51.94	52.00	52.06	52.12	52.18	52.24	52.30	52.36	52.42	52.48
63	52.54	52.59	52.65	52.71	52.77	52.83	52.89	52.95	53.01	53.07
64	53.13	53.19	53.25	53.31	53.37	53.43	53.49	53.55	53.61	53.67
65	53.73	53.79	53.85	53.91	53.97	54.03	54.09	54.15	54.21	54.27
66	54.33	54.39	54.45	54.51	54.57	54.63	54.70	54.76	54.82	54.88
67	54.94	55.00	55.06	55.12	55.18	55.24	55.30	55.37	55.43	55.49
68	55.55	55.61	55.67	55.73	55.80	55.86	55.92	55.98	56.04	56.10
69	56.17	56.23	56.29	56.35	56.42	56.48	56.54	56.60	56.66	56.73
70	56.79	56.85	56.91	56.98	57.04	57.10	57.17	57.23	57.29	57.35
71	57.42	57.48	57.54	57.61	57.67	57.73	57.80	57.86	57.92	57.99
72	58.05	58.12	58.18	58.24	58.31	58.37	58.44	58.50	58.56	58.63
73	58.69	58.76	58.82	58.89	58.95	59.02	59.08	59.15	59.21	59.28
74	59.34	59.41	59.47	59.54	59.60	59.67	59.74	59.80	59.87	59.93

(Table continued)

591

%	.0	.1	.2	.3	.4	.5	.6	.7	.8	.9
75	60.00	60.07	60.13	60.20	60.27	60.33	60.40	60.47	60.53	60.60
76	60.67	60.73	60.80	60.87	60.94	61.00	61.07	61.14	61.21	61.27
77	61.34	61.41	61.48	61.55	61.61	61.68	61.75	61.82	61.89	61.96
78	62.03	62.10	62.17	62.24	62.31	62.38	62.44	62.51	62.58	62.65
79	62.73	62.80	62.87	62.94	63.01	63.08	63.15	63.22	63.29	63.36
80	63.43	63.51	63.58	63.65	63.72	63.79	63.87	63.94	64.01	64.09
81	64.16	64.23	64.30	64.38	64.45	64.52	64.60	64.67	64.75	64.82
82	64.90	64.97	65.05	65.12	65.20	65.27	65.35	65.42	65.50	65.57
83	65.65	65.73	65.80	65.88	65.96	66.03	66.11	66.19	66.27	66.34
84	66.42	66.50	66.58	66.66	66.74	66.82	66.89	66.97	67.05	67.13
85	67.21	67.29	67.37	67.46	67.54	67.62	67.70	67.78	67.86	67.94
86	68.03	68.11	68.19	68.28	68.36	68.44	68.53	68.61	68.70	68.78
87	68.87	68.95	69.04	69.12	69.21	69.30	69.38	69.47	69.56	69.64
88	69.73	69.82	69.91	70.00	70.09	70.18	70.27	70.36	70.45	70.54
89	70.63	70.72	70.81	70.91	71.00	71.09	71.19	71.28	71.37	71.47
90	71.57	71.66	71.76	71.85	71.95	72.05	72.15	72.24	72.34	72.44
91	72.54	72.64	72.74	72.84	72.95	73.05	73.15	73.26	73.36	73.46
92	73.57	73.68	73.78	73.89	74.00	74.11	74.21	74.32	74.44	74.55
93	74.66	74.77	74.88	75.00	75.11	75.23	75.35	75.46	75.58	75.70
94	75.82	75.94	76.06	76.19	76.31	76.44	76.56	76.69	76.82	76.95
95	77.08	77.21	77.34	77.48	77.62	77.75	77.89	78.03	78.17	78.32
96	78.46	78.61	78.76	78.91	79.06	79.22	79.37	79.53	79.70	79.86
97	80.03	80.20	80.37	80.54	80.72	80.90	81.09	81.28	81.47	81.67
98	81.87	82.08	82.29	82.51	82.73	82.97	83.20	83.45	83.71	83.98
99	84.26	84.56	84.87	85.20	85.56	85.95	86.37	86.86	87.44	88.19

TABLE A.19. ORTHOGONAL POLYNOMIALS

a = 3

-1	+1
0	-2
+1	+1
2	6

a = 4

-3	+1	-1
-1	-1	+3
+1	-1	-3
+3	+1	+1
20	4	20

a = 5

-2	+2	-1	+1
-1	-1	+2	-4
0	-2	0	+6
+1	-1	-2	-4
+2	+2	+1	+1
10	14	10	70

a = 6

-5	+5	-5	+1	-1
-3	-1	+7	-3	+5
-1	-4	+4	+2	-10
+1	-4	-4	+2	+10
+3	-1	-7	-3	-5
+5	+5	+5	+1	+1
70	84	180	28	252

a = 7

0	-4	0	+6	0
+1	-3	-1	+1	+5
+2	0	-1	-7	-4
+3	+5	+1	+3	+1
28	84	6	154	84

a = 8

+1	-5	-3	+9	+15
+3	-3	-7	-3	+17
+5	+1	-5	-13	-23
+7	+7	+7	+7	+7
168	168	264	616	2184

a = 9

0	-20	0	+18	0
+1	-17	-9	+9	+9
+2	-8	-13	-11	+4
+3	+7	-7	-21	-11
+4	+28	+14	+14	+4
60	2772	990	2002	468

a = 10

+1	-4	-12	+18	+6
+3	-3	-31	+3	+11
+5	-1	-35	-17	+1
+7	+2	-14	-22	-14
+9	+6	+42	+18	+6
330	132	8580	2860	780

a = 11

0	-10	0	+6	0
+1	-9	-14	+4	+4
+2	-6	-23	-1	+4
+3	-1	-22	-6	-1
+4	+6	-6	-6	-6
+5	+15	+30	+6	+3
110	858	4290	286	156

a = 12

+1	-35	-7	+28	+20
+3	-29	-19	+12	+44
+5	-17	-25	-13	+29
+7	+1	-21	-33	-21
+9	+25	-3	-27	-57
+11	+55	+33	+33	+33
572	12012	5148	8008	15912

Answers to Most Odd-Numbered Exercises and All Review Exercises

CHAPTER 1

Exercises

1.1.1. 0.020, 0.019, 0.009, 0.034, 0.047, *E*.

1.1.3. a. 0.4.

   b. 0.216.

   c. 0.144.

   d. 0.288.

   e. 0.6.

1.1.5. a. $35/36$.

   b. $(1/36)^{24}$.

   c. $1 - (35/36)^{24}$.

   d. $0.5086/(1 - 0.5086) = 1.035$.

1.2.3. a. Theoretical results are insufficient; he wants to prevent cases of paralytic polio.

   b. The vaccine should be used.

1.2.5. a. H_0: $\pi = 0.5$.

   b. H_a: $\pi < 0.5$.

   c. 0, 1 with $\alpha = 0.109$.

   d. 2, 3, 4, 5, or 6 deaths.

   e. Do not reject H_0.

1.3.1. a. Survey.

   b. Survey.

   c. Experiment.

   d. Experiment.

1.3.3. His conjecture was based on a survey with no control of other variables.

Review Exercises

False:	1.4	
	1.6	1.12
	1.7	1.17
	1.8	1.19
	1.9	

Statistics for Research, *Third Edition*, Edited by Shirley Dowdy, Stanley Weardon, and Daniel Chilko. ISBN 0-471-26735-X © 2004 John Wiley & Sons, Inc.

CHAPTER 2

Exercises

2.1.3.   i. The English people
         ii. It is a subset of the population but it is not random
         iii. Obituaries of notable people are likely to be more detailed.

2.2.1. a. 2, 8, 1.
       a. 2, 1.
       b. 18, 43, 6, 3, 39.
       d. 8, 14, 20, 9.

2.2.5. a. The numbers of the 10 for the sample are 8, 39, 16, 11, 37, 22, 2, 3, 33 and 21
       b. i. Sample proportion is $7/10 = 0.70$
          ii. Sample average is 28.4

2.3.1. a. Continuous numerical.
       b. Nominal.
       c. Nominal.
       d. Nominal.
       e. Continuous numerical.
       f. Discrete numerical.
       g. Nominal.

2.3.3. a. Female, male.
       b. Less than 3, 3, more than 3.
       c. Blue-eyed, not blue-eyed.

2.3.5. a. Ordinal scale because the symbols he used are ordered
       a. If scores are classified as lower case or upper case letters, the odds of an upper case score are 3.5 times as large for the child of a skilled father.
       b. One way is with a two by two table with Skilled or Unskilled as rows and low score (lower case letter) or high score (upper case) as columns.

2.4.1. a. $1/6$.
       c. $4/6, 3/6, 1/6, 3/6$.

2.4.3. a. 1.
       b. $1/4$.
       c. $1/4$.
       d. $3/4$.
       e. $3/4$.

2.5.1. a: 1, $1/2$.
       b: 7, 2.
       c: 1.6250, 0.2969.

2.5.3. a. $p(0) = 0.94, p(5) = 0.03, p(10) = 0.02, p(25) = 0.01$.
       b. 0.60.
       c. No.
       d. 8.64.
       e. 0.97.

2.5.5. a. 2.5.
  b. $(a + b)/2$.

Review Exercises

False: 2.4   2.13
      2.5   2.15
      2.6   2.18
      2.9   2.19
      2.10

CHAPTER 3

Exercises

3.1.1. a. 1/5.
  b. 2/5.
  c. 3/5.
  d. 1.
  e. 0.
  f. 0.
  g. 4/5.
  h. 3/5.
3.1.3. a. 90/1024.
  b. 918/1024.
  c. 376/1024.
3.1.5. a. 25/7776.
  b. 1/1296.
3.1.7. a. 1.
  b. 3.
  c. 1.
  d. 10.
  e. 5.
  f. 4.
3.1.9. a. 0.11.
  b. 6.6×10^{-5}.
  c. 3.6×10^{-7}.
3.1.11. a. 1.6, 0.96.
  b. 1.6, 0.96.
3.1.13. 32.
3.1.15. a. 1/12.
  b. 1/6.
  c. 1/144.
3.1.17. a. 1/2.
  b. 1/32.

3.1.19. a. There are nine choices for the first station, eight for the second, and so on, and the total number of possibilities is the product.

b. 362,880.

c. 90,720.

3.1.21. a. 252.

b. 1.

c. 1/252.

d. The examiner might inadvertently indicate the pictures of the dead subjects.

3.1.23. a. $10!/(2!)(8!) = (10)(9)/2 = 45$

$(9 + 1) + (8 + 2) + (7 + 3) + (6 + 4) + 5 = (10)(9/2) = 45$

3.2.1. a. 0.000.

b. 0.000.

c. 0.904.

d. 0.238.

e. 1.000.

f. 0.548.

3.2.3. a. H_0: $\pi = 0.30$.

b. H_a: $\pi \neq 0.30$.

c. 0.053.

d. Accept H_0; the game may be operating as desired. He must assume the players are random.

3.2.5. a. 0.417.

b. Increase the sample size.

3.2.7. a. Twenty or fewer miles per gallon, more than 20 miles per gallon.

b. H_0: $\pi = 0.70$. (π is the proportion of Type B cars that average more than 20 miles per gallon.)

c. H_a: $\pi > 0.70$.

d. Type II. Use a large sample size.

3.2.9. a. 0, 1, 2, 11, 12, ..., 20.

b. 10, 11, ..., 20.

c. 0.176.

d. iv.

3.2.13. a. H_0: $\pi = 0.20$, H_a: $\pi > 0.20$.

b. 9, 10, ..., 25.

c. No, $P = 0.108 > \alpha$.

3.2.15. a. H_0: $\pi = 0.70$, H_a: $\pi \neq 0.70$.

b. 0, 1, ..., 10 or 18, 19, 20.

c. Discouraged because H_0 is rejected with the evidence in the direction of less than 70%.

3.2.17. a. H_0: $\pi_M = 0.50$, H_a: $\pi_M \neq 0.50$.

b. 0, 1, ..., 6 or 19, ..., 25.

e. No, 16 is not in the region of rejection.

3.3.1. (1) 0.229, 0.591.

    (2) 250, 90.

    (3) 0.263, 0.382.

    (4) 0.816, 0.897.

    (5) 0.164, 0.511.

    (6) 100, 17.

    (7) 0.046, 0.083.

    (8) 29, 0.90.

    (9) 500, 0.99.

    (10) 8, 0.25, 0.55.

3.3.3. a. $0.25 \leq \pi \leq 0.55$.

    b. $0.236 \leq \pi \leq 0.583$.

3.3.5. a. 0.14.

3.3.7. a. 0.52.

    b. $0.456 \leq \pi \leq 0.583$.

3.3.9. a. $0.471 \leq \pi \leq 0.588$.

    b. i. $H_0: \pi = 0.495$.

      ii. Acceptance.

      iii. 0.01.

3.4.1. a. $H_0: \pi = 0.50$, $H_a: \pi \neq 0.50$.

    b. 0.422.

    c. Do not reject H_0.

    d. Independence.

3.4.3. a. $H_0: \pi = 0.50$, $H_a: \pi < 0.50$.

    b. For $n = 25$, if $y \leq 8$; for $n = 50$ or 100, if 0.50 is not in the one-sided $CI_{0.95}$ for the upper bound found in A.5b and A.5c, respectively.

Review Exercises

False:
3.2	3.15	3.25
3.4	3.16	3.26
3.5	3.18	3.27
3.6	3.21	3.29
3.7	3.22	
3.9	3.23	
3.12	3.24	

CHAPTER 4

Exercises

4.1.1. a. Chironomid flies.

    b. 0.21.

4.1.3. a. Flaws.

     b. 2.5.

     c. 0.1336.

4.1.5. a. 0.082.

     b. 0.205.

     c. 0.918.

     d. 1, 2, 3, 4.

4.1.7. a. For $\lambda = 0.25$, $p(0) = 0.7788$, $p(1) = 0.1947$, $p(2) = 0.0243$, $p(3) = 0.0020$, $p(4) = 0.0001$, $p(5) = 0.0000$, ... For $\lambda = 0.50$, 1.00, and 10.00 use Table A.7.

4.2.1. a. 12.

     b. 0.0513.

     c. H_0: $\lambda = 12$ per 3 milliseconds, H_a: $\lambda > 12$ per 3 milliseconds. Reject H_0 if $P \leq \alpha$. Accept H_0. There is no evidence that the level is higher than 4 per millisecond.

4.2.3. H_0: $\lambda = 10$, H_a: $\lambda < 10$; reject H_0. There is evidence of a reduction.

4.2.5. a. H_0: $\lambda = 1$ per 100 cells.

     b. H_a: $\lambda > 1$ per 100 cells.

     c. 0.0190.

     d. 0.7787. It seems necessary because the probability of four or more basophils is 0.2213.

4.3.1. $1.4 \leq \lambda \leq 6.0$, $1.1 \leq \lambda \leq 6.7$.

4.3.3. $0.0158 \leq \lambda \leq 0.0527$.

4.3.5. a. i. Knowing she was watched could affect her behavior

       ii. They would likely not be independent

       iii. The probability of boredom could increase with length of time.

     b. For 16 half-minute units, CI $0._{80}$: $6.2213 \leq \lambda \leq 15.4066$. To obtain the interval of the estimate on a per minute basis divide L and U by 8.

     c. If the parameter to be estimated is the friend's boredom during that specific lecture, it is valid but not ethical for no one wants to be watched without permission.

4.4.1. 0.0047.

4.4.3. H_0: $\lambda = 4$, H_a: $\lambda < 4$; reject H_0 if $y = 0$ or 1. H_0 is accepted. No evidence of a reduction in the proportion of defective sets.

4.4.5. $\lambda \leq 9.0$.

Review Exercises

False:	4.1	4.6	4.13
	4.3	4.8	4.15
	4.4	4.9	4.16
	4.5	4.10	

CHAPTER 5

Exercises

5.1.1. a. 18.475.

    b. 2.156.

    c. 95.023.

    d. 0.05.

    e. 0.975.

    f. 18.307.

    g. 42.796.

    h. 5.

5.1.3. $\chi^2 = 9.336$, no evidence that the table is not random.

5.1.5. $\chi^2 = 1.240$, 75% of the plants may be red-flowering. We assume the nongerminating seeds would have produced the same proportion of plants with red flowers.

5.1.7. $\chi^2 = 53.427$, $P < 0.005$, there is evidence of a preference. The counts indicate a preference for the economy issue. This assumes that those who did not respond have similar views to those who did respond.

5.1.9. a. H_0: $\pi_1 = 9/16$; $\pi_2 = 3/16$; $\pi_3 = 3/16$; $\pi_4 = 1/16$.

    H_a: *At least one inequality.*

    b. $\chi^2 = 9.418$, the genes are probably not on different chromosomes.

5.2.1. $\chi^2 = 8.342$, b(y; 4, 0.40) may be the correct distribution.

5.2.3. $\chi^2 = 0.0222$, this may be from a binomial distribution.

5.2.5. $\hat{\lambda} = 0.5246$, $\chi^2 = 2.52$, this seems to be from a Poisson distribution.

5.3.1. H_0: *Both groups have the same pattern of colds.*

    H_a: *The groups differ with respect to colds.*

    $\chi^2 = 4.63$, the serum does not appear to be effective in preventing colds.

5.3.3. $\chi^2 = 2.67$, no evidence that the drug is related to a higher incidence of birth defects; homogeneity.

5.3.5. a. H_0: $\pi_A = \pi_B = \pi_C$. (π_i is the proportion of dead black files for each insecticide.)

    b. 9.210.

    c. $\chi^2 = 1.49$.

    d. Greater than 0.05.

    d. Do not reject H_0; the insecticides are equally effective.

5.3.7. a. H_0: The attractiveness of women is independent of city where seen.

    H_a: The attractiveness of women depends on city where seen.

    b. Chi-square $= 8.791$ (P-value $= 0.0123$).

    c. 55 of 200 were attractive, so odds $= 55/(200 - 55) = 0.379$.

5.4.1. a. prospective.

    b. relative risk $= (120/200)/(155/300) = 0.6/0.5166 = 1.16$.

    c. odds ratio $= (120/80)/(155/145) = 1.5/1.07 = 1.403$.

5.4.3. a. observational.

    b. relative risk $= (10/138)/(3/168) = 4.058$.

    c. odds ratio $= (10/128)/(3/163) = 4.245$.

5.5.1. a. $H_0: \pi_1 = \pi_2 = \pi_3 = \pi_4 = 0.50$.

    b. H_a: *at least one* $\pi_i \neq 0.50$.

    c. $\chi^2_{0.05,3} = 7.815$.

    d. $\chi^2 = 14.133$, the growth is different for different species, D grows fastest.

5.5.3. a. H_0: *Aggressiveness rank is independent of greediness rank.*

    b. The categories for both rows and columns are "above the median" and "below the median."

    c. $\chi^2 = 4.000$; there is evidence of an association between aggressiveness and greediness.

Review Exercises

False:

5.1	5.9
5.3	5.11
5.5	5.13
5.6	5.15
5.7	5.16
5.8	5.20

CHAPTER 6

Exercises

6.1.1. 70.

6.1.3. a. 2.0.

    c. 2.0.

    d. $\hat{\mu} = \bar{y} = 2.0$.

6.2.1. a. 6.

6.2.3. a. 1.68.

    b. 1.68.

6.2.5. b. 3, 9, 3.

    c. 1.5, 1.5, 2.8.

6.2.7. $\mu = 0.238$, $\sigma = 0.740$; $\mu \pm 2\sigma$ is -1.242 to 1.718, which contains 0.941 of the data; $\mu \pm 3\sigma$ is -1.982 to 2.458, which contains 0.972 of the data.

6.3.1. b. $22/3$, $38/9$.

6.3.5. c. 65.

    d. 65.

    e. 3.33.

    f. 1.67.

6.4.1. a. 5.25, 1.75, 1.05, 0.66.

    b. 5.25, 1.25, 0.45, 0.

Review Exercises

False: 6.1   6.11
             6.2   6.13
             6.4   6.14
             6.5   6.15
             6.8   6.19
             6.9

CHAPTER 7

Exercises

7.1.1. a. 0.818.
   b. 0.499.
   c. 0.382.
   d. 0.010.
   e. 0.500.
   f. 0.943.
   g. 0.124.
   h. 0.445.
   i. 0.318.
   j. 0.046.
   k. 0.002.
7.1.3. a. 0.933.
   b. 67.2 to 132.8.
   c. 120.8.
   d. 95.0.
7.1.5. a. 0.001.
   b. 0.159.
7.1.7. $\chi^2 = 10.789$, critical value 12.592, the sample seems to come from a normal distribution.
7.2.1. a. 1.64.
   b. -1.64.
   c. 2.33.
   d. -2.33.
   e. 2.58.
   f. -2.58.
7.2.3. a. H_0: $\mu = 19.3$.
   b. H_a: $\mu < 19.3$.
   c. $z \leq -1.64$ or $y \leq 18.808$.
   d. 0.359.

7.3.1. a. 0.106.

    b. $(0.106)^5$.

    c. 0.003.

7.4.1. 0.023.

7.4.3. a. 0.50, 0.371.

    b. 0.50, 0.159.

    c. i. $H_0: \mu = 90$, $H_a: \mu < 90$.

      ii. $\bar{y} \leq 83.44$.

      iii. 0.739.

7.4.5. a. 3.24.

    b. 2.06 to 10.55.

    c. $\chi^2 = 31.32$; do not reject H_0. The variance may be 3.0.

    d. 5.16 to 6.44.

7.5.1. a. i. 0.369.

      ii. 0.302.

      iii. 0.378.

    b. i. 0.147.

      ii. 0.174.

    c. The continuity correction is more important for small samples.

7.5.3. a. 25%.

    b. $H_0: = 0.25$, $H_a: \pi \neq 0.25$.

    c. $z = 2.47$; reject H_0. The disorder appears to be genetic.

7.5.5. a. 0.64.

    b. 0.0023.

    c. 0.55 to 0.73.

    d. There is evidence that people can distinguish because $\pi = 0.50$ is below the confidence interval.

7.5.7. $z = -2.21$; there is evidence of undercounts.

7.5.9. a. $H_0: \phi = 1$, $H_a: \phi > 1$.

      $z = 0.539/0.236 = 1.65$; reject H_0.

7.6.1. a. 25.5, 25.5, 25.5, 25.5.

    b. 20.825, 10.412, 6.942, 5.20ζ.

7.6.3. $z = -2.60$, the scrubbers reduce particulate emissions.

Review Exercises

False:	7.1	7.9
	7.2	7.11
	7.3	7.12
	7.4	7.14
	7.8	7.19

CHAPTER 8

Exercises

8.1.1.   a. 2.764.

      b. -2.764.

      c. 2.365.

      d. -2.365.

      e. 2.807.

      f. -2.807.

8.1.3.   a. $8000.

      b. 10,240,000.

      c. 2.00.

      d. Between 0.025 and 0.05.

8.2.1.   a. 1000.

      b. 100.

      c. 786.9 to 1213.1.

8.2.3.   a. 1.7.

      b. 54.7 to 61.7.

8.2.5.   a. 3.7 to 4.7.

      b. $\mu \le 4.7$.

8.2.7.   a. H_0: $\mu_d = 0$, H_a: $\mu_d > 0$.

      b. $t = 2.00$; H_0 is rejected. There is evidence of improvement on the second test.

8.2.9.   a. The design removes extraneous variability introduced by soil conditions, climate, and farming methods.

      b. 3.0.

      d. $t = 3.236$, reject H_0.

      e. There is evidence that the seed company's claim is correct.

      f. 2.24 to 3.76. H_0 is rejected because 2.0 is not in this interval.

8.2.11. a. 44.8.

      b. H_0: $\mu_d = 0$, H_a: $\mu_d \ne 0$.

      c. $t = 1.7$; do not reject H_0. There is no evidence of a difference in weight gain.

      d. -1.0 to 6.3. Since this interval contains 0 the null hypothesis is accepted.

8.3.1.   a. 105.

      b. H_0: $\mu_U = \mu_R$, H_a: $\mu_U > \mu_R$.

      c. $t = 2.30$; reject H_0. Urban pollution is higher.

      d. $\mu_U - \mu_R \le 24.7$.

8.3.3. -3.439 to -0.561.

8.3.5. $t = 1.80$; reject H_0. There is evidence that those who finish on time score higher. However, since this was obtained from a survey without control for other factors, it should be applied cautiously.

8.3.7.   a. -2.18 to -0.02.

    b. Since the interval does not contain zero, there is evidence of inequality. However, the evidence is weak because 0 is close to the upper limit -0.02.

8.4.1. a. 6.538.

    b. 4.886.

    c. 2.328.

    d. 0.430.

    e. 0.132.

8.4.3. $F = 4.00$; reject the hypothesis of equal variances. Use the t' test for means, $t' = -2.50$ with $v = 14$; reject H_0. There is evidence of a difference in the mean resin content.

8.4.5. a. $F = 0.444$; do not reject H_0. There is no evidence of different variances.

    b. 0.111 to 2.449.

8.5.1. a. The differences may not be normal.

    b. H_0: $\mu = 0$, H_a: $\mu > 0$.

    c. $z = 2.85$; reject H_0. There is evidence of a harmful effect.

8.5.3. $z = 2.19$; reject H_0.

Review Exercises

False:	8.1	8.9
	8.3	8.10
	8.4	8.11
	8.5	8.13
	8.6	8.14
	8.7	8.16
	8.8	

CHAPTER 9

Exercises

9.1.1. c.

9.1.3. Days per pound.

9.1.5. a. 180.

    b. 18.

    c. $\hat{y} = -208 + 18x$.

    d. 80.

9.1.7. a.   i. Positive

     ii. Yes.

     iii. 19.2.

     iv. Minutes per staff hour, per patient.

     v. $\hat{y} = -1 + 19.2x$.

     vi. 18.2, 95.0.

b. i. Negative
   ii. Not intuitive prior to the survey.
   iii. $\hat{y} = 19.15 - 1.43x$.
   iv. $x = 10$.

9.1.9. a. 68.
   b. 68.5.
   c. 0.50.

9.1.11. $\hat{y} = 53.69 + 0.6187x$.

9.2.1. c. i. $H_0: \beta = 0, H_a: \beta > 0$.
   ii. 1.895.
   iii. $t = 6.0$; reject H_0. There is evidence that increase in study is linearly related to higher grades.

9.2.3. a. Fish per hour.
   b. Fish.
   c. Fish.

9.2.5. a. i. $H_0: \beta = 0$. There is no linear relationship between time spent on patient care and patient load.
   ii. Time would seem to increase as number of patients increases.
   iii. 2.132.
   iv. $t = 16.0$; reject H_0. There is a linear relationship between time spent on patient care and patient load.

   b. i. $H_0: \beta = 0$.
   ii. It is not clear prior to the survey whether the relationship is positive or negative.
   iii. $t = -5.39$; reject H_0. There is a linear relationship; time for reports decreases as patient load increases.

9.2.7. a. 0.14.
   b. 0.14.
   c. i. $H_0: \beta = 0$.
   ii. Radioactivity disappears over time.
   iii. -2.353.
   iv. $t = -1.750$; do not reject H_0. There is no evidence of a linear relationship.

9.3.1. a. 40.
   b. 9.
   c. 72 ± 3.5.

9.3.3. a. $H_0: \beta = 0$.
   b. $H_a: \beta \neq 0$, since it is not obvious whether a larger number of fillings in the previous two years indicates that there will be little left to do or a very fast decay rate.
   c. ± 2.306.

9.3.5. a. 3.0.
   b. 1.44.
   c. $H_0: \beta = 0, H_a: \beta > 0, t = 1.00$; do not reject H_0. There is no evidence of a linear relationship.

d. 20.06.

e. 20.06 ± 0.78.

f. 19.7.

g. 19.7 ± 0.83.

h. Parts d through g are invalid because there is no linear relationship.

9.3.7. a. 0.00112 ± 0.00428.

b. 0.844 ± 0.262.

c. $0.79 < E(y|x = 50)$.

d. $0.657 < y$.

e. No, there is no evidence of a linear trend.

9.4.1. a. $-1, +1$.

b. 10, 1.

c. $-0.9, +0.4$.

d. Significant; nonsignficant.

9.4.3. $t = 4.0$; reject H_0. Length explains a significant portion of the variability in weight.

9.4.5. a. 2.

b. 1.

c. 2.

d. 1.

9.4.7. -0.990 to -0.651.

9.5.1. a. $r_s = 0.98$.

b. i. $H_0: E(r_s) = 0, H_a: E(r_s) > 0$.

ii. 1.645.

c. $z = 2.94$, reject H_0; there is a positive association.

9.5.3. a. $r_s = 0.58, r = 0.54$.

b. $z = 1.924$ for Spearman's test; accept H_0. The tests agree.

9.6.1. $\Sigma b_r x_i = b_r \Sigma x_i = (\Sigma y_i / \Sigma x_i) \Sigma x_i = \Sigma y_i$

9.6.3. a.

Estimates	Intercept	Slope
Least Squares	-0.656	0.889
Difference	-6	1
Ratio	0	0.875

Review Exercises

False: 9.2   9.12

9.5   9.14

9.6   9.15

9.7   9.17

9.8   9.20

9.11

CHAPTER 10

Exercises

10.1.1. $72/5 = 42/5 + 30/5$.

10.1.3. a. $F = 11.65$; reject H_0. There is a difference in the mean heights of the two groups.

b. $t = -3.41$; reject H_0.

c. $(t_{0.025,6})^2 = F_{0.05,1,6}$.

10.2.1. $F = 1.55$, H_0 is not rejected. There is no significant difference among the diets.

10.2.3. H_0: $\mu_A = \mu_B = \mu_C$, $F = 5.14$; reject H_0. There is at least one difference among the mean lifetimes.

10.2.5. H_0: $\mu_A = \mu_B = \mu_C$; this does not appear to be true from the graph. 3.885. $F = 216.7$; reject H_0. The mean amount of vitamin C differs for at least two of the methods.

10.2.7. $F = 23.56$; reject H_0. There is evidence of different mean weights at different locations.

10.2.9. a. $a = 7$, $n = 5$, total degrees of freedom $= 7(5) - 1 = 34$.

b. Trial.

c. Normality, independence, equal variances.

d.

df	SS	MS
6	330	
28	644	23

e. H_0: $\mu_1 = \mu_2 = \cdots = \mu_7$, H_a: At least one inequality.

f. $F_{0.05,6,28} = 2.445$. $F = 2.39$; do not reject H_0. There is no significant difference among the insecticides.

10.2.11. a. δ_h, ξ_{hi}.

b. $\Sigma_h \, \delta_h = 0$, ξ_{hi} is IND($0, \sigma^2$).

c. $h = 1, 2, 3 = a$. $i = 1, 2, \ldots, 5 = n$.

d. $F = 4.00$; reject H_0. There is a significant difference among the mean weight-bearing capacities.

10.3.1. a.

df	SS	MS
4	2392	
		180

c. Yes, F is significant.

d. $\bar{y}_C$ $\bar{y}_A$ $\bar{y}_D$ $\bar{y}_B$ $\bar{y}_E$.

10.3.3. a. $F = 2.88$; accept H_0.

b. No, F is not significant.

c. No significant differences.

10.3.5. a.

$$\bar{y}_I \qquad \bar{y}_{II} \qquad \bar{y}_{III}$$

b. 3.682. $F = 6.25$, which exceeds the critical value.

c. $H_0: \mu_3 = (\mu_1 + \mu_2)/2$, $H_a: \mu_3 \neq (\mu_1 + \mu_2)/2$.
Critical value 10.4, $\bar{y}_3 - (\bar{y}_1 + \bar{y}_2)/2 = 15$, so the yield with III is significantly different from the average of I and II.

10.4.1. There is a significant difference between the home type and the industrial type, $F = 9.68$.

10.4.3. a. $H_0: \mu_1 = \mu_2 = \cdots = \mu_6$, H_a: At least one inequality.

b. $F = 7.12$; reject H_0.

c. The placebo is significantly different from the analgesics.

e. 14%.

f. Pain relief is obtained more quickly with aspirin in any form than with the placebo.

10.5.1. a. 6.48 to 10.82.

b. -7.40 to 4.60.

c. -17.65 to -6.35.

d. 3.65 to 13.35.

10.5.3. a. $F = 4.0$; reject H_0.

b. 4.97 to 19.03.

c. 37.66 to 42.34.

d. -10.06 to $-1.94.1x(\text{^numListCount})$

e. 4.60 to 15.40.

10.6.1. $3.0045(6.36) = 19.109$. The value for Tukey's procedure is 18.2; since a larger difference is required for the Bonferroni procedure, it is statistically more conservative.

10.7.1. a. $H_0: E(\bar{r}_i) = 13/2 \ for \ i = 1, 2, 3$;
$H_a: E(\bar{r}_i) \neq 13/2 \ for \ some \ i = 1, 2, 3$.

b. 5.991.

c. $H = 7.269$; reject H_0.

d. There is a significant difference between alloys A and C; C lasts longer than A.
10.6.3. $H = 6.269$; there is evidence of a difference.

10.7.3   a. A nonparametric procedure is preferred when the data are not normal but the other conditions for ANOVA are satisfied.

b. $a_1 = 1$, $a_2 = -2$, $a_3 = 1$, $H = 5.65$. Reject H_0; B is significantly different from the average of A and C; B lasts longer.

Review Exercises

False: 10.2   10.10   10.16
         10.5   10.11   10.17

10.6	10.13	10.19
10.8	10.14	10.20
10.9		

CHAPTER 11

Exercises

11.1.1.   a. Fixed.
  b. Random.
  c. Random.
  d. Fixed.
  e. Fixed.

11.1.3.   a. $F = 3.09$; reject H_0: $\sigma_A^2 = 0$. There is evidence of significant variability among families.
  b. $r_I = 0.41$.
  c. Families with three brothers.
  d. Obesity is a characteristic of some families.

11.1.5.   a. REM.
  b. H_0: $\sigma_A^2 = 0$.
  c. $F = 19$; reject H_0.
  d. 0.90.
  e. Ten percent of the variability is due to the lab technique, and this may not be reliable enough for medical decisions.

11.2.1.   a. $F_{/\max} = 19.75$; reject H_0. There is at least one inequality among the variances.
  b.

σ_{NY}^2	σ_{SK}^2	σ_{LN}^2	σ_{CD}^2	σ_{DA}^2	σ_{RN}^2

11.2.3. $F_{\max} = 7.4$; do not reject H_0.

11.3.1.   b. 43.65 versus 1.58.

11.3.3.   a. Square root.
  b. Points seem random.

11.3.5.   b. F is significant; LSD indicates all transformed means are significantly different.

Review Exercises

False: 11.2   11.9   11.15
   11.4   11.12   11.17
   11.6   11.14   11.18

CHAPTER 12

Exercises

12.1.1.   a. i. The cock effect. Random.

          ii. The hen effect. Random.

      b. $F = 0.05$; do not reject H_0. There is no evidence of significant variability due to males.

12.1.3.

C   A   E   B   D

B or *D* should be purchased.

12.1.5.   a. 6, 7, 5.

      b. R-SQUARE $= 0.504467$ or 50.4%.

      c. $MS_a/MS_b = (32.333/5)/(75.413/36) = 3.087$.

      d. 105.840.

12.2.1.   b. Among hybrids $F = 38.98$; reject H_0.
Among locations $F = 5.82$; reject H_0.

      c. Yes.

      d. Yes.

      e. RC-3   DBC   FR-11   BCM
Any hybrid except RC-3 should be used.

12.2.3.   b. Fixed.

      c. Random.

      d. $H_0: \alpha_1 = \alpha_2 = \alpha_3 = \alpha_4 = \alpha_5$.

      e. Among models $F = 3.59$; reject H_0.
Among cities $F = 2.59$; do not reject H_0.

      f. Yes.

      g. Since Type I error is not serious, use Fisher's least significant difference.

      h.

D   B   C   A   E

C, *A*, and *E* get the best mileage.

      i. No.

      j. 17%.

12.2.5.   a. 4, 5.

      b. 1.2.

c.

$\bar{y}_1$	$\bar{y}_2$	$\bar{y}_3$	$\bar{y}_4$

12.3.1. b. For covers $F = 0.94$; do not reject H_0.
  For newsstands $F = 2.92$; do not reject H_0.
  For weeks $F = 1.29$; do not reject H_0.

  c. The mean sales among covers do not differ.

  d. Without this design, 125 repetitions of the experiment would be necessary.

12.3.3. c. For weeks $F = 0.22$; do not reject H_0.
  For days $F = 0.32$; do not reject H_0.
  For operations $F = 0.35$; do not reject H_0.

  e. Weeks are random, days are fixed, and operations are fixed.

  f. None of the effects analyzed contribute significantly to differences in the number of unsafe incidents.

12.3.5. SS_e would have zero degrees of freedom, so MS_e does not exist.

12.4.1. a. Fixed.

  b. Fixed.

  c. For diets $F = 12.6$, for jogging $F = 69.1$, for interaction $F = 1.6$.

  e. Yes.

  f. Yes.

  g. No.

  h. Use Fisher's least significant difference to locate the best diet and the best amount of jogging. Either a high protein or a high carbohydrate diet should be combined with two miles of jogging.

12.4.3. a.

Source	df	E(MS)	F
Plant species		$\sigma^2 + 5\sigma_{AB}^2 + 25\sigma_A^2$	5.125
Hillside	4	$\sigma^2 + 5\sigma_{AB}^2 + 30\sigma_B^2$	5.200
P × H		$\sigma^2 + 5\sigma_{AB}^2$	6.667
Error	120	σ^2	

  b. $6.667 > F_{0.05,20,120} = 1.662$ so there is a significant interaction.

  c. $\hat{\sigma}_A^2 = 13.2$, $\hat{\sigma}_B^2 = 11.2$, so species contributes more to the total variability.

12.5.1. b. All effects fixed.

  c. $F = 11.49$; reject H_0.

  d.

$$SS_a = 1{,}302.2 \qquad SS_b = 351{,}939.7 \qquad SS_c = 112{,}266.8$$
$$SS_{ab} = 2{,}572.8 \qquad SS_{ac} = 2{,}002.5 \qquad SS_{bc} = 15{,}366.5$$
$$SS_{abc} = 7{,}927.5 \qquad SS_e = 44{,}800.0$$

e. $E(MS_a) = \sigma^2 + bcn\Sigma\alpha_i^2/(a-1)$
$E(MS_b) = \sigma^2 + acn\Sigma\beta_j^2/(b-1)$
$E(MS_c) = \sigma^2 + abn\Sigma\gamma_k^2/(c-1)$
$E(MS_{ab}) = \sigma^2 + nc\Sigma\Sigma\alpha\beta_{ij}^2/(a-1)(b-1)$
$E(MS_{ac}) = \sigma^2 + nb\Sigma\Sigma\alpha\gamma_{ik}^2/(a-1)(c-1)$
$E(MS_{bc}) = \sigma^2 + na\Sigma\Sigma\beta\gamma_{jk}^2/(b-1)(c-1)$
$E(MS_{abc}) = \sigma^2 + n\Sigma\Sigma\Sigma\alpha\beta\gamma_{ijk}^2/(a-1)(b-1)(c-1)$
$E(MS_e) = \sigma^2$

f. Only the nitrogen levels and phorphorus levels are related to significant differences. There are no interactions.

12.5.3. a. Seed treatment (A), fixed.
Male (B), random.
Female (C), random.

b. F for Treatments 5.48; reject H_0.
F for Crosses 17.75; reject H_0.
F for $T \times C$ 13.00; reject H_0.

c. $SS_m = 26.09$, $SS_f = 13.93$, $SS_{mf} = 45.11$.

d. $SS_{tm} = 1.14$, $SS_{tf} = 29.34$, $SS_{tmf} = 31.93$.

e.

Source	df	F
Treatment (A)	1	no exact test
Male (B)	3	$MS_b/MS_{bc} = 1.74$
Female (C)	3	$MS_c/MS_{bc} = 0.93$
$A \times B$	9	$MS_{ab}/MS_{abc} = 0.11$
$A \times C$	3	$MS_{ac}/MS_{abc} = 2.76$
$B \times C$	3	$MS_{bc}/MS_e = 15.66^*$
$A \times B \times C$	9	$MS_{abc}/MS_e = 11.09^*$
Error	32	

f. 31%.

g. Because of the significant interactions which reverse the effects of scarification, the treatment has different effects on different crosses; scarification cannot be recommended in general.

12.6.1.

Source	df	F
Whole Units		
Wash temperature	1	80.34*
Brands	3	31.14*
Whole unit remainder	3	
Subunits		
Dry temperature	2	117.22*
Wash temp. × Dry temp.	2	17.51*
Subunit remainder	12	

12.7.1.   a. $y_{ijk} = \mu + \alpha_i + \beta_{ij} + \gamma_k + \alpha\gamma_{ik} + \varepsilon_{ijk}$

μ: the overall mean

α_I: fixed effect of ith level of Gender

β_{ij}: random effect of ijth experimental Unit

γ_k: fixed effect of the kth level of Target

$\alpha\gamma_{ik}$: The interaction effect between ith level of factor Gender and the kth level of factor Target.

b.

i.

Source	df	SS
Whole Units		
Gender	1	0
Units	6	264
Subunits		
Target	2	58,413
Gender × Target	2	37
Subunit remainder	12	302

ii. Rsquare $= 0.995$

c. Because the SS for Gender are zero, F $= 0$ and the P-value $= 1$.

d. i. Average time of males $= 180.75$ average time for females $= 177.25$

$$t = \frac{180.75 - 177.25}{\sqrt{\frac{2}{4}\left(\frac{302}{12}\right)}} = 0.987$$

ii. $t = \dfrac{180.75 - 180}{\sqrt{\frac{1}{4}\left(\frac{302}{12}\right)}} = 0.299$

12.7.3.   a. $y_{ijk} = \mu + \alpha_i + \beta_{ij} + \gamma_k + \alpha\gamma_{ik} + \varepsilon_{ijk}$

μ: the overall mean

α_I: fixed effect of ith level of time of buring

β_{ij}: random effect of ijth core

γ_k: fixed effect of the kth level of Depth

$\alpha\gamma_{ik}$: The interaction effect between ith level of factor Burning and the kth level of factor Depth.

c.

Source	df	SS
Whole Units		
Burning	2	3.010
Cores	3	0.390

Source	df	*SS*
Subunits		
Depth	2	3.6633
Burning × Depth	4	0.5567
Subunit remainder	6	0.8200

Review Exercises

False: 12.2   12.11
         12.4   12.13
         12.6   12.16
         12.9   12.17

CHAPTER 13

Exercises

13.1.1.   b. (3, 7), (4, 8), (5, 7).

    c. $\bar{x}_\cdot = 4$.

    d. $\hat{y}_{1j} = 1 + 2x_{1j}$, $\hat{y}_{2j} = 2x_{2j}$, $\hat{y}_{3j} = -3 + 2x_{3j}$.

    e. (4, 9), (4, 8), (4, 5).

    f. Increase. Order is changed.

13.1.3.   (1) *e*

    (2) *g*

    (3) *h*

    (4) not indicated

    (5) *c*

    (6) *f*

    (7) *d*

13.2.1.   c. $F = 4.93$; reject H_0. The adjusted alloy averages are significantly different.

13.3.1.   a. 4.

    b. $F = 33.78$; reject H_0. The slope is not zero.

13.3.3.   Yes.

13.4.1.   b. 0.80.

    e. adj $\bar{y}_{1\cdot} = 22.4$, adj $\bar{y}_{2\cdot} = 18.0$, adj $\bar{y}_{3\cdot} = 22.6$.

    f. $21.52 \le \mu_1 \le 23.28$, $17.26 \le \mu_2 \le 18.74$, $21.72 \le \mu_3 \le 23.48$.

    g. 18.0   22.4 22.6

13.4.3.   a. 4950.45, 76.92, 0.73, 50.84, 48.25.

    b. Birthweight.

c. To reduce the variability in the experimental groups.

d. No; P value equals 0.3971.

e. There are only two groups.

Review Exercises

False: 13.1   13.7   13.11
         13.2   13.10   13.12
         13.3   13.18
         13.5   13.19

CHAPTER 14

Exercises

14.1.1. a. $\begin{bmatrix} 13 & 27 \\ 24 & 25 \end{bmatrix}$

         b. $\begin{bmatrix} -30 & 10 \\ 54 & -18 \end{bmatrix}$

         c. $\begin{bmatrix} 16 \\ 64 \end{bmatrix}$

         d. $\begin{bmatrix} -1 \\ 50 \\ 7 \end{bmatrix}$

14.1.3. $\begin{bmatrix} 5 & -2 \\ -2 & 1 \end{bmatrix}$

14.2.1. a. $\begin{bmatrix} 10 & 20 & |4| & 1 & 0 \\ 20 & 40 & |2| & 0 & 1 \end{bmatrix} \rightarrow \begin{bmatrix} 1 & 0 & |12.4| & 4.1 & -2.0 \\ 0 & 1 & |-6.0| & -2.0 & 1.0 \end{bmatrix}$

         b. $F = 36.15$; reject H_0.

14.2.3. a. 0.7124.

         b. $F = 18.59$; R^2 is significant.

         c. Reject.

         d. $\hat{y} = 5852.06 - 2.563x_1 + 1.224x_2$.

         e. Decreased by -2.563.

14.3.1. a. $-3.518 \le \beta_1 \le -1.608$.

         b. $0.588 \le \beta_2 \le 1.860$.

         c. $t = -5.722$; reject H_0: $\beta_1 = 0$.
            $t = 4.099$; reject H_0: $\beta_2 = 0$.

         d. $1685.60 \le E(y \mid x_1 = 2000, x_2 = 860) \le 1871.80$.

         e. $1679.10 \le y \le 1878.30$.

14.3.3. a. $-0.1375, 0.2856$.

b. $-0.6413 \le \beta_1 \le 0.3663$.
    $-0.2904 \le \beta_2 \le 0.8616$.

14.5.1. a. i. 0.8645.

ii. 0.8602.

b. The model containing Oxygen and Depth is the better model.

14.5.3. a. $SSR/S_{yy} = 0.8811, 0.8613$.

b. i. 0.64%.

ii. 1.28%.

c. 2.0644.

d. i. The model containing only acres is best.

ii. $F = 1.0897$, $F_{0.05,2,19} = 3.522$; the reduction is not significant.

14.6.1. a. $\hat{y} = -71.6 + 48.5 \log x$. H_0: $\beta = 0$ is rejected with $t = 4.155$. There is a linear relationship.

14.6.3. b. i. -0.342.

ii. -0.998.

iii. -0.996.

c. i. She expects increased cooking time to reduce the number of salmonella colonies.

ii. $t = -15.42$; reject H_0.

d. i. 4.500.

ii. 2.852 to 7.099.

iii. Since $ae^{bx} = 0$ is impossible, solve $ae^{bx} = 1$. More than 19.4 minutes are required for an expected survival of zero.

14.7.1. a. 1.9401, -0.1125; both terms contribute significantly.

b. Yes, $F = 6.2$.

c. 43.02.

14.7.3. a. $F = 5.78$; reject H_0. There is a significant difference among fertilizers.

b. The linear and quadratic trends are significant.

c. From the group totals it seems to be included.

d. $R^2 = 0.683$ for the quadratic model.
    $R^2 = 0.684$ for the cubic model.

14.8.1. a. $\hat{\phi} = 1.403$.

b. $CI_{.95}$: $0.977 < \phi \le 2.106$.

e. 1 is in the confidence interval.
    This supports the hypothesis that ϕ is equal to 1.

f. The alternative hypothesis of interest in Exercise 7.5.8 is $\phi > 1$.

14.8.3. a. Galton's null hypothesis is that β is equal to 0, i.e. brewing time is unrelated to the probability of bitter tea.

b. $e^{-1.7849} = 5.959$. This is the multiplicative increase in the odds for bitter tea given a 1 minute more of brewing time. Increase is significant; P-value < 0.0001.

c. The predicted probability of bitter tea when the brewing time is 8 minutes is .19. The predicted probability of bitter tea when the brewing time is 9 minutes is .586. Don't brew the tea longer than 8 minutes.

Review Exercises

False:	14.1	14.10	14.16
	14.2	14.11	14.17
	14.3	14.12	14.19
	14.6	14.15	14.20
	14.7		

Index

Statistics for Research, *Third Edition*, Edited by Shirley Dowdy, Stanley Weardon, and Daniel Chilko.
ISBN 0-471-26735-X © 2004 John Wiley & Sons, Inc.